Grzimek's ANIMAL LIFE ENCYCLOPEDIA

P9-CDM-087

Grzimek's
ANIMAL LIFE
ENCYCLOPEDIA

Editor-in-Chief

Dr. Dr. h.c. Bernhard Grzimek

Professor, Justus Liebig University of Giessen
Director, Frankfurt Zoological Garden, Germany
Trustee, Tanzania and Uganda National Parks, East Africa

VAN NOSTRAND REINHOLD COMPANY

New York   Cincinnati   Toronto   London   Melbourne

First published in paperback in 1984

Copyright © 1970 Kindler Verlag A.G. Zurich

Library of Congress Catalog Card Number 79-183178

ISBN 0-442-23037-0

Printed in Federal Republic of Germany

Van Nostrand Reinhold Company Inc.
135 West 50th Street
New York, New York 10020

Van Nostrand Reinhold Company Limited
Molly Millars Lane
Wokingham, Berkshire RG11 2PY, England

Van Nostrand Reinhold
480 Latrobe Street
Melbourne, Victoria 3000, Australia

Macmillan of Canada
Division of Gage Publishing Limited
164 Commander Boulevard
Agincourt, Ontario M1S 3C7 Canada

16   15   14   13   12   11   10   9   8   7   6   5   4   3   2   1

EDITORS AND CONTRIBUTORS

Editor-in-Chief
DR. DR. H.C. BERNHARD GRZIMEK
Professor, Justus Liebig University of Giessen, Germany
Director, Frankfurt Zoological Garden, Germany
Trustee, Tanzania and Uganda National Parks, East Africa

DR. MASAO KAWAI
Primate Research Institute, Kyoto University   KYOTO, JAPAN

DR. ERNST F. KILIAN
Professor, Giessen University and Catedratico Universidad Austral, Valdivia-Chile   GIESSEN, GERMANY

DR. RAGNAR KINZELBACH
Institute for General Zoology, University of Mainz   MAINZ, GERMANY

DR. HEINRICH KIRCHNER
Landwirtschaftsrat (retired)   BAD OLDESLOE, GERMANY

DR. ROSL KIRCHSHOFER
Zoological Garden, University of Frankfurt a.M.   FRANKFURT A.M., GERMANY

DR. WOLFGANG KLAUSEWITZ
Curator, Senckenberg Nature Museum and Research Institute   FRANKFURT A.M., GERMANY

DR. KONRAD KLEMMER
Curator, Senckenberg Nature Museum and Research Institute   FRANKFURT A.M., GERMANY

DR. ERICH KLINGHAMMER
Laboratory of Ethology, Purdue University   LAFAYETTE, INDIANA, U.S.A.

DR. HEINZ-GEORG KLÖS
Professor and Director, Zoological Garden   BERLIN, GERMANY

URSULA KLÖS
Zoological Garden   BERLIN, GERMANY

DR. OTTO KOEHLER
Professor Emeritus, Zoological Institute, University of Freiburg   FREIBURG I. BR., GERMANY

DR. KURT KOLAR
Institute of Ethology, Austrian Academy of Sciences   VIENNA, AUSTRIA

DR. CLAUS KÖNIG
State Ornithological Station of Baden-Württemberg   LUDWIGSBURG, GERMANY

DR. ADRIAAN KORTLANDT
Zoological Laboratory, University of Amsterdam   AMSTERDAM, THE NETHERLANDS

DR. HELMUT KRAFT
Professor and Scientific Councillor, Medical Animal Clinic, University of Munich   MUNICH, GERMANY

DR. HELMUT KRAMER
Zoological Research Institute and A. Koenig Museum   BONN, GERMANY

DR. FRANZ KRAPP
Zoological Institute, University of Freiburg   FREIBURG, SWITZERLAND

DR. OTTO KRAUS
Professor, University of Hamburg, and Director, Zoological Institute and Museum   HAMBURG, GERMANY

DR. DR. HANS KRIEG
Professor and First Director (retired), Scientific Collections of the State of Bavaria   MUNICH, GERMANY

DR. HEINRICH KÜHL
Federal Research Institute for Fisheries, Cuxhaven Laboratory   CUXHAVEN, GERMANY

DR. OSKAR KUHN
Professor, formerly University Halle/Saale   MUNICH, GERMANY

DR. HANS KUMERLOEVE
First Director (retired), State Scientific Museum, Vienna   MUNICH, GERMANY

DR. NAGAMICHI KURODA
Yamashina Ornithological Institute, Shibuya-Ku   TOKYO, JAPAN

DR. FRED KURT
Zoological Museum of Zurich University, Smithsonian Elephant Survey   COLOMBO, CEYLON

DR. WERNER LADIGES
Professor and Chief Curator, Zoological Institute and Museum, University of Hamburg   HAMBURG, GERMANY

LESLIE LAIDLAW
Department of Animal Sciences, Purdue University   LAFAYETTE, INDIANA, U.S.A.

DR. ERNST M. LANG
Director, Zoological Garden   BASEL, SWITZERLAND

DR. ALFREDO LANGGUTH
Department of Zoology, Faculty of Humanities and Sciences, University of the Republic   MONTEVIDEO, URUGUAY

LEO LEHTONEN
Science Writer   HELSINKI, FINLAND

BERND LEISLER
Second Zoological Institute, University of Vienna   VIENNA, AUSTRIA

Volume 3

MOLLUSKS AND ECHINODERMS

Edited by:

BERNHARD GRZIMEK

OTTO KRAUS

RUPERT RIEDL

ERICH THENIUS

ENGLISH EDITION

GENERAL EDITOR:
George M. Narita

SCIENTIFIC EDITOR:
Erich Klinghammer

TRANSLATOR:
Renate Geist

SCIENTIFIC CONSULTANTS:
R. Tucker Abbott
David L. Pawson
Clyde F. E. Roper

ASSISTANT EDITORS:
Peter W. Mehren
John B. Brown

PRODUCTION DIRECTOR:
James V. Leone

ART DIRECTOR:
Lorraine K. Hohman

EDITORIAL ASSISTANT:
Karen Boikess

INDEX:
Suzanne C. Klinghammer

CONTENTS

For a more complete listing
of animal names, see systematic classification or the index.

1   The Mollusks

Phylum: Mollusca, by
L. V. Salvini-Plawen
and R. Tucker Abbott

All of us are familiar with at least some members of this large phylum of animals, known as the mollusks, whether it be seashells from the ocean's shores or snails from our gardens and local ponds. Lovers of shellfish foods are well aware of the deliciousness of clams and scallops, and in some countries baked squid and octopus or broiled snails are favorites of many gourmets. Hardly any one who has vacationed along the tropical seashores is not aware of the great diversity of attractive shells sometimes washed up on the beaches. The products of mollusks, in the form of lustrous mother-of-pearl and iridescent abalone shells are well known to admirers of jewelry, buttons, and other forms of shellcraft. Precious pearls originate from oysters and other mollusks. Some fresh-water snails of tropical countries transmit diseases fatal to man, and, hence, are of considerable public health concern.

Mollusks are greatly diverse in structure and represented by many curious forms, from the highly active squids to the slow and sluggish snails, from the huge 180-kg (400-pound) *Tridacna* clams to the minute woodland snails of the highest mountains. Yet all of these creatures have one or more basic morphological and embryological features that unite them into the phylum Mollusca.

Distinguishing characteristics

The mollusks are coelomate invertebrates, that is, animals lacking a backbone and possessing an internal cavity or coelom that is usually reduced in most mollusks to small pericardial, renal, and gonadial cavities. The animals are basically bilateral, with each side being the same, and not radial, as is found in the circularly arranged echinoderms, such as the starfish. Adult mollusks vary in body length from 1 or 2 mm to over 22 m, as in some deep-sea squid. Most mollusks produce a hard exterior shell, although many, such as the *Sepia* squid and the *Aplysia* seahare, produce internal ones; a few are without shells, such as the garden slugs and the sea nudibranchs. The body is without true segments such as are found in the annelid worms. The ventral side is usually developed into a muscular locomotory organ. Characteristic of the mollusks is a fleshy fold or cape known as the mantle which not only surrounds most of the body but produces the shelly skeleton. The mantle forms a protective cavity which usually contains paired gills, or ctenidia, the anus, the excretory and genital ducts, and sometimes the foot. In all classes, except the bivalvia, the foregut bears a toothed, chitinous ribbon, the radula. This typically molluscan organ is used to rasp and tear food. The heart, located posteriorly or centrally, acts as a pumping organ for the open circulatory system of primitive hemolymph. There is no well-developed brain, but there are several highly organized ganglia. The central part of the nervous system consists of a circumesophageal ring formed by a dorsal cerebral ganglion and a ventral labial commissure, as well as several pairs of large commissures innervating the foot and organs in the mantle cavity. Many mollusks, particularly the lung-bearing land snails, the pulmonates, are hemaphrodites with each individual bearing organs of both sexes. Other mollusks are dioecious, having the sexes in separate individuals. Eggs develop predominantly by spiral cleavage into first a trochophore larva with a ring of hairlike cilia around the oral cavity, and second, in many marine forms, into a free-swimming larval veliger stage.

The phylum shows a wide evolutionary diversity, and its success is attested to by the great number of living species and the great variety of its chosen habitats. Mollusks are found in the deepest parts of the ocean, in all bodies of fresh water, including arctic pools and thermal springs, and from tropical swamps to the limit of vegetation on the highest mountains; many are parasites on and within other invertebrate animals.

Great diversity of form

The phylum Mollusca has been variously subdivided over the years by various schools of thought, but the classificatory scheme for the larger groups appears to have reached a reasonable degree of stability. The following subdivisions are generally accepted today:

Class 1, Aplacophora. Class 2, Polyplacophora (Amphineura). Class 3, Monoplacophora. Class 4, Gastropoda. Class 5, Scaphopoda. Class 6, Bivalvia (Pelecypoda of some authors). Class 7, Cephalopoda.

At one time the Brachiopoda were included among the mollusks. More recently, in the original German edition of this encyclopedia, Luitfried von Salvini-Plawen, an outstanding expert on solenogasters, presented the following divisions: 1. Subphylum Aculifera Hatscheck, 1891, with three classes: the Solenogastres, the Caudofoveata, and the Placophora. 2. Subphylum Conchifera Gegenbaur, 1878, with five classes: the Tryblidiacea (Monoplacophora of other authors), the Gastropoda, the Scaphopoda, the Bivalvia, and the Cephalopoda. C. R. Stasek (1972, p. 40) raises Placophora von Ihering, 1876, to subphylum rank.

From a phylogenetic point of view, the mollusks can be included with the SPIRALIA, or animals characterized by an embryonic spiral cleavage, such as the flatworms, the threadworms, the kamptozoans, the sipunculid worms (see Vol. I), the Arthropoda (see Vols. I and II), the following

Form and phylogeny

Phoronida (see Chapter 8), and the mollusks themselves. The larval forms of these groups demonstrate that the mollusks are not phylogenetic successors to the arthropods. In all liklihood, the ancestor of the mollusk was a platyhelminthiclike creature. The arthropods were doubtlessly another side branch of the molluscan ancestor which changed radically due to a different mode of life.

As has already been mentioned in Vol. I, the term "worm" is not a very useful one in zoological comparisons. One can only understand phylogenetic relationships by comparing embryological and larval developments, musculature systems, nervous systems, and locomotory modes. At first glance, the mode of locomotion reveals a relationship between the flatworms, or platyhelminths, and the mollusks, in that both usually glide along without noticeable body contractions and often on a base of mucus. An earthworm, on the other hand, stretches and contracts like an accordion.

Versatility of the mollusks

The diversity of mollusks is expressed in many ways, including not only basic modes of locomotion, but also in the great differences of body size. Many solenogasters, snails, and adult clams barely exceed a length of 1 or 2 mm, while some kinds of squid almost rival the length of the largest vertebrates. Between these extremes the mollusks are represented by a wide variety of forms and tolerances to various habitats. There are bivalves and snails that are firmly cemented all their adult lives to a hard substrate; there are slow-moving garden slugs and river clams, and there are elegant pteropod "sea butterflies" and swift squids that criss-cross through the open oceanic spaces. Some land-snails can survive years of desert heat in a state of estivation. Some gastropods live in thermal springs where the temperature rises to 40°C (104°F), while others can readily survive the winter frozen into the ice cover. Some mollusks live in the perpetual darkness of caves, some under immense pressure at the bottom of the ocean, and others exist in the thin air above the 5,000 m mark in the Himalayan Mountains; others are internal parasites within the intestinal tracts of various animals belonging to other phyla. The octopus is one example of evolutionary development in habits and actions which is comparable to that of certain insects, and is surpassed only by that of the vertebrates.

Common characteristics

What are the common characteristics to all of these diverse molluscan forms? Which properties supposedly characterized the hypothetical "archeo-mollusk" according to our present-day conception of phylogeny (see Color plates, pp. 65–66)? The molluscan body is soft and has no regularly organized internal skeleton, except in the case of the *Sepia* squids, or cuttlefish. The body is neither truly segmented nor surrounded by articulating appendages, such as one finds in insects or crustaceans. Mollusks are particularly distinguished by their soft, flesh mantle, a cape-like organ capable of secreting the calcareous external, or rarely internal, shell.

The foot, a broad creeping disc or pointed flexible prong, varies greatly among the various classes of mollusks. In the chitons, limpets, and some tectibranchs it is separated from the mantle by a deep, broad groove encircling the animal's body. The groove usually contains a number of featherlike gills, or ctenidia. In conjunction with the gills, and within this groove or within a pocket-shaped mantle cavity, there may be the exits for the gonoducts, excretory organs, the intestine, and one of the gills may be highly modified into a chemoreceptive organ known as the osphradium.

Mollusks are also characterized by a unique rasping tongue, the radula, which generally serves in obtaining and breaking up food. In early evolutionary stages, the intestine probably was a straight tube with lateral pouches. Present-day molluscan forms, however, usually possess long and convoluted intestines. The foregut includes various salivary glands, and there are almost always glandular digestive diverticula which branch off the midgut. Digestion takes place in these areas, and in a region posterior to the stomach, known as the digestive gland. The radula is absent in all bivalves and a few parasitic gastropods.

Like the arthropods (see Vols. I & II), the mollusks possess a true heart, although in the tusk shells, or Scaphopoda, it is greatly reduced. It is located in the posterior third of the body above or around the rectum, circulating the body fluid through the body (this hemolymph is comparable to the blood and the lymphatic fluid of the vertebrates). Usually, however, true blood vessels are missing. The blood, therefore, circulates between the connective tissue spaces of the haemocoel, or primitive body cavity, and the body organs (open circulatory system). The heart, consisting of a ventricle and usually two auricles, is located in an elongated indentation along the dorsal midline, protectively enclosed in a pericardium which may have evolved originally to support the heart's pumping activity. The anterior portion of this pericardial cavity is filled with reproductive cells, and in present-day forms this part exists as an independent organ, the gonad. These body cavities are considered separate coelomic structures, associated with a pair of coelomoducts. In the majority of cases the pericardial ducts serve as excretory ducts for the "kidneys." The total complex is known as the urogenital apparatus.

The nervous system consists of a ganglionlike brain with four main nerve cords running posteriorly from it. A circumesophageal ring innervates the buccal region. The presence of sensory organs depends on the developmental phase, habitat, and living habits of the animal. The musculature of the mollusks is usually very well developed in order for the animal to carry out the mechanical functions of life. In light of phylogeny and function, the muscular strands (dorso-ventral musculature) which intertwine the mantle and foot are particularly important. Within this phylum there is a continuous series of muscle concentration depending on the development and function of the mantle and the foot. In the early mollusks one encounters serially arranged muscle pairs. In the transitional

Foot and mantle cavity

The radula and the alimentary canal

Circulation and the body cavity

Nervous system and musculature

stages there is a trend toward the fusion of these muscle strands into a single muscle bundle on each side of the animal's body. All these organs are surrounded by mesenchyme which is chiefly responsible for the soft body composition of the mollusks.

Reproduction

Larval development

Most mollusks exhibit sexual reproduction, many employing copulation. Others expel the sex products directly into the sea where the sperm find and enter the egg. The embryological processes of the various mollusks can be very revealing. The size of the eggs and the subsequent development into the adult stage varies greatly among mollusks. Yet despite the great diversity (plasticity) in the mollusks, spiral cleavage is a feature common to all except the highly specialized cephalopods (see Chapter 7). In most marine forms this results in a typical free-swimming larva characterized by a tuft of cilia and a ciliary girdle (see Fig. 8-25). In such primitive representatives as the solenogasters and the nut clams, the embryo proper is still enclosed by a protective layer of cells (yolk larva; see Color plate, p. 26); in other forms this cell layer has become reduced to a disc with long cilia (trochophore; see Color plate, p. 26) or has been modified into saillike velar lobes (veliger; see Color plate, p. 26). During metamorphosis the larvae loose their swimming and flotation organs, sinking to the bottom to complete their development and produce the adult shell.

Phylogeny of the mollusks

On the basis of fossilized remains we can trace to differing degrees the phylogenetic development of the various molluscan groups. However, our knowledge is limited to those forms which had shells that could be fossilized. These would be the shelled mollusks. Fossil records dating from the Cambrian (600 to 500 million years ago) serve as evidence for the geological antiquity of the molluscan phylum, but they do not illuminate the origin nor the phylogenetic interrelationship of the various molluscan classes. It was formerly assumed that the mollusks originated from the annelids, an assumption based only on the evidence of fossilized monoplacophorans. However, this theory has been largely rejected. The body structure of a newly discovered living monoplacophoran, *Neopilina galatheae* (see Chapter 4), has revealed a form of semi-segmentation (metamerism) that weakens the case that mollusks evolved from the annelids.

The origin of the molluscan phylum must go back to Precambrian times, more than 600 million years ago. Only minute calciferous structures with circular broad chambers, i.e., genus *Wyattia*, have been found from this period. According to Taylor, these forms may have been of molluscan origin. Therefore, while the appearance and habitat of the ancestral mollusk is still open to speculation, we can reconstruct a hypothetical primitive mollusk (see Color plates, pp. 65–66).

Only molluscan fragments have been described from the Cambrian. Some of these fossils can be classified with present-day classes (i.e., gastropods and cephalopods). One group, the Hyoliths (Hyolithida), was

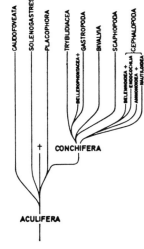

Fig. 1-1. Relationship between recent and some fossil (+) mollusks.

once thought to be tiny mollusks which were usually planktonic in the manner of the sea butterflies (Pteropoda; see Chapter 5) with which they were originally classified. These forms have only been recorded from the Palaeozoic era, and their club-shaped calciferous shells, which were not subdivided into chambers and measured only a few centimeters, are their only remaining legacy. In cross section these structures appear as elliptical or roundish triangles. They were equipped with a lid. That they were mollusks is questionable.

Other molluscan shells from the Cambrian undoubtedly corresponded in their structural design to those of the Cephalopoda (see Chapter 7). These shells possessed a chambered structure with a siphuncle which helped to maintain the animal's balance in the water. These shelled cephalopods, now extinct, included the genus *Volborthella*, an example from the Lower Cambrian, and the genus *Plectronoceras* from the Upper Cambrian. However, their shells were not spirally coiled in one plane as in *Nautilus* (see Chapter 7) but were straight or only slightly coiled. Cephalopods with these types of shells were common in the Palaeozoic in a great variety of species and forms. Here only the genera *Orthoceras*, *Gomphoceras*, *Phragmoceras*, *Endoceras*, and *Ascoceras* will be mentioned. They belong to the subclass Nautiloidea (see Fig. 1-1). This group gave rise to the completely coiled shells of nautiluses which originated in the early Palaeozoic and which still enjoyed worldwide distribution in the Lower Mesozoic. During the course of the Upper Mesozoic the number of nautiloid species dwindled steadily and only a few survived into the Tertiary and Recent times. The present-day nautiloid species can, therefore, be considered as true "living fossils." The nautiloid characteristic of the shell was already established in the Lower Mesozoic, and there is evidence of the living genus *Nautilus* from the early Tertiary.

Ammonoids

The ammonoids (subclass Ammonoidea) were another group of shelled cephalopods which were even more diverse in form, represented by a greater number of species than the nautiloids. The shells of the ammonoids also evolved from straight to coiled shapes. This example of parallel evolution is associated with a more advantageous position during swimming. The term Ammonoidea is derived from the Egyptian god, Ammon, who had ram horns as adornment on his head. The ammonoids lived from the Devonian up to the end of the Cretaceous (from 400 to 65 million years ago), although their ancestral forms, the Bactritida, which possessed straight shells, were already prevalent in the Ordovician and Silurian (500 to 425 million years ago). These representatives serve as the best "index fossils" of the Mesozoic. "Index fossils" means that certain species occur only in specific geological layers or formations of which they are characteristic. Some ammonites are known popularly as "snake stones" or "golden snails" depending on their configuration and state of preservation. The shells of the ammonoids can be differentiated from the nautiloid ones on the basis of their usually complex suture patterns (see

▷
Dondice banyulensis (nudibranch) is found in the Mediterranean. The body surface is covered with bizarre appendages (the cerata).

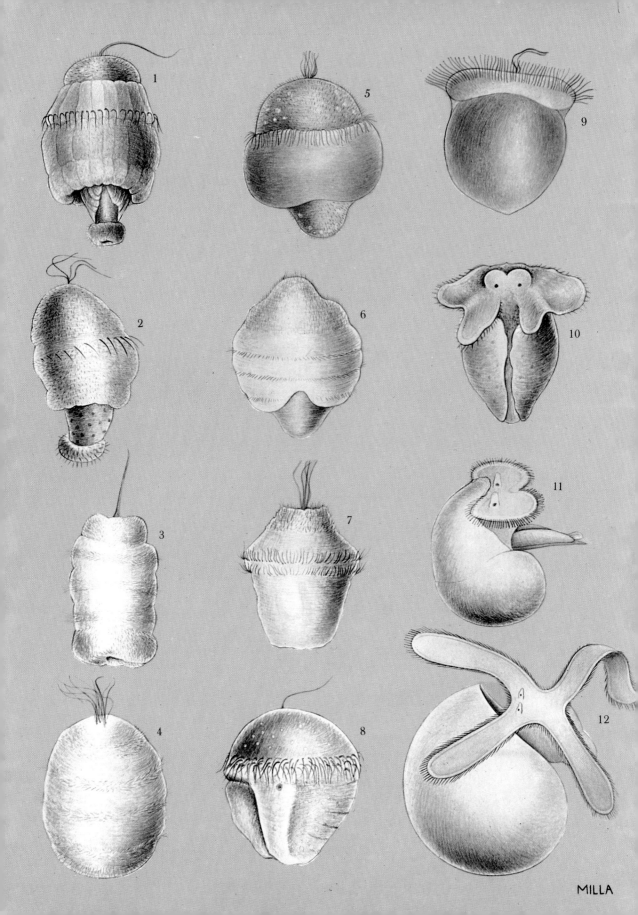

MILLA

Chapter 7), the consistent marginal siphuncle, and the horny or calcareous lids (aptych) which usually consisted of two parts. Certain geological layers in the Alps of Europe contain only these lids but no ammonoid shells. This was probably because after the death of an ammonoid, the argonite-containing shell with its gas-filled chambers drifted away, leaving behind the calcite aptych. During the transformation of loose sediment into solid stone (diagenesis), only the calcite lid was preserved. However, the soft parts of the ammonoids have also become known, and it turned out that despite their exo-skeleton the ammonoids are more closely related to the dibranchiate cephalopods than to the genus *Nautilus*. The ammonoid radula is similar to that of the living dibranchiates. The ammonoids also had an ink sac and eight to ten tentacles.

During the Lower Triassic (230 to 180 million years ago) the ammonoids reached their first phylogenetic peak. Toward the end of the Triassic the ammonoids became extinct except for the Phyllocerata. This group in turn gave rise to the ammonoids (Neoammonoides) of the Jurassic and Cretaceous periods. These forms flourished until the Upper Cretaceous (about 65 million years ago). In the meantime—particularly in the Cretaceous—numerous forms with aberrant shells evolved (the so-called heteramorphous ammonoids), for example the genera *Turrilites*, *Scaphites*, *Baculites*, and *Nipponites*. The Cretaceous period also included gigantic forms, such as *Pachydiscus seppenradensis* with a shell measuring over 2.5 m in diameter. All the ammonoids became extinct near the end of the Upper Cretaceous.

Belemnites

The belemnites (subclass Belemnoidea) are another group of cephalopods that became extinct. They existed during, and possibly before, the Carboniferous (approximately 300 million years ago) but did not flourish until the Jurassic and Cretaceous. Their body shape is reminiscent of modern dibranchiates but the internal skeleton consisted of a chambered phragmacone and a calcified rostrum, features indicative of the fact that they evolved from the same stem as the ammonoids, namely the straight-shelled cephalopods.

The monoplacophores or gastroverms (class Monoplacophora; see Chapter 4), living members of which were discovered only a few years ago, are another group of mollusks going back to the Palaeozoic. Judging by their distribution and thick shells, these formerly common animals must have inhabited the shallow seas.

Some of the geologically oldest and most primitive snails included mollusks with bilaterally symmetrical coiled shells and an aperture. These forms, the bellerophonts (suborder Bellerophontina), possibly were the common ancestral group to all snails (see Chapters 4 and 5).

Asymmetrical coiled shells are also known from the Upper Cambrian. These groups only began to flourish with the snails (class Gastropoda; see Chapter 5) during the Cenozoic. The gastropods are the most successful group of mollusks, for they have invaded practically every conceivable

◁
Larvae of mollusks: Yolk larvae: 1. *Nematomenia banyulensis*; 2. *Neomenia carinata*; 3. *Yoldia limatula*; 4. *Nucula proxima*; 5. *Epimenia verrucosa*; 6. *Dentalium dentale*. Trochophore type: 7. *Patella vulgata*; 8. *Ischnochiton*; 9. *Dreissena polymorpha*; 10. *Gasteropteron rubrum*; 11. *Nassarius*; 12. Veliger of *Murex ramosus*.

ecological niche. The geologically oldest pulmonate snails (genus *Anthracopupa*) have been recorded from the Lower Permian of North America.

It has not yet been possible to prove conclusively that the bivalves (class Bivalvia; see Chapter 6) originated in the Cambrian. However, the Babinkacea were already present during the Palaeozoic era. These forms can only be differentiated from other bivalves by their several pairs of abductor muscles. Although one cannot regard these bivalved animals as direct ancestral forms to the true clams, or Bivalvia, they nevertheless serve as their precursor. These bivalve precursor forms and the Monoplacophora can be derived from a common ancestral form. Like the gastropods, the bivalves did not reach their phylogenetic peak until the Cenozoic era.

The rudists (of the bivalve order Hippuritoida) of the Cretaceous possessed greatly modified shells. One of their valves was cemented to the substratum. It had a cup or gobletlike shape. The other valve was flat and served as a lid. In this feature the rudists deviated somewhat from the usual bivalve design.

Proving the existence of the scaphopods (class Scaphopoda; see Chapter 6) and the chitons (class Placophora; see Chapter 2) in deposits of Cambrian origin is also difficult. Forms found from Palaeozoic deposits do not differ basically from their living representatives today. Some fossil remains from the Cambrian (genus *Matthevia*) were identified as placophores and were placed in a separate class, but additional evidence is necessary for a final placement. The oldest placophores (Palaeoloricata) lacked the second lowest shell layer (articulamentum) of today's chitons.

It is difficult to estimate the number of living species of mollusks because of varying opinions of the morphological and biological limits of a so-called species. Estimates have ranged as high as 125,000, but more recent reappraisals set a lower figure of about 60,000 living species.

Approximately 60,00(species of mollusks

It is impossible to discuss all of these molluscan species in one volume, just as it was with the many insect species, because the mollusks are the second largest animal phylum, outnumbered only by the arthropods. The living mollusks are classified into classes on the basis of differences in the mantle and associated shell structures, the pallial cavity, the foot, and the formation of other body parts. The Monoplacophora, Aplacophora, and Polyplacophora include the most primitive and exclusively marine forms. The Cephalopoda, the Bivalvia, and Gastropoda contain many highly developed and specialized forms.

2   The Solenogasters and Chitons

Solenogasters and
Chitons, by L. v.
Salvini-Plawen and
R. Tucker Abbott

Three groups of primitive mollusks have been variously classified as either gastropods, or snails, or have been treated as separate classes, or even as non-mollusks. In the original German edition of this encyclopedia, L. von Salvini-Plawen combined these groups into the subphylum Aculifer with three classes: Solenogasters, Caudofoveata, and Placophora. At one time they were classified as the Amphineura. Today, they are generally accepted as two classes: 1. APLACOPHORA, the small vermiform, deepsea solenogasters. 2. POLYPLACOPHORA, the eight-plated, marine chitons, also known as Placophora and Loricata. In both these groups the mantle covers the entire animal and may secrete a shelly or leathery cuticle which partially or entirely covers the body. Calcareous spicules are embedded into this integument. The BL varies from 0.5 mm to over 30 cm. They lack cephalic eyes and tactile and gravity receptors (statocysts). The nervous system consists of lateral nerve cords with a suprarectal commissure. The intestinal tract always lacks a proto-style. The anus is always found just beyond the subterminal body section. They are strictly marine.

Members of these classes are represented on the color plates of this volume (see Color plates, pp. 35 and 36) and show that these creatures do not meet our accustomed image of a mollusk. Although they represent only a small number of the mollusks, they are of great significance in phylogenetic studies. In a way they are "relict groups." In addition to numerous similarities in the finer structural details, all are similar because they lack a head region with eyes and statocysts. The four large nerve cords are still connected by numerous commissures and there is a suprarectal commissure.

The solenogasters, or "worm mollusks," included two phylogenetically distinct relict groups which, however, did not share many common characteristics, except for the vermiform body form and certain primitive features noted in Chapter 1. Any apparent similarity in this case is due to parallel evolution; therefore, the solenogasters and caudofoveates are regarded as separate subclasses by many scientists.

The SOLENOGASTERS (subclass Neomeniida or Solenogastres, see Color plates, pp. 65/66) are characterized by a laterally compressed body. The mantle is entirely covered with a cuticle and calcareous spicules. The BL of adults may vary between 1.5 mm and 30 cm. The foot is vestigial and is represented by a median ridge in a small ventral groove. The mantle cavity (pallial cavity) is posterio-ventral. There are no true ctenidia, however. There are gill-like structures (filaments, leaflets, and papillae) which may aid in respiration. Terminal sensory organs are often vestigial or are multiple. The midgut is usually associated with several successive lateral pouches. The dorso-ventral musculature between the mantle and the foot consists of strands. True gonoducts are almost always vestigial. Gonodal cells are discharged via the coelomoducts. The animals are hermaphroditic, and fertilization is internal. They are free-living on the bottom of the sea or live as predators of other invertebrates, usually on corals and hydroids. According to present-day knowledge, there is only one family Neomeniidae) in the subclass Solenogastres with fifty genera and 115 species.

"And this worm is called a mollusk!" an observer once exclaimed at the rather rare sight of a solenogaster squirming in a petri dish. Only upon close examination does one notice two peculiarities which definitely characterize this creature as a mollusk. The entire animal is covered with a dense coat of shining spicules or scales, and has a continuous ventral groove, or an elongated line. This groove contains the highly reduced molluscan foot. In the solenogasters this foot is further reduced to varying numbers of longitudinal folds. The animal's mode of life may explain this feature of extreme reduction of the foot. Solenogasters still use this structure for locomotion, but with the aid of cilia. The musculature of the body is not the sole means of locomotion. Consequently, one can speculate that during the course of evolution the locomotory mode was not altered but rather that newly evolved life habits required a greater degree of agility. The animals either slid or wound themselves along. This in turn favored a narrowing of the body.

The total reduction of the ctenidia, structures usually thought of as a molluscan characteristic, can also be explained on the basis of the solenogasters' mode of life. Since the animals inhabit the floor of the ocean, they are in constant contact with a changing water supply. At certain regions of the body, sufficient gas exchange takes place through the skin. This made the gills less necessary, so they disappeared without any loss in respiratory efficiency. However, several species evolved secondary gill structures, such as filaments, leaflets, or papillae on the posterior edge of the mantle cavity, which function solely as respiratory organs.

With the evolution of the cylindrical body shape, the mantle cavity became limited to the posterior portion. This shift also resulted in the mantle cavity becoming located partially within the body proper. Here

Class: Solenogasters

Distinguishing characteristics

Fig. 2-1. Calcareous spicules of the solenogasters and chitons: 1 and 2. Solenogasters, and 3. Polyplacophora.

Fig. 2-2. "Cupid's arrow" of the solenogaster *Genitoconia rosea*.

Fig. 2-3. Various radulae of the solenogasters (only one transverse row is illustrated).

Fig. 2-4. Distribution of *Neomenia carinata* in Europe.

it gave rise to glandular gonoducts or commonly was modified into a copulatory organ. In *Genitoconia* and other genera. excitatory organs (copulatory spicules or "cupid's arrows") function in conjunction with the copulatory organ, which consists of either cuticular or calcareous material just like the secretions of the mantle. This copulatory spicule occurs only in the solenogasters (and a few gastropods), and is functional during these hermaphroditic animals' copulation. The actual gonoducts are almost always reduced. Instead, the eggs and sperms are discharged via the ducts of the pericardium (coelomoducts), which open into the anterior section of the mantle cavity.

The oral cavity is associated with a buccal cavity, often of trunklike appearance. The buccal cavity possesses an eversible gustatory subradular organ which functions as a food taster. The buccal cavity usually contains a bipartite radula which commonly is shaped like a pair of pincers or some form derived from this. In many aspects this radula is reminiscent of that of the snails. In many suctorial solenogasters species the teeth are often reduced. These species probably secrete glandular substances which break down the host's tissues, which are then sucked up by the predacious solenogasters. Two well-known species are *Rhopalomenia aglaopheniae* and *Nematomenia banyulensis* (see Color plate, p. 35). Both feed on hydrozoans. *R. aglaopheniae* (L 10–35 mm) are found along the European coastlines at depths of 50 to 100 m where they are always associated with the hydrozoan species *Lytcarpia myriophyllum*. The reddish *N. banyulensis*, approximately the same size as *R. aglaopheniae*, feeds upon various hydrozoans (see Vol. I). The nematocysts which are also sucked in pass harmlessly through the intestine without discharging. Probably some kind of glandular secretion inactivates the "weapons" of the prey (see Chapter 5). *Nematomenia flavens* (BL 4 cm; see Color plate, p. 35) found in the Mediterranean and the *Nematomenia corallophila* (BL 2 cm) found near Algiers behave very similarly. The latter species can erect the whitish scales on the body, thereby mimicking the coloration of the red coral (*Corallium rubrum*).

However, the free-living predacious species of solenogasters, for example *Genitoconia rosea* (L 2–3 cm; see Color plate, p. 35) of the northeastern Atlantic Ocean, or *Eleutheromenia sierra* (L 1 cm) of the Mediterranean, both of which possess bipartite radula, often also prey on hydrozoans. Only a few species, including *Proneomenia sluiteri* (L 1–15 cm), from northern Europe, seem to be omnivorous. Although the ventral groove is extremely small, the animals use it as their sole means of locomotion. They always move along on the sediment on the ocean's floor. Only one burrowing form has been discovered. This is the stout *Neomenia carinata* (L 1–3 cm; see Color plate, p. 35), found along Europe's seacoast. It burrows with the trunk and searches for microscopic creatures. *Biserramenia psammobionta* (L 2–3 mm), on the other hand, inhabits sandy niches. Solenogasters move by the ciliary action of the foot together with the mucus secreted by numerous glands which are concentrated at the anterior end of the groove above the extensible ciliary groove. All benthic species, for example *Dorymenia vagans* (L 6 mm) or *Pruvotina impexa* (L 1 cm; see Color plate, p. 35), both from the Mediterranean, glide along and in this aspect as well as in their external appearance are often strongly reminiscent of certain nemertines (Nemertini; see Vol. I).

Since the solenogasters are usually carnivorous, their alimentary canal is always straight. Usually there are lateral caeca. The muscular strands leading to the foot are located between these digestive pouches. In many species a prominent longitudinal muscular strand which facilitates the protective rolling-in of the groove has evolved in both sides of the pedal groove.

The freely deposited eggs undergo spiral cleavage and then develop into free-swimming larvae. In most of the forms which have been studied, a well-developed, large-celled prototroch or test surrounds the embryo proper (yolk larva). The embryo grows out posteriorly in a conical shape (invaginal cone). The prototroch is either absorbed or is cast off during metamorphosis. This type of development takes place in *N. carinata*, *R. aglaopheniae*, *N. banyulensis*, and other species. The comparatively large southeast Asian *Epimenia verrucosa* (L 30 cm; see Color plate, p. 35), in contrast, is characterized in the early stages by a hat-shaped pretrochal region. This type of larva is transitional, and is a precursor of the mollusks that are characterized by a trochophore larva (see Color plate, p. 26). In *E. verrucosa* and also *Halomenia gravida* and *Provotina providens*, which occur in the Pacific Ocean, brood care is provided by pouches within the mantle.

Solenogasters commonly inhabit great depths of water. Since these creatures are usually of small size they are often overlooked or are not properly identified because of their external appearance. The spiny cuticle serves as protection against smaller carnivores such as polychaetes, fish, or other predators which hunt along the bottom of the sea. The calcareous spicules embedded in the skin serve as a shield against the action of the discharged nematocysts of the hydrozoans. *Forcepimenia protecta* (L 2 mm), which preys on hydrozoans in the Red Sea, was examined and its entire mantle was found to be covered with discharged nematocysts.

Unfortunately, little is known about these remarkable animals because for decades they have not been investigated. Our knowledge about the distribution patterns of individual species is particularly scarce. For instance, *Strophomenia indica* (see Color plate, p. 35) has been found in Indonesia as well as in Naples. It would be of great value if zoologists would devote more time to this unusual relict group of mollusks.

The subclass CAUDOFOVEATA (see Color plate, p. 35) is characterized by an elongated body covered entirely by cuticle and scales secreted by the mantle. The range in length is from 0.5 to 14 cm. A uniform or lobed pedal shield (burrowing and sensory plate) is present posteriorly or

Embryological development

Fig. 2-5. Ctenidia of the solenogaster, *Chaetoderma nitidulum*.

The caudofoveate solenogasters

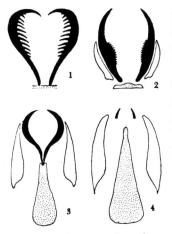

Fig. 2-6. Retrogression of the radula in the caudo-foveate solenogasters (in each case one transverse row is illustrated). 1. *Scutopus ventrolineatus*, 2. *Prochaeto-derma*, 3. *Falcidens*, 4. *Chaetoderma nitidulum*.

Fig. 2-7. North Atlantic distribution of *Falcidens crossotus*.

around the mouth. The mantle cavity is located posteriorly. It contains a pair of true bipectinate ctenidia. The terminal sense organ is always unpaired. The midgut is usually associated with a ventral digestive caecum. The dorso-ventral musculature between mantle and ventral side consists of numerous, but usually reduced, pairs of muscle strands. The gametes are discharged via the pericardial ducts (coelomoducts). True gonoducts are absent. The sexes are separate. Fertilization is external. The animals burrow in the sediment of the ocean floor. There are three families—Limifossoridae, Prochaetodermatidae, and Chaetodermatidae—with six genera and about sixty species.

Again we are faced with a group of mollusks about which little is known except for a few species from well-explored regions. The caudo-foveate solenogasters were formerly classified as only one family, Chaeto-dermatidae, or the "worm mollusks." However, closer studies of the body structure show that the caudofoveate solenogasters evolved as a separate and independent group, which might be regarded as one of the earliest side branches of the molluscan stem.

The uniqueness of the caudofoveate solenogasters is particularly evident in the armorlike plate located posteriorly to or surrounding the oral cavity. This structure has the dual function of a perch during burrowing or as a digging organ. There is no true foot. The whole body is cylin-drical. The only distinguishing external feature is the bell-shaped posterior mantle cavity which houses a pair of ctenidia. The ancestors of the caudofoveate solenogasters were probably flat and broad. During the course of time the body became laterally rounded which eventually resulted in the total loss of the molluscan foot. This loss can be explained on the basis of the animal's particular mode of life. The caudofoveate solenogasters are burrowers in the bottom sediment of the sea, and the groping and searching activity of the burrowing head section of the flatter ancestral form with its creeping foot must have required a twisting of the body. This meant that the posterior end, with the ctenidia, protruded from the sediment as the animal burrowed to greater depths. The posterior part of the foot thus lost its usefulness. Gradually a progressive reduction of the creeping foot occurred. Total absence of the foot at the anterior end coincided with the adaptation of the new body form, and with the mode of burrowing by erectile tissue. According to S. Hoffmann, the pedal shield, consisting of the burrowing and sensory plates just posterior of the mouth cavity, probably is the last remnant of the most frontally located foot section.

Scutopus ventrolineatus (L 10–35 mm), from the North Sea and other oceans, already illustrates reduction of the foot from posterior to anterior. A "scar line" is clearly discernible on the anterior part of the body but not in the posterior section. The ctenidia of the present-day burrowing forms were a necessity, but these organs moved to the posterior end along with the mantle cavity.

Certain degenerations have taken place in the Caudofoveata in correlation with the particular body form and mode of life. For example, gonoducts of the paired unisexual gonad were lost. Here, as in the solenogasters, the gametes are discharged via the pericardium and coelomoducts into the mantle cavity. This cavity, housing a pair of bipectinate ctenidia, can be sealed completely by a sphincter muscle, a condition similar to that of the solenogasters except for the manner in which the two frontal sections of the mantle penetrate into the body. Those cause the formation of short glandular grooves or cavities (mucous grooves or mucoid ducts) into which the coelomoducts open. The common reduction of the musculature between mantle and central abdomen, representing the former foot, is correlated with the newly evolved body shape. Only *Scutopus ventrolineatus* still possesses groups of muscular strands in addition to a longitudinal muscle (rolling-in musculature) on both sides of the plane of symmetry. The three-layered epidermal muscle trunk is particularly significant in burrowing. Here, unlike the solenogasters, this muscular layer is well developed.

The animal's mode of life has also had its impact on the feeding apparatus. As a rule, the primitive limifossorids (family Limifossoridae) are still characterized by a strongly developed bipartite radula. In the prochaetodermids (family Prochaetodermidae) fewer transverse rows are noticeable and there is a trend toward the formation of various platelike structures. In the chaetodermids (family Chaetodermidae) the radula has been reduced to only one or two pairs of pincers, as in the genus *Falcidens*, or has atrophied to a few denticles. *Chaetoderma* has lost the radula altogether, replacing it with platelike structures which obviously function only as crushing organs (see Fig. 2-6). The alimentary canal consists of a short cylindrical stomach, usually a long, unpaired lateral caecum which passes ventrally to the midgut, and a long, straight hindgut.

All Caudofoveata inhabit the sediment at the bottom of the sea. The animals may partially protrude from a tube and often may expose the mantle bell, surrounded by long protective spicules, to the surface. In this position the ctenidia are greatly extended to facilitate breathing. Almost all species investigated so far also burrow parallel to the surface of the sea bottom, for example the *Limifossor talpoideus* (L 6–12 mm; see Color plate, p. 35), from Alaska. It is essential that both ctenidia are well circulated for respiration. This problem has been solved by the pumping action of a powerful heart. The heart beat in *Scutopus ventrolineatus* is usually nine to twenty-one beats per minute; in *Falcidens sagittiferus* twenty-five to thirty; in *Falcidens crossotus* (L 10–25 mm; see Fig. 2-7) twenty-seven to thirty-two beats per minute.

Because of their mode of life these elusive creatures are rarely found. Their bodies are very resistant to crushing in the unsettled bottom sediment. They are well adapted to their environment by the dense, scaly

▷
Solenogasters: 1. *Neomenia carinata*; 2. *Nematomenia banyulensis*; 3. *Nematomenia flavens*; 4. *Strophomenia indica*; 5. *Pruvotina impexa*; 6. *Epimenia verrucosa*; 7. *Genitoconia rosea*.
Caudofoveata: 8. *Chaetoderma nitidulum*; 9. *Falcidens loveni*; 10. *Limifossor talpoideus*.

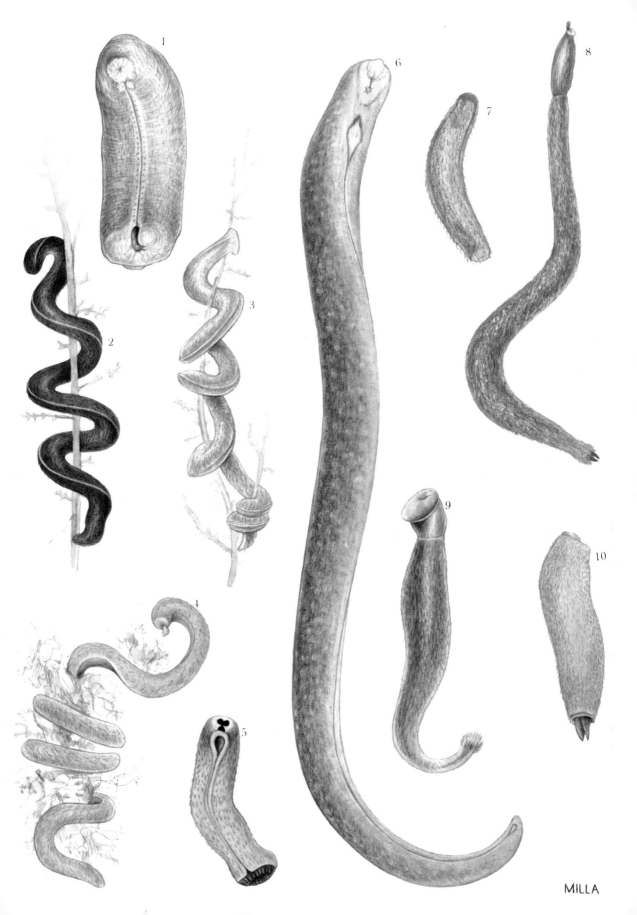

MILLA

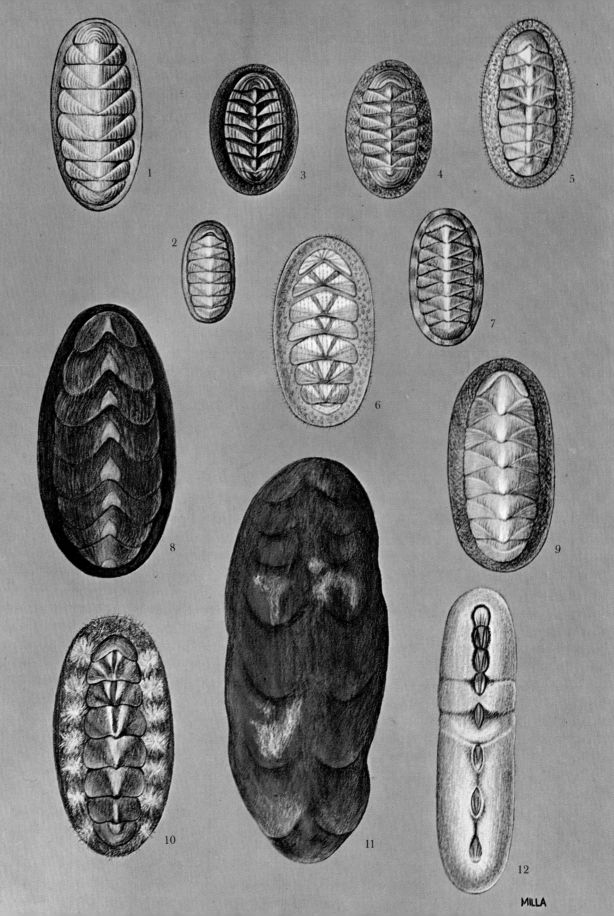

MILLA

◁
Polyplacophora (chitons):
1. *Lepidopleurus cajetanus*;
2. *Lepidopleurus cancellatus*;
3., 4. *Ischnochiton varians*;
5. *Callochiton laevis*;
6. *Nuttalochiton hyadesi*;
7. *Lepidochitona cinerea*;
8. *Placiphorella vestita*;
9. Mediterranean chiton
(*Chiton olivaceus*);
10. *Acanthochiton fascicularis*;
11. Giant Pacific chiton
(*Amicula stelleri*);
12. *Cryptoplax larvaeformis*.

Class: Polyplacophora

skin-covering and the prevalent grayish-brown coloration which matches the surrounding habitat. The Caudofoveata are found in their vertical J-shaped burrows, rarely in horizontal ones, but unlike many other burrowing animals they are unable to back out of their hiding places. They penetrate the sediment by alternately stretching and contracting the frontal part of the body. In order to move to a new location they must burrow in a frontal direction, probably because their scales point posteriorly. Under certain circumstances they may bury themselves totally again after changing location. The entire burrowing process, until the animal has completely disappeared beneath the ocean floor, takes twenty to thirty minutes in *Falcidens crossotus*, from the North Atlantic, and five to ten minutes in the Mediterranean *Falcidens gutturosus* (L 5–15 mm). In other species it usually takes more than thirty minutes.

The Caudofoveata feed on microorganisms such as algae, or protozoans which they pick up from the surrounding sediment. The prey is crushed between the pharyngeal plates of the radular apparatus and from there finds its way into the alimentary canal.

Intestinal samples have shown that Caudofoveata species with well-developed radula also feed on this algal food source. In *Prochaetoderma raduliferum* (L 3–4 mm), from the Mediterranean, the radula extends out of the gaping mouth. Perhaps in this case the radula serves as a broom, sweeping up the food, and the pedal shield functions as a testing organ, picking out the edible particles.

Only a few of the European species are widely distributed. *Chaetoderma nitidulum* (L 1–8 cm; see Color plate, p. 35), *Falcidens crossotus*, and *Chaetoderma canadense* (L 3–5 cm) are found from North America to the Mediterranean. Yet here, too, scientific studies have long been neglected. Probably many more species exist than we are aware of, with wider ranges of distribution than we presently know. The six living northern European and four Mediterranean species probably do not represent the true range of diversity for the Caudofoveata in these regions. There is still a wealth of material to be gathered about the animal's embryological development. The development of the externally fertilized egg and the larval formation have not yet been observed. In summary, even our knowledge about the biology of the Caudofoveata is still extremely fragmentary.

The CHITONS (class Polyplacophora) usually have strongly dorsoventrally compressed bodies. The central portion of the dorsal mantle is covered by a longitudinal series of eight overlapping calcareous plates which are encircled and sometimes partially covered by the mantle or girdle (perinotum). The mantle is a muscular region which is heavily cuticularized with spicules and scales. In length, adult chitons vary from 3 mm to over 30 cm. The foot is broad. The mantle cavity, a groove, houses 6–88 pairs of ctenidia arranged in groups. The osphradium is vestigial. The mouth is anterior to the foot. There are a pair of pharyngeal glands and midgut digestive glands which open into the stomach. The structure of the glands is characteristic for the various groups of chitons. The hindgut opens into the most posterior part of the pallial groove. The paired nephridiopores and gonopores are lateral to the anus. The nephridia show a great degree of branching. The shape of the nephridia differs in the various chitons. The animals are dioecious. Fertilization is external. The dorso-ventral musculature consists of 16 pairs of strands. The chitons mainly eat algae on rocky surfaces. There are two main orders of chitons: 1. Paleoloricata, an extinct, primitive group. 2. Neoloricata, existing from the Carboniferous to the Recent and containing three large suborders: 1. Lepidopleurina, 2. Ischnochitonina, 3. Acanthochitonina, with approximately 1,000 species.

The chitons characteristically have eight movable valves on the animal's dorsal surface. The shells are encircled by a mantle which is a muscular skin fold bearing calcareous spicules and scales, a characteristic feature for the entire class. Other characteristic features are the broad, elongated, oval-shaped foot and the pallial groove that encircles it. All these body structures meet our conception of the archeo-mollusk (see Color plates, pp. 65/66). Of course the eight overlapping shell plates are unique to the chitons. These dorsal shell plates evolved at a latter time period, probably as a protective measure against adverse environmental conditions. Correspondingly, the musculature between mantle and foot becomes concentrated into eight pairs of muscular strands.

All the chitons are marine, most inhabiting the littoral zone and especially the intertidal region. Only a few species are found in deeper waters, with some living at depths of over 1,150 m. Few are laterally elongated. The articulation of the shell plates and the prominent longitudinal muscular strand permit the animal to roll itself up like an armadillo, which provides protection against injuries and small enemies. Under normal conditions, however, the chiton is firmly pressed against the substrate with the foot and ventral surface of the mantle with the aid of the dorso-ventral muscles. This adhesion prevents desiccation of the animal during receding tides. It is sometimes easier to pull a chiton apart than to detach it from its substrate. To obtain an undamaged specimen it is necessary to surprise the chiton with the quick thrust of a knife.

The radula is fairly uniform throughout the entire group. The structure consists of transverse rows of seventeen teeth each. The radular complex serves primarily to scrape algae off the rocks in the coastal zones.

The LEPIDOPLEURIDS (suborder Lepidopleurina) are characterized by shell plates that are not interlocked and lack serrated edges. Each of the eight shell plates is subdivided into an anterior and posterior area. This primitive group consists of three families including the lepidopleurids, the hanleyids, and the choriplacaphores.

There are several interconnected features which demonstrate the special status of these primitive chitons. The most noticeable characteristic

Suborder:
Lepidopleurina

Fig. 2-8. Anatomy of a chiton (without the digestive system). 1. Gonad, 2. Mantle (pallial) cavity, 3. Mouth opening, 4. Ctenidia, 5. Nephridium, 6. Anus, 7. Pericardium, 8. Gonoducts.

Fig. 2-9. Geographical distribution of *Lepidopleurus asellus*.

Fig. 2-10. Geographical distribution of *Lepidochitona cinerea*.

Suborder:
Ischnochitonina

is the rather primitive construction of the shell plates. The tight connection between the dorsal plates and musculature is still absent. The articulamentum, the lower layer of the shell, which developed during the later stages of chiton phylogeny, is represented only by two apophyses in shell plates two to eight. The family LEPIDOPLEURIDS (Lepidopleuridae) includes the smallest forms of chitons, as well as most of the deepsea forms. Among the few species inhabiting shallow waters, *Lepidopleurus asellus* from northwestern Europe and *Lepidopleurus medinae* from Tierra del Fuego have highly specialized diets. Both species are from 1 to 2 cm in length, and inhabit the sandy sea floors where they feed on siliceous algae. *Lepidopleurus intermedius*, which measures only 4 to 5 mm, has thus far been recorded only in the Adriatic Sea. This species inhabits the spaces between the coarser shell sand. Frequently these species, as well as the equally small *Lepidopleurus cancellatus* of the North Atlantic (see Color plate, p. 36), escape detection in their rolled-up position when they have been dislodged from the substrate. Additionally, these chitons are well camouflaged against the background, a condition evident in the otherwise conspicuously colored *Lepidopleurus cajetanus* (L 3 cm; see Color plate, p. 36) from the Mediterranean and Portugal.

Most lepidopleurid species inhabit deep water. They are characterized by their small, inconspicuous shapes measuring up to 2 cm. The depth record is held by 8-mm-long *Lepidopleurus benthus* which was found at a depth of 4,200 m in the North Pacific. This species along with *Lepidopleurus belknapi* (L 10 mm), which was found at 1,890 m depth, and the antarctic coastal species *Hemiarthrum setulosum* of the family Hanleyidae are unusual in possessing only six or seven pairs of ctenidia, a characteristic considered primitive. The hanleyids include the widely distributed but inconspicuously muddy-colored *Hanleya hanleyi* (L 1–2 cm) which feeds on sponges.

The suborder ISCHNOCHITONINA is characterized by the distinctly edged notches on most of the shell plates. The eight shell plates are subdivided into an anterior and a posterior region. The broad girdle is embedded with calcareous formations of great variety. There are nine families, including the true chitons (Chitonidae), the mopaliids (Mopaliidae), and the ischnochitons (Ischnochitonidae).

This diverse group includes small as well as large forms. A particularly highly developed characteristic distinguishes several species, for example *Ischnochiton varians* (see Color plate, p. 36), and *Lepidochitona cinerea* (L 2 cm; see Color plate, p. 36), which varies greatly in its coloration. Occasionally even without the aid of a magnifying glass one can detect small dully pigmented to darkish glossy dots on the dorsal side of the shell plate. These are perforations containing the terminal caps of highly sensitive nervous epidermal strands (esthetes). Each esthete consists of megaloesthetes (central cores) with numerous branching microesthetes penetrating through the two upper layers of the shell plate. In *L. cinerea* there are 14 to 17 microesthetes. These sensory organs serve as light and shade receptors. Placophores are predominantly photonegative. Several species even possess blue eyes on the shell plates! The most highly developed evolutionary stage of these organs is represented by sensory cells consisting of a lens surrounded by pigment cells. Such eyes are present in the chitonid genus *Acanthopleura*, specifically the species *A. spiniger*.

One investigator has claimed that *L. cinerea* practices brood care, but there is still some controversy concerning this species. Yet it has clearly been shown that *Lepidochitona raymondi, Ischnochiton imitator,* and *Ischnochiton hewitti* provide care for their young. *L. raymondi* is the only hermaphroditic species. All other chitons are reported to be dioecious. The gonads are usually fused into a single, unpaired tube with a gonoduct on each side. In *Notochiton mirandus* and *Nuttalochiton hyadesi* (see Color plate, p. 36) the two gonads persist, although pressed very closely together. *N. hyadesi* even lacks an aorta. Two other species of this suborder which practice brood care are *Nuttalochiton thomasi* and *Middendorffia polii.* The fertilized eggs remain in the pallial groove until the larvae hatch. *Callistochiton viviparus* reportedly gives birth to fully developed juvenile chitons. *Chiton barnesi* and *Chiton nigrovirens* of the family Chitonidae carry their under-developed young within the pallial grooves.

The Polyplacophora are referred to as "chitons." The Greek word *chiton* sometimes meant a suit of armor plate. The true chitons are not well represented in the European region, but the MEDITERRANEAN CHITON (*Chiton olivaceus*; see Color plate, p. 36) is found rather frequently in its Mediterranean range of distribution. This species is colored in many shades. The color can vary according to age or the type of diet of the individual. Color changes are frequently brought about by algal growth, by other organisms which adhere to the shell, or by erosion of the surface layers of the shell plates. As a rule, chitons serve as a substrate for various organisms. The bivalve mollusk *Montacuta oblonga* has even been found on *Ischnochiton*. Chitons usually live for three to six years. Injuries and age-dependent deficiency symptoms incurred during the animal's life span may also contribute to its color changes. The life span for the WEST INDIAN CHITON (*Chiton tuberculatus*) is said to be up to twelve years.

Callochiton laevis (see Color plate, p. 36) is rather unusual among the European species because of its broad girdle. This form is more color constant than the Mediterranean chiton. *C. laevis* (L 2 cm) of the Mediterranean prefers to attach itself to the red *Peysonellia*, and has a red coloration for camouflage. This chiton has curved spicules which extend beyond the edge of the mantle into which they are embedded. The posterior margin of each shell plate is arranged with two lateral triangular rows of dark spots. These are the ocelli and the associated small lenses. In addition there may be scattered photosensitive esthetes. Even such a highly modified species as the large arctic *Placiphorella vestita* (see Color plate, p. 36) still

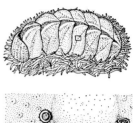

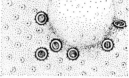

Fig. 2-11. Esthetes of *Acanthopleura spiniger.* The box from the upper illustration is magnified several times in the lower picture.

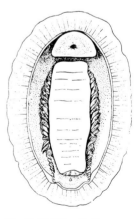

Fig. 2-12. West Indian chiton seen from the ventral side. The ctenidia are situated on either side of the central foot.

Fig. 2-13. *Placiphorella velata*, anterior mantle lobe raised to capture shrimp.

Suborder:
Acanthochitonina

Fig. 2-14. Geographical distribution in Europe of *Acanthochiton fascicularis*.

possesses esthetes although its dorsal plates, with the exception of a few pits, are entirely covered by the mantle.

James McLean has described some rather unusual anatomical peculiarities which appear only in the genus *Placiphorella* in the family Mopaliidae. One unusual feature of *Placiphorella* is its "head flap," an anterior extension of the girdle. This species also possesses a pre-cephalic mantle lobe with tentaclelike projections. James McLean, who studied the food gathering and other life habits of the species *Placiphorella velata* (L 2 cm) in Monterey Bay, California, tried to determine the advantages of these special anatomical structures. "At the beginning the animals were observed under water (in the field) where they conspicuously projected the reddish-colored undersides of the head flap out of crevices and rocky niches. At the slightest touch they would quickly clamp this flap against the rocky substratum. This same type of behavior was also observed in the aquarium. The pre-cephalic tentacles remained pressed against the substratum while the head flap was raised. In the aquarium, living amphipods were offered to *Placiphorella* with forceps, and the head flap immediately clamped down. In this manner this chiton captured amphipods up to 6 mm long, and digested them within one hour. Thus it seems that the chitons, which commonly feed on algae, also include some species with predacious habits. *Chaetopleura papilio*, and a few species of the genus *Mopalia*, for example *Mopalia hindsi*, reportedly feed on food of animal origin. According to McLean's observations *Placiphorella velata* seems to be a regular trapper.

The order ACANTHOCHITONINA is comprised of chitons in which the marginally matched shell plates are more or less covered by the mantle. The eight shell plates are subdivided into four areas. There is only one family: Acanthochitonids (Acanthochitonidae).

The tendency of the mantle to overgrow the shell plates laterally to varying degrees, already noted for *Amicula* and its related forms, is found in all acanthochitonids. The ACANTHOCHITONIDS (genus *Acanthochiton*) show another special characteristic. When glancing at the European species *Acanthochiton fascicularis* (see Color plate, p. 36), which is found up to the hightide line, one notices nine pairs of tufts of bristles at the sides of the shell plates. These tufts are highly sensitive to touch. *Acanthochiton communis* (L 2.5 cm), found in the Mediterranean, along the Portuguese coast, and in the Indo-Pacific, can only be distinguished from the European species by its smaller size. In the GIANT PACIFIC CHITON (*Amicula stelleri*; see Color plate, p. 36) the shell plates are covered by the mantle from all sides. This brick-colored species from the northern Pacific Ocean is the largest chiton, measuring up to 33 cm. This giant is completely covered with small calcareous spicules. In this respect it resembles the solenogasters. However, this aberrant similarity is only due to convergent evolution. Formerly the genus *Cryptoplax* was considered, because of its elongated body, as a transitional link between the chitons and solenogasters. Such similarities serve as fine examples of convergent evolution brought about by similar modes of life, as we have already noted in other features of the solenogasters and Caudofoveata.

Cryptoplax larvaeformis (see Color plate, p. 36), which reaches a length of 5 to 10 cm and which inhabits the Pacific Ocean, and other representatives of this genus, according to the words of L. Plate, "live in holes and rock crevices. They can stretch to extraordinary lengths, up to one foot, whereby the animal's body contours around the substrate. If a rocky cleft is very narrow only the animal's front end will penetrate. The rest of the body will remain outside as a thickly swollen appendage and is subject to tearing at the contact point." *Cryptoplax* also creeps along the surface of coral reefs. The extraordinary development of the mantle musculature, the high degree of contractibility, and the reduction of the shell plates are adaptations to a burrowing mode of life.

With the exception of the chitons that practice brood care or bear living young, the eggs are usually deposited into the water as two gelatinous strings. The externally fertilized eggs develop into free-swimming trochophore larvae. The trochophore possesses a pair of eyes and a single pedal gland. Both of these structures disappear at a later stage. The eight dorsal shell plates are recognizable at a very early stage. After only a few hours the prototroch is shed and the larva sinks to the bottom.

3   The Shelled Mollusks

The Shelled Mollusks,
by L. v. Salvini-
Plawen and
R. Tucker Abbott

All those mollusks that either bear a shell (concha) or have a shell-bearing ancestor also share many other characteristics. For this reason we have discussed them under one chapter. In the original German edition, von Salvini-Plawen applied the subphylum name Conchifera to this group, but that name, used first by Lamarck in 1818, was used up into the Victorian era for the Bivalvia and Brachiopoda. The most conspicuous structure of these groups is the shell. The shell or calcareous mantle formation is usually large and uniform but in several groups it is reduced or is totally absent. However, the larval stages reveal that this conspicuous feature does develop in all mollusks at some time. When discussing the mollusk shell one cannot speak of a "skeleton." Although the shells are solid structures they are nevertheless merely secreted deposits. The shells in these higher mollusks lend support and protection to the animals' bodies but are not a life-essential feature, as becomes clear from the numerous species whose shells have been reduced.

Distinguishing
characteristics

The internal organs are frequently concentrated in a dorsal pouch. The visceral mass is covered only by the mantle which is not embedded with calcareous spicules. Eyes and tentacles occur frequently and statocysts are always present. The longitudinal cords of the nervous system are connected by a subrectal commissure. The shape of the foot varies greatly. The mantle cavity is frequently rotated due to torsion but is rarely reduced. The alimentary canal often contains a crystalline style (see Chapters 5 and 6). The midgut gland is paired in the embryological state. In many cases the anus does not open posteriorly. The remaining five classes: 1. Gastroverms (Monoplacophora); 2. Snails (Gastropoda); 3. Tusk shells (Scaphopoda); 4. Bivalves (Bivalvia); 5. Squid and octopods (Cephalopoda). There are approximately 70,000 living species.

When looking at illustrations of snails and bivalves one is overcome by the wealth of shapes and colors of their shells. Shells of the mollusks enjoy great popularity, mainly due to their often very beautiful colors and frequently bizarre calcareous formations. Conchs, tritons, and cowries have filled curio cabinets and natural product stores and were actively bartered; shells have long been presented in museums and private collect-ions. However, these interests were usually limited to the pretty mollusk shells. Today, scientific inquiry takes into account the animals themselves, their characteristics and life habits.

The great diversity of expression in the calcareous shell illustrates the vast variety of the animals themselves. This shell structure is not only evident as a cap over the smallest of glassy, delicate snails and the large, massive bivalves of over one meter, but also in terrestrial snails and the squid cephalopods with their internal cuttlebone (horny or calcareous shell of the cephalopods).

The shell is secreted by invaginated shell glands on the dorsal side. However, subsequent increase in the shell surface usually is deposited by the glandular region of the mantle margin. Thickening is caused by the secretion of the cellular layer of the dorsal epidermis. As a rule there is only a narrow portion of the mantle margin that remains sensory and is not covered by the three-layered shell. Starting with the innermost layer, the shell itself consists of the nacreous layer which frequently is atrophied but which basically can result in pearly formation (see Chapter 6) in most forms. However, since pearl formation is a lengthy process, they are only found in certain long-living species. The central prismatic layer is sur-rounded by the organic outer layer, the periostracum (Fig. 3-1). In this respect the three-layered shell corresponds exactly to the dorsal shell plates of extinct chitons which also consisted of three layers.

The developmental process from the egg to the adult is different in the more highly evolved forms, as for example in the cephalopods, terrestrial snails, and fresh-water groups than in the remaining mollusks. Generally, however, spiral cleavage and larval stages without body segmentation are distinguishing characteristics of embryogenesis. In the bivalves the larva develops in three stages. The major classification of mollusks is based mainly on the characteristics of the shell, mantle, and foot. When taking these features into consideration one can usually differentiate between the representatives of the individual classes just on the basis of their external appearance.

The shell

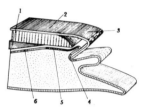

Fig. 3-1. Formation of the shell (schematic cross-section of the mantle margin of a bivalve): 4. Mantle epithelium which secretes the nacreous layer (1); 5. Mantle epithelium which secretes the prismatic layer (2).

▷
Archaegastropoda (from Europe): 1. Thick topshell (*Monodonta turbinata*); 2. Fresh-water nerite (*Theodoxus fluviatilis*); 3. Painted topshell (*Calliostoma zyziphinus*; compare Color plate, p. 74); 4. Blue limpet (*Patella caerulea*; compare Color plate, p. 74); 5. *Diodora italica*; 6. *Haliotis tuberculata* form *lamellosa* (compare Color plate, p. 74).

MILLA

MILLA

4 The Gastroverms

Class: Mono-
placophora, by
L. v. Salvini-Plawen
and R. Tucker Abbott

◁

Taenioglossa: 1. European
apple snail (*Viviparus
contectus*) (freshwater);
2. Needle whelk (*Bittium
reticulatum*) (marine);
3. *Gourmya vulgata* (brackish water); 4. *Turboella
inconspicua* (L 1.8 mm,
marine); 5. Round-
mouthed snail (*Pomatias
elegans*) (land); 6. Common
periwinkle (*Littorina
littorea*; compare Color
plates, pp. 71 & 74)
(marine); 7. Trellis wentletrap (*Epitonium clathrus*)
(marine); 8. *Serpulorbis
arenaria* (marine); 9., 10.
Adanson's worm-shell
(*Vermetus adansoni*)
(marine); 11. *Entoconcha
mirabilis* (marine parasite);
12. Common slipper shell
(*Crepidula fornicata*)
(marine).

One may recognize in the simple, limpet-shaped gastroverms (class Monoplacophora), one of the most primitive groups of today's living mollusks, numerous characteristics which are transitional between the chitons (see Chapter 2) and the gastropods (see Chapter 5). In the main, they have a single cap-shaped to spoon-shaped shell which is bilaterally symmetrical or longitudinally curved. There are three orders, all of which occur in the earliest fossil records of the Cambrian. Only the order Tryblidioidea, which was once common in epicontinental Cambrian Seas, has living representatives that now live at depths ranging from 4,000 to 6,000 m. The first living species, *Neopilina galatheae* (family Tryblidiidae), was not collected until 1952, and was finally described in 1957.

The Monoplacophora possess a body which is elongated along the longitudinal axis. The mantle covers the entire animal and secretes a uniform patelliform shell. The length of the shell is about 2 mm to over 35 mm. The foot is flat. The pallial cavity encircles the foot and bears five or six pairs of ctenidia. The stomach is entered by two large ventral "liver" glands and also contains a crystalline style (see Chapters 5 and 6). The dorsoventral musculature between the mantle and the foot consists of eight to ten pairs of muscular strands. The animals are dioecious. They are marine and live at great depth. At present only five species are known.

The fact that the entire animal is covered by a mantle suggests that the Monoplacophora are the most primitive shelled mollusks living today. The mantle secretes a thin shell which extends over the entire dorsal side of the animal. The shell is patelliform and untorted. The apex is located on the anterior point of the conically growing shell (see Fig. 4-3). The ventral surface of the animal reveals a circular to oval-shaped foot and the anteriorly positioned mouth, with associated tentacles and a velum. The pallial groove which encircles the foot contains five or six ctenidia on each side. Each ctenidium consists of a stem bearing leaflets on only one side. Six nephridiopores are also found in the mantle groove, in addition to the anus which is located on a papilla and terminates posteriorly.

The internal structure of a gastroverm is of particular scientific interest since it is phylogenetically highly significant. Prominent features of the digestive system are the radular membrane with its eleven transverse rows of teeth, the pharyngeal pouch and the convoluted midgut with its digestive diverticula, and the two large "liver" glands. These structures are reminiscent of those found in the chitons. The muscular arrangement is even more informative: eight pairs of well-developed pedal retractor muscles are serially arranged between the mantle of the shell and the margin of the foot. The first and seventh muscle pairs are still distinctly divided. Thus we re-encounter the sixteen pairs of dorso-ventral muscle strands of the chitons (see Chapter 2) which are concentrated into ten strands on each side as a result of the uniformly fused shell.

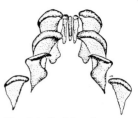

Fig. 4-1. Radula of *Neopilina galatheae*.

The nervous system consists of a kind of "brain" and four longitudinal cords without segmented ganglia. There are serially repeated latero-pedal connectives but the nerves given off from the outer edge of the lateral nerve cord to the mantle are somewhat irregularly arranged.

The excretory-reproductive system contains two pairs of gonads which are situated below the intestine and midgut gland, and also six pairs of nephridia (metamerism). The heart consists of one pair of ventricles which probably developed from a rudimentary sac that divided during the course of evolution. There are also two auricles. The blood is conducted via the true aorta to the cephalic region. The Monoplacophora are dioecious but do not possess copulatory organs. The eggs are fertilized externally in the water. Larval shells (protoconchs) were observed in *Neopilina galatheae* and *Neopilina bruuni*, which in the former species are spirally coiled (exogastric). Perhaps free-swimming larvae do occur in some Monoplacophora.

Fig. 4-2. Anatomy of *Neopilina*: 1. Gonad, 2. Mouth opening, 3. Nerve cords. 4. Nephridia, 5. Ctenidia, 6. Anus, 7. Pericardium, 8. Dorso-ventral musculature.

This brief discussion of the body structure suggests that in many aspects the gastroverms represent a transitional form between the chitons and the gastropods. The first zoologists who studied the newly discovered living gastroverms assumed that these animals were serially segmented (coelomic metamerism) and were therefore derived from segmented animals (annelids; see Vol. I). However, this concept proved to be incorrect since the serially repeated individual organs were not interconnected and there was no definite proof of a previous existence of a segmented secondary coelomic cavity.

Fossil Monoplacophora

The few living specimens of Monoplacophora, which caused such a stir in zoological circles, were preceded by a great number of well-known fossil representatives. Some of these possessed patelliform shells while others were characterized by spirally coiled shells. The trend of these fossils towards a progressive fusion of the dorso-ventral musculature into a few muscular strands (seven to two) seems to indicate a relationship to the extinct Bellerophontacea (see Chapter 1) and thereby to the gastropods (see Chapter 5). At this point special reference should be made to the extinct species *Sinuitopsis acutilira*: its rolled-up shell has three pairs of

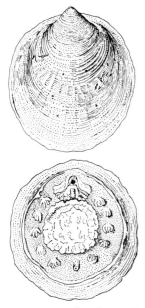

Fig. 4-3. Dorsal and ventral surface of *Neopilina galatheae*.

muscular scars. In addition there is also a raised depression (sinus) on the posterior oral margin which up to now had been known only in the Bellerophontacea which had one pair of muscles. In these coiled snails (see Chapter 5) this sinus is present as a slit. Thus *Sinuitopsis* demonstrates that the anteriorly spiral shell of the Bellerophontacea, on the one hand, and the raised depression or the slit, on the other, were already present before the intestines of the gastropods underwent torsion.

Living *Neopolina* have been found at depths of 2,500 to 6,500 m since the initial discovery in 1952. Subsequently, researchers on board the *Vema* discovered a second species (*Neopilina ewingi*; L 15 mm). In 1960 several *Neopilina* were captured in the Pacific Ocean off the coast of Baja California. This species, which measures only 2.5 mm, was named *Neopilina veleronis*. Finally, in 1967 R. Menzies disclosed the discovery of the species *Neopilina bruuni* and *Neopilina bacescui* which were found in the deep-sea trench off the coast of Peru. Following these discoveries in the deep-sea trenches of the Pacific coast of the Americas, *Neopilina* was also reported from the Old World on 22 February 1967. These specimens were dredged from a depth of 3,500 m from the Gulf of Aden, and showed great similarity to *Neopilina galatheae*. These specimens can probably be regarded only as representatives of a geographical subspecies (*Neopilina galatheae adenensis*).

The *Neopilina* species which inhabit the muddy bottom of the deep oceans are obviously detritus feeders. They feed on numerous protozoans including radiolarians, diatoms, and foraminiferans. The intestinal content of *Neopilina ewingi* also revealed spicules from sponges and sea urchins. If one compares the findings of the *Galathea* with those of the *Vema*, it becomes evident that there are far more species inhabiting the abyssal depths of 3,300 to 3,700 m than depths of 6,000 m. It is amazing how many animal groups are still represented at the greater depths, for example the annelids, bivalves, crustaceans, hydrozoa, polychaetes and ophiuroids, solenogasters, Caudofoveata, and gastropods. On the other hand, the scaphopods, decapods, pantopods, bryozoans, and sea urchins are apparently missing from depths beyond 4,000 m.

These somewhat sparse yet informative findings about the ecology of *Neopilina* provide us with at least a few clues and concepts about the mode of life of the animal kingdom at these depths which had been assumed to be devoid of living creatures. Scientists eagerly await additional research done on animals from the ocean's greatest depths. Probably many surprises remain to be uncovered.

5 The Gastropods

The average layman living far inland generally regards the snails (gastropods) as the characteristic representatives of the shelled mollusks. Yet because of its high degree of evolutionary development, this animal class deviates from the general molluscan body plan in many respects. Nevertheless, land and sea snails are undoubtedly the most familiar of the molluscan forms. This also becomes evident from the fact that the concept "snail" occurs quite frequently in everyday word usage. For instance, there is the proverbial "snail's pace." In many countries, spirally coiled baked goods are referred to as snails. Housewives are annoyed with "salad snails" in their vegetables. Finally, some "ESCARGOT" snails are treasured ingredients in the culinary arts.

Class: Gastropoda, by L. v. Salvini-Plawen and R. Tucker Abbott

The SNAILS (class Gastropoda) usually have a distinct cephalic region. As a rule, the visceral mass is asymmetrical and spirally coiled, and there is a univalve shell which, however, can demonstrate all stages of reduction and can also be elongated. Adults may range in size from 1 mm to over 60 cm. The foot is flat and usually elongated in a taillike manner. It may be broadened into flaplike projections; sometimes it is reduced. The alimentary canals of some forms contain a crystalline style. The musculature between the mantle and the foot (dorso-ventral musculature) is usually concentrated as an unpaired spindle muscle on the right side or is reduced. Gastropods are either dioecious or hermaphroditic. Fertilization is usually internal. Snails are found in marine, fresh-water, and terrestrial habitats. There are three subclasses: 1. Prosobranchia ("fore-gilled" snails) with two orders; 2. Opisthobranchia with eight orders; 3. Pulmonata with two orders. Altogether there are approximately 40,000 living species.

Distinguishing characteristics

To people living inland, the most familar forms of gastropods are either the garden snails and slugs or the pond and aquarium snails. To those living near the sea, a host of periwinkles, conches, and whelks are usually known to the casual observer of life along the coast. Even at a glance one can recognize that a snail possesses a true head with eyes, and a creeping foot topped by a spirally coiled shell. The snail is able to retract its body almost completely into this shell. In approximately one percent of all species the shell is absent or only poorly developed in the adult stage, and is therefore of no protective value. The asymmetrical body plan of the snail is often not conspicuous to the average observer. Not only is the shell often coiled conically to one side but the entire visceral mass which is encircled by the mantle is also not symmetrical. The mantle cavity, with all its organs, does not surround the foot posterially in a horseshoe shape but is turned anteriorly towards the head. This turn-about or torsion of the visceral complex and the shell by 180 degrees, and all its consequences, differentiate the gastropods from all other mollusks. For this reason, of course, the gastropods are a long way from being the characteristic type for the phylum.

Torsion of the visceral mass

Several different hypotheses exist about what caused torsion. Whatever the cause, the far-reaching effects of these changes can be recognized in every gastropod. The creeping sole of the foot is elongated in a taillike fashion. Usually the right muscular strand between the mantle and the foot is present while the opposing strand is vestigial. Only a few members of the suborder Pleurotomariina and a few other forms still possess remains of this muscle. In the Opisthobranchia and Pulmonata there is a gradual unpairing of the osphradia, auricles, nephridia, gonads, and mid-gut glands due to partial detorsion. These organs became relocated from the original left side to the right side. Only the gonads are still situated on the right, often discharging via parts of the degenerated nephridium.

Shell forms

The diversity of the body plan becomes evident first by the extraordinarily great variety of shells which serve as standards for classification purposes. In a typically coiled shell one distinguishes between the overall form (be it conical, cap-shaped, plate-shaped, towered, etc.), the degree of whorling (wide or narrow), and whether it has an umbilicus, which may be wide or narrow. The shell opening is known as the aperture, and the early tip is the apex.

The previously mentioned trend of a reduction of the shell, most pronounced in the pulmonates, influences not only the body arrangement per se but also the entire body structure. It becomes evident, however, that these shell-less snails are also typical gastropods if one looks at the "one-sidedness" of their internal organs.

The creeping sole of the foot

The creeping sole, which is endowed with pedal glands, permits the animal to move along mainly in an uninterrupted gliding motion. The basommatophores are aided in their locomotion by the ciliary propulsion of the creeping sole. Usually, however, there are additional muscular locomotory waves which run along the foot. The locomotory waves may start at the posterior end of the foot (retrograde, as in the chitons; see Chapter 2) or at the anterior part (direct) or, finally, also on both sides of the foot (shuffling or alternating). Many species of gastropods may swim with the aid of lateral flaps (parapodia) which permit a type of "wing beating" through the water.

There is great variation in the radula of the various groups. This great diversity permits the snails to explore all possible sources of food. The radula is absent in only a few groups or species. The alimentary canal is frequently prolonged into a proboscis anteriorly in association with a powerful suctorial pharynx. This pharynx, just like the diagonal peri-cardial muscle fibers, and frequently also the pedal flaps, is made up of striated muscles. The mid-gut is often equipped with typhlosolar ridges. Those species which feed on minute particles and which usually exist as "ciliary feeders" possess a groove instead of these ridges. This groove contains the crystalline style which is formed from mucoglobulin (see also bivalves, Chapter 6).

Striated musculature

The reproductive apparatus is frequently highly complex but is always unpaired and located on the right side. In the less complex cases, primarily the dioecious Prosobranchia, the reproductive system consists of a gonad with a gonoduct which in certain forms leads into the transformed right nephridium. The males frequently possess a gland and a copulatory organ (penis). In the females one finds glandular albuman and mucoid cells along the uterus. The hermaphroditic Pulmonata possess several sex organs in addition to the hermaphroditic gonad (ovotestis) with a gonoduct and a copulatory organ. This duct may be separated sexually. The female section is often divided into an oviduct; and the bursa copulatrix accessory appendages, such as an albumen gland, mucous gland, seminal receptacle, and fertilization chamber, are often also present. Copulation in these species is usually mutual and reciprocal with every animal simultaneously active as male and female. The terrestrial snails of the family Helicidae inject delicate calcareous darts ("love arrows") into their opposite partner.

The eggs are deposited as a mass surrounded by gelatinous material, or in capsules, or singly. The eggs of most marine gastropods undergo spiral cleavage and then develop into swimming larvae of the trochophore and veliger type (see Chapter 1). This line of development is characteristic for most invertebrates. The terrestrial and fresh-water gastropods, on the other hand, release eggs which are rich in yolk and quite often young individuals emerge fully developed. Some forms practice brood care or are ovoviviparous.

The only group which can be regarded as the direct ancestors of the modern gastropods is the extinct suborder Bellerophontacea (see Chapter 1). These fossilized forms were still symmetrical, with an anteriorly coiled (exogastric) shell and paired spindle muscles. Torsion caused rotation of the visceral mass and resulted in the asymmetrical structure and posteriorly coiled (endogastric) shell of the gastropods. Only these turned-about forms can be regarded as the true representatives of the gastropod class.

The systematics of the gastropods are based on the differences in the

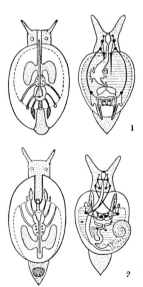

Fig. 5-1. Comparison of the hypothetical anatomy of the extinct Bellerophontacea (1) and the living Strepto-neura gastropods (2).

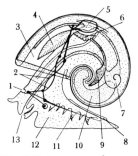

Fig. 5-2. Anatomy of a streptoneurous gastropod: 1. Eye, 2. Crossed-over nerve strands, 3. Mantle (pallial) cavity, 4. Ctenidia, 5. Pericardium, 6. Nephridia, 7. Midgut, 8. Operculum, 9. Gonad, 10. Digestive gland ("liver"), 11. Pedal cirri, 12. Foot, 13. Mouth opening.

Fig. 5-3. Shell of Beyrich's slit-shell, *Pleurotomaria beyrichi* (L 10 cm).

Fig. 5-4. Distribution of the Eastern Atlantic abalone. 1. (*Haliotis tuberculata*), and its Mediterranean form, 2. *H. t. lamellosa*.

pallial complex. Most scientists differentiate between "fore-gilled snails," "back-gilled snails," and "lung snails." Some scientists use the structural reorganization of the nervous system with the cerebral ganglia, buccal ganglia, and two lateral nerve ends, and divide the gastropods into the subclasses Streptoneura and Euthyneura.

The "fore-gill snails" (subclass Prosobranchia) usually possess a well-developed shell and corresponding operculum. The mantle cavity is on the anterior side of the shell. The gills (ctenidia) are in front of the pericardium. The nervous cords are situated laterally and are twisted (crossed-over). They are usually dioecious and marine. There are two orders: 1. Archaeogastropoda; 2. Caenogastropoda. There are approximately 20,000 species.

Most marine gastropods with well-developed shells belong to this subclass. Usually they are characterized by thick-walled shells with an operculum. In museums these snails are often conspicuous because of their colorfully shimmering and bizarrely shaped shells. In certain regions, particular marine snail shells were also incorporated into legends and traditions. Snail shells were symbols for death and rebirth in pre-Columbian Mexican cultures. The coiled form of the shell symbolized life, growth, and degeneration.

The operculum (compare Fig. 5-8) which is borne upon the dorsal surface of the metapodium (foot) is a secretion of horny or calcareous materials. In many species the operculum fits very closely inside the aperture. The central part of this structure, called the nucleus, originated as the operculum in the larval stages. Subsequent concentric or spiral rings are deposited around this nucleus as the snail grows. Occasionally there is also a projection on the underside of the operculum which partially serves for muscle attachment and partially as a hinge with the shell margin.

The previously mentioned torsion has resulted in the relocation of the mantle cavity to the anterior rim of the shell. Consequently the paired or occasionally unpaired ctenidia came to lie in front of the pericardium. For this reason this group is referred to as "fore-gilled snails" (Prosobranchia).

The order Archaeogastropoda contains prosobranches usually possessing two auricles, and occasionally two ctenidia and two nephridia. The ctenidia (present in all but a few genera) are bipectinate, with filaments alternating in two sides of the axis and free at the front end. There is no siphon or probscis. The males are without prostate and penis. The nervous system is not concentrated, and the pedal cords are ladderlike. The inner layers of shell in the majority of the genera are nacreous. There are numerous families arranged into twenty-one superfamilies having both fossil and living representatives. There are approximately 3,000 living species, nearly all of which are marine.

The Archaeogastropoda include some of the most primitive gastropods known. This is borne out by the fact that these gastropods still have two auricles and a cephalic mantle cavity which in some forms still houses two true sets of ctenidia. The many forms of the superfamily Pleurotomariacea, particularly, have retained this primitive condition. In yet another aspect the Pleurotomariacea also seem to be the ideal link to the extinct Bellerophontacea. Shells of the family Pleurotomariidae have a slit band in the anterior section which extends along the roof of the mantle cavity back to the anus. Waste products are thereby conducted dorsally and not towards the cephalic region. The few living species which belong to this group can be regarded as "living fossils" from the Lower Cambrian (500 million years ago). Today these forms are found at depths below 100 m in the Caribbean, off South Africa, and in east Asian oceans. At this point the *Pleurotomaria beyrichi* (see Fig. 5-3) from the oceans around Japan and the *Pleurotomaria adansoniana* of the West Indies should be mentioned. The latter can have a shell which measures more than 20 cm and usually is of a brilliant orange-yellowish with red flames. The shell surface is finely beaded. *Scissurella costata* and the slightly larger *Scissurella crispata* (L only 4 mm) from the North Atlantic are very similar to these large-shelled forms but are very much smaller in size, white in color, and still possess part of the left shell muscle.

The ABALONES (family Haliotidae), with a brilliant iridescent nacre on the inner surface of the shell, deviate considerably from the usual gastropod form. The few whorls of the shell are flattened and rapidly increase in diameter so that the largest part of the shell consists of the last part of the whorl ("ear-form"). The European representatives of this family is the ORMER (*Haliotis tuberculata*) which is characterized by knotty longitudinal shell ridges. It is distributed from the English Channel to western Africa. *Haliotis tuberculata lamellosa* (see Color plate, p. 45, and Fig. 5-4) is only a form of the former species that is characterized by transverse shell ridges. Both forms usually measure 5 to 7 cm. The largest species occur along the Pacific coast of the United States, in northern Japan, and southern Australia, where the animals are commercially fished for their delicious meat. The iridescent shells are used extensively in making jewelry. Most of the approximately seventy species are found in cool waters. Some members of this group attain shell lengths of over 20 cm and live to an age of 10 to 13 years. Aside from the characteristics already mentioned, abalones are further unique in that the slit band is present only as a series of small holes. During further growth abalones develop additional holes in a curved line, and the "retired" holes are sealed over.

The abalones inhabiting the intertidal zone to a depth of about 50 m scrape algae off rocks. All abalone species have adapted to this ecological niche by developing a broad suction foot with a correspondingly large shell. The light-shuning animals attach themselves to shady parts of the rock with this broad foot. The suction force of this foot is more than 4,000 times that of the animal's body weight. Just as in the

▷
Marine Taenioglossa:
1. Tiger cowrie (*Cypraea tigris*; see Color plate, p. 72); 2. *Carinaria mediterranea* (pelagic); 3. *Atlanta peroni* (pelagic); 4. Pink conch (*Strombus gigas*); 5. Pelican's foot (*Aporrhais pespelecani*); 6. Giant tun shell (*Tonna galea*); 7. Trumpet shell (*Charonia tritonis*); 8. *Sinum leachi*; 9. *Naticarius stercusmuscorum* (see Color plate, p. 74); 10. Large helmet (*Cassis cornuta*; compare Color plate, p. 71).

MILLA

scissurelids, the left muscle is still present as a weakly developed bundle.

The KEYHOLE LIMPETS (family Fissurellidae) are even more modified in their external appearance. Their body is completely symmetrical and in other respects they also seem to resemble the true limpets. A closer examination of these animals, which can usually be found underneath stones at shallow depths, reveals a different picture. For example, the apex of the 13-mm-long shell of the SLIT LIMPET (*Emarginula huzardi*) of the Mediterranean is slightly rolled posteriorly, and the slit on the anterior margin is conspicuous. This short, ribbonlike opening functions in conducting waste products away from the head and pallial regions.

In *Puncturella noachina*, distributed from the Arctic to the Mediterranean, and in North America, the anus and the slit notch in the shell move anteriorly during continued growth. This phenomenon also takes place in the keyhole limpets (genus *Fissurella*; see Fig. 5-5). The genus *Diodora*, which includes *Diodora italica* (see Color plate, p. 45) and *Diodora graeca*, is closely related. They possess the shell opening at the apex. Other members which belong to this family are *Fissurella nubecula*, of the Mediterranean, and *F. barbadensis* of the Caribbean. In these species the apical perforation is slightly elongated and shaped somewhat like a keyhole.

All members of the suborder Pleurotomariina possess a rhipidoglossate radula (see Fig. 5-9) consisting of a central tooth, a few lateral teeth, and numerous marginal teeth. Many species are herbivorous while others evidently ingest sponges.

The LIMPETS (family Patellidae), on the other hand, have a docoglossate radula (see Fig. 5-6), although they graze algal growth from rocks and cliffs in a manner similar to the abalones and keyhole limpets. The central tooth is frequently degenerated but the lateral teeth are strongly developed with recurved serrate ends. The marginal teeth are less well developed and there are usually only three of them. The limpets differ from the Pleurotomariina because of their powerful radula which is usually several times longer than the animal's body. The limpet's cap-shaped exterior is very similar to the keyhole limpets. This phenomenon has to be viewed as parallel evolution which came about by similar modes of life. The limpets, together with the families Acmaeidae and Lepetidae, are classified in a separate superfamily, Patellacea or Docoglossa, characterized by a patelliform shell. The dextral embryonic shell is shed in the juvenile stage but is retained in forms which feed on the seaweed *Laminaria*. In the PELLUCID LIMPET (*Patina pellucida*; L 1–2 cm) of Europe, which is vividly marked with blue iridescent bands in its juvenile stages, one can recognize the rolled-in shell at the apex. As do the abalones and keyhole limpets, limpets lack an operculum in their adult stage.

The individual species of the limpets proper (genus *Patella*), found in great numbers on every coastline except North America's, along rocks and cliffs in the intertidal zone, are very difficult to differentiate since their sizes, which may vary from 3 to 6 cm, fluctuate greatly depending

◁
Marine Prosobranchs:
1. Atlantic dogwinkle (*Nucella lapillus*); 2. *Ceratostoma erinaceum*; 3. *Conus marmoreus* (compare Color plate, p. 73); 4. *Oliva flammulata*; 5. Black olive snail (*Oliva maura*); 6. Murex dye shell (*Trunculariopsis trunculus*); 7. *Mitra cornicula*; 8. Waved whelk (*Buccinum undatum*; compare Color plate, p. 71); 9. *Hinia reticulata*.

on their age. The color and form of the shell generally determine the identity of a species. The COMMON EUROPEAN LIMPET (*Patella vulgata*; see Color plate, p. 74) is distributed from Norway to the Mediterranean. The BLUE LIMPET (*Patella caerulea*; see Color plate, p. 45) is widely distributed throughout the Mediterranean region. *Patella rustica*, found from the Mediterranean to the Bay of Biscay, prefers locations along the high-water line which are rich in oxygen and exposed to light.

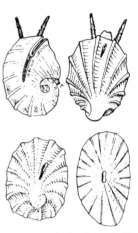

Fig. 5-5. Shift of the shell perforation during the course of development in the keyhole limpets.

As are many other marine gastropods, some limpets are always found at the same location. Each limpet occupies a slight depression exactly fitting its body contours, which it hollows out with the aid of some acidic secretions. Neither sun radiation during low tide, the dangerous desalinating effects of rainfall in the surrounding waters, nor the most powerful surges can endanger the limpet because of its ability to lock the external environment out completely. Nevertheless, limpets are able to creep. As do many other snails, they move along with the aid of muscular undulations of the foot, even if only for four to five hours at night. Some species characteristically creep to the left and make an arc-shaped path of one meter in diameter at the most. They usually return on their own track and by daybreak they are back in their own "bed." In the limpets, particularly the ones that inhabit the intertidal zone, this homing behavior and faithfulness to one site seems, according to recent investigations, to be due to some chemical-tactile stimuli. Some gastropods, therefore, always return to their own specialized mucous sites that are secreted by the pedal glands. This is a case of true territorial marking. Additionally there are further orienting aids which have not yet been investigated, and we do not know if we are dealing with measurement of distance, magnetic orientation, or gravity. Tactile as well as chemico-sensory input (perhaps with the pedal margins) helps guide the limpet back to its exact former position in the hard substratum. The tentacles along the mantle margin have tactile sensory abilities. These tentacles also help to control growth of the shell and to adjust it to the substrate. Adjustment can occur within a few hours or take up to a day.

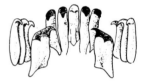

Fig. 5-6. Docoglossate radula of the common *Patella* limpet.

In contrast to the Pleurotomariina, the true limpets, with simple epithelial cup eyes, and the eyeless Lepetidae (for example *Propilidium ancyloide* from the European coasts) do not have ctenidia in their very small pallial cavity. Another outstanding feature of the limpets is the powerful, horseshoe-shaped shell muscle which enables the animals to withstand a pull up to 15 kg. Other representatives of the Patellina have similarly well-developed musculature with the exception of the thin-shelled *Propilidium ancyloide*, which still retains the embryonic whorls in its shell. The Acmaeidae possess a pectinibranch and occasionally also cordlike external mantle gills (branchial cordon). One of these species, the TORTOISE-SHELL LIMPET (*Acmaea virginea*; shell L 1 cm) of Europe, is less well known, since it is usually found only at depths of over 10 m along European coasts. The brownish spotted *Acmaea testudinalis* (shell L 3–4 cm), distri-

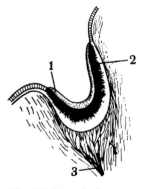

Fig. 5-7. Eye of a limpet (*Patella*): 1. Optic cup, 2. Retina, 3. Optic nerve.

buted throughout the circumpolar boreal regions, on the other hand, is not only conspicuous because of its design but also because of its size.

Gastropod groups of flat patelliform shells without an operculum also include the families of the Cocculinidae and Lepetellidae; in the case of the Cocculinidae, however, frequently only the shells are known. It becomes evident that these two families are related with the previously mentioned groups because of their rhipidoglossate radula, the horseshoe-shaped muscle, and the respiratory organ which is usually present as a folded monopectinate gill. Several forms, for example *Addisonia paradoxa*, distributed in the North Atlantic and Mediterranean, also have secondary gills originating from the margin of the mantle cavity. Since the gastropods of both these families are primarily deep-sea dwellers, they lack eyes. Interestingly, these forms, in contrast to most other Archaeogastropoda, are hermaphroditic.

The suborder Trochina represents the most extensive group of Archaeogastropoda. The trochids lack a slit band and have only one ctenidium. Due to the general effects of detorsion, the right ctenidium is already reduced, as in all the following snails. The conical shells of the trochids are favorite collector's items because of their frequently brilliant coloration and nacreous shimmer on the inside. Certain species, for instance *Tectus niloticus*, with shells which measure from 10 to 20 cm, are often used in the manufacture of buttons and arm bracelets.

Calliostoma topshells

Among the European trochids, the PAINTED TOPSHELL (*Calliostoma zyziphinus*; see Color plates, pp. 45 and 74) occurs quite frequently although it is found below the tidal zone. Its shell, which has a wavy reddish-brown design, can reach lengths of 15 to 20 mm and in the Mediterranean even 35 mm. Like most trochids, the *C. zyziphinus* has a longitudinally divided foot (ditaxic locomotion). The trochid snails feed on algae and organic detritus; one North Atlantic species feeds on the polyps of hydroids. *Calliostoma laughieri* (L 1-1.5 cm), found at shallow depths along southern European coasts, is conspicuous because of its deep-red body and darkly waved shell. Since shell coloration can vary greatly in this gastropod, it actually has earned the nickname "colorful top" even though olive-brown with grayish-blue marbling is predominant. The so-called GRAY TOP (*Gibbula cineraria*; L 15-20 mm), a somewhat less attractive species, is found in the upper coastal zones of northern and western Europe and also the Mediterranean. The slightly larger *Gibbula magus* is found in the Mediterranean below depths of 10 m but its range of distribution goes up to the Shetland Islands. *Gibbula divaricata* and *Gibbula tumida* of northern Europe will be mentioned. Both species are found at lesser depths. All these trochids are quite abundant, and great quantities of their shells are often washed up on the beaches.

Gibbula adansoni (formerly known as *adriatica*) is a favorite item around the Adriatic Sea. These gastropods, which barely measure 1 cm, are very common in the algal zone of the coast, and are gathered in great quantities. In north Adriatic coastal towns they are polished down to the nacreous layer and are made into jewelry, especially necklaces. The THICK TOPSHELL (*Monodonta turbinata*; L 2-3 cm; see Color plate, p. 45) of the Mediterranean is one of the most abundant trochid species among the area's approximately eighty indigenous species. Most visitors to these rocky coasts are familiar with *Monodonta* snails, since they come up to the high water line and at low tide are seen on stone buildings in the harbor. *Jujubinus exasperatus* is regularly found from the Mediterranean to England, and the Azores, occurring among algal growth, in particular the shrublike brown algae (genus *Cystoseira*), along coastal cliffs.

The families Stomatellidae, Turbinidae, Skeneidae, Phasianellidae, and Orbitestellidae are not represented by quite so many species as the previous families. Among this group the heavy and massive shell of the GREEN TURBAN (*Turbo marmoratus*; L 15 to 20 cm) is an impressive exhibition piece. The foot bears a large, white, calcareous operculum. In its Indo-Pacific habitat this gastropod's body serves as food for the natives and the pearly shell is used in shellcraft. The similarly large-shelled genus *Astraea* has a furrowed, calcareous operculum and a rugose exterior. Many species occur in tropical waters. *Astraea rugosa* is an inhabitant of the Mediterranean and coastal waters of Portugal. Its brownish-red body is occasionally sold in fish markets. The shell measures 4 to 5 cm and is partially covered by spines.

European representatives of the family Skeneidae, which have corneous, multispiral operculå, on the other hand, are generally very small. One of these, *Tubiola nitens* (L 8-9 mm), is regularly found along European coasts. The European PHEASANT SHELL (*Tricolia pulla*) measures 1 cm and has a calcareous operculum. Like almost all trochids, *T. pulla* possesses long cephalic tentacles and a tentaculate epipodium. There is a median longitudinal cleft in the foot sole, which enables the animal to move relatively quickly, alternating on one or the other pedal side. This form is frequently found at shallow depths on sandy or algal substrate. It feeds on seaweeds which when digested may influence its shell's color.

The last superfamily of the Archaeogastropoda is the Neritacea, represented in Europe by 20 species. Judging by their great variety of habitats, they seem to be a very diverse group. They are characterized by secondary gills or a pulmonary sac, absence of the right nephridium, and a shell without the nacreous layer. Generally they still possess two auricles. The genus *Nerita* is well represented along tropical rocky shores. The shelly operculum has a pronglike peg for better attachment to the foot. The genus *Neritina* has invaded brackish and fresh waters, and some species ascend mangrove trees.

The family Neritopsidae is represented by one living species in the Indo-Pacific and one in the Caribbean region. The family Neritidae occurs in European areas but only in the form of the EMERALD NERITE (*Smaragdia viridis* (L 8 mm), distributed from the Mediterranean to the West Indies.

Turban shells

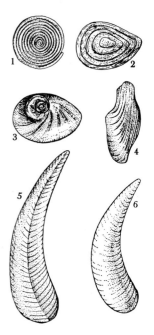

Fig. 5-8. Opercula: 1. Multispiral (*Gibbula cineraria*), 2. Concentric (*Viviparus viviparus*), 3. Paucispiral (*Littorina littorea*), 4. Unguiculate (*Nucella lapillus*), 5. Sickle-shaped (*Strombus*), 6. *Conus*.

The FRESHWATER NERITES (genus *Theodoxus*), with eighteen European species, are well-known members of the European animal world. All these forms, however, live in fresh water. *Theodoxus meridionalis* and the north German subspecies of the common freshwater nerite penetrate the brackish zones of rivers. These intertidal forms are somewhat smaller than their close relatives in the rivers.

Fig. 5-9. Rhipidoglossate radula of the European nerite, *Theodoxus fluviatilis*.

The FRESHWATER NERITE (*Theodoxus fluviatilis*; L approximately 1 cm; see Color plate, p. 45) is typified like all *Theodoxus* snails by a strongly flattened shell with a wide body whorl and low spire. The operculum, with a thornlike process, is red on the periphery and has a beautiful dark-red network. *T. fluviatilis* is widely distributed from Spain to the British Isles and from Lapland to the Caspian Sea, also occurring in the entire Danube region and the Ukraine. In Germany they are replaced by other species. Thus *Theodoxus danubialis* is found from Kelheim in the Danube onward, but is also distributed around the Baltic Sea and the Ukraine. This form is characterized by oblique zig-zag lines on a light background, and an operculum with a seam. *Theodoxus transversalis* occurs upstream from Ingolstadt in the Danube region, and in the Dnieper area. *T. transversalis* is characterized by three bands running parallel to the whorls, and by the operculum with a seam.

Among all other *Theodoxus* snails, in part limited to specific regions such as Spain or the islands off the Baltic coast, *Theodoxus prevostianus* deserves special mention. This gastropod is a vestige from the Tertiary period and is found in warm waters of the Alps region between Hungary and the Black Sea. It particularly frequents thermal springs, for example near Voslau, Austria. *Theodoxus subterrelictus* (L barely 5 mm) is another snail which occupies an unusual habitat. It can also be regarded as a European relic from the Tertiary. The shell of this snail is colorless with tilelike overlapping ribs. It inhabits subterranean springs and water caves near Metkovic, Yugoslavia.

Terrestrial forms with "lungs"

The only Archaeogastropoda of the land family Hydrocenidae in the European region is *Hydrocena cattaroensis*. It has lunglike vascularized tissue in the mantle cavity. In this form and in the few other species of this family, the oxygen is not taken up with the aid of ctenidia or similar structures. Respiration is carried out by finely interconnecting networks of blood channels. In the section of this chapter dealing with the pulmonates we shall discuss this system more thoroughly. In *Hydrocena* the right auricle is also degenerated. This form is found on rocks and walls near Cattaro in Dalmatia. There it inhabits moist crevices and cracks.

Numerous species of the family Helicinidae have become decidedly terrestrial. Most of them have a chitinous operculum with an exterior calcareous layer. Some even inhabit trees. Their range of distribution is limited to the South and North American and the Indo-Pacific regions.

The last of the Archaeogastropoda is the family Titiscaniidae which is represented by the single species *Titiscania limicina*. This marine snail is distributed throughout the Indian Ocean. It does not have a shell nor an operculum and therefore is the only free-living naked gastropod of the Prosobranchia. However, it still possesses a small pallial cavity with a single ctenidium behind its head.

The second order of the Prosobranchia is the Caenogastropoda. They are characterized by a single auricle, ctenidium, and nephridium. The ctenidium has only one row of filaments. The system is usually attached to the roof of the mantle cavity (ctenobranchy). In a few exceptions the pedal ganglia are developed into nerve cords. The nacreous layer of the shell is absent. There are two suborders based upon radular characters: 1. Taeniglossa, 2. Stenoglossa. (These two suborders were at one time called Mesogastropoda and Neogastropoda, respectively.) Altogether there are approximately 20,000 species, making this the largest gastropod order. The gill system that is attached to the roof of the pallial cavity has filaments on only one side. This condition is termed pectinibranch. They have only a single auricle and nephridium. This condition was already present in a few forms of the Archaeogastropoda and can be regarded as a transitional stage. The Caenogastropoda represent the most predominant shelled gastropods of the marine environment. Specializations, such as dimorphism in sperms, occur in representatives of sixteen families. The Caenogastrapoda have taken a wide variety of adaptive forms. This order not only includes some of the largest living gastropods, which reach lengths of over 60 cm (two feet), but also fresh-water and terrestrial inhabitants, free-swimming planktonic creatures, and even true internal parasites.

*Order:
Caenogastropoda*

In the suborder Taenioglossa the radula, with a few exceptions (e.g., family Capulidae) possesses more than three, usually seven, teeth in each tooth row. There are about seventy-eight families in sixteen superfamilies and approximately 11,000 species of Taenioglossa.

*Suborder:
Taenioglossa*

This suborder, which has an extremely rich variety of forms, is called Mesogastropoda. The first superfamily, CYCLOPHORACEA, consists mainly of terrestrial forms which have retained a very primitive characteristic: the pedal ganglia are not elongated into cords but are circular masses with nerves. Their mode of life is always adapted to a particular ecological condition. Thus the terrestrial families like the Cyclophoridae, Poteriidae, Pupinidae, and Cochlostomatidae have lost their gills but instead have a well-developed vascular net. Numerous species of the three families mentioned first live in the Americas and southeast Asia. Others, for example *Craspedopoma lyonetianum*, are found on the Azores and Canary Islands. The Cochlostomatidae, on the other hand, are frequently represented in central and southern Europe by *Cochlostoma septemspirale*. It is found among leaves and stones on limestone ground. Numerous other species of the cochlostamatids are primarily found in southern Europe. *Cochlostoma henricae*, however, does penetrate into the Austrian region of the Alps.

Fig. 5-10. The taenioglossate radula has seven teeth per transverse row as in the hydrobiid, *Peringia ulvae*.

The apple snails

Usually the Viviparidae and Ampullariidae are treated as a separate superfamily (VIVIPARACEA). These forms are always aquatic, and therefore have retained their gill. The European *Viviparus contectus* (see Color plate, p. 45) occur in almost all stagnant waters rich in vegetation. The tapering mantle cavity has the attached gill on the left side. The intestine and anus, and, in the female, the oviduct are on the right side. As in several other species of this group, a special structure is present: the floor of the mantle cavity bears a longitudinal raised, reddish ridge (epitaenia) which divides the cavity into left and right halves. Along the base of this ridge runs a well-developed mucous groove. This unique arrangement serves as a filter for suspended organic particles that enter with the respiratory current. The water current enters on the left side and is conducted posteriorly via ciliary action. Gas exchange takes place in the gill. The deoxygenated water proceeds anteriorly outside. This respiratory current is filtered free of food particles and microscopic animals by the gill leaflets which entrap the food with mucus. The outgoing respiratory current flows out along the mucous groove to the right of the snout. Periodically the animal excretes an accumulated ball of mucus. The river snail is a herbivore as well as a detritus feeder. The animal not only scrapes algal growth with the radula but also feeds via the mucociliary feeding method. This feeding mode is of particular significance with a low oxygen content when the snails have to stay near the surface where they are not able to scrape food with the radula.

Like the *Viviparus contectus*, the RIVER SNAIL (*Viviparus viviparus*), Europe's second most common species, is ovoviviparous. These two species are commonly confused with each other. These forms produce two kinds of sperm. There is the ordinary flagellate sperm (eupyrene type) and a food-filled spermatocyte (apyrene). The ovoviviparous fresh-water snails retain the eggs in the lower part of the oviduct until the offspring hatch as fully developed individuals. One can regard this as a type of brood care since the young undergo the entire larval development within the safety of the maternal body. The river snail, with its bluntly rounded shell apex, unlike the *V. contectus*, is almost always found in moving water. One of the common Chinese species, *Viviparus malleatus*, was introduced to California in 1890 and has spread throughout the rest of the U.S.A., where it is also known as the apple snail and is raised in aquaria.

The Ampullariidae are limited to the tropics. They also possess a longitudinal ridge (epitaenia) at the floor of the mantle cavity, which divides this cavity into two halves. One of these representatives is *Pomacea gigas*. Just like the APPLE SNAILS (e.g., *Pomacea scalaris*, formerly genus *Pomus*), the South American ampullariids are favorite snails for aquarium fans. In this artificial environment the animals feed solely with the aid of the radula. The ampullariids have a ctenidium on the right side and a pulmonary sac on the left side. Thus this respiratory system seems to be adapted to an aquatic as well as a terrestrial existence. In addition, these

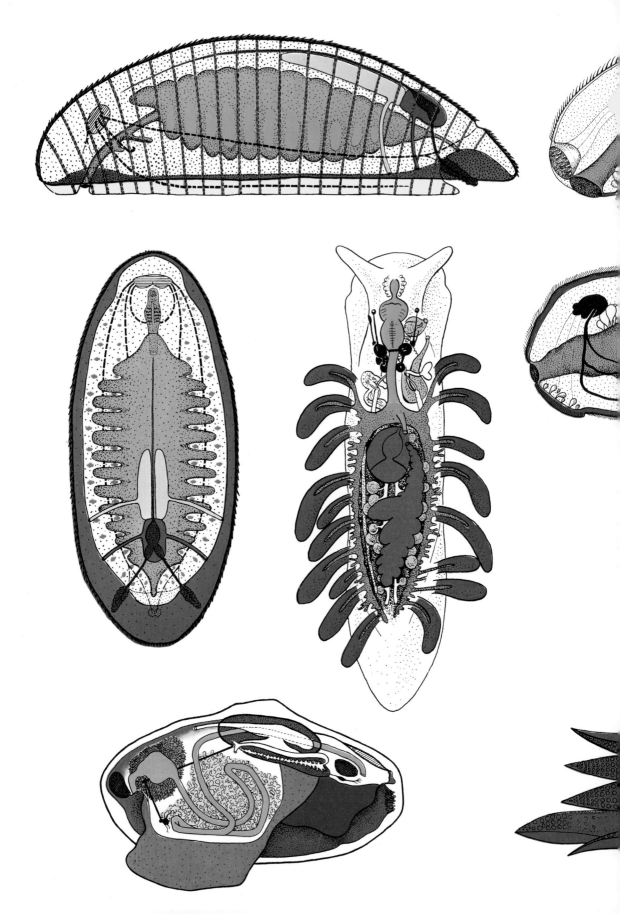

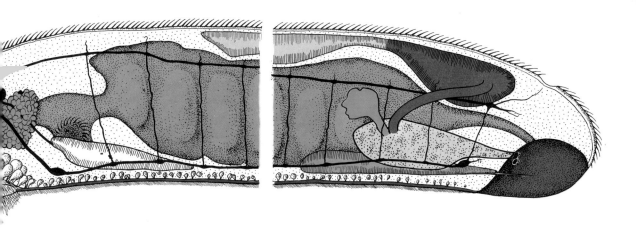

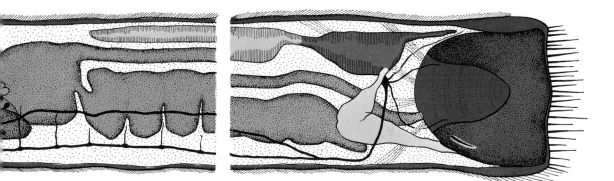

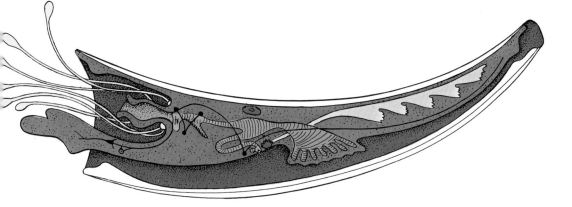

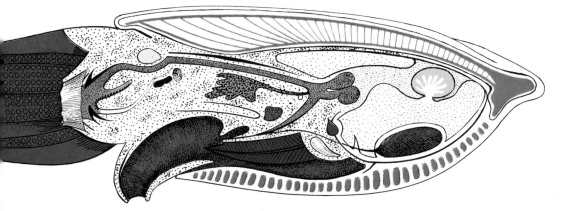

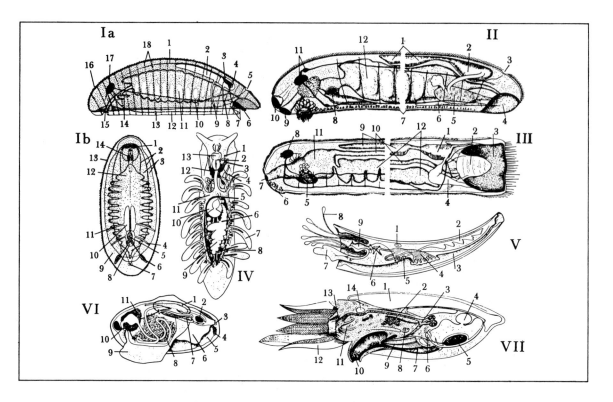

Molluscan Anatomy:

I. a. Lateral view of a reconstructed archeo-mollusk: 1. Dorsal blood sinus, 2. Gonads, 3. Heart, 4. Pericardial sac, 5. Terminal sensory organ, 6. Ctenidia, 7. Mantle (pallial cavity), 8. Coelomoduct (later nephridium), 9. Gonoducts, 10. Lateral nerve cords, 11. Midgut, 12. Sole of foot, 13. Abdominal (ventral) nerve cords, 14. Radula, 15. Oral opening, 16. Cuticle with calcareous bodies, 17. Brain mass, 18. Dorso-ventral musculature. I. b. Dorsal view of reconstructed archeo-mollusk: 1. Brain mass, 2. Lateral and ventral nerve cords, 3. Dorso-ventral musculature, 4. Pericardial duct (later nephridium or coelo-

moduct), 5. Heart, 6. Ctenidia, 7. Terminal sensory organ, 8. Pericardial sac, 9. Mantle (pallial cavity), 10. Gonads, 11. Gonoduct, 12. Midgut, 13. Cuticle with calcareous bodies, 14. Radular apparatus with double-rowed radula.
II. Anatomy of Soleno-gasters: 1. Gonad, 2. Pericardial sac, 3. Terminal sensory organ, 4. Mantle (pallial) cavity, 5. Gonoduct, 6. Gonopore with copulatory spicule, 7. Pedal groove, 8. Radular apparatus, 9. Mouth opening, 10. Oral sensory organ, 11. Nervous system, 12. Midgut.
III. Anatomy of Caudofoveata: 1. Pericardial duct, 2. Ctenidia, 3. Mantle (pallial) cavity, 4. Mucous

tract, 5. Radular apparatus, 6. Pedal shield, 7. Mouth, 8. Nervous system, 9. Gonads, 10. Hindgut, 11. Midgut, 12. Midgut caecum.
IV. Anatomy of an euthyneurous opisthobranch gastropod (family Stiligeridae): 1. Eye, 2. Foregut, 3. Penis, 4. Oviduct, 5. Hindgut, 6. Nephridium, 7. Cerata, 8. Midgut gland, 9. Hermaphroditic gland, 10. Pericardial sac, 11. Stomach, 12. Upper and lower esophageal nerve masses, 13. Foregut.
V. Anatomy of a scaphopod (elephant's tusk): 1. Pericardial sac, 2. Gonad, 3. Mantle (pallial) cavity, 4. Midgut gland, 5. Nephridium, 6. Radular apparatus, 7. Burrowing foot, 8. Captacula, 9. Oral bulb.

VI. Anatomy of a freshwater eulamellibranch clam (*Anodonta cygnaea*): 1. Pericardial sac, 2. Retractor muscle, 3. Exhalent opening, 4. Inhalent opening, 5. Mantle (pallial cavity), 6. Gill, 7. Nephridium, 8. Gonad, 9. Foot, 10. Mouth opening, 11. Midgut gland.
VII. Anatomy of a cephalopod (*Dibranchiata*; compare p. 203): 1. Shell (cuttlebone), 2. Midgut gland, 3. Midgut (stomach), 4. Gonad, 5. Ink sac, 6. Heart, 7. Mantle (pallial) cavity, 8. Nephridium, 9. Ctenidium, 10. Pallial funnel, 11. Radular apparatus, 12. Head tentacles, 13. Jaws, 14. Cephalic capsule (cartilage).

animals possess a long left nuchal lobe which draws in air from above the water. Since these snails can reach sizes of over 50 cm and are very tasty, they are favorite food items, particularly in eastern Asia. They may live from five to ten years.

The VALVE SNAILS, superfamily VALVATACEA, which consist of numerous species which are divided into numerous geographical subspecies, enjoy a wide range of distribution in the fresh waters of the northern hemisphere. In these forms one can still witness a completely formed ctenidium which often protrudes extensively. The long tentaculate mantle process on the opposite side probably corresponds to the right gill system. The COMMON VALVE SNAIL (*Valvata piscinalis*), barely 5 mm long, is the most common European form of this group. It was introduced into the Great Lakes of North America in the 1880's. This snail is found in almost all stagnant or slowly flowing waters of Europe with the exception of those in the most southerly regions. It is also present in Asia Minor and deep into Siberia, where it inhabits the mud. Depending on its particular environment this form can vary extraordinarily. The hermaphroditic VALVE SNAILS are unmistakably unique with their anteriorly two-lobed foot, and long penis which is to the right of the head. There are six species in North American fresh water, the commonest being the golden-yellow, 4-mm-wide *Valvata tricarinata* that occurs in shallow lakes and streams throughout most of Canada and the northern United States. *Tropidina macrostoma* is intermediate in its shell form. It inhabits swamps, sloughs, and ditches. The *Borysthenia naticina* is found only in eastern Europe. Like the river snail, this form is ovoviviparous, and the young leave the maternal body as fully developed individuals.

The operculate land snails

Most representatives from the LITTORINACEA are marine. Many have become terrestrial like the European operculate snails (family Pomatiasidae) and the Central and South American family Chondropomidae. The ROUND-MOUTHED SNAIL (*Pomatias elegans*; L 10–15 mm; see Color plate, p. 46) prefers the warmer spots of central Europe, and is primarily present in the Mediterranean region. One can find these snails between fallen leaves and among rocks. The animal is rather conspicuous because of its unique locomotory method. The two halves of the foot, divided by a deep longitudinal groove, are raised alternately and pushed forward, much as one would shuffle forward in a potato sac.

This "trend for the conquest of the land" is already present in the more primitive PERIWINKLES (family Littorinidae) which are present in great numbers on rocks and cliffs of seacoasts. Rocky cracks and niches are particularly favorite places where the animals are closely crowded together. One of these forms, the COMMON PERIWINKLE (*Littorina littorea*; see Color plates, pp. 46 and 71), is distributed throughout the entire North Atlantic from the 43rd degree latitude upwards. It appeared in New England in the 1850's and spread south to Maryland by 1959. This snail is characterized by a partially reduced ctenidium in addition to a lunglike roof of the pallial cavity. The microscopic egg capsules float freely through the water for two or three weeks before hatching. The ROUGH PERI-WINKLE (*Littorina saxatilis*) is ovoviviparous and thereby practices "brood care." The SMOOTH PERIWINKLE (*Littorina obtusata*; see Color plate, p. 74), however, which is present in New England and the western region of the Mediterranean, does not show any specializations in the pallial complex. *L. obtusata* undergoes direct development without a larval stage. *Littorina neritoides*, on the other hand, resembles the common periwinkle in these characteristics. Such similarities would be evident if the individual species were occupying respective habitats in the tidal zone. In reality, however, *L. neritoides* is usually found in the spray-water zone. The ovoviviparous rough periwinkle is present in the high-water zone while the smooth and the common periwinkles settle near the low-tide mark. The common periwinkle sometimes does go up higher. The common periwinkle is usually easily recognized by its decidedly dark coloration. The three smaller species of periwinkles, which only measure 1–2 cm, can often only be differentiated on the basis of their habitat or by the special character-istics in their form. The frequently observed faithfulness to one location in the littorinids is probably due to the inherent "light compass." The animal's U-shaped creeping tracks always fall in the direction of the sun and then back towards the home location. Additionally, these creatures are able to differentiate between the oscillation of polarized light. The common periwinkle was a favorite food centuries ago, and is still eaten today.

Homing habits

The East Indian *Cremnoconchus conicus* is also a member of the littorin-ids. Like its closely related species, it inhabits the moist cliffs of the ocean. Its ctenidium, however, is not degenerate. Members of the family Lacunidae are strictly marine inhabitants. *Lacuna vinca* and *Stenotis pallidula* and a few other species occur in the algal zones of both sides of the North Atlantic from the shoreline to a depth of approximately 20 m.

The RISSOACEA are closely related to the Littorinacea. Shells belonging to this superfamily are generally minute and seldom measure over 5 mm. Here too a trend toward a fresh-water or even terrestrial habitat is evident. Only the major family (Rissoidae), which was named after the well-known zoologist G. A. Risso (1777–1845), and the vitrinellids (Vitrinellidae) have remained marine. Of the exceptionally numerous rissoids which are present in Europe, only the three major indigenous genera shall be repre-sented here by the extremely variable *Turboella inconspicua* (see Color plate, p. 46), and *Alvania reticulata*, which is sculptured with netlike ribs and spirals. Another variant of this family is *Rissoa violacea*, which has a purple band. This snail is found close to the tidal zone along European coasts. Like the round-mouthed fresh-water snails, hydrobids, and almost all bivalves, the rissoids are characterized by a stomach with an anterior evagination in which special digestive enzymes are secreted. These digestive substances are mixed with mucus into a more or less solid rod

Crystalline style

(crystalline style). This rod is constantly worn out during food digestion, and is constantly rebuilt. Such crystalline styles are confined to herbivorous gastropods that feed on minute creatures via ciliary currents. The rissoids are therefore usually limited to the algal zones of the coast.

Assiminea grayana, a representative from the family Assimineidae, which lives among the grasses and herbs on the muddy bottom of the shore line, also possesses a crystalline style. This species is thought to have been introduced to Europe from China in the early 1800's. In other aspects, however, this form is adapted to a terrestrial existence. Clear evidence for this condition are the absence of fully developed ctenidia and this animal's presence along the edge of brackish water. These snails rarely enter the water. The POINT SHELLS (family Aciculidae) are strictly terrestrial. In Germany they occur underneath autumn leaves. This group also includes the SMOOTH POINT SHELL (*Acicula polita*), which only measures 3 mm and has a reddish-brown shell.

The European HYDROBIIDS (family Hydrobiidae), represented by two species, are quite common in the sandy-bank region of the North Sea. Since 1883 *Potamogyrus jenkinsi*, which was introduced from New Zealand, has increased sporadically in this region. The two indigenous species, *Peringia ulvae* and the SPIRE SHELL (*Hydrobia ventrosa*), live in Europe's moist sandy-banks region. The *Hydrobia ventrosa* are brackish-water forms which seek out neither fresh water nor the open ocean. *Lithoglyphus naticoides*, on the other hand, inhabits sluggishly flowing streams and lakes in central and eastern Europe. Its last whorl, which is greatly enlarged, may reach a size of 7 to 9 mm. Another aquatic form, *Bythinella austriaca*, and its related forms, are found in the eastern Alps. SPIRE SHELLS (*Amnicola steini*) are frequent in northern central Europe, and in North America there are a dozen representatives, including *Amnicola limosa* and *A. integra*.

Genera *Lartetia*, *Paladilhia*, and *Paladilhiopsis* are characteristically inhabitants of subterranean springs or underground water. They are represented by a great variety of forms since each group lives isolated in a limited environment. Thus many separate species originated. The only European exception is *Paladilhia bourguignati*, which has a subterranean distribution in limestone substrate from Provence to Holland. All these cave snails, as typical cave dwellers, have vestigial eyes and are usually colorless. An example of this is *Lartetia quenstedti*, which is often present in the river caves of the Swabian Alps. Frequently only the shell of the cave snails is found, because usually only this part of the animal comes to light. Nevertheless, new species are occasionally discovered in old wells, shafts, or water-filled tunnels. Often only a few dozen individuals make up one species. The Lartetien well in Klingenberg-on-the-Main was even named after a snail species. In 1865 a living cave snail, *Lartetia rougemonti*, was discovered in the well beside the anatomy building at the University in Munich. Nowadays, however, this place of discovery is inaccessible. The description of the "dwarf cave snail" is indicative of the small size of all these snails. *Paladilhiopsis geyeri*, which is distributed in springs from Wachau to Vienna, has a shell with a height of only 2.5 mm. Several species of the fresh-water genus *Oncomelania* in the Orient, southeast Asia and the Philippines are the first intermediate host of parasitic diseases fatal to man such as schistosomiasis, caused by a blood fluke. Control of these diseases is partially obtained by destroying the snail host.

Truncatella subcylindrica is a representative of the TRUNCATELLIDS (family Truncatellidae). It measures 5 mm, prefers warm temperatures, and is frequently found on the beach above the spray-water zone underneath stones or moist sand. Although this form is a land dweller, so to speak, it still possesses a normally developed gill. *Bithynia tentaculata* and *Bithynia leachi* inhabit standing waters. These two members of the BITHYNIIDS (family Bithyniidae) have an extraordinarily wide distribution with a few gaps. *Bithynia tentaculata* even penetrates into Siberia and was introduced by man to America in the 1870's.

The VITRINELLIDS (family Vitrinellidae) are quite similar to the valvatids in shell shape and formerly were even grouped close together. The vitrinellids again bring us back to the marine representatives of the Rissoacea. They are characterized by a protruding gill, long snout, and two right parapodia. The family Tornidae is represented in Europe by *Tornus subcarinatus*, which possesses a deeply sculptured shell. These snails live embedded in mud underneath big rocks. For this reason they are seldom seen although they are almost always just below the low-tide line. *Circulus striatus* was also formerly regarded as a relative to the valvatids. This form, contrary to many reports, does not occur very frequently. *C. striatus*, like other Rissoacea, has a long snout and two parapodia.

In more recent times four families have been included in the superfamily RISSOELLACEA. These forms can be considered as being most closely related to the Rissoacea. They already show a few characteristics which are indicative of the euthyneurous gastropods (subclass Pulmonata). The origin of this subclass, which will be discussed in more detail later, is undoubtedly rooted in the hermaphroditic forms. In other aspects these hermaphrodites still have a rather uncomplicated and slightly specialized body structure. C. R. Boettger has described the European coastal genus *Omalogyra*, which is a member of this superfamily, as being generally dioecious. The animal first produces sperms and then eggs from the same gonad. This condition is known as protandry. The first individuals to hatch, in the summer, produce only sperm for a short period. Then, if conditions are optimal, eggs will be produced and a hermaphroditic situation will exist for a longer period, during which sperms and eggs are formed simultaneously. Such conditions may have persisted in the ancestors of the euthyneurous snails.

In other aspects *Omalogyra* (family Omalogyridae) is also quite similar to the Cephalaspidea, the ancestral group for the opisthobranch snails, not considering, of course, the special adaptations which are unique to all

▷
Left, from top to bottom: Orange-mouthed scorpion conch (*Lambis crocata*; L 8 cm) Indo-Pacific. Periwinkles (*Littorina littorea*) (compare Color plates, pp. 46 and 74). Capsules of a buccinid snail showing internal veliger larvae and nutritive eggs. Right, from top to bottom: *Buccinum humphreysianum* (L 9 cm; compare Color plate, p. 56) Europe. Bull-mouth helmet (*Cypraecassis rufa*; L 12 cm; compare Color plate, p. 55) Indo-Pacific. Spiny dye murex (*Murex brandaris*; L 8 cm) Europe. Egg capsules laid by several *Murex brandaris*. Europe.

▷ ▷
Various cowrie snails (family Cypraeidae) of the Indo-Pacific and Ovulidae of the West Indies. Above: *Cypraea maculifera* on a *Tubipora* organ pipe coral. Lower left: *Cyphoma gibbosum* (L 3 cm). West Indies. Middle: *Cypraea erosa* (L 3 cm). Indo-Pacific. Right below: Tiger cowrie. Indo-Pacific. (*Cypraea tigris*; L 6 cm; see Color plate, p. 55).

▷ ▷ ▷
Above, from left to right: *Harpa ventricosa* (L 7 cm). Indo-Pacific. Cone (*Conus distans*; compare Color plate, p. 56). *Hydatina velum* (L 4 cm). Indo-Pacific. Below: Mediterranean umbrella snail (*Umbraculum umbraculum*; see Color plate, p. 103).

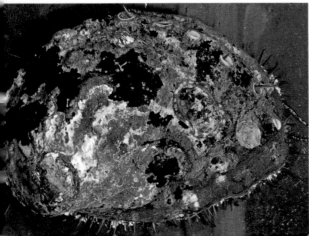

species of this superfamily. The degeneration of the ctenidium in the Rissoellacea is undoubtedly a unique feature associated with living conditions in the tidal zone. The North Atlantic representative, *Skeneopsis planorbis* (L 1.5 mm), has hardly more than 8 to 10 lamellae on the roof of the mantle cavity. The slightly larger *Cingulopsis fulgida* only has 3 or 4 gill filaments. *Rissoella glabra*, with two columellar muscles, and the previously mentioned *Omalogyra atomus*, hardly reveal any traces of the gill. Further adaptations of some species to their particular modes of life are the development of a large pedal gland with a continuing deep groove and the direct development of the eggs without larval stages. Other specializations of the Rissoellacea may be based on the fact that they belong to the smallest prosobranch snails, and feed on minute algae. The representatives of the genus *Omalogyra*, for example, rip algae off with the powerful central teeth of their radula and then suck out the plant juices.

The conical to disklike shells of the SUNDIAL SNAILS (superfamily ARCHITECTONIICACEA), which have beautiful designs of colored stripes and patterns, are favorite collector's items, for example the Indo-Pacific SUNDIAL (*Architectonica perspectiva*). Numerous species live on various soft corals. One of these is the southern European *Philippia hybrida*, which barely measures 2 mm. Some representatives of the genus *Heliacus*, for example the West Indian *Heliacus cylindricus*, are mainly found on crusted anemones (Zoantharia; see Vol. I) on whose tissues they feed. The representatives from the mathildids (family Mathildidae) have biologically similar behavior to the last group. They possess a tower shell, for example the southern European *Mathilda quadricarinata*, and a sinistral embryonic shell.

The CERITHIACEA are a similarly comprehensive superfamily like the Rissoacea. In Europe they are represented by not fewer than seventeen families. Forms outside Europe include the aquatic Syrnolopsidae of eastern Africa, the Abyssochrysidae of the deep sea, the closely related fresh-water Pleuroceridae and Thiaridae, the marine Planaxidae, which is adapted to warm coastal zones, and the marine Modulidae.

The subsequent group is the TURRET SNAILS (family Turritellidae). The European TURRET SHELL (*Turritella communis*) possesses a shell of up to 19 whorls, which measures up to 5 cm. The operculum is covered by bristles. This form is a detritus feeder and consequently has a crystalline style. These snails are extremely common in the muddy bottoms of all European oceans but they are usually buried. The quite similar but slightly smaller *Archimediella triplicata*, which are distributed in the Mediterranean and connecting Atlantic, is conspicuous because of the two or three raised longitudinal strips on the outside rings. The swamp-living *Siliquaria obtusa* is included in the family Siliquariidae. *Siliquaria obtusa*, which is found in the Mediterranean, is characterized by an uncoiled shell with a longitudinal slit. The shell resembles a worm tube.

The true WORM SNAILS (family Vermetidae) are rather unusual animals,

◁
Left, from top to bottom: Common European limpet (*Patella vulgata*; L 4 cm; compare Color plate, p. 45).
Painted topshell (*Calliostoma zyziphinus*; L 3 cm; see Color plate, p. 45). Europe.
Naticarius stercusmuscarum (L 6 cm; see Color plate, p. 55). Europe.
Kamchatka abalone (*Haliotis kamtschatkana*; L 8 cm; compare Color plate, p. 45). Alaska.
Right, from top to bottom: *Lucapina crenulata* (L 8 cm). California.
Smooth periwinkle (*Littorina obtusata*; L 1 cm; compare Color plate, p. 71, and Color plate, p. 46 on *Fucus*). North Atlantic.
Calliostoma annulatum (L 3 cm). California.

for only the juvenile shell is reminiscent of the usual snail form. They are sessile and attached to the substrate. The vermetids possess an irregularly coiled shell, and through the progressive degeneration of the foot they seem more to resemble the tube-building polychaetes (see Vol. I) than the true snails. It is due to the efforts of the scientist Adanson in 1757 that the vermetids came to be recognized as mollusks, and the type-specific species was therefore named after him (*Vermetus adansoni*; see Color plate, p. 46). *Serpulorbis arenaria* (see Color plate, p. 46), which is most frequent in Germany, reaches a size of over 10 cm and in cross section measures 10 to 15 mm. The shell of the smaller *Bivonia triquetra*, which is present on the south European coast, is usually less bizarre. In cross section the shell appears triangular. The whorls are characterized by a knotted line which winds along the spirals. The sedentary habits of these peculiar worm snails has resulted in a great change in the mode of feeding. Worm snails are mucociliary feeders. Their pedal gland discharges mucous strings which form a net which is used to trap microscopic creatures and organic substances in the water. Periodically the vermetid grasps such a mucous string with its radula and pulls it into its mouth.

Fig. 5-11. High-spired shell of *Mathilda quadricarinata* (L 3 cm). Europe.

The very small CAECIDS (family Caecidae) are now known to belong to the superfamily Rissoacea. These tiny 3 mm snails live on sandy bottoms, and at first have a coiled shell with flat spirals. During subsequent growth of the shell, the juvenile snail shell is rejected and a septum is formed at the site of the rupture. The largest European representative of this family is *Caecum trachea*, measuring 5 mm, which is easily recognized by its spiral shell. *Caecum glabrum* measures only 2 mm. In places it is far more common than the former species. Its shell is smooth. *C. glabrum* is the only streptoneurous species known today that inhabits the spaces between coarse sand along the European coasts. There are dozens of species elsewhere in the world.

Fig. 5-12. *Caecum glabrum* (L 2 mm) and juvenile shell. Europe.

The families Thiaridae, Melanopsidae, and Pleuroceridae were formerly grouped with the Melaniidae in the broad sense. Planaxidae and Modulidae follow these three families. It seems obvious that some of the aquatic thiarids and marine planaxids reproduce by parthenogenesis since no males have yet been found. A brood pouch serves as a protection for the progeny. In contrast, the aquatic melanopsids and pleurocerids are dioecious and lay eggs. The few species which represent the melanopsids in the Old World prefer warmer regions. *Fagotia acicularis* and *Fagotia esperi*, which occur in Hungary and the Balkan peninsula, inhabit warm springs and water of similar temperatures. *Amphimelania holandri*, distributed from the Vienna basin to the southeast, and *Amphimelania parvula*, distributed from the Carpathian Mountains to the Balkans, are found in similar habitats. The pleurocerids are very numerous in species and individuals in the rivers of North America.

The shells of *Pirenella conica*, from the brackish waters or saline lakes

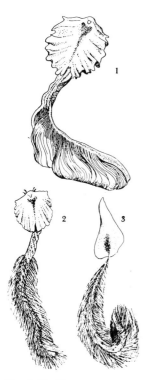

Fig. 5-13. Giant spermato-phore (spermatozeugma) with tail of attached sperms of *Cerithiopsis tubercularis* (1), *Epitonium clathrus* (2), *Janthina janthina* (3).

Fig. 5-14. Violet snail with bubble float (see p. 78).

of the Mediterranean, and *Cerithidium submamillatum*, which is one-fourth as small (L 5 mm) and is distributed from the Black Sea to Biscay, already closely resemble the true HORN SHELLS (family Cerithiidae). The NEEDLE WHELK (*Bittium reticulatum*; see Color plate, p. 46), which is quite common in the algal zones of all European coasts, belongs to the cerithiids. Other representatives are the *Gourmya vulgata* (see Color plate, p. 46), which can measure over 8 cm and is found from the Mediterranean and the southeastern Atlantic, and the much smaller *Gourmya rupestris* from southern and southwestern Europe. The latter species inhabits muddy substrates. *Cerithiopsis tubercularis* (L 7 mm), on the other hand, occurs in the algal zones of swamps. Generally it seems that numerous cerithiids are adapted to feeding in swamp vegetation. The uniqueness of the cerithiids also becomes evident if one considers the diversity of the sperms. The cerithiids have giant sperm which are sperm carriers. The marine family Triphoridae deviates from the usual shell form in that in many species the shell winds to the left (sinistral). Sinistral whorling of the shell was responsible for the scientific name of *Triphora perversa*.

The WENTLETRAPS and VIOLET SNAILS (Ptenoglossa) are quite an interesting group from the biological point of view. This group includes the WENTLETRAPS OR STAIRCASE SHELLS (family Epitoniidae) and the VIOLET SNAILS (family Janthinidae) because both families share many similarities in the more general and detailed aspects of their body structures. W. Ankel reported that despite the totally different shell of the TRELLIS WENTLETRAP (*Epitonium clathrus*; see Color plate, p. 46), its radula has a similar structure and the same function as that of the VIOLET SNAIL (*Janthina janthina*). The jaws and the characteristically arranged salivary glands and nervous system also have a similar basic plan. In both genera a pallial gland discharges a purplish secretion of unknown significance. In addition, these two groups are characterized by a most peculiar type of sperm formation which is unique in the entire animal kingdom. Both produce giant sperms by the same method. They produce normal sperms that become attached to these sperm carriers (giant sperm; see Fig. 5-13). Finally, the species of both genera are specialized to take up similar foods. Both live on hydrozoans, the wentletraps preferring sea anemones (Actiniaria) and the violet snails preferring floating jellyfishes and ctenophores.

Yet despite all these similarities, both families have basically different modes of life. The wentletraps, which include the already mentioned Trellis Wentletraps and the HOTESSIER'S WENTLETRAP (*Opalia crenata*) from the Mediterranean and the western Atlantic, as well as *Cirsotrema communatum*, are bottom dwellers and locate their food with the aid of sensory stimuli. The violet snails are pelagic. They drift along upside-down on the ocean surface (see Fig. 5-14) with a self-constructed foam float. These snails are blind and are without statocysts. They are only able to trap prey which accidentally gets caught in the float. The violet snails have lost the ability to crawl normally because their foot has been specialized for the function of producing the float. The operculum is also degenerate. In order to construct a foam float the animal captures an air bubble from the water surface with its spoonlike foot, the propodium, covers it with mucus, and attaches it to additional bubbles along the sole of the foot. The violet snails, which are dioecious in some species, sometimes also use the float as a raft to which an extremely large number of eggs (up to 2-1/2 million) are attached to the float in more than 500 capsules. Following the free-swimming larval stage (veliger type; see Chapter 1) the juveniles begin to construct their own floats. Like a few other snails (e.g., sea hares) and most cephalopods, violet snails discharge a colored substance when disturbed. As the name indicates, the violet snails discharge a violet-colored substance.

The ACLIDIDS (family Aclididae) are the last group of the superfamily EULIMACEA which are characterized by a distinct radula which is very similar to the one in the Ptenoglossa. A European representative of this family is *Aclis supranitida*. The EULIMIDS (family Eulimidae) parasitize echinoderms. With the aid of a proboscis they suck the body pieces of their living hosts. In the genus *Leiostraca* and the species *Leiostraca subulata*, which lives along European coasts, only a few small, pointed radular teeth are present. In related species, however, the teeth as well as the jaws are missing. *Eulima incurva* parasitizes brittle stars in a manner similar to that of *Eulima perminima*, *Balcis polita*, and others which are frequent forms along European coasts.

The STILIFERIDS (family Stiliferidae, see Color plates, pp. 65–66) and the subsequent three families are true parasites. They lack a radula. The stiliferids and the eulimids are grouped under the Aglossa (tongueless). Within this group there is progressive specialization from external to internal parasitism. All members of this group either live in or on echinoderms (see Chapters 12 and 13). These snails are characterized by a "false mantle" which is a disclike or collar-shaped flaplike projection of the foot, which is located at the base of the usually long proboscis. This projection extends up to the tentacles in the hermaphroditic European *Pelseneeria stylifera* and partially encircles the shell. This species still possesses small eyes, and the foot is still used as a creeping organ. *Pelseneeria profunda*, which is a parasite of the deep-sea sea urchin in the Atlantic, on the other hand, has only a very small foot, and degenerate eyes and tentacles. *Gasterosiphon deimatis* (see Fig. 5-15) is a true internal parasite of a deep-sea sea cucumber (*Deima blackei*).

This species is described by H. F. Nierstrass in the following manner: the adult animal has an oval body which is extended by two processes. One of these, the siphon, is short and punctures the skin; the other is long, consisting of two different parts, and opens into the ring canal of the host. The snail is enveloped in a gigantic secondary mantle, which in this case is a true extension of the proboscis. The animal has retained distinct whorls but has lost the shell. The vestigial foot is present in the form of

▷
Bulloidea: 1. *Acteon torn-atilis*; 2. *Haminea hydatis*; 3. Common bubble shell (*Bulla striata*); 4. *Philine quadripartita*; 5. *Scaphander lignarius*; 6. *Retusa truncatula*; 7. *Doridium depictum*; 8. *Microhedyle glandulifera* Thecosomata (pteropods): 9. *Clio pyramidata*; 10. *Cavolinia tridentata*.

MILLA

MILLA

two longitudinal lateral flaps. Otherwise the tentacles with the eyes, the mantle, the ctenidia, heart, and kidneys are absent.

The Japanese *Paedophoropus discoelobius* (L 3–5 mm) parasitizes the water-vascular system of a sea cucumber (Polian vesicle of *Eupygurus pacificus*). This parasite, which does not have a secondary mantle, is dioecious. Its larvae mature in a pedal cavity which serves as a brood chamber. This larval form still has a shell with two whorls and an operculum, although this as well as the anus is lost in the adult. The tubular species *Entocolax ludwigi*, from the Bering Sea, and *Entoconcha mirabilis* (see Color plate, p. 46), from the European coasts, have greatly reduced internal organs. They are dioecious. The dwarf males of this species live within the secondary mantle. They also attack sea cucumbers. The host of the latter species, the sea cucumber *Leptosynapta*, is frequently found along European coasts. *Enteroxenos ostergreni* (shell L up to 10 cm) parasitizes the intestinal tract of the sea cucumber *Stichopus*, which is distributed along Norway's coast. This parasitic species, as well as *Entoconcha* and *Entocolax*, is dioecious, and does not have an intestine. The interior of this parasite is merely a large cavity which is connected with the intestinal wall by a ciliary canal. The cavity also contains the egg balls and the highly modified "dwarf male."

The superfamily HIPPONICACEA brings us back again to the familiar shelled snails with radulae. This superfamily encompasses the VANIKORIDS (family Vanikoridae) of the Indo-Pacific and Caribbean, the FALSE CUP-AND-SAUCER (*Cheila equestris*), which is distributed in the tropical oceans, and *Fossarus costatus* (shell L 6 mm), which occurs in the Atlantic and the Mediterranean. The CALYPTRAEACEA are a smaller superfamily. The group includes the CAP SHELLS (family Capulidae), which are without an operculum and are conspicuous because of their cap-shaped form, rather in contrast to the northern *Trichotropis borealis*. The LARGE CAP SHELLS (*Capulus hungaricus*; L 5 cm), regularly found in the North Atlantic and rarely in New England, are filter feeders, eating microscopic animals and therefore also possessing a crystalline style. The parasitic cap shells, on the other hand, have an elongated pharynx shaped as a proboscis, with which they suck up the body fluids of the host. They do not have a radula. Capulids have been found as fossils from the Silurian age. The Indo-Pacific *Thyca* lives on starfishes. *Thyca stellasteris*, which is found only in India, lacks a proboscis. It is the only species in this group of which the males are much smaller than the females. The foot in these snails is barely functional and is displaced by a special suction disc. The alimentary canal is often simplified in the capulids.

The CALYPTRAEIDS (superfamily CALYPTRAEACEA) contain several forms that are noted for their protandric hermaphroditism. They do not have an operculum but have an internal septum which separates and supports the visceral mass. The circular CHINESE HAT (*Calyptraea chinensis*; L 2 cm) otherwise deviates little in its external form from capulids. The well-

◁
Basommatophora (pond snails): 1. Great pond snail (*Lymnaea stagnalis*); 2. Dwarf pond snail (*Galba trunculata*); 3. Bladder snail (*Physa fontinalis*); 4. Bog or marsh snail (*Galba palustris*); 5. Ear pond snail (*Radix auricularia*); 6. Whirlpool ram's horn (*Anisus vortex*); 7. River limpet (*Ancylus fluviatilis*); 8. Great ram's horn (*Planorbarius corneus*); Stylommatophora (of Europe); 9. *Zebrina detrita*; 10. *Ena montana*; 11. Amber snail (*Succinea putris*).

known SLIPPER SHELL (*Crepidula fornicata*; see Color plate, p. 46) of the eastern U.S.A. and introduced to Europe is a typical mucociliary feeder which is harmful to oyster colonies. Just like the Chinese hat this gastropod also has a crystalline style characteristic for this mode of feeding. Mainly, however, the slipper shell has become well known because of its conspicuous method of reproduction: sexually mature animals which creep around are always males. Sooner or later they become sessile, often on females. This unique condition is known as consecutive hermaphroditism and results in the well-known "tower chains" which consist of females at the bottom and sexually mature males at the top. Fertilization of the females is brought about by a wandering male and not by one of its "own" chain.

The CARRIER SHELLS (family Xenophoridae), with *Xenophora conchyliophora* from the Caribbean, belong to the superfamily STROMBACEA. These snails camouflage themselves by cementing broken shells and stones to their own shells. The stromboids are characterized by a greatly enlarged body whorl which in the adult form flares out like a wing. The foot is long and very muscular, and mounted with a sharp, sickle-shaped operculum. This group also includes the mucociliary-feeding family Struthiolariidae, from the Indo-Pacific, and the well-known PELICAN'S FOOT (*Aporrhais pespelecani*; see Color plate, p. 55), which is frequently found on muddy bottoms of European oceans where it feeds partially on substrate particles, and partially by the mucociliary method. The step-wise locomotory mode of the *A. pespelecani*, which is rather unusual for a gastropod, is brought about in the following manner: first the shell is lifted and the animal pushes itself frontwards; then the foot underneath the shell is raised and brought forward. The Caribbean PINK CONCH (*Strombus gigas*; see Color plate, p. 55) moves in a similar manner, although these highly active animals can manage a type of jump of half a body length by anchoring and suddenly stiffening the posterior part of the foot. Occasionally there is pink pearl formation in *S. gigas*. This gastropod, which has good eyes and is capable of directional vision, is feared because of its razor-sharp operculum. If disturbed in its habitat close to the beach, it can beat about wildly and inflict deep wounds into a wader. They are vegetarians and feed on red algae. *Lambis chiragra*, which lives in the Indo-Pacific region, possesses six heavy shell processes. It was formerly used by the natives as a dangerous fist weapon comparable to brass knuckles.

The LAMELLARIACEA are an interesting superfamily, certain forms of which still possess two columellar muscles. It includes the genera *Lamellaria*, *Velutina*, and *Trivia*. Some of them are found on tunicates (see Chapter 18), for example *Velutina velutina*, or the widely distributed *Lamellaria perspicua*, whose shell is completely overgrown by the mantle. *Velutina flexilis*, on the other hand, feeds on hydrozoans (see Vol. I). *Erato voluta* is found on tunicates as well as on corals. *Trivia monacha* and *Trivia*

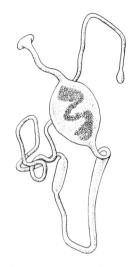

Fig. 5-15. *Gasterosiphon deimatis*, a true internal parasite.

Pelican's foot

Pink conch

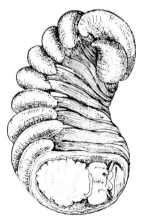

Fig. 5-16. Stacked (copulatory) chain of slipper shells (*Crepidula fornicata*).

Fig. 5-17. *Trivia monacha* on a colonial ascidian; right, egg capsules. (After Fretter and Graham, 1962).

Fig. 5-18. *Simnia patula* on *Alcyonium*. (After Fretter and Graham, 1962).

Panther cowrie

Money cowries

europaea live on colonial tunicates. They deposit their egg capsules, which are characterized by a neck, into the unicates, in a fashion similar to that of other species. The majority of the lamellariids and eratoids (families Lamellariidae and Eratoidae) are also characterized by a special larval type (i.e., echinospira). When the larval shell develops, the outer layer separates and a space of various dimensions appears which is filled with tissue fluid. The outside of this shell, which can measure up to 2.5 mm, is often covered with spicules or is sculptured. This type of shell acts like a float and allows the larva a prolonged planktonic existence. The previously mentioned cap shell also has an echinospira larva.

The COWRIES (superfamily CYPRAEACEA) resemble the Lamellariacea biologically as well as in the mantle flaps which cover the major part of the shell and the respiratory siphon which is a tubelike elongation of the mantle cavity. These shells are well-known collector's items. This group includes all European OVULIDS (family Ovulidae) which without exception live on various corals. Some species, such as *Simnia spelta* deposit color pigments in their mantle, which are obtained from digested horny corals. After a few weeks of such feeding the snails are protectively colored. The geographically adjacent *Simnia patula* lives on sea fans and hydroid polyps (see Vol. I). The southern European PEDICULARID (*Pedicularia sicula*), which is only half the size of the previous species, attaches itself to horny and stony corals and even adjusts itself to the contours of the host.

The European COWRIES (family Cypraeidae), on the other hand, are free-living on muddy or hard substrates where they prey on corals and sponges. Typical examples are the two southern European species: *Cypraea pyrum* (L 3 cm) and the larger, highly variable *Cypraea lurida*. These two species as well as the PANTHER COWRIE (*Cypraea pantherina*) already had special significance in times of antiquity. They were carried as amulets against infertility and venereal disease. On Cyprus the shell of the panther cowrie was dedicated as *concha venerea* to Aphrodite, the goddess of love. The name of the genus, "Cypraea," also originated in this context. Even today, girls still string these shells into necklaces and use them as decoration although the original meaning of these amulets has presumably been lost. The beautifully colored TIGER COWRIE (*Cypraea tigris*; L 8 cm; see Color plates, pp. 55 and 72) and the GOLDEN COWRIE (*Cypraea aurantium*) are also very well known.

Among the Indo-Pacific cowries the shining, yellow *Cypraea moneta* (L up to 25 mm) is quite famous. Formerly, its shells were used for monetary purposes in China, Japan, and India. The scientific term, "moneta," reflects this use. These shells reached Europe long ago via the trade routes. Even up to the 19th Century these snails had monetary value. *Cypraea annulus* played a similar role in the former states of Siam and Bengal while *Cypraea moneta* was used as money at the upper Coe River in southwestern China, elsewhere in the Orient, and in central Africa.

All these cowrie shells and many other species as well have attained strong cultural significance in various regions of the world. In ancient Egypt, for instance, shells were put in graves. Gastropod shells have been used as mystical symbols, various types of jewelry, diverse working tools, and they have even been stylized in ideograms, identifying sentences dealing with concepts of trade, economics, and money. In many of these frequently beautifully designed shells, only the body whorl is visible since it grows around the spire. In living animals the shells are often not visible at all because the broad mantle flaps cover almost the entire shell. Also the mantle flaps are able to secrete calcareous substances themselves. It is interesting that porcelain received its name from the snail and not vice versa. M. Schilder reported: "When Marco Polo brought back the first porcelain to Europe at the end of the 13th Century his countrymen were reminded of a similarity between this material and the familiar 'porceletta' or 'porcellana' snail shells. They transferred this name to Marco Polo's imported material."

The HETEROPODS (superfamily Atlantacea) introduce us to gastropods that are adapted to a free-swimming existence in the high seas and appear near coastal lines only during fall or winter storms. The animals swim on their backs and are completely transparent. The anterior section of the heteropod foot is laterally strongly compressed; a suction cup makes up the remainder of the foot. The compressed section of the foot, if turned dorsally, serves as a starting motor when rapidly beaten. These snails include the ATLANTIDS (family Atlantidae) which possess a very thin, keeled shell plus an operculum. Among this group one finds the species *Atlanta peroni* (see Color plate, p. 55), which is quite common in the Mediterranean, and *Oxygyrus keraudreni*. Both species measure only a few millimeters. Since all heteropods are pelagic, these species are dependent on drifting food which they may partially detect with the aid of their stalked eyes. They also have complicated statocysts. With the suctorial proboscis and the grasping radula, these snails commonly devour jellyfishes and other free-swimming animals which they can overpower. Even the stately *Carinaria mediterranea* (L up to 30 cm; see Color plate, p. 55) preys on larger animals. In addition, this snail swims very well with the aid of its broad propodium and the metapodium which has been modified as a steering fin. In this species as well as in all CARINARIIDS (family Carinariidae) the shell exists only as a small hatlike structure that fits over the visceral mass which is surrounded by the secondary gills. Finally, in the PTEROTRACHEIDS (family Pterotracheidae), mantle and shell are completely degenerated. Externally they greatly resemble the carinariids. It is said that the pterotracheids emit phosphorescent light when disturbed and that they thereby contribute to the illumination of the ocean. Among this family are the species *Pterotrachea coronata* and *Firoloida desmarestia* which frequently occur in the Mediterranean. In this family, only the males have a suction cup on the "keel." Tentacles are present only in the males of the genus *Firoloida*.

Superfamily:
Atlantacea

Moon shells

The MOON SHELLS (superfamily NATICACEA) inhabit soft substrates where they usually burrow themselves into the mud. They obtain oxygenated water via a respiratory siphon. They prey on other mollusks, both gastropods and bivalves. The propodium can be made turgid by lymph, and during digging it acts as a type of "plow." If a moon shell has captured a prey, the victim is at first "corroded" with a secretion from an organ underneath the oral cavity. Subsequently the powerful radula enlarges the corroded hole in the shell until the proboscis along with the radula can be inserted completely and the victim broken up. The BANDED MOON SHELL (*Lunatia alderi*; L 2 cm) winds mucous strands around its victim, thereby immobilizing it; then the gastropod captor bores holes into the shell at 90-degree angles. Related species, such as *Naticarius stercusmuscarum* (see Color plate, p. 74) from southern Europe and *Sinum leachi* (see Color plate, p. 55) from the Indo-Pacific region, pierce captured bivalves and snails in a similar manner. Some naticids deposit their spawn in the characteristic form of a sandy collar. In the BROWN MOON SHELL (*Lunatia catena*) the spawn is arranged in a conical manner, while in *L. alderi* it is circular.

Some representatives of the last superfamily, the TUN SHELLS (Tonnacea) of this large suborder Taenioglossa have also entered the history of civilization. Generally these snails are very large. They possess a respiratory siphon, a powerful foot, and a handsome shell. They prey mainly on echinoderms and bivalves. Although *Oocorys sulcata* from the depths of the Atlantic measures only 4 cm we can already notice the trend towards a strong development of the shell in the PRICKLY HELMET (*Galeodea echinophora*; L 5–11 cm), which lives at shallow depths in the Mediterranean. The Indo-Pacific HELMETS (family Cassidae) are favorite collector's and exhibition items. The LARGE HELMET (*Cassis cornuta*; see Color plate, p. 55; compare Color plate, p. 71), for example, was a favorite household decoration. *Cypraecassis rufa* is still today carved into famous cameos. The related family Cymatiidae includes the TRITONS (genus *Charonia*). The typical TRUMPET SHELL (*Charonia tritonis*; see Color plate, p. 55) occurs in the Indo-Pacific. Another species which can reach a size of 40 cm is found in the northwestern region of Africa. It is the largest gastropod in the African-European area.

Triton's Trumpet

The triton's shell was used as a war trumpet in the South Seas. A blowhole was carved into its side so it could be used like a German flute. The Romans also made use of the "Buccina." It was used to call the citizens to war. In this case, however, the TRUMPET SHELL (*Charonia lampas*; L 30 cm) was utilized. Even today trumpet shells are occasionally used in the Pacific and Mediterranean as a signal horn. Here the blowhorn is not carved into the side of the shell but the tip is filed to produce an opening. During the time of the Rococo the helmet shells were used in a variety of ways as decorations. *Bursa scrobiculator* (family Bursidae), which measures about 8 cm, is found on muddy flats of moderate depths in the

Mediterranean. The well-known TUN SHELL (*Tonna galea*; see Color plate, p. 55) of the family Tonnidae occupies the same habitat. It can reach a respectable size of 25 cm. Like all tun shells, this form is carnivorous and also has a special weapon to kill its prey. Its predatory nature is recognizable on the basis of the long and movable proboscis with the terminal suction cup. The animal produces a solution containing 2% to 4% sulfuric acid in its two large buccopharyngeal glands. This is a unique occurrence in the entire animal kingdom. The gastropod injects this corrosive secretion into its victim by means of the two jaw plates which have been modified into grooved hooks and the prey either becomes paralyzed or is killed. Then the snail rips large chunks out of the victim with its radula and eats them. The tun shells represent a final link in the chains of this suborder, with their adaption to a carnivorous existence, similar to the cones in the following suborder, Stenoglossa (or Neogastropoda).

Sulfuric acid in the animal kingdom

The suborder Stenoglossa represents the most recent branch of the prosobranch gastropods. The radula usually possesses only two or three tooth plates per transverse row. As a rule the shell has a long siphon. These forms are almost always marine carnivores or scavengers. Fossil Stenoglossa are only known from the Upper Cretaceous and they have therefore been referred to as Neogastropods. Their are eighteen families within four superfamilies and a total of approximately 5,000 species.

Suborder: Stenoglossa

With the ROCK SHELLS (superfamily Muricacea) we again encounter a group of snails which at one time were very famous and were highly valued by man. In ancient times these gastropods were the source of the well-known purple color. The muricids are characterized by their frequently drawn-out siphon. Their most notable feature, however, is the hypobranchial gland in the mantle cavity. Although it is found in most aquatic, shelled gastropods, exposing the secretion of the muricid's gland to sunlight gives it a color ranging from yellow to purple. The Phoenicians from Tyre (now in Lebanon) were already aware of the property of this "purple gland." The tribes around the Mediterranean utilized the muricids, thereby adding a new dimension to the world of fashion. Purple was a desired dye for garments worn by Roman caesara and senators, the tribunals, and rich citizens.

The extraction of purple dye

In those days *Trunculariopsis trunculus* (see Color plate, p. 56), from the Mediterranean and the Lusitanian region, the SPINY DYE MUREX (*Murex brandaris*, see Color plate, p. 71), from the Mediterranean, and *Thais haemastoma*, from the North Atlantic and Mediterranean, were referred to as *purpura*. *T. trunculus* produces intermediate colors such as yellow, green, blue, red, and, in the final stage, light purple. The secretions of *M. brandaris* and *Thais haemastoma*, however, turn violet in sunlight. Naturally, one single snail produces only a tiny amount of dye. Consequently, thousands upon thousands of snails were killed for their dye. Proof of this slaughter are the gigantic shell deposits in Tyre, Sidon, and other places, as well as the famous Monte Testacea (shell mountain) near Tarentia. There are still a few regions in the Mediterranean where the purple dye is extracted from these gastropods. Formerly the ATLANTIC DOGWINKLE (*Nucella lapillus*; see Color plate, p. 56) was used for similar purposes in Brittanny and Norway, especially to mark laundry. In Central America the species *Purpura patula* was used by the Indians to dye cotton threads. The famed purple gowns worn by cardinals are still reminiscent of this formerly priceless dye. Today, however, purple dye is produced by chemical methods.

All these rock shells prey on other living animals or feed on carrion. They press their prey to the bottom with their foot and then devour it. Usually they feed on bivalves and other gastropods (see Vol. I). Some species, such as *Ceratostoma erinaceum* (L 6 cm; see Color plate, p. 56), drill a hole into the valves with the aid of the radula and then insert the long proboscis. Others, for example *M. brandaris*, pry open the bivalve with a strong projection in the middle of the outer lip of the aperture.

The great variation of the rock shells is of interest and is probably due to genetic factors since the chromosomes are dimorphic. The beautiful *Murex pecten* or VENUS COMB MUREX from the Indo-Pacific should be mentioned. The latter species has a shell which measures 10 to 15 cm and has an elongated siphonal canal. Its shell has four rows of dozens of thorns.

The juvenile forms of the MAGILIDS (family Magilidae) strongly resemble the rock shells, although they do not have a radula, and are almost always parasites on hydrozoans. Most forms live on corals, for example the Mediterranean-Lusitanian *Coralliophila lamellosa* (L 40 cm). Some magilids, like the Indo-Pacific *Magilus antiquus*, burrow themselves into the calcareous layers of the corals by chemical means. Some even become enclosed by the growing corals.

Fig. 5-19. Rachoglossate radula of *Buccinum undatum*.

One of the largest marine superfamilies is the BUCCINACEA, which, among several families, includes the DOVE SHELLS (family Columbellidae). A few of them live on other animals as ecto-parasites, if only in or on sponges. Occasionally *Mitrella scripta* is found in the Mediterranean. *Columbella rustica* (L 2 cm) is twice as large as the previous species, and inhabits the algal coastal regions of southern Europe. A close relative of it is the COMMON DOVE SHELL (*Columbella mercatoria*; L 2 cm) from the Caribbean.

Fig. 5-20. Distribution of the waved whelk, *Buccinum undatum*.

The well-known WAVED WHELK (*Buccinum undatum*; L 10 cm; see Color plate, p. 56, and Fig. 5-20) is a pure scavenger. This easily recognizable form is widely distributed in the North Atlantic and in northeastern United States. The species belongs to the very comprehensive family of WHELKS (Buccinidae). Just like all creatures that are active as "sanitation workers on the ocean" in that they feed on animal corpses, the waved whelk is endowed with excellent chemoreception. This sense is made more efficient by the siphon which is pointed into the direction of the water current. North Sea fishermen like to use the marbled body of the waved whelk as bait and often also to enrich their own menu. The whelks lay numerous egg capsules piled up into irregular clumps. Each capsule contains up to 1000 eggs but only approximately ten actually undergo full development. The others serve as "nutritive eggs." The animals hatch as fully developed young snails from the capsule. Formerly the fishermen used the empty capsules as "sea soap" to clean their hands. Hermit crabs frequently take possession of empty whelk shells.

Of the numerous whelks in the European region the following species are quite common: *Cantharus d'orbignyi* (L 2 cm), which inhabits the Mediterranean, *Pisania striata* (L 3 cm), which is found in the algal zones and also in harbors where it often settles in dense concentrations, and *Buccinulum corneum* (L 6 cm), which is found on stony substances. The shells of the SPINDLE SHELL (*Neptunea antiqua*; shell L over 15 cm) were formerly used as whale-oil lamps. This species is distributed throughout the northeastern Atlantic. The northern genus *Colus*, with *Colus gracilis* (L 7 cm), also belongs here.

The CONCHS (family Melongenidae) have evolved numerous gigantic forms. *Syrinx aruanus*, from northern Australia, measures up to 60 cm from the siphonal canal to the apex. It is one of the largest shelled gastropods known, exceeded in size only by the sea hare, *Aplysia*.

If one places a piece of fresh meat or carrion on the flat muddy bottom in the tidal zone, in addition to whelks, a great number of yellowish-brown snails will be attracted which will approach the prey with an extended siphon. These are widely distributed from Norway to the Black Sea. They are called *Hinia reticulata* (L 3 cm; see Color plate, p. 56) and belong to the MUD NASSAS (family Nassariidae). These animals are able to smell a substance from a distance of thirty meters. When *H. reticulata* is resting, one can only detect its presence by the siphon which protrudes out of the mud and which is pointed into the direction of the water current. *H. reticulata* is characterized by an interesting flight reaction when it comes within touching distance of a starfish. The snail will tip the shell anteriorly and release the foot from the substrate, thereby causing the shell aperture to lie dorsally. Subsequently it will stretch out its foot, turn laterally, and settle on the substrate again. The animal can escape faster by repeating this maneuver than by creeping away. Related species, for example the Indo-Pacific *Nassarius arcularius*, are adapted to opening bivalves in a manner similar to that of some rock shells. These carnivorous snails lie in wait until the bivalve victim opens its shells slightly. *N. arcularius* will rapidly tip the shell frontwards and push the snout margin between the valves so that the long proboscis can be inserted.

In contrast to the egg-shaped mud nassas, the TULIP SHELLS (family Fasciolariidae) are more reminiscent of the elongated shells of the buccinids. In Florida and the West Indies the fusiform TRUE TULIP (*Fasciolaria*

▷
Stylommatophora (land snails):
1. Red slug (*Arion rufus*; see Color plate, p. 94); 2. Black slug (*Arion ater*; see Color plate, p. 94); 3. *Cochlodina laminata* (compare Color plate, p. 94); 4. *Limax maximus* (see Color plate, p. 94); 5. *Deroceras agreste* (compare Color plate, p 94); 6. *Daudebardia rufa*; 7 and 8. Grove snail (*Cepaea nemoralis*; compare Color plate, p. 94); 9. *Achatina achatina* (see also Color plate, p. 94); 10. Copse snail (*Arianta arbustorum*); 11. *Helicella obvia* (compare Color plate, p. 100); 12. *Perforatella incarnata*.

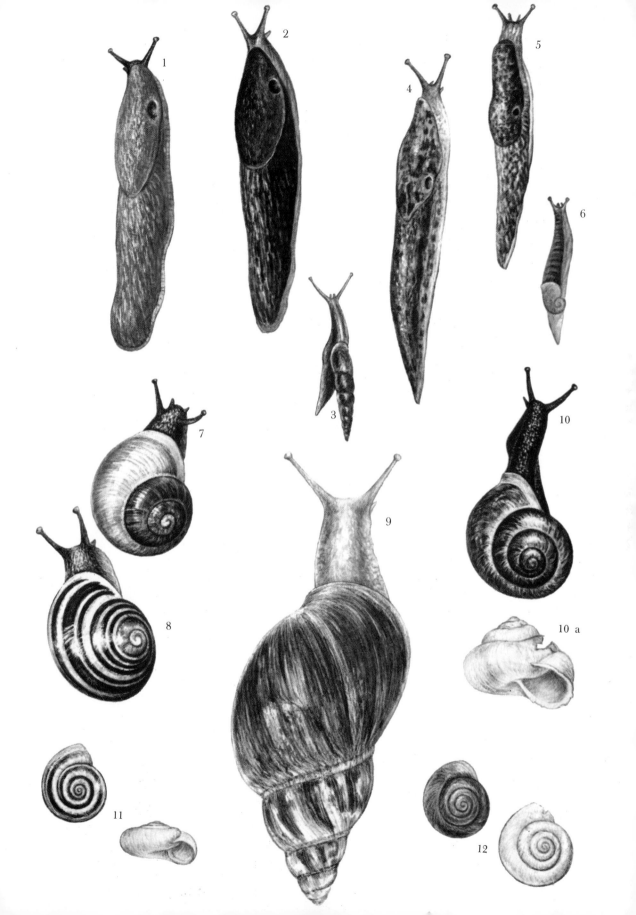

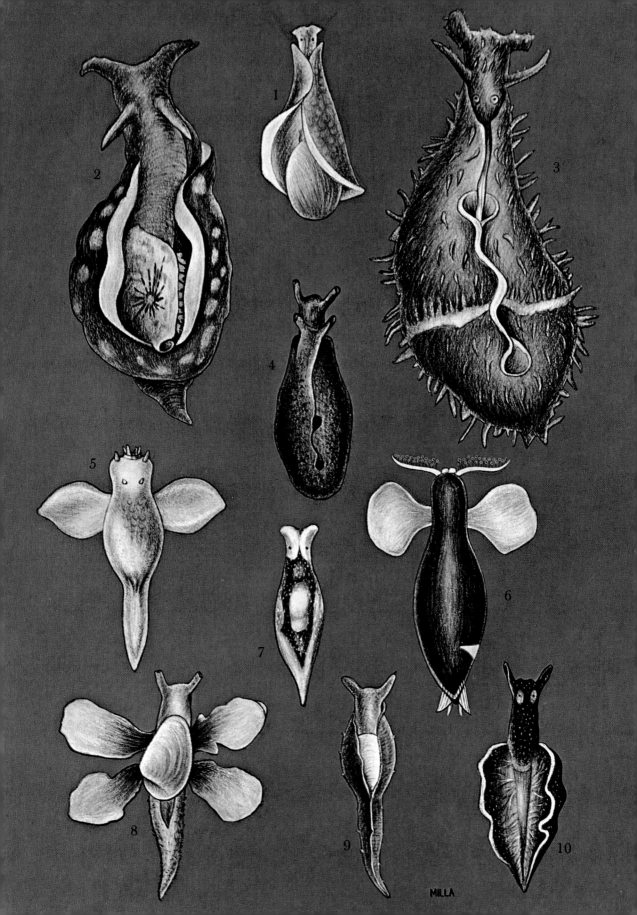

MILLA

Aplysiacea: 1. Paper bubble shell (*Akera bullata*); 2. Sea hare (*Aplysia depilans*; compare Color plate, p. 122); 3. *Dolabella scapula*; 4. *Aplysiella virescens*. Gymnosomata (pteropods): 5. *Clione limacina*; 6. *Pneumodermon violaceum*. Sacoglossa: 7. *Limapontia nigra*; 8. *Lobiger serradifalci*; 9. *Oxynoe olivacea*; 10. *Elysia viridis*.

Volutes

tulipa; L 8 cm) lives in shallow water among marine algae and sand, and feeds on bivalves. It lays attractive clumps of chitinous urn-shaped egg capsules which are attached to rocks and dead shells. *Fasciolaria hunteria* from Florida and *F. lignaria* from the Mediterranean, and the larger *Buccinofusus berniciensis* (L 9 cm) are familiar representatives of this group. All species possess a relatively bulky shell unlike, for example, the slender shell with the elongated siphonal canal of *Fusinus rostratus* (L 7 cm) from the Mediterranean.

The various OLIVE SHELLS (family Olividae) are truly beautiful. They are not found in European oceans. Their attractively colored olive-shaped shells are favorite items for collectors. The foot which covers the shell from both sides is often uniquely decorated. Olive shells are primarily tropical gastropods, but the LETTERED OLIVE (*Oliva sayana*; L 5 cm) is abundant in sand in the temperate region of the southeastern U.S.A. *Oliva flammulata* (L 3 cm; see Color plate, p. 56) inhabits the west African coast, and the doubly large BLACK OLIVE SHELL (*Oliva maura*; see Color plate, p. 56), the Indo-Pacific region. One of the best-known representatives is the TENT OLIVE (*Oliva porphyria*; L over 10 cm) from the Pacific side of Central America.

The VASE SHELLS (family Vasidae), which are quite similar in shape to tulip shells, except that the columella has three to five strong folds, are represented in the central European area by *Metzgeria alba* (L 25 mm). The largest species of WEST INDIAN CHANK (*Turbinella angulata*; L 12 cm) lives in shallow water in the Bahamas, Cuba, and Mexico. Special mention should be made of the SACRED CHANK (*Turbinella pyrum*; L 8 cm) which is the sacred shell of the East Indian god Vishnu. In Bengal these snail shells are cut into arm bracelets and are used as fertility symbols. The HARP SHELLS (family Harpidae) are represented by about eleven species, most of which come from the Indo-Pacific. However, *Harpa ventricosa* (see Color plate, p. 73), from the Indian Ocean, is well known as a collector's item. It can voluntarily drop off the last third of its foot. It feeds on crabs. The VOLUTES (family Volutidae), most of which are without an operculum, are represented by about 180 worldwide species. There is the familiar MUSIC VOLUTE (*Voluta musica*) from the lower Caribbean Sea. Its scientific name is based on the design of the shell which resembles lines of the staff. The Iberian *Cymbium olla* (L over 10 cm) is ovoviviparous and is equally conspicuous with its projecting body whorl. Usually these animals are burrowed in the sand of the beach. In the Gulf of Mexico and southeastern U.S.A. the attractive, brown-spotted JUNONIA (*Scaphella junonia*; L 5 cm) is greatly sought after by shell collectors.

Next in line are two families which also are carnivorous. The CANCELLARIIDS (Cancellariidae) includes the European *Narona pusilla* (L 6 mm), and the COMMON NUTMEG (*Cancellaria reticulata*; L 2 cm) which inhabits the area from Florida and the Caribbean, both of which do not possess an operculum. The MARGINELLIDS (Marginellidae) are also without an operculum. These small, shiny shells inhabit sandy strips of beach along warm oceans of the world. The largest and most beautiful come from western Africa. The external shape of these forms is greatly reminiscent of the cowries but frequently the marginellids are extremely tiny. *Gibberulina clandestina* (L not quite 2 mm) is regularly found in algal regions and underneath stones on rocky coasts of southern Europe.

The following superfamilies, the CONACEA and MITRACEA, are included in the Stenoglossa because of the number of radular teeth per transverse row in the radula. In other aspects the Conacea deviate greatly. They are characterized by venom glands and a radula which is modified into harpoonlike teeth. The MITER SHELLS (superfamily MITRACEA) are the least modified conids. They still have a rachoglossate radula and therefore were formerly put close to the volutes. The shape of the mitrid shell, which is usually without an operculum, is strongly reminiscent of a bishop's miter. One of the Indo-Pacific species is called the EPISCOPAL MITER (*Mitra mitra*; L 4 cm). *Mitra cornicula* (L 2 cm; see Color plate, p. 56) is found in the Mediterranean and the adjoining Atlantic. It inhabits the deeper algal zones along the shore and also inhabits sponges. The slightly larger *Mitra ebenus* has a similar mode of life and range of distribution. There are many species in the warm waters of the Indo-Pacific and Caribbean.

Miters

Within the superfamily CONACEA, the families of TURRID SHELLS (Turridae and Thatcheriidae) are externally quite similar to the Conidae and formerly both were grouped under one family. Yet there are vast differences in the structure of the radula in these two groups. The turrid shells have tall, conical shells but with a peripheral slot in the outer lip similar to that in the primitive archeogastropod, the Pleurotomariidae. There are either seven radular teeth per transverse row as in the European species *Spirotropis carinata* and *Clavus maravignae* or five radular teeth as in the northwest African species *Perrona nifat*. In other species only two rows of marginal teeth are present. Most of these turrids still have an operculum and, in addition, their partially modified radulae are attached to the base by a membrane.

Turrids

The radula of the turrids is similar to those found in the conids and in the AUGER SHELLS (family Terebridae) in being modified into grooved or hollow-needle or stilettolike teeth which serve as injection needles for venom. For this reason these gastropods are known as Toxoglossa. One of these is the BLUE OENOPOTA (*Oenopota turricula*; L 6–25 mm) which is found at great depths in the Azores and western Europe; in northern Europe and the Arctic it comes up to the 20-m line. *Cythara albida* and *Cythara taeniata* (L barely 1 cm) are regularly and frequently found on soft substrates in the Mediterranean. *Mitrolumna olivoidea*, with a reticulate sculptured shell, on the other hand, primarily inhabits sponges. *Mangelia attenuata* (shell L 15 mm) has an extensively elongated shell aperture. This form is common at lower ocean depths.

▷
Archidoris tuberculata has a retractile respiratory organ (below) and two rhinophores at the anterior (top). (See p. 130).

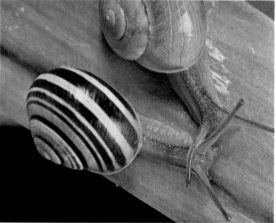

Fig. 5-21. Distribution of
Conus ventricosus.

The euthyneurous
gastropods

◁
Stylommatophora (land
snails):
Left, from top to bottom:
Achatina zebra (compare
Color plate, p. 89);
Garden snail (*Cepaea
hortensis*; compare Color
plate, p. 89);
Black slug (*Arion ater*; see
Color plate, p. 89);
Red slug (*Arion rufus*; see
Color plate, p. 89).
Right, from top to bottom:
Trochulus unidentatus, and
two door snails (family
Clausiliidae; compare
Color plate, p. 89);
Amber snail (*Succinea
putris*; see Color plate,
p. 80);
Great slug (*Limax maximus*;
see Color plate, p. 89);
Deroceras laeve (compare
Color plate, p. 89).

The closely related family Thatcheriidae is represented in today's seas by the beautiful, pagodalike THATCHER TURRID (*Thatcheria mirabilis*; L 6 cm) from deep water off Japan.

The well-known and frequently feared CONES (family Conidae) possess two single stilettolike teeth per transverse row on the radula. Their beautifully designed shells are prize exhibits in any collection. In Europe these gastropods are represented by the MEDITERRANEAN CONE (*Conus ventricosus*; L 2–5 cm). This animal, which greatly varies in its coloration, lives among algae along rocky shores where it preys on a variety of animals. Like all worm-eating representatives of this family, *C. ventricosus* is characterized by an extraordinarily long, extensible proboscis which is invaginable. In other conids the effect of the venom is of primary importance in the capture of prey. The venom has a totally paralyzing effect on the attacked animal, be it a fish or a snail.

The cones often burrow in the sand. The proboscis is pushed towards the victim where it stiffens and suddenly injects the venom dart into the prey. Then the victim is drawn into the proboscis. Only one stiletto (dart) is injected at a time. The resultant injections by certain species are unusually severe, for instance in the MARBLED CONE (*Conus marmoreus*; see Color plate, p. 56) and *Conus geographus* from the Indo-Pacific. The venom has caused the death of several people.

The EUTHYNEUROUS GASTROPODS are now considered as two subclasses—the OPISTHOBRANCHIA and the PULMONATA. Some workers use the subclass EUTHYNEURA to embrace the Opisthobranchia and Pulmonata, and further divide these gastropods into a total of thirteen orders. The shell is less prominent and is either greatly reduced or absent. The operculum is primarily limited to the larval stage. The mantle cavity usually contains only one secondary plicate gill or a vascular net which is usually situated on the right side. The lateral nerve cords are usually uncrossed (euthyneurous) and concentrated anteriorly. The bolsters of the radula are associated with strong muscles. They are almost always hermaphroditic. They are distributed in the oceans, on land, and in fresh water. There is an approximate total of 15,000 living species.

The euthyneurous gastropods encompass a series of groups which have followed different evolutionary lines. All groups are characterized by the absence of the crossed-over lateral nerve cords. This may have been caused by detorsion of the mantle cavity to the right side or (as in the pulmonates) by a prominent concentration of the main ganglia around the pharynx and esophagus. This not only resulted in the mirror symmetry of the nervous system but also of the external form of these snails. However, this is only external and apparent, for, as presumed descendants of the Caenogastropoda, the euthyneurous gastropods still possess the same unsymmetrical arrangement of the internal organs as their ancestors. Those groups which are characterized by a detorsioned mantle cavity also show a trend towards a reduction of the shell and mantle cavity.

Their ctenidium is replaced by a plicate gill which originated from the folding of the former gill septum. This plicate gill, in turn, is frequently degenerated and sometimes is even replaced by other body appendages.

The subclass OPISTHOBRANCHIA consists of mainly marine gastropods that are detorted, rarely torted, with a reduced shell that is sometimes internal or absent, with a reduced mantle cavity along the right side or entirely absent, usually without an operculum, with one auricle always posterior to the ventricle, and with one hermaphroditic reproductive system. There are thirteen orders.

Some forms of the order CEPHALASPIDEA are still very primitive. The mantle cavity is usually detorted. The shell is predominately spiral and exposed. The head is characterized by tentacles and rhinophores and usually has a broad head-shield. The mantle cavity is on the right. As a rule a plicate gill is present. Occasionally there are parapodia. They are almost exclusively marine. There are five superfamilies. There are approximately 1,500 species.

Order: Cephalaspidea

The BULLACEA possess a characteristically broad head which was caused by the fusion of the lower (anterior) tentacle and the upper rhinophore. Many species of this group have a special elongated sensory organ which is located on the right side between this cephalic shield and the foot. This is the organ of Hancock and functions for the detection of scent and taste. As in most marine euthyneurous gastropods, the eyes play only a subordinate role.

As in many streptoneurous gastropods, the Bullacea produce an eyeless veliger larva (compare Color plate, p. 26) which hatches out of the egg after a brief embryonal period. This larva possesses a small spiral shell with an operculum even in those species which are later shell-less. After the free-swimming stage, during which the veliger feeds independently on minute creatures, the larva sinks to the bottom and undergoes metamorphosis. Only a few species, from the genera *Runcina* and *Retusa*, undergo direct development without a larval stage. As a rule the Bullacea (in the narrow sense) possess a distinct shell and a foot with a prominent sole. Parapodia suitable for swimming are found only in the family Gastropteridae. The Cephalaspidea inhabit sedimentary substrates. There are six superfamilies with seventeen families.

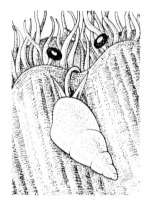

Fig. 5-22. *Odostomia eulimoides* parasitizing *Pecten* (see p. 101).

Within the superfamily ACTEONACEA we find not only the most primitive species of the Cephalaspidea but of the euthyneurous gastropods as a whole. These primitive representatives are the ACTEONIDS (family Acteonidae). Their lateral nerve cords are still crossed over and the animals are able to withdraw into a relatively large shell (L 2.5 cm). They still have an operculum. The mantle cavity is anterior. Thus the auricle lies in front of the ventricle. The penis is not evertible. *Acteon tornatilis* (see Color plate, p. 79) is found in all European oceans, often occurring on sandy bottoms and the so-called "sea lettuce" (*Ulva*). This species uses its cephalic shield for burrowing and plowing. With the aid of the shield the animal usually constructs short channels below the sandy surface. *Lissacteon exilis* (L 5 mm), from the deeper zones of the northern Atlantic and the Mediterranean, is very similar. The families ringiculids (Ringiculidae) and hydatinids (Hydatinidae) also belong to this same group. The ringiculids which inhabit the warmer oceans possess globose shells. They are usually smaller than the acteonids. *Ringicula auriculata* (family Ringiculidae) inhabits the Mediterranean. In this form one can detect the first external modifications, for example the loss of the operculum. Nevertheless, *R. auriculata* can still completely withdraw into its shell like the hydatinids. The hydatinids (family Hydatinidae) are immediately recognizable by the two head-shield appendages and two or four cephalic flaps. These forms, for example the Indo-Pacific *Hydatina physis*, are found in the tropics and already possess an evertible penis.

Most representatives of the superfamily Bullacea are still able to withdraw completely into their shell. However, only the BUBBLE SHELLS (family Bullidae) have a relatively solid shell. In the European species, COMMON BUBBLE SHELL (*Bulla striata*; L 2.5 cm; see Color plate, p. 79), the shell is marbled with purple spots. This species is easily recognized by its conspicuous cephalic shield, and is a regular inhabitant of the sea grass meadows and sandy stretches of the European coast lines.

The frequent and much smaller ATYIDS (family Atyidae) have thin-walled shells, and the foot is laterally broadened into so-called parapodia. Some species, for instance *Haminea hydatis* (L 15 mm; see Color plate, p. 79) and *Haminea navicula* (L 25 mm), which are distributed throughout western Europe and the Mediterranean, largely cover their transparent shell with these parapodia. Only the posterior section of the sunken-in spire is still visible. The closely related *Atys diaphana* is much less common in the Mediterranean, and the parapodia do not cover the more solid shell.

The RUNCINIDS (family Runcinidae), which belong to the superfamily PHILINACEA are very tiny; some are without a shell. Parapodia are absent. These highly developed and partially specialized animals have a concentrated nervous system and a true ctenidium which is extended posteriorly. They undergo direct embryonal development without a larval stage. *Runcina hancocki* (L 4 mm), which belongs to this group, occurs quite frequently along the coasts of the North Atlantic and Mediterranean.

The equally small PHILINOGLOSSIDS (family Philinoglossidae) inhabit the spaces between the ocean sand. Some workers make them the separate order PHILINOGLOSSOIDEA. The species *Pluscula cuica* still has a small vestigial shell. It is ovoviviparous, either bearing live young or having the embryos highly developed within the egg. The vestigial shell is absent in *Sapha amicorum* from the Red Sea, in *Philinoglossa remanei* from the Mediterranean, and *Philinoglossa helgolandica*. The last species has an elongated and flat, almost rectangular appearance. It has been found quite frequently along European coasts in sand at depths of two to thirty meters.

Fig. 5-23. *Runcina hancocki*, a marine cephalaspidean.

The next family to follow is the SHELLED RETUSIDS (family Retusidae), because of the structure of their reproductive organs. They are small gastropods without parapodia which inhabit deep sedimentary substrates. There they probably feed on soft microscopic creatures since they lack a radula as well as jaws. Some of the larger representatives of this group are the species *Retusa trunculata* (see Color plate, p. 79) and *Retusa obtusa* (L 3–5 mm), which are quite common in European oceans. The latter species does not have a free-swimming larval stage.

The members of the superfamily PHILINACEA are unable to withdraw into their somewhat elongated and thin shell, which is extensively covered by the mantle. This group includes the carnivorous SCAPHANDRIDS (family Scaphandridae), for example *Scaphander lignarius* (L 3 cm; see Color plate, p. 79), which is common along European coasts. In northern Europe this species is found in the shallower waters whereas in the Mediterranean it is frequently present at great depths. The species feeds primarily on scaphopods (Scaphopoda). The most familiar species of the philinids (family Philinidae) is *Philine quadripartita* (L 5 cm; see Color plate, p. 79). It is frequently found on sedimentary bottoms along European coasts while *Philine catena* (L 10 mm) inhabits the sand, and its juvenile forms occupy the spaces among the sand.

The externally similar AGLAJIDS (family Doridiidae), which also possess an internal shell, inhabit the warmer oceans in contrast to the philinids. They are not carnivorous and do not possess a radula or chewing plates within the stomach. *Doridium depictum* (L 6 cm; see Color plate, p. 79) from the Mediterranean is easily recognizable by its deep red coloration with white spots. Like all species of this family, it has a mantle which is extended posteriorly into flaps. The conspicuous BATWING SEA-SLUG (*Gasteropteron rubrum*; L 2–4 cm) has a very small internal shell which often is very difficult to remove. It is the only species of this suborder that swims with its strongly developed parapodia. However, the swimming motions are still rather jerky and undirectional. This sporadic fluttering is caused by beating of the lateral parapodia above and below the body. The "wing span" of this animal may be from 3 to 5 cm. *G. rubrum* is of a deep red color with bluish spots and a bluish-yellow margin around the parapodia. These gastropods are distributed from western India to the Adriatic Sea and also are found in the Caribbean. In places they are quite numerous at depths of thirty to eighty meters on muddy substrates. Aside from the conspicuous external form and the plicate gill, *G. rubrum* is also characterized by a tentaculate process of the mantle cavity (palliocaecum). The function of this structure is unknown. The same process is also found in the acteonids, the genera *Haminoea*, *Scaphander*, and *Doridium*, the chilinids, certain Planorbidae, and akerids. This indicates a phylogenetic relationship.

The superfamily diaphanoids (DIAPHANACEA) is comprised mainly of only minute snails which are rarely more than 5 mm. The NOTODIAPHA-

▷
The edible snail or "Escargot" (*Helix pomatia*).
Below, from left to right: copulation, egg deposit, freshly hatched snail.

NIDS (family Notodiaphanidae) inhabit the Indian Ocean; the DIAPHANIDS (Diaphanidae; e.g., *Diaphana glacialis* and *Newnesia antarctica*) mainly prefer colder oceans or the deep sea. Some species, however, are also distributed in the Italian part of the Mediterranean, for example *Diaphana minuta*.

One can derive the ACOCHLIDIIDS (order Acochlidioidea; L 1–4 mm) from these tiny species with their distinct yet transparently thin shells. The acochlidiids lost their shells, probably as an adaptive feature in response to their specialized habitat in the spaces among the sand granules. Among these forms one also finds the only cephalaspids which have penetrated into fresh water in the tropics, *Acochlidium amboinense* of Indonesia and the Palaus. The acochlidiids do not contain many species but are well represented in Europe and frequently are separated into their own group since they do not have a cephalic shield. One of the most widely distributed species is *Microhedyle milaschewitchii* (family Microhedylidae), which is quite easily recognized among the European species since it has only one pair of tentacles. *Microhedyle glandulifera* (see Color plate, p. 79) and *Hedylopsis spiculifera* both have two pairs of tentacles. The former species has strongly developed tentacular glands. *H. spiculifera* has calcareous bodies throughout its body, which is not an uncommon occurrence in sand dwellers.

The little-known species *Cylindrobulla fragilis* (superfamily CYLINDRO-BULLACEA), which is distributed in the Mediterranean and along the Portuguese coast, seems, according to more recent investigations, to be interrelated in certain aspects with the saccoglossans.

We now come to the order ENTOMOTAENIATA. In recent times the PYRAMIDELIDS (superfamily Pyramidellacea) were also placed in a similar special position. Formerly they were grouped together with the prosobranch eulimoidids. Comparative studies were able, however, to disclose the true phylogenetic position of these parasites. These animals are characterized by their sinistral embryonal shell and a dextral solid adult shell and pharynx which has been modified into a long proboscis with a suction cap. The pharynx lacks a radula but possesses a stiletto-like modified jaw with a salivary gland canal which corresponds to the jaws of the other snails. The animals use this piercing organ to penetrate the tissues of their prey, in order to inject the glandular secretions and to suck up the victim's body fluids. *Odostomia eulimoides*, which is very common along European coasts, preys on *Pecten* just like *Odostomia conoidea*, while *Eulimella laevis* parasitizes sponges and ascidians. *Turbonilla elegantissima*, on the other hand, preys on polychaetes.

The order THECOSOMATA is characterized by a partially internal shell. The sole is positioned off center. The parapodia have developed into prominent winglike structures. The body is watery and transparent. They are swimmers of the open high sea. There are two superfamilies with six families.

Glandular secretions

◁
Helicella itala and *Cochlicella* of Europe (compare Color plate, p. 89) enter into a state of estivation during the hot hours of the day.

Formerly the thecosomatids were classified together with the PTERO-PODS (Pteropoda). There is a similar trend in the Thecosomata as in the heteropods towards a stepwise reduction of the sinistral shell and the concommitant loss of the gill. The winglike enlargement of the propodium is indicative of a special adaptation towards a pelagic existence. All the cosomatids feed on minute planktonic creatures. Some species like *Spiratella helicina* and *Clio pyramidata* (see Color plate, p. 79) form huge swarms in the colder oceans and constitute an important food source for the beard whale (see Vol. XI). They belong to the superfamily SPIRATELLACEA.

Several species (*Peracle reticulata, Procymbulia valdiviae* and related forms) of another superfamily (PERACLIDACEA) still have a whorled shell, but the more highly developed CYMBULIIDS (family Cymbuliidae) lack a true shell, having in its place an internal pseudo-shell (pseudoconch). Finally, the genus *Desmopterus* is made up of forms that are free-swimming, shell-less snails without a mantle cavity. One representative, *Desmopterus papilio*, lives in the Mediterranean.

The CYMBULIIDS (family cymbuliidae) do not have gills, heart, nor kidneys, but a well-developed pharyngeal weaponry is present in the genus *Cymbulia*, e.g., in *Cymbulia peroni* from the Mediterranean. Radula and jaws are absent in the genera *Gleba* and *Corolla*. All species of the cosomatids are equipped with a more or less prominent proboscis for sucking up plankton. Like many cephalaspids they possess gizzard plates for crushing hard-shelled food.

The order SOLEOLIFERA is an aberrant group of naked, sluglike gastropods that have been the cause of much dispute concerning their phylogenetic relationship. They include the superfamilies ONCHIDIACEA and VERONICELLACEA. They are characterized by a detorsion of the mantle cavity, lack of a shell, gills, parapodia, and cerata. The respiratory cavity is either on the right side or is opisthopneumate. The head usually has one or two pairs of tentacles and stalked eyes. They are terrestrial or live in the intertidal region near the ocean. There are three families with approximately 200 species.

In the beginning the soleoliferans, especially the terrestrial and the marine onchidiids, were classified as Stylommatophores because of their stalked eyes and the gill-less respiratory cavity. However, stalked eyes are present not only in Stylommatophores but also in Thecosomata, Gymnosomata, and indicative of the Cladohepatica. With the exception of a few primitive species, the tentacles are only invaginable in the onchidiids. This is in contrast to the Stylommatophores which can only retract the tentacles, but is similar to the tentacles of the Thecosomata. The detorsion of the mantle complex during embryological development and the concentration of the nervous system in the Stylommatophores as well as the Soleolifera is due to convergence and therefore does not betray any phylogenetic relationship. Even the large pedal gland and the

▷
Notaspidea gastropods: 1. *Umbraculum umbraculum* (see Color plate, p. 73); 2. *Tylodina perversa.* Europe; 3. *Oscanius membranaceus* (see Color plate, p. 119); 4. *Pleurobranchaea meckeli.* Europe. Nudibranchia: 5. *Glaucus marinus* (compare Color plate, p. 121), a tropical pelagic species; 6. *Hero formosa* (L 15 mm). Northeastern Europe; 7. *Polycera quadrilineata* (L 18 mm). North Atlantic; 8. *Hexabranchus marginatus* (L 6 cm; compare Color plate, p. 119) of the Indo-Pacific.

Order: Soleolifera

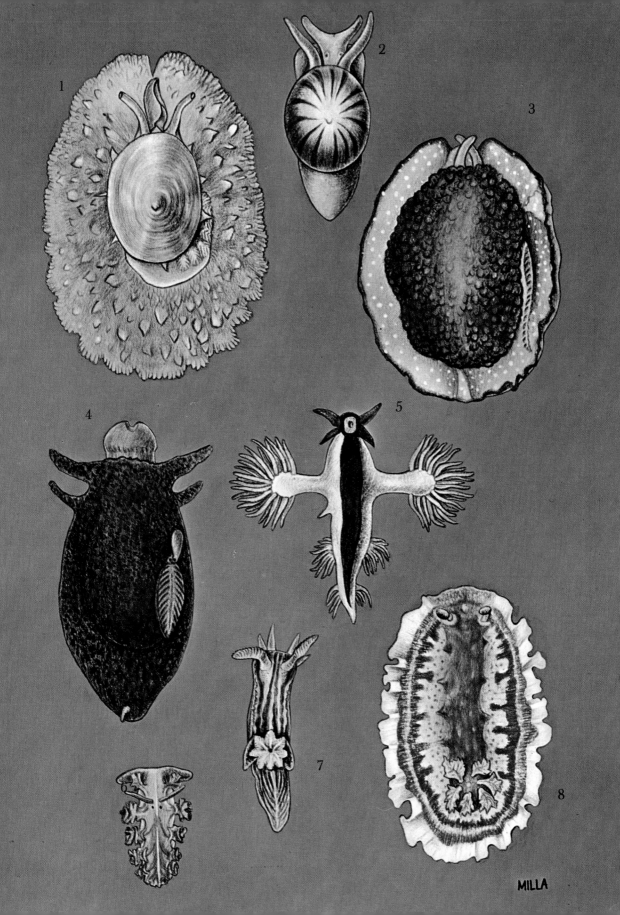

MILLA

MILLA

jaw structure are specific adaptations to terrestrial life and diet and cannot serve as clues to systematics, since this is a highly specialized group. It is probable that the individual families of the Soleolifera arose from a common root together with the Cephalaspidea and basommatophores, but then followed different evolutionary paths.

The superfamily VERONICELLACEA consists of two families. In the RATHOUSIID SLUGS (family Rathousiidae), the respiratory cavity is still located close to the anterior end of the right side. The female gonophore also opens into this cavity. These snails are strictly carnivorous. They possess a jawless extensible proboscis. *Rathousia leonina* is found in China. Representatives of the genus *Atopos*, which are carinate on the dorsal side, are distributed from India to northern Australia. *Atopos semperi* (L 2–3 cm; see Color plate, p. 113) occurs on the island of Mindanao in the Philippines. The rathousiids are more closely related with the VERONICELLIDS (family Veronicellidae). The latter inhabit the tropics, where they are primarily found underneath rocks, rotten wood, and leaves. They are nocturnal and are rarely seen by man. The usually elongated animals feed on vegetation. *Angustipes plebejus* (L 4–6 cm; see Color plate, p. 113) is distributed from Australia to the Fiji Islands, and also in Brazil. *Vaginulus taunaysi* (see Color plate, p. 113), however, is limited to Brazil. Members of both families are characterized by cross ridges along the foot (soleolae). The animal's scientific name is based on this feature.

The ONCHIDIIDS (superfamily Onchidiacea) take up a somewhat unusual position since they lack soleolae and only have one pair of tentacles. Only until recently have the onchidiids been classified with the Opisthobranchia. *Onchidiella floridana*, which inhabits rocky crevices, is amazingly faithful to one locality. Just like the limpets, it always returns to its original spot after searching for food during the low tide period. Almost all onchidiids are oval and more or less flat snails which often possess specialized organs on their wart-covered dorsal side. These organs may be bunchlike respiratory organs, as in the genus *Onchidium*, or dorsal eyes with receptors pointing away from the source of light, as in the genera *Onchidium*, *Platevindex*, and others (Fig. 5-36). In *Onchidina australis* and *Platevindex montana* the respiratory cavity is more or less positioned on the right side of the body. In the veronicellids this cavity is opisthopneumonate (at posterior end). *Platevindex montana* (L 10–25 mm) is one of the few species of this family which have given up a marine existence. This species inhabits the mountainous countryside of the Philippines. *Onchidium typhae* (see Color plate, p. 113) behaves somewhat similarly. It is found among leaves along the banks of East Indian rivers. However, most species live in or close to the tidal zone. There the animals feed on algae. They include the extremely widely distributed *Onchidium verruculatum* (Fig. 5-37), the southeastern Pacific *Onchidiella chilensis* (see Color plate, p. 113), and the Indonesian *Platevindex granulosa* (see Color plate, p. 113). The only European species, *Onchidiella celtica*

Nudibranchia: 1. *Facelina drummondi* (L 3 cm; compare Color plate, p. 120). Atlantic; 2. *Spurilla neapolitana* (L 4 cm; see Color plate, p. 120). Atlantic; 3. *Duvaucelia gracilis* (L 10 mm). Mediterranean; 4. *Peltodoris atromaculata* (L 5 cm). Mediterranean; 5. *Glossodoris gracilis* (L 4 cm; compare Color plate, p. 119). Mediterranean. Sacoglossa (for comparison): 6. *Thuridilla hopei* (L 2 cm; see Color plate, p. 119). Mediterranean.

(L 10–25 mm), is distributed in the tidal zone from Cornwall to Sicily.

Within the order RHODOPACEA there is but one species formerly thought to be a nudibranch. This small snail, *Rhodope veranyi* (see Color plate, p. 113), which barely measures 4 mm and lacks shell, tentacles, jaws, radula, and heart, was also formerly classified by some investigators with the Turbellaria (see Vol. I) and then later as a marine nudibranch. Recent studies of the organ structure and the nervous system revealed that this snail is closely related to the Soleolifera, and it was therefore included with the latter group in a separate family (Rhodopidae). *R. veranyi* is not commonly found. It inhabits the spaces between the sand granules in undisturbed stony and sandy coastal places, where it feeds primarily on sponge larvae known as *Trichoplax* (see Vol. I).

The order ANASPIDEA well-known as aplysiid sea hares, is a unique evolutionary branch of the euthyneurous gastropods. Formerly they were classified as "tectibranchs" (Tectibranchia) along with the Cephalaspidea and the Notaspidea. The anaspids are characterized by detorsion of the mantle cavity, and a shell that is largely covered by the mantle. The shell may be small or absent. A plicate gill may be present or absent. The head is usually endowed with a pair of free tentacles and a pair of curled rhinophores. The foot possesses broad parapodia used in swimming. The mantle cavity is on the right side. They are exclusively marine. There is one superfamily—APLYSIACEA. There are approximately 100 species.

It is probable that the anaspids evolved from the Philinoidea. The anaspids are well characterized by their free tentacles, the position of the nervous circumenteric ring behind the esophagus, and the lateral lobes adapted for swimming.

The superfamily APLYSIACEA is the only group of the anaspids. They are characterized by well-defined but usually small shells that are covered by the mantle. The foot is developed into a true creeping sole. Parapodia are present. They inhabit coastal regions covered by rich vegetation. There are two families.

The PAPER BUBBLE SHELL (*Akera bullata*; L up to 5 cm; see Color plate, p. 90) is a member of the family Akeratidae, which is widely distributed along European coasts, and shows in its appearance and the indicated cross-over of the lateral nerve cords a certain similarity to the Cephalaspidea; however, the remainder of the nervous system still betrays that the paper bubble shell (family Akeratidae) must be classified with the Aplysiacea. These snails always have their shell covered with broad parapodia. The fragile, thin, globular shell encircles the entire visceral mass, but the animal is unable to withdraw into its shell. When the broad parapodia are flapped together over the animal's back, they form a funnel which opens posteriorly. Water is pressed out by their quick muscular contraction, propelling the animal forwards. They are particularly active in the spring, when one can see them go head

Order: Anaspidea

Fig. 5-24. Distribution of the lake limpet (*Acroloxus lacustris*).

Fig. 5-25. Distribution of the fresh-water limpet *Ancylus fluviatilis* in Europe.

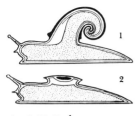

Fig. 5-26. Stylommato-
phora, 1. With shell,
2. Naked snail (slug).

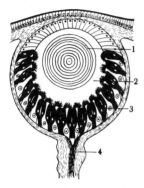

Fig. 5-27. Direct eye of the
Helix snail (longitudinal
section): 1. Lens,
2. Secretory body,
3. Retina, 4. Optic nerve.

Order: Gymnosomata

first, with the visceral mass enclosed in the shell which resembles a
barrel.

The EUROPEAN SEA HARE (*Aplysia depilans*; L often over 20 cm; see
Color plate, p. 90) and its relations (family Aplysiidae) are also able to
swim via the back stroke. They form a tube when flapping the parapodia
together; the animal is quickly propelled forward by a true wave of
muscular contraction which lasts 1.2 seconds. The sea hares possess only
a very small, flat inner shell which usually is barely visible. The genus
Dolabella, with the gigantic *Dolabella dolabella* (L up to 40 cm; see Color
plate, p. 90), one of the largest euthyneurous gastropods, still has a
calcified small shell. Other representatives from this family, however,
lack any trace of a shell in the adult stage. This is exemplified in the green-
colored *Phyllaplysia depressa* (L 4 cm) from the Mediterranean. All sea
hares, including *Aplysia fasciata* (L 30 cm), *Aplysia rosea* (L 8 cm), and
Aplysiella virescens (L 3 cm; see Color plate, p. 90), from the Mediterranean, are able to expel a milky to brownish-violet secretion from their
mantle cavity if disturbed. The ancient Romans had assumed for this
reason that the sea hares were poisonous. In fact, however, this reaction is a
harmless protective measure just as in the violet snails and the Dibranchiata (see Chapter 8). In the Society Islands in the Pacific, *Dolabella
termidi* (L 13 cm) is even eaten.

Sea hares are vegetarians and are usually found in the algal zones along
the coast lines. Their body coloration is adapted to this habitat. During
the spring one can often find masses of juvenile sea hares in these areas.
It is quite a sight to observe a chain of sea hares copulating; a group of up
to twelve animals swims in a row along the sea bottom. The largest known
living gastropod is the Californian sea hare which reaches a length of
75 cm and a weight of 16 kg.

The shell-less order GYMNOSOMATA (formerly a group of the Pteropoda) is even more highly evolved as pelagic creatures than the Aplysiacea, which are already well adapted for swimming. In the Gymnosomata
the foot has been transformed into a swimming organ. The animals
possess special prey-catching devices which are an adaption to their
carnivorous existence (adaptations which are not found in any other
gastropod group). There are seven families.

The species *Laginiopsis triloba*, from regions of the Azores, and *Anopsia
gaudichaudi*, from the Indo-Pacific, are distinct from the rest of the Gymnosomata because they followed separate evolutionary paths quite early
in their development. In the first species the buccopharyngeal armature
and radula are degenerated, and the animal catches its prey with the aid
of the long proboscis. *Anopsia*, on the other hand, can still retract the
head into the mantle, and catches its prey with the aid of the radula.
This species is also notable for internal brooding. Only after the mother
has died will her body-wall burst open and the fully developed juveniles
be released.

All other Gymnosomata possess modified parapodia which also serve as flapping swim organs. These structures range in shape from being more or less cylindrical to spindle-shaped. Aside from the radula, the special buccopharyngeal armature consists of two or more invaginable hook sacs which aid in prey-catching. In some forms, for instance *Thalassopterus zancleus* (L 2 mm), from the Mediterranean, these hook sacs may be reduced. The Gymnosomata swim on the surface when looking for prey. They seem to de diurnal although the eyes, located on the posterior tentacles, are partially reduced. The CLIONES (*Clione limacina*; L 3–4 cm; see Color plate, p. 90) of the family Clionidae are a representative of this order, and are periodically found in great swarms in boreal and arctic waters. This species feeds exclusively on *Spiratella helicina*, but in turn it and *S. helicina* are the major food source for the baleen whales (see Vol. XI).

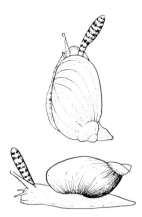

Fig. 5-28. Amber snail with *Leucochloridium* sporocytes in its right tentacle.

The smaller NOTOBRANCHIDS (family Notobranchaeidae) are quite similar in appearance to the cliopsids (family Cliopsidae), which are characterized by a long proboscis. Both groups possess a secondary plicate gill on the posterior end in contrast to the Gymnosomata, which lacked respiratory organs. A true plicate gill is perhaps only found in the PNEUMODERMATIDS (family Pneumodermatidae). The structure of this gill and that of the equally posterior secondary plicate gill are barely distinguishable. The pneumodermatids are characterized by suckered arms in the buccopharyngeal region which were already indicated in the clionids as so-called buccal cones. The purplish-violet *Pneumodermon mediterraneum* (L 1–2 cm) is equipped with a pair of club-shaped organs, each endowed with six to nine suction cups. The animals extend these suckered arms, the proboscis, and the hook sacs when capturing their prey. *Pneumodermopsis ciliata* (L 1 cm; Fig. 5-39) is similarly equipped; the suckered arms, however, are very short, which has resulted in two transverse rows of suction cups. In addition, there is a median tentacle located on the proboscis. This species, which is also violet, is sporadically much more common than *Pneumodermon. P. mediterraneum* is found in the North Atlantic and in the Mediterranean up to the Adriatic.

Fig. 5-29. Distribution of the slug family Athoracophoridae.

The sacoglossans (order Sacoglossa), which are not represented by many species, have produced an amazing array of polymorphic forms. The sacoglossans are characterized by detorsion of the mantle cavity, a shell which is thin-walled, oval, or bivalved, or absent. The shelled species possess a ctenidium. The number of tentacles varies (two, one, or none). Forms without shells possess cerata. The foot occasionally has one or two parapodia. A suctorial buccal pharyngeal apparatus without jaws is always present. The anterior end of the radula with a single longitudinal row of teeth is located in a blind pouch (ascus), in which used and worn teeth accumulate. There are three superfamilies: 1. Juliacea; 2. Oxynoacea; 3. Elysiacea. There are approximately 200 species.

Despite the great diversity of sacoglossans, they are all characterized

Order: Sacoglossa

Fig. 5-30. *Aneitia sarasini*, at left, dorsal side; at right, the ventral side.

Fig. 5-31. *Athoracophorus bitentaculatus*. New Zealand.

Superfamily Elysiacea

Fig. 5-32. *Limax maximus* laying eggs (see p. 132).

by a single-rowed radula with the associated ascus. The marine sacoglossans feed by sucking out the juice from algal filaments, which they pierce with one radula tooth. These frequently very small snails are almost exclusively limited to the algal zones along the coast, where they crawl about along the sea bottom or the entangled plants.

The superfamily JULIACEA is comprised of shelled forms with long lateral nerve cords that in some cases are still crossed over. There are two families: 1. Arthessids (Arthessidae), able to withdraw completely into the shell; the lateral nerve cords are crossed over. Little is known about this family. 2. Juliidae, characterized by two shell valves (bivalvelike).

The bivalves of the Julias are not an indicative factor that these gastropods are a transitional form to the bivalves. A study of the larval development of a few species, for example *Berthelinia limax* from Japan, has shown that the left valve of the Julias represents the actual shell. The larva possesses a normal coiled shell with an operculum. The right valve develops as the adult shell becomes stronger and less symmetrical; eventually two valves are formed. Occasionally, as in *Berthelinia limax* or in *Berthelinia chloris* (L 1 cm) from California, the coiled embryonal shell is still present in the spire of the left valve.

Until 1959 no living representative of this family had been found, merely the two valves. It is therefore not surprising that these forms were classified as bivalves. In 1959 the Japanese scientists S. Kawaguti and D. Baba discovered the Julias. Just like the sacoglossans, the Julias feed on certain algae to which they are matched in color. *Julia japonica* feeds on the *Caulerpa ambigua*. Following cross-copulation the eggs are laid on specific algae in bunches of over 2,000. Only approximately one dozen species have been found in Madagascar, Japan, Australia, the Caribbean, and along the Pacific coast from California to Peru. Very few of these have been studied thoroughly. The genus *Midorigai* can be differentiated from *Berthelinia* by its disc-shaped embryological shell.

The superfamily ELYSIACEA encompasses the remainder of the sacoglossans, which are quite diverse in shape. The shell is present or absent. The nervous system is concentrated. There are five families.

One of the European forms, *Oxynoe olivacea* (see Color plate, p. 90) from the Mediterranean and belonging to the superfamily OXYNOACEA, is easily recognized by its elongated taillike foot. Its body proper and the thin shell is covered by the parapodia. The animal, which is uncommon, lives on algae of the genus *Caulerpa*. *O. olivacea*, just like *Elysia viridis* (L 1–2 cm; see Color plate, p. 90), which is commonly distributed along European coasts, has only one pair of rolled-up rhinopores. This is a feature which serves as a good criterion for distinguishing the *Elysia* snails (superfamily ELYSIACEA) and also for many other sacoglossans. The colorful *Thuridilla hopei* (L 5–15 mm; see Color plate, p. 104), from the Mediterranean, inhabits seaweed and algal growth just like other species of this family. It is able to swim short distances between plants with the aid of the undulating flapping of its parapodia. *Bosellia mimetica* (L 5–10 mm), from the Mediterranean, on the other hand, is almost always limited to shallow waters where the alga *Halimeda tuna* grows. This species is not only camouflaged in its color but also in its body shape relative to the surrounding environment. When *Bosellia mimetica* is resting, the net of blood vessels which is visible through its skin also gives it an additional leaflike appearance. In this species the male gonads develop before the female gonads (protandry).

Fig. 5-33. European distribution of *Testacella haliotidea*.

An additional family (Stiligeridae) is characterized by the prominent concentration of the nervous system. This group encompasses the LOBI-GLRIDS which are characterized by a long foot, ear-shaped shell, and two pairs of long, free parapodia used as swimming organs. *Lobiger serradifalci* (see Color plate, p. 90) is common in the seaweed meadows of the Mediterranean. It feeds on the green algae of the genus *Caulerpa*. All other families are characterized by one pair of tentacles at the most, and also by the absence of the shell and parapodia. Instead they are usually endowed with numerous cerata. In the POLYBRANCHIDS (Polybranchidae) the cerata are leaf-shaped, and in the STILIGERIDS (family Stiligeridae) they are usually club- or thread-shaped. The stiligerids are therefore easily mistaken with Eolidoidea from the order Nudibranchia, but the latter possess two pairs of tentacles. It has not yet been determined with certainty if the North American species *Olea hansineensis* should be included here, although it has been placed in the family Oleidae.

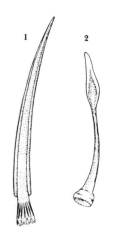

Along European coastlines the stiligerids, particularly *Stiliger vesiculosus* (L 5–10 mm), are quite common. *S. vesiculosus* is found in the algal zone as well as harbor regions. It feeds on the spawn of other naked snails. The species *Hermaea bifida* (L 15–25 mm) and *Placida dendritica* (L 10–15 mm) are less common inhabitants of seaweed and algal growth regions. *Ercolania viridis* (L 5–10 mm) is limited to green algae in the Mediterranean, where it is sporadically distributed. This form is easily recognized by its numerous bottle-shaped cerata. Occasionally these uniformly green animals have blue spots, but only on their cerata. According to L. Schmekel these blue spots become blurred after a short period of hunger and disappear altogether with prolonged fasting. Special attention should be focussed on *Alderia modesta* (L 8–10 mm), which is the only sacoglossan that has adopted an amphibiotic existence. This species inhabits the grass-like growth of the green alga *Vaucheria*, and frequently also occurs outside the water. *A. modesta* is distributed along European coasts, in the Baltic Sea up to Helsinki, and along the Atlantic and Pacific coasts of North America.

Fig. 5-34. "Love darts" of the edible *Helix* snail (1) and *Arianta arbustorum* (2).

The LIMAPONTIIDS (family Limapontiidae) are naked snails lacking cerata. They can therefore be mistaken for certain Cephalaspidea but also for flatworms. In fact, the Northeastern Atlantic *Limapontia nigra* (L approximately 5 mm; see Color plate, p. 90), which lacks tentacles, was described as a "worm" several times in all works on zoology. *L. nigra*

Fig. 5-35. Land snails, *Cylindrus obtusus*, from Europe; L 10 mm.

has also penetrated the Baltic Sea and the Mediterranean. That this creature is a snail is demonstrated by the asymmetrical arrangement of its body organs and the presence of the radula. *Limapontia nigra* inhabits algal growth or dwells underneath stones, and is often difficult to find. The approximately equal-sized *Acteonia corrugata* is distributed from the North Sea to Biscay. In contrast to the *Limapontia*, it is endowed with a pair of rolled-up tentacles.

Order: Notaspidea

The NOTASPIDS (order notaspidea) are a small group of euthyneurous snails, some of which have a flat patteliform shell which may be covered; some have no shell. This group is significant because it serves as a starting point for the Nudibranchia. Notaspidea are characterized by a groove-like mantle cavity on the right side, which always contains a ctenidium. Parapodia are lacking. There are similarities in the concentration of the nervous system of the Notaspidea and Nudibranchia. Often these two groups are classified as a single order. The Notaspidea are marine. There are two families with approximately 150 species.

The UMBRELLA SHELLS (Umbraculidae) are characterized by snails that have a flat umbrellalike shell. *Tylodina perversa* (2.4 cm; see Color plate, p. 103), which is found in the Mediterranean and on the coast of Portugal, is still somewhat protected by its shell. This form inhabits the brownish-yellow sponge *Verongia aerophoba* below a depth of 10 m. It feeds on this sponge and also takes on its color. In contrast, *Umbraculum umbraculum* (L up to 15 cm; see Color plates, pp. 73 and 103) seems to be more reminiscent of a wandering sponge topped by a little hat, for it has an extremely enlarged and thickened foot with a warty surface.

Special protective measures

In the PLEUROBRANCHIDS (family Pleurobranchidae) this spongelike appearance is even more prominent. These forms usually take up a lot of water in the tissues of the foot and the so-called "sail" of the joined frontal tentacles. If the animal is disturbed, it expels the body fluid and literally "shrinks." The brilliantly orange *Berthella auriantiaca* (L up to 3 cm) even rolls up in case of danger. *Berthella* has a relatively large shell which, however, is invisible because it is surrounded by the mantle as in all pleurobranchids. *Oscanius membranaceus* (L 6 cm; see Color plate, p. 103) is less noticeable for its yellowish-brown coloration than for its circular shape. Its body is covered by warts, and the foot is entirely covered by the mantle. The fragile, small internal shell (L 20 mm) is silhouetted as a dorsal shield. These snails prey on other animals. *Oscanius* feeds on ascidians (see Chapter 18). When in danger this snail releases a powerful acidic mucus. Fish that swallow pleurobranchids spit them out immediately. This is also the case with *Philine quadripartita*, sea hares, *Elysia*, holohepatids, and various lamellariids. *Pleurobranchaea meckeli* (L up to 10 cm; see Color plate, p. 103), which are also carnivorous inhabitants of the deeper zones of the Mediterranean and adjoining Atlantic, are capable of "swimming." The presence of this snail endangers all edible creatures in its vicinity. Neither worms nor mollusks are safe from *Pleurobranchaea*, and even conspecifics are not spared. These naked animals are characterized by an exposed gill which is not covered by the mantle.

The NUDIBRANCHS (order Nudibranchia) are highly diverse and multi-formed opisthobranchs. They are characterized by detorsion of the mantle cavity. Mantle cavity, shell, and parapodia are absent. Respiration takes place via the epidermis or with the aid of respiratory projections on the dorsal surface. There are usually two pairs of tentacles on the head; the rhinopores are often retractable into sheaths. Cerata are absent or present. All species are marine, despite the erroneous report of *Ancylodoris baicalensis* which probably came from European seas and not Lake Baikal, USSR.

The marine nudibranchs, which are bilaterally symmetrical externally, are sometimes overlooked or difficult to find because they are frequently quite small. The total loss of the mantle cavity has been compensated for in some groups by the formation of respiratory projections and cerata on the dorsal surface. Formerly the nudibranchs were divided into the Holohepatica and Cladohepatica; however, this classification did not correspond to the true phylogenetic relationships. The present classification is based on the difference of several external characteristics (tentacles, etc.), the arrangement of the digestive organs, and the often highly branched nephridium. There are four suborders: 1. Doridoidea; 2. Dendronotoidea; 3. Arminoidea; 4. Eolidoidea. There are forty-eight families and approximately 1000 species.

The doridacean nudibranchs are characterized by a dorso-ventrally flattened form and a feathery secondary gill partially or entirely surrounding the anus. Generally, calcareous spicules are present. The left midgut gland is compact but the right one is degenerated. There are five superfamilies and sixteen families.

Except for a few representatives which include *Doridoxa ingolfiana* from Greenland, the superfamily Doridoidea encompasses forms that are endowed with highly branched respiratory organs that surround the posteriorly located anus. These external gills almost always have the same color as the rhinopores, which appear ringed. Frequently the branched respiratory organs can be retracted. This is demonstrated in the European *Archidoris tuberculata* (L up to 10 cm), *Peltodoris atromaculata* from the Mediterranean, and in the species from the Indo-Pacific genus *Hexabranchus*, which are capable of swimming. This includes the brilliantly red *Hexabranchus marginatus* (see Color plate, p. 103). The GLOSSODORIDS (family Glossodoridae) protect their gills in a similar fashion. It is probable that the coloration of these animals serves as a direct camouflage.

H. R. Haefelfinger, whose accounts we follow here, has intensively studied the coloration of these snails. When diving, he noted that the coloration of the marine animals was influenced by the quality of the

Order: Nudibranchia

Suborder: Doridoidea

▷
Nudibranchs: 1. *Tethys fimbria* (L 20 cm). Atlantic; 2. Frond dorid (*Dendronotus frondosus*; L 6 cm). North Atlantic; 3. *Phylliroe bucephala* (L 3 cm). Atlantic and Mediterranean. Soleolifera: 4. *Onchidium typhae* (L 2 cm). East Indies; 5. *Onchidiella chilensis* (L 3 cm). South America; 6. *Atopos semperi* (L 3 cm). Philippines; 7. *Onchidium peroni* (L 6 cm); 8. *Vaginulus taunaysi* (L 4 cm). Brazil; 9. *Platevindex granulosa* (L 3 cm). Indonesia; 10. *Angustipes plebejus* (L 4 cm). Southwestern Pacific; 11. *Rhodope veranyi* (L 4 mm).

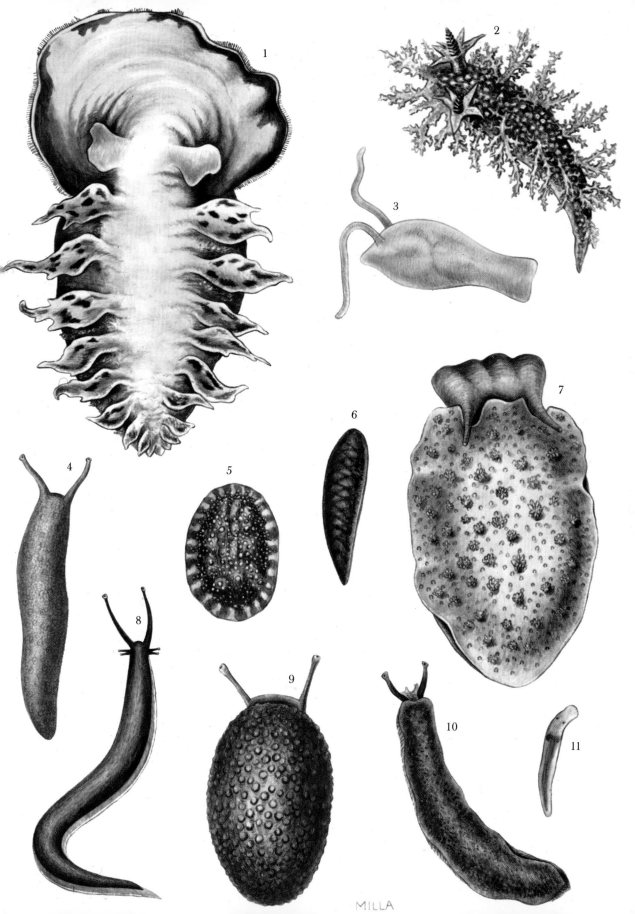

MILLA

◁
Eolidioidea: Above:
Flabellina affinis (juvenile;
L 10 mm). Mediterranean.
Below: *Facelina rubrovittata*
(juvenile; compare p. 104).

water. The various colors that we admire in these animals under daylight rapidly loose their intensity in increasing depths of water. The brilliant reds become paler, the blues become darker, the yellows dirty-looking, and finally one notices only various shades of grayish-blue color. The most flashy color shades can be just as effective as camouflage as inconspicuous pale shades. Especially while diving, it repeatedly becomes evident how difficult it is to recognize snails in their own habitat.

According to Haefelfinger the splendid coloration has real significance in the life of the snail, although it still remains a mystery why there is such diversity of design within one family, genus, or even within a single species. The genus *Glossodoris*, which is represented by numerous species in the Mediterranean, serves as an example for extraordinary differences in color and design. In some forms juvenile coloration and design remain constant for life. *Glossodoris luteorosea*, for example, is always characterized by irregular golden-yellowish, white, and dark pink spots framed against a pale reddish background, and *Glossodoris tricolor* is always dark blue with white stripes. On the other hand, there are species in which the adult form differs so greatly from the juvenile one that it is not surprising that they were often regarded as separate species. The design alternation in *Glossodoris gracilis* (see Color plate, p. 104) is the most intricate. The juvenile snails already possess a distinct yellow line on the dorsal surface and a relatively broad white median line which gradually deposits yellow pigment. Later, additional yellow lines form between the mantle line and dorsal shield margin. The median line splits, and in the adult only a fine yellow net of lines remains, covering the entire dorsal surface and the flanks of the animal. The whole thing becomes even more confusing since the juvenile stages in *Glossodoris tricolor* and *Glossodoris gracilis* are conspicuously similar while the adults show distinct differences.

In *Polycera quadrilineata* even the various age groups are colored and designed differently, and each stage has been mistakenly described as a different species. Although it is known that this species inhabits algal growth, one is not certain of its exact source of food. Other glossodorids are characterized by very specific food requirements. For example, *Glossodoris tricolor* feeds only on the tissues of the sponge *Ircinia*, and the large *Peltodoris atromaculata* is usually found on the sponge *Petrosia ficiformis*. Aside from their specialized food requirement, the Doridoidea are also characterized by mesenchymal calcareous spicules. *Polycera quadrilineata* (L 1–3 cm; see Color plate, p. 103), which occurs along European coast lines, has only a few embedded calcareous spicules. In other species these spicules, however, serve as an excellent protection against enemies. *Jorunna tomentosa* (L 2–4 cm) is characterized by curious dorsal sensory organs which are reminiscent of spice cloves. These structures are called caryophyllids, and in *Rostanga rubra* (L 10-15 mm) and several related species they are surrounded and supported by spicules.

Spiculated
nudibranchs

Several representatives of this suborder, for example *Aegires punc-*

tilucens (L 13–20 mm), which is covered with papillose warts, are not able to retract the posterior external gill into a pouch. Some of these forms are ectoparasites on other animals. *Okenia elegans* feeds on ascidians. Many species of the genus *Lamellidoris* feed on bryozoans (see Vol. I), and are therefore predominantly found on seaweed. The two big species of the genus *Dendrodoris* from the Mediterranean, *Dendrodoris limbata* (L 5–7 cm) and *Dendrodoris grandiflora* (L 5–19 cm), are sporadically found underneath stones and on sponges. The rare PHYLLIDIIDS (family Phyllidiidae), which have respiratory organs that are located as a lamellar ring below the mantle, and the previous group, are classified as Porostomata. They are suctorial forms lacking a radula, and they are able to protrude their narrow proboscis.

Representatives of DENDRONOTACEAN NUDIBRANCHS (suborder Dendronotoidea) usually possess a paired row of cerata and a laterally positioned anus. Usually gills are lacking. The midgut gland is either compact or highly branched; the right gland is smaller. The head is usually endowed with coiled rhinophores which can be retracted into their own sheaths. There are ten families.

Suborder:
Dendronotoidea

The most primitive form of this group is *Duvaucelia gracilis* (L 1 cm; see Color plate, p. 104) from the Mediterranean. This species lacks gills, and has, along each dorsal ridge, a row of cerata. The rhinophores are larger than the cerata and can be withdrawn into tubelike sheaths. The cephalic shield or vail is also characteristic of this species. Subsequent species that follow *Tritonia* and *Duvaucelia* in external appearance are *Marionia blainvillea*, *Hancockia uncinata*, *Lomanotus genei*, and *Doto coronata*. The leaflike cerata of these forms are reminiscent of certain sacoglossans. Some of these species are highly specialized in their food requirements. *Tritonia hombergi* and other species of this genus are always found on alcyonarians. The species of the genus *Lomanotus*, which swim by body undulations, inhabit sponge growth. *Hancockia* and *Doto* feed on various kinds of hydroids. *Hancockia* even stores the nematocysts in terminal pouches of the cerata. The FROND DORID (*Dendronotus frondosus*; see Color plate, p. 113) also feeds on various kinds of hydroids.

Special feeding habits

Tethys fimbria (L 20–30 cm; see Color plate, p. 113), from the Mediterranean, lacks jaws as well as a radula and is characterized by a frontal vail. This form scoops up all sorts of prey on which it feeds, ranging from microscopic creatures to small fish. *T. fimbria* can swim for longer periods by undulating its flat body. *Scyllaea pelagica* (L 2–4 cm) is pelagic, as indicated by its scientific name. It is found on drifting sargassum weed on the open sea where it feeds on ctenophores and other hydroids. It has the same color as the seaweed in which it lives. *Phylliroe bucephala* (L 3–4 cm; see Color plate, p. 113) is also pelagic. It is characterized by a laterally flattened body which has an axelike or fishlike appearance. It is colorless and almost transparent and lacks a proper foot. Its only appendages are the long tentacles. The eyes and radula are reduced. As in the genus

Rhodope, *P. bucephala* is so highly modified to its habitat that it hardly looks like a snail. *P. bucephala* moves in a snakelike fashion in accordance with its body shape. As a juvenile this snail parasitizes the medusa of *Zanklea costata* (see Vol. I) which it progressively sucks empty during its growth period until finally the prey merely hangs like an appendage from the vestigial foot. The adult animal also feeds primarily on hydroids, particularly on siphonophores (see Vol. I) and other medusae.

Suborder: Arminoidea

The ARMINACEAN NUDIBRANCHS (suborder Arminoidea) encompass rather heterogeneous forms. They may lack or possess cerata. As a rule the head is endowed only with the retractile rhinophores without sheaths. The midgut gland is branched. There are three superfamilies and nine families.

Only a few arminacean nudibranchs are found in the oceans around Europe. The European genus *Arminia* lacks any body appendages. The anterior end of this animal is only recognizable by the presence of the rhinophores. *Arminia maculata* (L 10 cm), which is found in the Mediterranean and along the Portuguese coast, is the only species of the five European representatives that is characterized by flat warts on its dorsal surface. In all other species, however, numerous leaflike respiratory projections are located on both sides between mantle and foot. The snails inhabit the deeper muddy sea bottoms. The considerably smaller *Janolus hyalinus* (L 5–10 mm) and *Antiopella cristata* (L 30 mm) probably feed on hydrozoans (see Vol. I). These species can easily be distinguished from the other species by the cerata which encircle the entire mantle. These forms are only occasionally found along western and southern European coasts. The North Atlantic *Hero formosa* (see Color plate, p. 103) has short, branched cerata on its dorsal side which, in contrast to the cnidosacs of *Hancockia*, and the subsequent eolidacean nudibranchs, do not store the nematocysts of the captured hydroids.

Suborder: Eolidoidea

The suborder Eolidoidea is characterized by numerous cerata which range from club-shaped to thread-shaped appendages. The cerata are usually arranged in several rows or in bunches. There are terminal cnidosacs for storing nematocysts. The animals are usually 5–35 mm in length. As a rule the head has two pairs of tentacles without sheaths and frontal vails. The paired midgut gland is branched but the right side is smaller. Occasionally the eyes are stalked. There are three superfamilies and thirteen families.

Storage of nematocytes

The eolidacean nudibranchs feed almost exclusively on various kinds of hydroids (see Vol. I). The latter's nematocysts are temporarily stored in the terminal sacs (cnidosacs) of the cerata. The undigested nematocysts migrate through the intestine and are then sorted by the epithelial lining of the cnidosacs. Periodically these undesirable food particles are expelled to the outside. This process can be regarded as a type of "excretion." The amazing fact is that the nematocysts do not discharge and therefore cannot harm the snail. Similarly, as in the many solenogasters (see Chapter 2) and other animal groups, the "weapons" of the hydroids remain without effect. The clown fish (see Vol. V) have also adapted to an existence among the tentacles of the sea anemones, without being harmed by the nematocysts. The possibility exists that the mollusks which feed on hydroids employ a secretion that makes them immune against nematocysts and which also camouflages them. There could also be another possibility that the gastropod discharges a prey-specific mucus from its foregut glands that prevents the prey from recognizing the foreign species and thereby prevents the triggering of nematocysts before and during the gastropod's feeding.

The reddish *Coryphella lineata* (L 1–2 cm) is a regular inhabitant of the algal zone of European coasts. *Coryphella verrucosa* is present in the North Sea. In both species, as well as in the antarctic noteolidiids (family Notaeolidiidae), the hindgut terminates laterally.

The pale yellowish *Eubranchus tricolor* (L 1–2 cm) is characterized by numerous violet papillae which are arranged in thirteen or fourteen rows, each with three to five papillae. Below each papilla tip is a yellow circle. This whole arrangement is conspicuous as well as beautiful. This snail is usually associated with the hydroid *Obelia dichotoma*, on which it feeds. One subspecies has also penetrated into brackish water, where it primarily feeds on the hydroid *Cordylophora*. *Tenellia ventilabrum* (L 7 mm) is also found in regions of low salinity content. For example, this form is distributed in the Baltic Sea among algae and water plants. The smaller species *Embletonia pulchra* and *Tergipes despectus* possess only two longitudinal rows of alternating club-shaped cerata. *Embletonia* lacks cnidosacs. This tiny snail inhabits the sedimentary ground and often also penetrates into the spaces among the sand.

The PSEUDOVERMIDS (family Pseudovermidae) are inhabitants of the spaces among the sand. Eight species are known from the Mediterranean and Atlantic Ocean and two from the Pacific. They deviate from the other eolidacean nudibranchs by lacking tentacles and a clearly defined foot. Like many other small marine naked snails, they move with the aid of cilia. Although over thirteen pairs can be fully developed, the papillae are only clearly visible in the species *Pseudovermis papillifer* (L 2–3 mm) from the Aegean and Adriatic Seas, and *Pseudovermis salamandrops* (L 2–6 mm) from Brazil. Usually the animals are able to reduce the cerata with the associated cnidosacs. It often takes close observation to detect the presence and number of cerata.

In contrast to *Trinchesia foliata* (L barely 5 mm), which vary greatly in color and which are regularly found among algae, the cerata are arranged in stalked bunches in the *Flabellina affinis* (L 1–2 cm) and the *Calmella cavolinii* (L 1 cm) from the Mediterranean. In the latter species these bunches can be brilliant brick-red depending on the food ingested. *Fiona pinnata* (L 1–5 cm), which is often found on drifting objects in the Atlantic and Mediterranean, is similarly affected by the color of the

Brackish-water species

▷
Left, from top to bottom (sea slugs):
Phyllidia uphilis (family Phyllidiidae), a nudibranch.
Hexabranchus imperialis (see Color plate, p. 103), an Indo-Pacific nudibranch.
Thuridilla hopei (see Color plate, p. 104), a sacoglossate snail.
Center, from top to bottom:
Batwing sea-slug (*Gasteropteron rubrum*), a cephalaspidea.
Superfamily Doridoidea:
Peltodoris atromaculata (L 5 cm). Mediterranean dorid.
Right, from top to bottom:
Oscanius membranaceus (see Color plate, p. 103), a pleurobranch.
Glossodoris purpurea (L 3 cm; compare Color plate, p. 104). Mediterranean.
Rose-shaped spawn of a dorid snail.

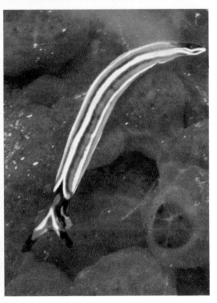
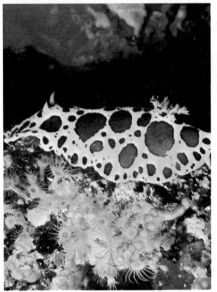
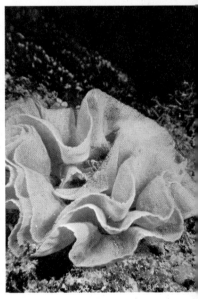

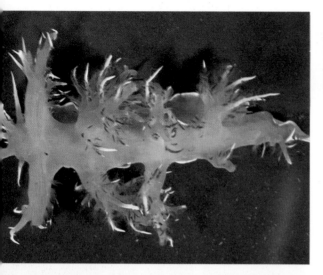

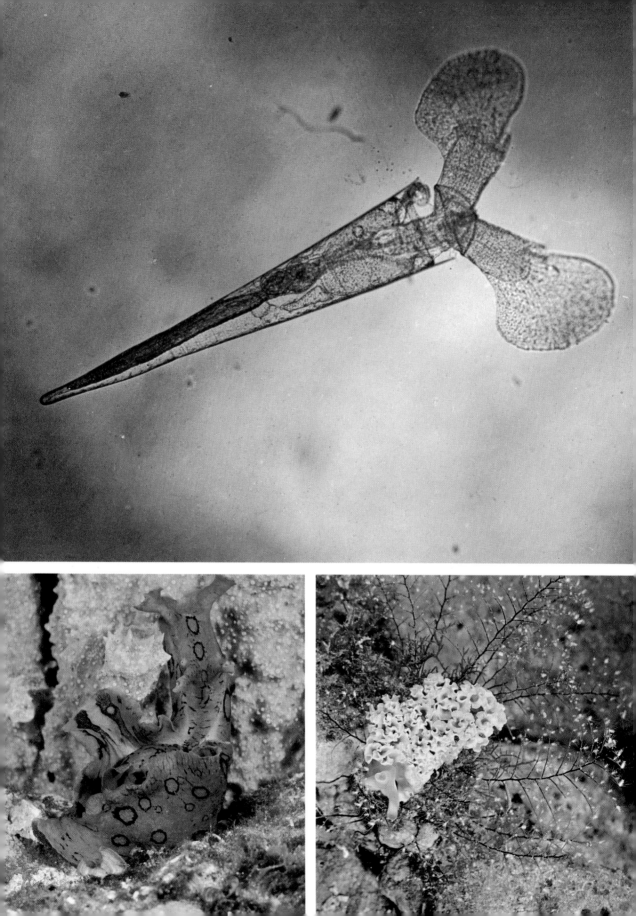

◁

Top:
Creseis acicula (L 1 cm), a pteropod, worldwide.
Left, bottom:
Aplysia dactylomela (L 10 cm; compare Color plate, p. 90). West Indies.
Right, bottom:
Tridachia crispata, a West Indian sacoglossate.

◁ ◁

Glaucus (bottom; compare Color plate, p. 103), a pelagic nudibranch, eating a *Porpita porpita* (above). The photo is magnified approximately 5 X.

◁ ◁ ◁

Left, from top to bottom:
Glaucus (compare Color plate, p. 103);
Spurilla neapolitana with string of spawn (see Color plate, p. 104);
Eubranchus tricolor.
Right, from top to bottom:
Favorinus branchialis laying eggs;
Facelina coronata (compare Color plate, p. 104);
Dondice banyulensis (see Color plate, p. 25).

Order:
Basommatophora

hydroids it feeds on. *Calma glaucoides* (L 5–15 mm) strongly deviates in its mode of life. It is found under rocks in the algal zone along European coasts, where it feeds on fish spawn and eggs of other mollusks. In this form there has been a concurrent reduction in radular teeth, an absence of cnidosacs in the numerous lateral cerata, and no hindgut. In all other representatives of the superfamily Flabellinacea the anus is located on the dorsal side, but in *Calma* the hindgut ends as a blind sac.

The superfamily Eolidiacea possess a radula with a single row of teeth and an anus which is located on the right side of the dorsal surface. This group includes the GREY SEA SLUG (*Eolidia papillosa*; L 5–12 cm) from the North Atlantic and the common *Facelina drummondi* (see Color plate, p. 104). The latter species is covered by numerous long, thin cerata which cover the rather colorless body and give it a characteristic coloration. The very similar *Hervia peregrina* and *Hervia costai* (L 1–14 cm) are found in the algal regions of the coast line. They always inhabit hydroid colonies (*Eudendrium ramosum*). *Favorinus branchialis* (L 10–15 mm), which occurs along European coasts, is unique in that it feeds on the spawn of other marine naked snails. Usually *Favorinus* is brown but this can change depending on the food eaten. H. R. Haefelfinger describes *Spurilla neapolitana* (L 2–4 cm; see Color plate, p. 104): "If one feeds *Spurilla neapolitana* with their main food source, the sea anemone *Aiptasia mutabilis* (which is also fed on by the bluish iridescent *Berghia coerulescens*), this snail soon takes on a characteristic yellow to olive-brownish color. If *Spurilla* is fed with the actinian (*Actinia equina*), it changes to red within a matter of a few hours. This process is reversible, which means that the color pigment is not permanently stored in the body cells."

Glaucus marinus (L 1–3 cm; see Color plate, p. 104) is characteristic pelagic form of the open sea. This form possesses flat elongated cerata which are arranged in three pairs of bunches and enable the animal to float on the water surface. The shining blue ventral surface, which is pointed up, and the colorless dorsal surface, which faces down, are adaptive features to this pelagic way of life. *Glaucus* is also able to utilize gas or air for propulsion. Just like the violet snail (*Janthina*) or the *Phylliroe bucephala*, *Glaucus* is dependent on free-swimming hydrozoans. Even their reproductive habits are adapted to life on the open sea. *Glaucus* lays its spiral chains of eggs on the skeleton of siphonophores it has fed on. The violet snails and the *Glaucus* species form a close community with the siphonophores, and drift along as silvery-blue swarms—a "blue fleet," it was called by W. Ankel.

With the BASOMMATOPHORES (order Basommatophora) we start the discussion of the third well-known subclass, PULMONATES (Pulmonata), which embraces the true land dwellers as well as many fresh-water pond-snails. In the primitive basommatophores there is no torsion of the mantle cavity. The shell is always clearly present but an operculum sometimes is found. The ctenidium is degenerated. A secondary gill or well-developed vascular net functions as a lung. The eyes are beneath the cephalic epidermis and are not stalked. They are primarily fresh-water inhabitants, such as the planorbids and lymnaeids. A few species such as the siphonariids and ellobiids are found on the shore or coastal regions of the ocean. There are thirteen families within six superfamilies, with approximately 1,000 species.

The pulmonates are characterized by a newly evolved organ, the vascular net in the mantle cavity, which acts as a lung. In the basommatophores this structure is partially displaced by a secondary gill. The superfamily SIPHONARIACEA, which was formerly known as Thalassophila, embraces two related families: the siphonariids (Siphonariidae) and the trimusculids (Trimusculidae). Their body structure points to many connections with the Actinoidae among the cephalaspids. The most primitive representatives of this group are obviously the limpetlike Siphonariids, because they still possess part of the true ctenidium in addition to the pulmonary sac. Their flat shell without the operculum, and the almost circular pedal muscle, on the other hand, are adaptations for an existence in the tidal and shoreline zones on the ocean, for example in the species *Siphonaria pectinata* from southwestern Europe. The Trimusculids, which are also devoid of tentacles and possess a patelliform shell, for example the circular *Trimusculus garnoti* (L 1 cm) from the Mediterranean, are characterized by a muscle which is fully exposed anteriorly. The trimusculids are found in the tidal zones and the air spaces in cave ceilings. The AMPHIBOLIDS (superfamily Amphibolacea), which possess a normally coiled shell, are distributed from eastern Asia to New Zealand. They are the only pulmonates that still have an operculum. Their water-filled respiratory cavity, however, is devoid of a gill although they live submerged in brackish water and the mouth of rivers. There is but one family, the Amphibolidae.

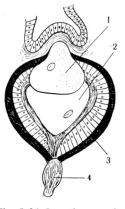

Fig. 5-36. Invaginate vesicular eye of *Onchidium* (see p. 105): 1. Lens, 2. Lens cell, 3. Retina, 4. Optic nerve.

The most primitive fresh-water pulmonates are undoubtedly the South American CHILINIDS (family Chilinidae). They still have crossed-over lateral nerve cords. The flattening of the whorled shell, as it is found in *Chilina fluctuosa*, was probably cause for the splitting off of several other families. The New Zealand LATIIDS (Latiidae) belong to one of these families. These forms are endowed with patelloid shells with an internal septum. One member, the photophobic *Latia neritoides* (L barely 1 cm), inhabits swiftly flowing creeks. It is the only known fresh-water animal that is luminescent. In other aspects the chilinids are the prototype for the lymnaeids and lancids and probably also for the ACROLOXIDS (superfamily Acroloxacea, family Acroloxidae), which are rather a group apart. The LAKE LIMPET (*Acroloxus lacustris*; L up to 7 mm; see Fig. 5-24), which belongs to this last group, is widely distributed throughout Europe. In this species the body openings (e.g., anus, gonopore) are situated on the right side. Some structural characteristics of the acroloxids point to the relationship with the chilinids, latiids, and lym-

Fig. 5-37. Distribution of *Onchidium verruculatum* of the Indo-Pacific.

Family: Physidae

naeids, although the unique development of their eggs is indicative of the specialized development of these animals.

The POND SNAILS (superfamily Lymnaeacea, family Lymnaeidae) are generally well known in North America and Europe and are represented by several species and subspecies. The relatively hard shell of the GREAT POND SNAIL (*Lymnaea stagnalis*; L often over 5 cm; see Color plate, p. 80) is subject of great variations in form and appearance depending on the biotope. For example, the shell becomes shorter in strong water currents or becomes indented around the aperture in forms that inhabit the reeds and rushes. These adaptations may be caused by purely external influences or may be dependent on the type of food. The Lymnaeidae are omnivorous, feeding on animal as well as plant food. The EAR POND SNAIL (*Radix auricularia*; see Color plate, p. 80) and the WANDERING SNAIL (*Radix peregra*) show similar variations. The latter species is even found at water currents which flow 30 cm per second in glacial brooks and in layers of water where the temperature is up to 30°C. The DWARF POND SNAIL (*Galba trunculata*; L 1 cm; see Color plate, p. 80) acts as intermediate host for the liver fluke of sheep and is therefore of great economic significance. They are predominantly present in ditches, meadow ponds, and also wet grass, where grazing animals pick them up and thereby get infected. In Volume I we discussed in detail these unique and fascinating life processes. In contrast to the previously mentioned species, the dwarf pond snail and the BOG or MARSH SNAIL (*Galba palustris*; see Color plate, p. 80), which is twice as large as the dwarf pond snail, are characterized by a brown shell which is not subject to variations in its form.

The patelloid LANCIDS (family Lancidae), from North America, succeed the previous group in their body structure. *Lanx patelloides* is one of its best-known species. The usually delicately shelled BLADDER SNAILS (family Physidae) are quite similar to the lancids, although they possess a sinistral shell and the body openings are also on the left side, as is common for the entire superfamily ANCYLACEA. The broad tentaculate mantle-margin covers the shell of the bladder snails. Touching these fingerlike mantle processes of the BLADDER SNAIL (*Physa fontinalis*; see Color plate, p. 80) elicits an immediate defensive reaction. The animal repeatedly swings the shell from one side to the other and thereby shakes off smaller enemies. The warmth-loving *Physa acuta*, which originated from the Mediterranean and is present in central Europe, is found only in warm springs, for example in Baden near Vienna, in waste waters from industry, or in aquaria. The MOSS BLADDER SNAIL (*Aplexa hypnorum*), on the other hand, is a cold-adapted form which is rarely found south of the Alps.

Just like the wandering snail, the PLANORBS (family Planorbidae; see Color plates, pp. 326–327 in Vol. IV) are extremely adaptable. There is no water too polluted nor any pond with too little oxygen for them. However, they avoid swiftly flowing water since they cannot get a foothold on the moving substrate, because they have a high sinistral shell. The best-known European species among them are the GREAT RAM'S HORN (*Planorbarius corneus*; L up to over 30 mm; see Color plate, p. 80), and the RAM'S-HORN (*Planorbis carinatus*). The extremely flat WHIRLPOOL RAM'S-HORN (*Anisus vortex*; see Color plate, p. 80) and *Gyraulus laevis* have wide ranges of distribution. *G. laevis* penetrates into water layers up to 30°C. Since all planorbids have hemoglobin in their blood, they occasionally release a red drop. For this reason the great ram's horn was formerly also known as the "purple snail of the fresh water." The red pigment not only dissolves the oxygen in the blood but also binds it chemically. This process facilitates a more efficient use of the oxygen content of the water. In addition, there is a well-vascularized erectile skin flap in the respiratory cavity which serves as a "gill." On the basis of this respiratory equipment, the great ram's horn is particularly well adapted to unfavorable water conditions.

Of the seven subfamilies of the planorbids, the BULINIDS (Bulininae) and the FERRISS SNAILS (Ferrissiinae) are often separated into their own families. The bulinids have a primitive form which is still quite similar to the bladder snails. The bulinids contain many species which in the tropics act as intermediate hosts for the dreaded disease, schistosomiasis, a blood disease caused by the blood fluke *Schistosoma* (see Vol. I). The American ferriss snails originally were not present in Europe, but three species have been introduced. The phylogenetic relationship of the LAKE LIMPETS (family Ancylidae) is not entirely clarified yet. The RIVER LIMPET (*Ancylus fluviatilis*; L 5–10 mm; see Color plate, p. 80), which is widely distributed throughout Europe, is characterized by a caplike shell. It inhabits creeks and the shore zone of lakes, where it can be found closely pressed against rocks and thereby can withstand even the most vigorous water movement.

The ELLOBIIDS (superfamily Ellobiacea) occupy an amazing variety of habitats. They deserve our special interest since apparently the terrestrial pulmonates originated from this group. All species of the ellobiids have dextral, oval to cylindrical shells with apertures that are often folded or serrated. They are primarily terrestrial or live close to land. *Ovatella myosotis* is found on moist sandy beaches and brackish-water regions of western and southern European coast lines. This species is characterized by a pair of stumplike anterior tentacles on the snout. The highly diverse HERALD SNAIL (*Carychium minimum*), on the other hand, is found along river banks and in the marshy meadows of Europe and the Near East, and in the Alps up to elevations of over 1800 m. Representatives of the genus *Zospeum* are regular cave gastropods. Examples of this genus are *Zospeum alpestre*, which occurs in the eastern Alps, and *Zospeum speleum*, from the adjacent southeasterly Karsh plateau. Several of the larger species of ellobiids are limited to the Indian and Pacific Oceans. One of these species, *Ellobium aurismidae* (L 10 cm), which possesses a very solid shell, is rather more reminiscent of a prosobranch seashell. The OTINIDS (family

Hemoglobin

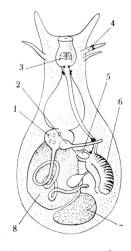

Fig. 5-38. Anatomy of an anaspid (sea hare): 1. Midgut (stomach), 2. Nephridium, 3. Pharynx with radula, 4. Penis, 5. Heart, 6. Plicate gill, 7. Ovotestes, 8. Midgut gland.

Herald snail

Otinidae) can be derived from the ellobiids. In Europe this family is represented by the species *Otina otis* which is located in the upper tidal zone (spray zone) from the British Isles to the western Mediterranean. They are the only group of the basommatophores which is incapable of withdrawing into its small, caplike shell.

Development

Only a few basommatophores, for example *Siphonaria* and *Amphibola*, develop a true larval form of the veliger type. The larvae of a few *Siphonaria* species undergo a free-swimming developmental stage. Generally, development begins within the egg. Just like the terrestrial pulmonates, fresh-water forms undergo direct development, although in some one can still differentiate well-developed or only weakly suggestive larval forms with ciliary girdles which generally possess protonephridia.

**Order:
Stylommatophora**

The STYLOMMATOPHORES (order Stylommatophora), which include the familiar shelled garden and tree snail, can be derived directly from the ellobiids of the basommatophores. The mantle cavity is detorted. The shell is partially reduced and is devoid of an operculum. There is always a vascular net which serves as a pulmonary sac. As a rule the head has two pairs of tentacles. The eyes are situated at the tip of the posterior tentacles. They are terrestrial without a free-swimming larval form. There are four suborders which are differentiated in part on the basis of position and structure of the nephridium: 1. Orthurethra; 2. Mesurethra; 3. Heterurethra; 4. Sigmurethra. There are approximately 10,000 species.

Just like the ellobiids among the basommatophores, the stylommatophores withdraw the soft part of their body into the shell. They do this by first pulling in the cephalic foot and then the end of the sole. The aperture is closed by the bulging mantle. The degree of the concentration of the nervous system equals that of the ellobiids. There are further similarities in other body structures such as the unpaired jaws, but the stylommatophores possess two pairs of tentacles on the head. The posterior pair bears the lensed eyes which have the photoreceptors which it orients towards the source of light. These optic tentacles are invaginable. The operculum is already absent during embryonal development. Nevertheless, these gastropods can effectively protect themselves against cold or desiccation; many forms secrete a winter operculum or an epiphragm which protects them against disturbances or small enemies.

It is likely that the three other suborders originated from the suborder Orthurethra. The nephridium is parallel to the heart. The ureter is straight and opens directly to the outside. The shell is always well developed. There are twelve families with approximately 2,000 species.

The most primitive stylommatophora are probably those found in the superfamily ACHATINELLACEA. Many of them live in trees, for example the ACHATINELLIDS (family Achatinellidae) which inhabit the tall forest growths of the Hawaiian Islands. The mode of reproduction is of interest since the PARTULIDS (family Partulidae) lay eggs containing highly developed embryos (ovoviviparity), while others such as the achatinellids

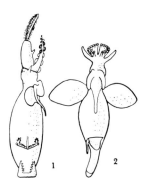

Fig. 5-39. 1. *Pneumodermon mediterraneum* with protrusible buccopharyngeal organs (lateral view), 2. *Pneumodermopsis ciliata* (ventral view). Both naked and pelagic.

(family Achatinellidae) bear live young. The closely related superfamily CIONELLACEA, on the other hand, just lays the usual eggs with under-developed embryos.

One of North America's and Europe's most frequent meadow snails is *Cionella lubrica* (L up to 7 mm). The elongated oval shells of this form as well as those of *Azeca menkeana* are difficult to find, particularly since the animals prefer to stay underneath the autumn leaves. Because of the shape of their shell, cionellids are frequently eaten by birds, particularly pigeons, since they mistake these snails for grain kernels. Within the super-family PUPILLACEA one finds the families Vertigidae, Orculidae, Chondrinidae, Valloniidae, Enidae, Pyramidulidae, and Pupillidae, which are collectively known as whorl snails. As a rule they are smaller than 1 cm. This group also includes the sinistral *Vertigo angustior* and the WHORL SNAIL (*Vertigo pygmaea*). Both inhabit meadows. *Orcula dolium* occurs underneath forest ground vegetation. *Abida frumentum*, the LARGE CHRYSALIS SNAIL (*Abida secale*), and *Chondrina avenacea* are all characteristic representatives from the calciferous Alps. Finally there are the MOSS SNAILS (*Pupilla muscorum*) and many additional related forms. All possess bulge-like to pointed, oval shells which frequently have folded apertures. *Orcula fuchsi*, found only near the Goller in lower Austria, is a relict from the ice ages.

Pupillid snails

The VALLONIDS (family Vallonidae) are characterized by small rounded shells with low spires. To some extent self-fertilization is evident as in the RIBBED GRASS SNAIL (*Vallonia costata*). The PRICKLY SNAIL (*Acanthinula aculeata*; L 2 mm) and the ROCK SNAIL (*Pyramidula rupestris*) show the typical spirally wound shells. All vallonid shells are widely umbilicate with more or less strongly defined ribs which in the prickly snail terminates with a thorn on each rib. The viviparous rock snails in particular are highly adaptable. This form is not only found in calcareous mountains at elevations up to 3,000 m but has occasionally also been detected on the Vorarl Mountain and the Kaiser Mountains and in cave entrances.

Vallonia snails

Zebrina detrita (L 1–3 cm; see Color plate, p. 80) is of economic significance since it serves as the first intermediate host for a sheep-liver fluke (see Vol. I). *Z. detrita* is frequent on limestone meadows. The sinistral *Jaminia quadridens*, from western Europe and the Mediterranean region, penetrates into the southern Tyrol and the Rhine region. The BULIN (*Ena montana*; see Color plate, p. 80) and two other species, on the other hand, are widely distributed.

In the suborder HETERURETHRA the nephridium is transverse to the heart along the posterior end of the lung. The ureter is bent. The shell is usually delicate or almost reduced. There are three families with approximately 300 species. Some workers believe this group may represent an aberrant opisthobranch stock.

In Europe this small group is best known by six of the AMBER SNAILS (family Succineidae). The larger species, for example the AMBER SNAIL

(*Succinea putris*; see Color plate, p. 80), are often mistaken for mud snails because the shells of these two forms are similar and also the former are the only Stylommatophora which live on the shore and marshy vegetation in the vicinity of water. The amber snail cannot always withdraw its entire body into the shell, partially because of the high water content in the tissues. The snail, therefore, falls prey to frogs, snails, ants, and other enemies. The amber snail, in particular, is the intermediate host for the trematode *Distomum macrostomum* (see Vol. I) which lives in birds. The sporocyte generation of this trematode, which is known as *Leucochloridium*, forms tubelike structures several centimeters long which, during the day, are situated within the snail's tentacle. These sporocytes, which shine as green and brown rings through the snail's tentacle, pulsate forty to seventy times a minute. This signal attracts primarily thrushes and other worm-eating birds, which feed on the simulated "worm" and are thereby infected.

The colorless thin shell of the AFRICAN AILLYIDS (family Aillyidae) is covered along its margins by the mantle. The ATHORACOPHORIDS or JANELIDS (superfamily Athoracophoracea) are elongated nude snails which measure only a few centimeters. They are characterized by a few shell remains which are buried in the dorsal skin, and only one pair of tentacles. Numerous blind tubules extend from the minute respiratory cavity of the Athoracophoridae into surrounding tissue. This respiratory provision is known as the "tracheate" type. *Athoracophorus bitentaculatus* inhabits trees and shrubs in New Zealand. *Aneitia sarasini* lives in similar habitats in New Caledonia. Other species are also found on neighboring islands and Australia.

In the suborder Sigmurethra, the short nephridium is situated near the posterior end of the lung. The ureter is usually bent back secondarily and ascends alongside the rectum to the outside. The shells of these animals range from well developed to greatly reduced. There are about forty-three families within eleven superfamilies, with more than 15,000 species.

Six families constitute the suborder MESURETHRA (superfamily CLAUSI-LIACEA) on the basis of the unique morphology of the excretory system. This suborder encompasses the cerions (family Ceriidae) from the West Indies, the megaspirids (Megaspiridae) from the tropics of the Southern Hemisphere, the corillids (superfamily Corillacea) from southern India, and the dorcasiids (Dorcasiidae) from South Africa. The stropho-cheilids (Strophocheilidae), which are distributed in South America, cause some damage to coffee plantations and other plants. Equally harmful is *Strophocheilus oblongus*, which measures 20 cm. Its eggs measure 5 cm, and resemble those of a small bird.

The best-known and most numerously represented family of the superfamily Clausiliacea in Europe and Asia are the DOOR SNAILS (family Clausiliidae). They are highly towered, slender snails which include not

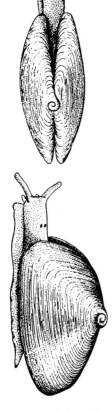

Fig. 5-40. *Berthelinia chloris* (L 1 cm), the "Bivalved Gastropod", dorsal view and seen from the left side

less than twenty-eight central European species with shell lengths from 5 to 20 mm. They are all characterized by a special aperture lid. Along the aperture of the sinistral shell are various lamellae and a piece attached by an elastic stalk. The animal is able to retract this calcareous piece or clausilium over the aperture so it fits exactly into the lamellar seam. Some of the more frequent species are the *Clausilia parvula* (L 10 mm), *Cochlodina laminata* (see Color plate, p. 89), and *Laciniaria biplicata*. In Germany *Delima itala* (L 15–20 mm) is only present in areas where grapes are cultivated. Just like the vine stocks, these snails love warmth, in contrast to other door snails which frequently prefer cool, moist places. The dextral *Cecilioides acicula* (L approximately 5 mm), which is widely distributed in central Europe, the Orient, Africa, and introduced to the U.S.A., is subterranean and lives underneath grass, stones, and roots, in areas rich in limestone. The whitish shell and the degeneration of the eyes are adaptive features to this subterranean mode of life. *Rumina decollata* (L 2–4 cm) is sporadically very common in the Mediterranean region and has been introduced to the American tropics and the U.S.A. It is found on stony substrate such as cliffs, rubble, or stone walls. The tip of the elongated conical shell of this snail is always broken off in the older animals, which results in a peculiar rounded cylindrical shell.

Within the suborder SIGMURETHRA, the flesh-eating SPIRAXIDS (family Spiraxidae), which are distributed throughout Central America, are considerably larger representatives of the door snails. Another genus, *Poiretia*, is also present in the Mediterranean. This form is characterized by a pair of large oral flaps beside the tentacles. This feature is also prominent in the carnivorous *Euglandina rosea* which is found in the southeastern U.S.A. The family Achatinidae even surpass the previous group in size. *Achatina achatina* (see Color plate, p. 89) can reach a size of more than 30 cm. Its shell can reach a length of more than 20 cm and a width of almost 10 cm. This giant snail is primarily found in the tropical rain forests from Guinea to Nigeria. It is nocturnal and can cause considerable damage to the plants it feeds on. This snail serves as food for the natives. *Achatina fulica* is feared because of the damage it inflicts to commercial plants. It has been introduced to parts of India, Japan, and Indonesia, via the trade routes. It is a regular "snail plague." In recent times this snail has even "conquered" Florida. In Germany *Achatina marginata* has occasionally been introduced. Natural enemies of the giant snails are planarians (Geoplanida; see Vol. I), beetles (family Lampyridae; see Vol. II), ants, crabs, toads, monitor lizards, and ground-dwelling birds. These predators feed either on the eggs, which measure 3.0 mm, or the snail itself. Another way to combat these snails is by carefully introducing two carnivorous snails, *Edentulina affinis* (L 3–5 cm) and *Gonaxis kibweziensis* (L approximately 2 cm), which specifically prey on the giant snails.

These two carnivorous snails, *E. affinis* and *G. kibweziensis*, belong

Giant African snails

to the superfamily STREPTAXIDS (Streptaxacea). Closely related are members of the RHYTIDACEA, which are also flesh-eating snails, but which are not found in Europe. The ACAVIDS (family Acavidae), on the other hand, are vegetarians. *Helicophanta magnifica* (L 10 cm, shell L 6 cm), from Madagascar, belongs to this group. The BULIMULIDS (superfamily Bulimulacea), which are divided into four families (Bulimulidae, Odontostomidae, Orthalicidae, and Amphibulimidae), will be mentioned later. Another group is the urocoptids (Urocoptidae). These gastropods primarily inhabit America, with sporadic distribution also in Africa and the southwestern Pacific. Their shells range in shape from conical to towered, and measure only a few centimeters. Gigantic forms do also occur in this group, as exemplified by the New Caledonia *Placostylus fibratus*.

Punctum snails

The family Endodontidae, which measure only a few millimeters, include the DWARF SNAIL (*Punctum pygmaeum*; L 3.5 mm). This form is the smallest of the European terrestrial snails. It has a flat, widely umbilicate shell, and is frequently found under autumn leaves, wood, or stones. Similar habitat is occupied by *Discus ruderatus*, *Discus rotundatus*, and *Discus perspectivus*, which lives only in the mountains. On the average all measure about 6 mm. Related forms of the superfamily ENDODONTACEA include groups which show a trend for the reduction of the shell. The southwestern Pacific OTOCONCHIDS (family Otoconchidae) possess only a delicate and flattened shell which is partially covered by the mantle. The well-known European SLUGS (family Arionidae) externally do not betray any sign of a shell. However, a shell vestige is still present underneath each mantle shield along the dorsal side of the anterior section of the body and to the right of the center of the visible breathing hole.

Every European garden owner or hiker must be familiar with the two indigenous species of slugs (family Arionidae). There is the RED SLUG (*Arion rufus*; see Color plate, p. 89), which, however, can also be brown or dark gray, and the BLACK SLUG (*Arion ater*; see Color plate, p. 89), which has juvenile forms that may also be yellowish, gray, or greenish. These animals, which can measure up to 15 cm, are frequently present in moist areas. Heavy rainfalls or thundershowers will bring these animals out of their hiding spots. The equally large DUSKY SLUG (*Arion subfuscus*; L 5–7 cm) feeds exclusively on mushrooms and is therefore found only in evergreen forests, dry deciduous stands, and also on the heather. The GARDEN SLUG (*Arion hortensis*) seems to be a regular "domestic animal" which is only occasionally encountered outside gardens and parks. Along with other European *Arion* species, these animals are more or less vegetarians. Only a few, for example *Arion hortensis*, could cause damage to man's crops. This snail can be a vector of various plant diseases, thereby influencing cultured crops.

The final stage of the regression of the shell is reached by the PHILOMYCIDS (family Philomycidae). There is no trace of a shell and the broad

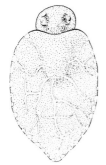

Fig. 5-41. *Bosellia mimetica* (L 2 cm), a Mediterranean sacoglossate.

mantle surrounds a wide, empty cavity. The philomycids are distributed throughout North America (for example *Philomycus carolinensis*), south-eastern Asia, and Indonesia.

Under the superfamily ZONITACEA, the CRYSTAL SNAILS (family Vitrinidae) are characterized by a transparent, glassy shell. The PELLUCID SNAIL (*Vitrina pellucida*) can barely withdraw into this shell. Complete withdrawal is impossible for *Eucobresia diaphana*. These snails cannot tolerate warm temperatures, and in the summer they usually are hidden. They are common in cool and moist regions, where they are always active. Even in the winter they do not hibernate. Generally the ZONITIDS (family Zonitidae) are small animals with flatly spiral shells. They frequently feed on plant substances. They include representatives of the genera *Vitrea, Nesovitrea, Zonitoides*, or *Aegopsis* from the eastern Alps. The genus *Aeginopella*, on the other hand, feeds primarily on animal matter. The species of the genus *Oxychilus* are also primarily meat eaters. The CELLAR SNAIL (*Oxychilus cellarius*), however, inhabits bushes, caves, greenhouses, and cellars, and has been introduced into North America.

Fig. 5-42. Cross-copulation in *Stiliger*.

Daudebardia rufa (see Color plate, p. 89) and *Daudebardia brevipes*, which are difficult to distinguish from one another, are exclusively carnivorous. They are characterized by very small, caplike shells. Hidden in loose, moist ground, they prey on worms, insect larvae, and snails. The external appearance of the *Daudebardia* leads to the familiar LIMACID SLUGS (family Limacidae), which have shells in various degrees of regression that are covered by the mantle shield. In contrast to the slugs, their pneumostome is located posterior to the mantle shield center. The MILACIDS (family Milacidae) are a transitional group including the quite common European representative *Milax rusticus* and *M. gagates* which are distributed in Europe and North America, and the less common *Aspidoporus limax*. All other European species of Limacidae possess a vestigial shell which is entirely enclosed by the mantle, for example the GREAT SLUG (*Limax maximus*; see Color plates, pp. 89 and 94), the ASH-BLACK SLUG (*Limax cinereoniger*; L up to 15 cm) and *Deroceras agreste* (see Color plate, p. 89); and *Deroceras reticulatum* (L 3–6 cm), which can hardly be distinguished externally from the latter species. *D. agreste* can cause considerable damage to crops and gardens. Many of these slugs have been introduced throughout North America and most temperate countries settled by European peoples. This group also includes *Deroceras laeve* (L 2–3 cm; see Color plate, p. 94), introduced to the United States and which is found in moist habitats, and *Lehmannia marginata* (L 6 cm), which frequently inhabits trees.

Some species from the TRIGONOCHLAMYDIDS (Trigonochlamydidae), which are indigenous to Iran and the Caucasus, are devoid of any shell. On the other hand, the SYSTROPHIIDS (Systrophiidae), which are now classified after the previous group, still possess a transparent, thin shell. The systrophiids are found in Central and South America. The TAWNY GLASS SNAIL (*Euconulus fulvus;* L 2–3 mm) is strongly reminiscent of the

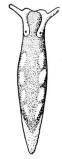

Fig. 5-43. *Acteonia corrugata* (L 6 mm). Eastern Europe.

Pyramidula rupestris, although it actually belongs to the sinistral *Rhinocochlis nasuta* and related forms. *E. fulvus* inhabits trees on Borneo, and *Thyrophorella thomensis* inhabits trees on the Isle São Thomé in the Gulf of Guinea. *E. fulvus* has a beaklike widening on the keeled shell aperture, and *T. thomensis* even has a flaplike shell extension on the upper part of the aperture which closes over it.

The superfamily OLEACINACEA is limited to America with the exception of the family Testacellidae with one species, the SHELLED SLUG (*Testacella haliotidea*; L 5–10 cm; see Fig. 5-33), which has infiltrated Germany from southwestern Europe and is found in American greenhouses. Like many other oleacinids, the testacellids are true carnivores which prey on earthworms underneath the loam. They possess small caplike shells.

The superfamily HELICACEA is the most highly evolved group of the Stylommatophores. Transitional forms which bridge the previously mentioned forms to the Helicacea are the American families Oreohelicidae and Camaenidae. These families, however, still lack the dart sac and digitiform glands and the "amorous dart" which characterizes the Helicidae. Of the BUSH SNAILS (Bradybaenidae), only the BUSH SNAIL (*Bradybaena fruticum*; L 2 cm) is present in Europe. It is found in deciduous forests, in brush, and also on moss-covered calcareous rocks. *Bradybaena similis* has been introduced in nearly all moist tropical areas of the world.

Because of its frequent occurrence in Europe, the HELICIDS (family Helicidae) are gastropods that have been thoroughly investigated, and we know much about their mode of life. Among these the EDIBLE SNAIL (*Helix pomatia*) is not only generally well known but has been extensively studied by experts. These snails inhabit vineyards and all regions which are not too moist, particularly brush. With the onset of winter its "biological clock" induces the edible snail to burrow itself into loose soil to a depth of up to thirty centimeters. The shell aperture becomes covered by an epiphragm, and thus the snail survives the cold season. *Helix pomatia* becomes active again with the warmth of spring, recuperates, and above all equalizes its loss of water content. Cross-copulation takes place during the moist days of May or June. The two partners of this hermaphroditic species erect their anterior parts of the body and firmly press together the foot soles. Each partner injects a calcareous copulatory dart into the other. This has a stimulatory effect. Edible snails produce sperm throughout the entire warm season but eggs for only a limited time.

P. Farb describes this process: "This feature probably protects the edible snails against self-fertilization during the months of their highest sexual activity. It is assumed that at least during part of this interval (egg deposit in July/August) the foreign spermatophores degenerate. This would avoid accidental self-fertilization of the eggs as they pass along the oviduct to the spermatheca." As in almost all stylommatophores, the fertilized eggs have a high yolk content. They measure 3 mm and have a calcareous shell. The snails lay their eggs into a dug-out hole in the

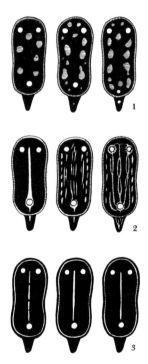

Fig. 5-44. Design formation in: 1. *Glossodoris luteorosea*, 2. *Glossodoris gracilis*, 3. *Glossodoris tricolor*.

ground. An edible snail can lay sixty to eighty eggs in one to two days. The fully developed young snails hatch after twenty-five to twenty-seven days. At first they possess delicate transparent shells. During the period until after the first hibernation in winter, the number of young snails is reduced sharply by ants, beetles, toads, and other predators. One can sometimes determine the age of a snail by the number of yearly "growth rings" in the shell.

One encounters *Perforatella rubiginosa* (L 5–7 mm) on marshy meadows and wet depressions. The well-known COPSE SNAIL (*Arianta arbustorum*; see Color plate, p. 89) also prefers highly moist spots in brush areas and herbaceous growth in forests. Some representatives of the genus *Helicella*, on the other hand, are inhabitants of arid regions. Some, including the HEATH SNAIL (*Helicella itala*), remain exposed outside even during the hot summer days. This form avoids desiccation by sticking to a branch and also by forming an epiphragm over the shell aperture. The heath snail, as well as *Helicella obvia* (see Color plate, p. 89) and almost all other *Helicella* species, possesses more or less distinct brown spiral bands. Both *H. itala* and *H. obvia* serve as intermediate host for the sheep-liver fluke (see Vol. I) and therefore are of economic importance.

Adaptations against desiccation

Snails that live in southern Europe and in Africa demonstrate still more astonishing heat-adapted features. The WHITE or SANDHILL SNAILS (*Theba pisana*) of Europe and California bunch up in rows along dried grass or bushes, where they sit during the hot hours of the day. In places the temperature may even go to over 60°C. These snails are active only during the night. The helicids are just as cold-resistant, for quite frequently they secrete a succession of calcareous lids in addition to the epiphragm. In several European species one can nicely identify the upper shell layer (periostracum), which is conspicuous because of its hairs, for example in the CHEESE SNAIL (*Helicodonta obvoluta*), *Trochulus villosus*, and other species. *Isognomostoma holosericum* is characterized by flat, platelike shells. One of the most common European species, the LAPIDARY SNAIL (*Helicogona lapicida*; L 2 cm), is a mountain snail which often escapes to the trunks of bushes and other deciduous trees during rain showers.

The genus *Cepaea* are particularly conspicuous because of the pretty bands in the shells. This group includes the GROVE SNAIL (*Cepaea nemoralis*; see Color plate, p. 89), introduced to the eastern U.S.A., the GARDEN SNAIL (*Cepaea hortensis*) of Europe and New England, *Cepaea vindobonensis*, and other species which often can only be differentiated by the design and structure of the shell aperture. The similarly shaped *Monacha cartusiana*, on the other hand, is easily recognized because of its grayish-white shell. Aside from the COMMON SNAIL (*Helix aspersa*), which has been introduced to many countries throughout the world, another quite unusual representative, *Cylindrus obtusus* (L 10 to 15 mm), should be mentioned. This snail has a pale gray shell which is cylindrical and is reminiscent of the whorl snail shell. *C. obtusus* is a regular relict which is limited in its distri-

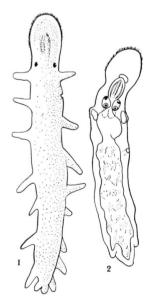

Fig. 5–45. 1. *Pseudovermis papillifer*, 2. *Pseudovermis schulzi* (both are seen from the dorsal view); these forms inhabit the spaces between wet beach sand.

bution to the Austrian calcareous Alps, to Schneeberg, and up to Dachstein at elevations of 1,100 to 2,500 m. This is the only species of the genus that was discovered in seventy-five different regions.

Even in ancient times the *Helix* snail was a favorite food item, and the edible snail also played a role in folk medicine. Recent investigations have demonstrated, in fact, that specific glandular substances from the edible snail, garden snail, and the larger arionid cause agglutination of certain bacteria, and therefore could be of therapeutic value against whooping cough, asthma, and other diseases. This might also explain the tonic effect which results from eating snails. Napoleon's soldiers carried canned edible snails as emergency rations during their great campaigns. The extract of 1,000 snails per man was supposed to be sufficient food for one week. Aside from the fastidious Frenchmen, people in southwestern Germany and many other regions also consider edible snails in a herb sauce a great delicacy. These snails are especially bred for the food industry.

Helicella and whorl snails also have special value for the collector, especially if the shell of an edible snail is sinistral, which is a great exception. It is then termed a "snail king." Certain Cuban bush snails are also valuable, from the collector's point of view, such as the brilliant *Polymita picta* which incorporates various colors in its shell depending on the particular habitat.

6 Scaphopods and Bivalves

Class: Scaphopods, by
L. v. Salvini-Plawen
and R. Tucker Abbott

An encyclopedia from the year 1751 dealing with natural science described various forms of a peculiar group of animals which were called "Dentalia" or "Entalia," toothed snails: "These are elongated, striped tubes that are blunted on both ends. The tubes, however, are somewhat longer and taper to a point on one end. Some assumed these to be teeth of certain fishes, and therefore gave them a corresponding name. However, these are not teeth but little tubes and bivalves which probably contain a worm of corresponding size."

These "toothed snails" form their own highly uniform class within the large phylum Mollusca. Today this group is referred to as TUSK SHELLS, TOOTH SHELLS, or SCAPHOPODA. Their shells vary in length from 2 mm to 15 cm. They are bilaterally symmetrical. The body is elongated. The mantle consists of two lobes which are fused at the ventral side, resulting in a tube with open ends. The shell is a tapering cylinder open at both ends. The head consists only of a proboscis. The base of the snout bears two pairs of clusters of protrusile filaments called captacula. Eyes and true tentacles are lacking. The burrowing foot, which lacks a creeping sole, is protruded by blood pressure. The foot and the organs of the mantle margin are located at the anterior end of the ventral side. Gills are lacking. There is only one pair of muscular strands which extends from mantle to foot. The heart lacks auricles and vessels, and the pericardium is small. The radula bears five rows of teeth and beneath it lies the non-protrusile taste organ (subradular organ). The muscular stomach is associated with two lobed midgut glands. The hindgut is short with an adjoining gland (rectal gland, the function of which is unknown). The simple nephridia are lobed, and are not connected with the pericardium but open into the mantle cavity behind the hindgut. The sexes are separate. The single unpaired gonad is strongly lobed, and usually without a gonoduct. There are two families: 1. Elephant's tusks (Dentaliidae); 2. Siphonodentaliidae. There are approximately 350 species.

In many aspects of their body structure the scaphopods are simplified

Distinguishing characteristics

Fig. 6-1 Geographical distribution of *Cadulus jeffreysi* (see p. 143).

mollusks. They are limited in their evolutionary development and are mostly sedentary. They lack gills and their circulatory system is degenerated. The animal burrows by extending the stumplike foot and plowing it deeply into the substrate. The foot swells due to blood pressure, anchoring the animal. Then the pedal muscle strands pull the body down. The animal burrows itself into sand or sandlike mud at an angle until only the tip of the shell protrudes out of the substrate. Gas exchange takes place through this projecting posterior end. The numerous captacula—in some species there are 130 on each side—emerge as two clusters from lobes in the head. The captacula have club-shaped pads at their ends. In the process of feeding, the captacula are extended into water-filled spaces among the loosely arranged substrate particles. These tiny cavities contain an amazing array of microscopic plants and animals. This so-called sand-space flora and fauna provides sufficient food for the scaphopod. The primary sources of food are foraminiferans (Foraminifera; see Vol. I), diatoms, and other protozoans. These particles are collected with the aid of sticky glands and are conveyed by the captacula to the mouth region. A ciliary pathway passes the food particles in the direction of the mouth. If the food source becomes exhausted in one locality, the animal leaves and burrows itself into a new spot.

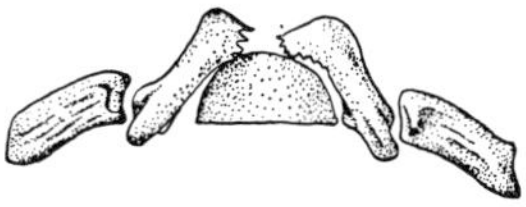

The radula within the proboscis is non-protrusile. The radula and the paired supports serve more as a crushing apparatus. The shape of the two first lateral teeth are particularly well suited for this crushing function. Before the food is crushed and passed into the alimentary canal, it is tested by the subradular organ underneath the radula for its palatability.

The scaphopod body plan, briefly discussed among the distinguishing characteristics, corresponds to its specialized mode of life and feeding. The transformation of the foot into a burrowing apparatus and its relocation to the anterior end of the animal is significant, as is the presence of the sensory organs in the mantle margin at the cephalic region. The osphradium, normally located in the vicinity of the gills, is found at the base of the scaphopod's foot, approximately at the body midline. The anus and nephridiopore are also found in this location. Due to the relocation of the foot, the actual body axis, from the mouth to the anus, as opposed to the original position characterized by the axis of the shell, is shifted by twenty to thirty degrees. On the basis of scaphopod larval development, we know that only the mantle and the open-ended shell maintain the original axis.

The ventral margins of the mantle, and subsequently the shell margins, fuse together in the larval stage. This fusion results in a mantle cavity that extends freely from the anterior to the posterior end. In some species this fusion is incomplete. Thus numerous species have notches or small slits which have remained open along the posterior part of the ventral shell. In *Fissidentalium plurifissuratum* (L over 6 cm) a series of holes is found in this position, and in the West Indian species *Fustiaria stenoschizum* (L 35 mm) there is a long cleft.

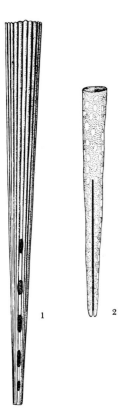

Fig. 6-2. A transverse row of scaphopod radulae.

Fig. 6-3. Shells of 1. *Fissidentalium plurifissuratum*, 2. *Fustiaria stenoschizum*.

The ciliated inner surface of the mantle has taken over the respiratory function of the scaphopod. The respiratory current is sucked in through the posterior opening of the mantle with the aid of ciliary action. Fecal and excretory products may enter the same way and are expulsed again through the same opening. Peculiarly, the hemocoele is in contact with the mantle cavity through an opening on each side of the anus and is thereby also connected to the external environment.

A unique feature of the scaphopods is the unpaired unisexual gonad, usually located in the posterior third of the body, which is without a gonopore. At the time of sexual maturity the gonad connects with the right nephridium. The wall of the nephridium is penetrated and the gametes are discharged into the nephridial cavity and the excretory duct, and finally into the mantle cavity and to the outside.

Fertilization of the ova and development of the trochophore larva occurs in the water. Just as in *Epimenia verrucosa* (see Chapter 2), this larva undergoes a transitional stage between the yolk larva and trochophore larva (see Chapter 1; and Color plate, p. 26). Later, an umbrellalike ciliated organ, the velum, forms at the anterior end. In the scaphopods the arrangement of the bi-lobed mantle, as well as the bilateral symmetry and the unique nervous system, suggest a great similarity to the bivalves. At a later stage the mantle and its secreted three-layered shell surround the animal in a tubelike fashion. During the course of development the shell becomes elongated anteriorly by new overlapping tube segments and at the same time it also becomes wider. In the ELEPHANT'S TUSK (*Dentalium*) the three-lobed foot is already evident at the very early stage as a bulge on the frontal margin of the animal. However, the larva sinks to the bottom after the velum has been discarded or has shrunk, and begins its burrowing mode of life.

The two scaphopod families are differentiated on the basis of characteristics in the burrowing foot and the radula. The European COMMON ELEPHANT'S TUSK (*Dentalium vulgare*; L 6 cm; see Color plates, pp. 65/66 and 139) represents the family of the ELEPHANT TUSKS (Dentaliidae). Like its relatives, this species is characterized by an elongated shell which constantly grows at the anterior end, often becoming broken off at the posterior end. The individual species of this family are usually easily recognized by the various patterns of longitudinal ribs, rings, reticulation, or other surface designs on the shell. The foot consists of a tip flanked by two large lateral lobes. Muscular action can shorten or retract the foot but not invaginate it. As in the majority of species, the shell of the common elephant's tusk is of a whitish color. Scaphopods are regularly found in the sea-bottom deposits of European oceans. *Fustiaria rubescens*, which measures 3 to 4 cm and is found in the Mediterranean, is one of the few forms that is brightly colored. Its shell is purplish-red. *Dentalium rectum* (see Color plate, p. 139), which measures 6 cm and is found in the eastern Atlantic, is characterized by distinct color patterns. *Dentalium entale*

▷
Scaphopods: 1. Elephant's tusk (*Dentalium vulgare*); 2. *Dentalium rectum*; 3. *Pulsellum lofotense*; 4. *Siphonodentalium vitreum*; 5. *Cadulus subfusiformis*. Protobranchs: 6. Common nut clam (*Nucula nucleus*; L 6 mm); 7. *Nuculana fragilis* (L 8 mm); 8. *Yoldia limatula* (L 5 cm). European Taxodont Bivalves: 9. Noah's Ark (*Arca noae*; L 6 cm); 10. Common bittersweet shell (*Glycymeris glycymeris*; L 6 cm).

▷ ▷
Leptodont Bivalves (all natural size): 1. Blue mussel (*Mytilus edulis*); 2. Bearded mussel (*Modiolus barbatus*); 3. Date-shell (*Lithophaga lithophaga*); 4. Hammerhead oyster (*Malleus malleus*); 5. *Musculus marmoratus*; 6. Great pearl oyster (*Pinctada margaritifera*); 7. Pen shell (*Pinna nobilis*; compare Color plate, p. 162).

MILLA

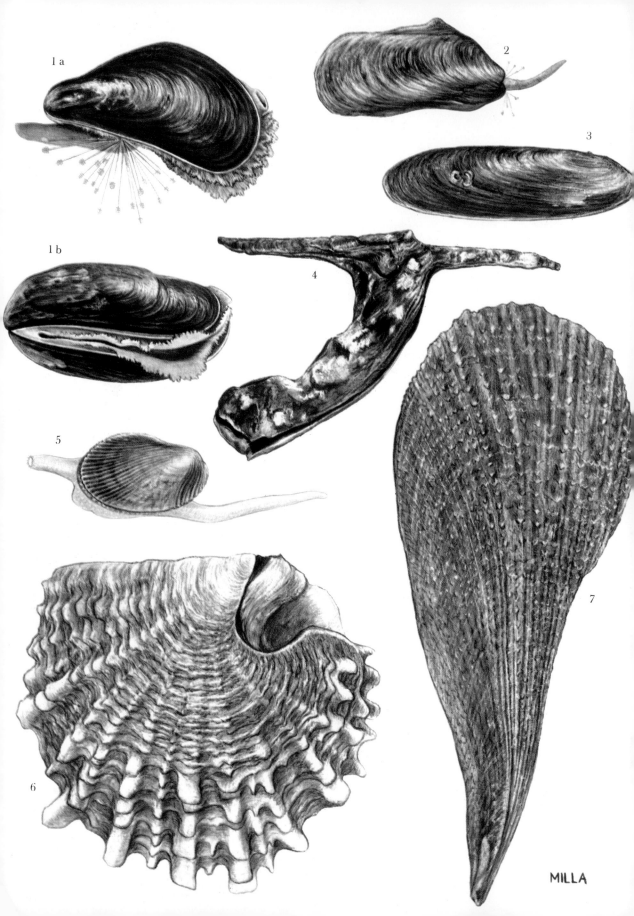

1 a

2

3

1 b

4

5

7

6

MILLA

MILLA

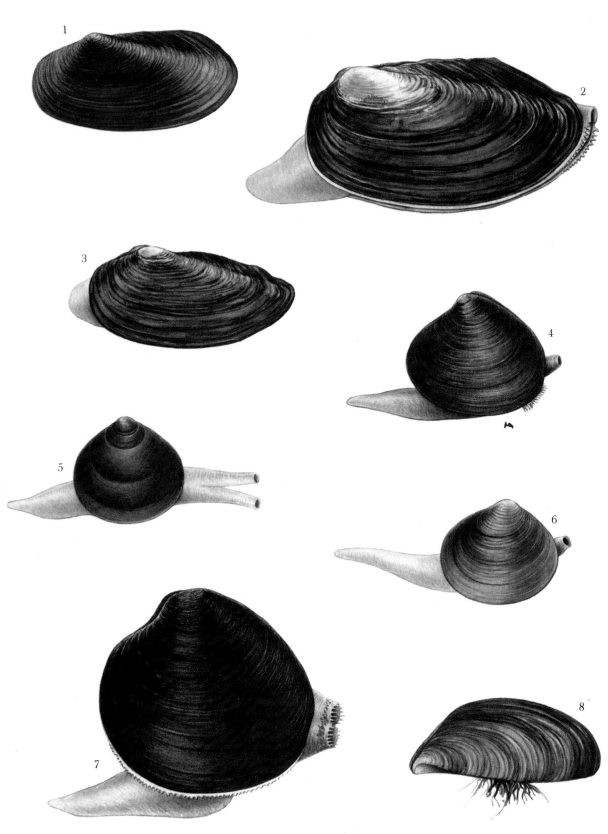

MILLA

(L 4 cm) from the North Atlantic and *Dentalium dentale* (L 3 cm) from the Mediterranean are both difficult to distinguish from the common elephant's tusk since they all are whitish in color. *Dentalium dentale* has a specialized diet. It feeds only on foraminiferans. The animal collects up to one dozen of these hard-shelled protozoans in its mouth, and then crushes and digests its prey. Other species probably are similarly adapted to a single source of food, although little information on this subject is available.

Even less is known about the mode of life of the second family, the SIPHONODENTALIIDS (Siphonodentaliidae). Their foot is characterized by a terminal disc or a simple tip without lateral lobes. In the latter case the foot is entirely evertible. Both midgut glands discharge into the stomach via the left duct. The gonad is located partly in the mantle. The shells are smooth or patterned by only a few ridges. Their shell form is somewhat similar to that of the elephant's tusk, as is exemplified by the pale-pink to bluish shell of *Entalina quinquangularis*, which is five-cornered in cross section. The shell structures of juvenile forms of the European species *Pulsellum lofotense* (see Color plate, p. 139) are easily mistaken for *Dentalium*. The usually very small species of the genus *Cadulus*, which enjoy worldwide distribution, possess shells which are more or less roundish, spindle-shaped. Like almost all scaphopods, they are also distributed on sedimentary ground down to great depths. They are represented by a variety of species. Although the siphonodentaliids usually measure only 2 to 15 mm, their frequently fragile shells are conspicuous because of their porcelain color and are therefore relatively easy to find. In European oceans the genus is represented by the two very similar species, *Cadulus subfusiformis* (see Color plate, p. 139) and *Cadulus jeffreysi* (see Fig. 6-1). The genus *Siphonodentalium*, with the North Atlantic species *Siphonodentalium vitreum* (L 10 mm; see Color plate, p. 139), serves as a transitional form between *Entalina* and *Cadulus* in respect to body shape and shell form.

The scaphopods are protected primarily by their shells. The openings can be tightly sealed by mantle bulges. Their burrowing mode of life is another protective measure. Occasionally these animals are extremely difficult to find. Nevertheless, the scaphopods have numerous enemies among the fishes, polychaetes, and gastropods; for example, the canoe shell (*Scaphander lignarius*) (see Chapter 5) has specialized in preying primarily on scaphopods. Generally, however, the animals live for several years. Unlimited growth can result in quite sizeable forms. The largest of all scaphopods, *Fissidentalium verneli*, from southeastern Asia, can reach a shell length of from 9 to 13.5 cm. Empty shells are utilized not only by hermit crabs, annelids, and Sipunculida but on occasion have been used as jewelry or money by North American Indians.

The BIVALVES (class Bivalvia) are one of the largest and best known of the various classes of shelled mollusks. They lack a proboscis and the radula. Adult shells vary from 2 mm to over 130 cm. Their body is usually

◁ ◁
Leptodont Bivalves (all natural size): 1. Pilgrim's scallop (*Pecten jacobaeus*; compare Color plate, p. 159); 2. Varied scallop (*Chlamys varius*); 3. Spiny oyster (*Spondylus gaederopus*); 4. Inflated file shell (*Lima inflata*; compare Color plate, p. 160); 5. False saddle oyster (*Anomia ephippium*); 6. Common European oyster (*Ostrea edulis*); 7. Japanese oyster (*Crassostrea gigas*).

◁
Schizodont Bivalves: Unionid river mussels. 1. Painter's mussel (*Unio pictorum*); 2. Swan mussel (*Anodonta cygnea*); 3. Pearl mussel (*Margaritifera margaritifera*). Heterodont Bivalves: 4. Northern Astarte (*Astarte borealis*; L 4 cm); 5. Common orb mussel (*Sphaerium corneum*; L 9 mm); 6. Pea mussel (*Pisidium amnicum*; L 5 mm); 7. Ocean quahog (*Arctica islandica*; L 6 cm); 8. Zebra mussel (*Dreissena polymorpha*; L 3 cm).

laterally compressed and entirely surrounded by the lateral mantle folds, which are greatly enlarged and covered by a longitudinally divided shell. In a few primitive species the foot is flat but as a rule it is developed as a burrowing organ which becomes erectile by blood pressure and is usually ax-shaped; the scientific name, Pelecypoda, refers to this description. The byssus gland, found in such sessile forms as the marine mussels, at the base of the foot, secretes sticky threads which become hardened in the water and serve as mooring lines to the substratum. The mantle cavity lies on both sides of the foot and contains a pair of modified and enlarged ctenidia. The stomach is always associated with a crystalline style. There are from zero to seven (usually one or two) pairs of pedal retractor muscles going from the mantle to the foot. The nervous system is essentially simple, and depending on its degree of evolutionary development is concentrated to varying degrees in the posterior end of the body. Bivalves are predominantly dioecious. Fertilization takes place in the water or in the mantle cavity. As a rule they are filter feeders. They are marine or live in the intertidal zone. There are four basic types of gills: 1. Protobranchia, 2. Filibranchia, 3. Eulamellibranchia, 4. Septibranchia. There are approximately 8,000 species.

The bivalves are one of the most peculiar groups of animals. A bivalve is a concept to almost everyone, and almost everyone has seen one. Yet who knows and understands why this two-shelled mollusk developed such a body structure? There is not recognizable head section, and therefore it is difficult for a non-expert to differentiate between the front and back of the animal. The fact that specific bivalves are found only in salt water or fresh water also makes it difficult to obtain more information about these animals. Although most bivalves are able to move freely, one only rarely sees them changing their position or locality. This limited amount of mobility, even in the mating season, easily gives these creatures the reputation of being uninteresting to the average nature lover. However, if one compares the body structure and life history of the bivalves with other mollusk groups, the bivalve's peculiar structure as well as its embryological development and mode of life become understandable and increasingly fascinating.

Just as the pages of a book are protectively enclosed by its cover, the bivalve's body is surrounded by the laterally enlarged mantle folds and the two shell plates it secretes. The equivalent to the spine of the book is the uncalcified ligament which holds the two shell valves together without influencing their movement against each other. Along the hinge line the two shell halves usually possess interlocking teeth. Muscular action is responsible for the tight and firm closure of the bivalve shell. In an earlier evolutionary stage there was an anterior and a posterior shell muscle in each valve. An adductor muscle consists of two kinds of fibers which perform for a rapid closing of the shells, and the smooth muscle fibers in the joint react slowly and can keep the mollusk closed for weeks with a

Class: Bivalves, by
L. v. Salvini-Plawen
and R. Tucker Abbott

Bivalve
characteristics

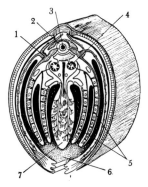

Fig. 6-4. Fresh-water bivalve (*Anodonta*) anatomy in cross section: 1. Auricle, 2. Ventricle, 3. Endodermis, 4. Kidney, 5. Ctenidia, 6. Mantle cavity, 7. Foot with intestine and gonads (sectioned).

minimum expenditure of energy. This two-valved structure contains the actual animal, which is usually completely covered. In a whole series of species, however, the valves gap open where the inhalent and exhalent siphons enter.

If one takes into consideration the life habits of the bivalve, it makes sense that they lost the entire buccal mass and radula in the course of their phylogenetic development. Most species are filter feeders and live on microscopic creatures which they suck in with the inhalent current. The exhalent current along with all the waste products is then expelled. In this type of bivalve the inhalent and exhalent openings are found at the most posterior end of the body. Frequently these openings are elongated tubes or siphons, similar to those of certain gastropods (see Chapter 5). Sometimes the siphons are separated, or they can be fused to a double tubular structure which sticks out from the bivalve's body.

Siphons

The shell develops first along the hinge line. This part, the umbo, often stands out as a conical elevation. During the process of growth, more or less concentric "growth rings" form below the umbo. Along the beach or inland water shores one can frequently pick up empty bivalve shells on which one can recognize the hinge, umbo, growth rings, and the scars of the adductor muscles, and on occasion one can also "read" the three layers of the shell if the leathery outer peristracum as well as the nacreous layer has been developed.

The foot and the mantle cavity are modified in a very characteristic manner. The morphology of the respiratory organs deviates greatly from that of the other mollusks. Only in the primitive protobranchs are the true ctenidia still evident. The majority of bivalves either possess the so-called filibranch gills, where each gill separately forms a "W" in section, or the eulamellibranch gill, where adjacent filaments are united by vascular cross connections. This two-sided, apparently double, gill of the eulamellibranchs is additionally fused to the adjoining foot by the tip of the ascending filament. In this manner the gills have separated the mantle chamber horizontally into an upper cavity which collects the exhalent water current and conducts it to the exhalent aperture. The inhalent current, which also contains food particles, enters the lower cavity via the action of millions of beating cilia which line the entire mantle chamber. The inhalent current is then pressed through the netlike gill spaces. In this process food particles become trapped on the external gill surface and are conducted on mucoid strands to the mouth opening which is surrounded by four labial palps. The outgoing current in the upper (exhalent) cavity also flushes out the waste products from the anus and nephridopore, which terminate there.

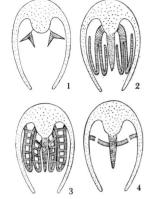

Fig. 6-5. Types of ctenidia: 1. Protobranch, 2. Filibranch, 3. Eulamellibranch, 4. Septibranch.

All bivalves are characterized by a special crystalline style within a separate section of the stomach, which is also found in certain gastropods and limpets (see Chapters 4 and 5). This structure is an adaptation to feeding on microscopic creatures. The rotation of this style is responsible for distributing digestive juices to all sides of the stomach. Concommitantly the finer food particles become separated from the usually coarser sand granules which also enter the stomach with the ciliary current. Along with the style enzymes, the lighter particles pass into the two digestive divertcula for digestion. The heavier particles, along with the glandular feces and mucus, are conducted to the midgut and finally out through the anus.

Since there is not a true head region in the bivalves, the mantle margin in contact with the external environment has taken over the cephalic functions. The mantle margin is endowed with a great number of sensory cells and, not uncommonly, also eyes. The sensory system is often extended into numerous tentacles or lobes which are more highly concentrated along the tips of the inhalent and exhalent currents. A special mantle nerve receives the stimuli of the sensory cells and sensory organs and conducts most information to the posterior ganglion (visceral ganglion). This paired ganglion also receives the sensory stimuli of the gills and the osphradium, which is located on the base of the gills.

The various eyes found in many of the species demonstrate a wide range of structural complexity. There is the simple eye of the vesicular converse type, which consists of pigment and a few light-sensitive cells. This eye, gathered in faceted groups, is similar to the compound eye of the arthropod, and consists of about 70 to 250 visual cells. However, highly developed eyes are also represented. The eyes of the scallops, for instance, contain two successive retinas. Yet even bivalves lacking eyes can perceive differences in light intensity and movement. Just as we detect the movement of a finger along our skin with the aid of our sense of touch, the bivalve also perceives a stimulus when light moves over the epidermis of the mantle.

The morphology of the eye

Bivalves are usually dioecious. As a rule, their gonads are located in front of or below the pericardium in the anterior part of the body. The gonads usually consist of numerous lobes or tubes. The gonoduct and the opening which empties into the upper mantle chamber are simple. Since neither a seminal receptacle nor other associated structures are present, the gametes are discharged directly to the outside via the exhalent current. In numerous species the progeny is protected in special brood chambers. Usually, however, the eggs are fertilized in the surrounding water. A larval development almost always follows. The embryos of the primitive protobranchs are oval and develop within a ciliated cellular test. This yolk larva (see Color plate, p. 26) developed into two evolutionary lines, as was stated by P. Chanley and others. One is the "Trochophore-veliger line" where the cellular ciliated test becomes modified into a propelling velum. In the other, the "glochidium-lasidium line," the test is lost or reduced. The glochidium and lasidium larvae may be considered as a modified veliger stage of the fresh-water bivalves adapted to a parasitic life. The trochophore and veliger larvae (see Chapter 1), on the other hand, are the predominant larval forms for all other bivalves.

The larvae

The filtering activity of the bivalves contributes to the cleanliness of the water since they remove detritus. The bivalve's role of "health officer" becomes significant and life-essential to man if one recognizes that tremendous numbers of bivalves often cover the floors of inland waters or parts of the ocean. Bivalves also play a considerable role as a food source to man. Their significance in the jewelry industry is well known. Oyster cultivation, pearl diving, and artificial growing of pearls are of great economic importance in certain countries.

However, in addition to these benefits, bivalves can also cause "harm" to man: in fresh water the parasitic glochidia can occasionally weaken the fish population. Various boring bivalves can cause great damage to wooden harbor installations or ship bottoms. In certain places bivalves have played a great, though undervalued, role in folklore for quite some time. Erich Thenius notes that it is not an uncommon occurrence for peculiar heart-shaped structures to become exposed in the rocky surface in the southern Tyrolean Dolomites and in the Dachstein Mountains in Triasbalken of the Alps. These structures are faintly reminiscent of cattle tracks. "They are usually found in great numbers and often cover entire layers of stratified rock. Sometimes these depressions measure several decimeters." Actually these are the remains of megalodonts which have been most frequently found in the Dachstein Mountains. The megalodonts were characteristic of the Upper Triassic period. They are often fossilized as stone nuclei which vary in cross section depending on size and form. The shepherds in the Alps usually regarded these fossils as "fossilized cow tracks." In certain places one still encounters the old interpretation of these fossils, which says that they are traces of wild chases which stormed over these parts long ago. These concepts go back to a time long before the introduction of Christianity.

The classification of the bivalves is based mainly on the characteristics of the entire animal class. These features are the gills, the hinge, and the musculature. Yet the present-day classification remains unsatisfactory since the individual related groups overlap in certain characteristics. Consequently, several taxonomic systems are used side by side. Each system has its advantages and disadvantages. In this discussion we follow a classification system based on the differences in gill structure. This simplifies comprehension of this class, for only one characteristic is considered. The classification into protobranchs, filibranchs, eulamellibranchs, and septibranchs, whereby the protobranchs are grouped into a special category, is somewhat substantiated by fossil records. On this evidence, bivalves can also be classified into two types of groups based on the differences in the hinge morphology. The protobranchs are characterized by a row of teeth (ctenodont) along the hinge margin, while the raylike (actinodont, primitively radial) tooth arrangement of the fossilized forms like *Babinka*, *Lyrodesma*, and others (see Chapter 1) served as the prototype for the hinge structure of all remaining bivalves—or at least for the Lucinacea and

Bivalves in folklore

Leptonacea within the Heterodonta. The special position of the proto-branchs among all other bivalve orders also becomes evident when examining the processes of its development.

In recent years, a classification combining the fossil families and the living forms has been adopted, especially in North America and England. While this encyclopedia follows the older classification based almost entirely on gill structure, we believe it best to give, in outline at least, the arrangement now standard in American treatises (R. C. Moore, 1969).

I. SUBCLASS PALAEOTAXODONTA. Includes the marine nuculoids which have taxodont hinge teeth, equivalved shells, and nacreous or crossed lamellar shell material.

ORDER: Nuculoida. Includes the Recent families Nuculidae, Nuculani-dae, and Malletiidae.

II. SUBCLASS CRYPTODONTA. Includes primitive, almost hingeless sol-emyoid clams. Contains a large fossil order, Praecardioida, and the Recent:

ORDER: Solemyoida. One Recent family: Solemyidae, a siphonate burrowing protobranch.

III. SUBCLASS PTERIOMORPHIA. Includes the byssate marine clams, such as the taxodont arks, the filibranch marine mussels, the scallops, and the oysters.

ORDER: Arcoida. The major living family include the Arcidae, Limo-psidae, and Glycymerididae, all of which have taxodont teeth.

ORDER: Mytiloida. The two major families are the marine mussels, Mytilidae, and the pen shells, Pinnidae.

ORDER: Pterioida. Contains the living wing oysters, Pteriidae; the hammerhead oysters, Malleidae; the scallops, Pectinidae; the spiny oysters, Spondylidae; the jingle shells, Anomiidae; the file shells, Limidae; and the true edible oysters, Ostreidae; as well as several other minor living and fossil families.

IV. SUBCLASS PALAEOHETERODONTA. Primitive clams with strong hinges; usually nacreous within. The living orders include:

ORDER: Unionoida. The fresh-water mussel families Unionidae, Mute-lidae and Etheriidae.

ORDER: Trigonioida. Includes the living Trigoniidae of Australia.

V. SUBCLASS HETERODONTA. The advanced eulamellibranch marine clams with porcellaneous, crossed-lamellar shell structure. Contains the majority of species of living marine clams.

ORDER: Veneroida. Clams having thick shells with well-developed hinge teeth. Includes numerous families and subfamilies among which are the Lucinacea; the jewel boxes, Chamacea; the small, often parasitic clams, Leptonacea; the Carditacea; the true cockles, Cardiacea; the Mactracea; the razor clams, Solenacea; the tellins, Tellinacea; the Venus clams, Veneracea; and others.

ORDER: Myoida. Marine clams with fragile shells, long siphons, degener-ate hinges. Includes the soft-shell clams, Myidae; the basket clams,

Corbulidae; the piddocks, Pholadidae; and the shipworms, Teredinidae.

VI. SUBCLASS ANOMALODESMATA. Fragile marine clams, with weak hinge having a spoonlike fossate; lateral teeth absent; internally nacreous; gills usually septibranch or eulamellibranch.

ORDER: Pholadomyoida. Includes the superfamilies Pholadomyacea; the Pandoracea; the Poromyacea; and the watering-pot clams, Clavagellacea.

From this point we return to the older classification system.

Order: Protobranchia

The PROTOBRANCHS (order Protobranchia) are primitive forms which possess true ctenidia. The hinge margin usually consists of numerous teeth which are more or less of the same size. Both adductor muscles are still present and are usually equally well developed. Inhalent and exhalent siphons may be present or absent. They have a yolk larva with a thin test (see Chapter 1). There are four living families, including: 1. Nuculidae, 2. Malletiidae, 3. Solemyidae, 4. Nuculanidae. There are approximately 600 species.

The common nut shell

The protobranchs are the most primitive representatives of today's living bivalves. In certain aspects the anatomy of their body is quite illuminating, and provides a better understanding of the deviating molluscan nature of the bivalves. Let us examine as an example, the COMMON NUT SHELL (*Nucula nucleus*; see Color plate, p. 139) which measures only slightly more than 1 cm and is distributed along European coasts from the North Sea to the Aegean. Its labial palps are conspicuously elongated. When the animal is feeding, these tentaculate processes are stretched out of the shell and are drawn in again with a supply of unicellular organisms and algae as well as organic debris (detritus) from the sea bottom. Finally these food particles are conducted along the ciliary food groove to the mouth in an "assembly line" manner. This feeding mechanism is far less complex than in the remaining bivalves, but is prevalent in all forms closely related to the nut shells which are included in the NUCULIDS (family Nuculidae). The nut shells were already present in the Devonian period (380 to 320 million years ago).

The pair of bipectinate ctenidia, which characterizes all representatives of this order as well as the majority of the mollusks discussed so far, also plays a role in the feeding mechanism. In the nut shells, however, and in the toothless SOLEMYIDS (Solemyidae), the gills only participate to a small extent since the inhalent current is filtered for food particles prior to passing through the gills. The remaining protobranchs (compare *Nuculana fragilis*; see Color plate, p. 139) are far more "advanced" in their mode of feeding since they developed inhalent and exhalent siphons. The yellowish *Nuculana pella*, found in the Mediterranean and Atlantic, and *Yoldia limatula* (see Color plate, p. 139), which, like the nut shells, also develops a yolk larva (see Color plate, p. 26), belong to the more "progressive" forms with siphons.

The brownish upper shell layer (periostracum) which protrudes

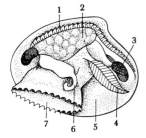

Fig. 6-6. Anatomy of a *Nucula* nut shell: 1. Hinge, 2. Gonad, 3. Adductor muscle, 4. Ctenidia, 5. Mantle cavity, 6. Labial palps, 7. Foot.

beyond the thin shell on the larva margin is clearly recognizable in *Solemya togata* which measures 3 to 4 cm. The Nordic *Yoldia arctica* achieved a certain fame because the precursor of the Baltic Sea was known as the "Yoldia Sea." During the last ice age (approximately 12,000 to 10,000 years ago) the Baltic Sea broke open and, along with the salt water, marine animals migrated into this basin. One of these immigrant species was *Yoldia arctica*, which increased greatly in its population numbers, and became a particularly prominent segment of the animal fauna in this newly arisen salt-water ocean.

"Yoldia Sea"

The usually numerous, almost equal-sized teeth along the hinge margin are a primitive characteristic feature of the protobranchs. All protobranchs have a flat, broad foot which is strongly reminiscent of the creeping sole of the chitons and gastropods, although these forms already move in the manner of most bivalves. The foot is extended and burrows itself into the substratum; then it swells due to blood pressure and anchors itself; finally the body is also pulled down. The life habits of these bivalves provide favorable conditions for certain other organisms to settle on their shell surface, since the bivalves burrow themselves into the soft sandy mud of the ocean. Nut shells found at depths under fifty meters are often associated with all sorts of commensals but particularly hydroid colonies (see Vol. I), as is found on *Nucula sulcata* from the North Atlantic.

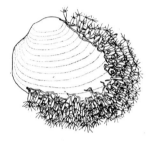

Fig. 6-7. *Nucula sulcata* with adhering (epizoic) polyps of *Neoturris pileata*.

The larval development of the protobranchs also verifies the group's status as a primitive order. Just as in the solenogasters (see Chapter 2), the embryo develops within a large-celled test. This yolk larva (see Color plate, p. 26) constitutes the free-swimming stage. During the course of further development, the larva becomes modified into an adult and the test is thrown off. The protobranchs give us an insight about what forms of bivalves existed before evolutionary radiation gave rise to all the other bivalves, because of their primitive features in general body morphology and in larval development.

The gill filaments of most FILIBRANCHS (order Filibranchia) are elongated into threadlike structures which bend upwards at the sides (exceptions: file shells and oysters). The hinge structure is diverse. As a rule the anterior adductor muscle is degenerated or is absent. Siphons are always absent. There are two suborders: 1. Taxodonta, 2. Leptodonta. There are approximately 1,000 species.

Order: Filibranchia

The Taxodonts are usually characterized by numerous similar teeth along the hinge. As a rule the shell is equivalve, and not nacreous. There are seven living families, the main ones being: 1. Ark shells (Arcidae), 2. Bittersweet shells (Glycymerididae), 3. Limopsidae, and 4. Philobryidae. There are approximately 250 species.

Suborder: Taxodonta

Fossilized ARK SHELLS (family Arcidae) showed that these forms were already present in the Lower Ordician about 500 to 450 million years ago. Just like the protobranchs, the ark shells still exhibit numerous primitive characteristics. Their two adductor muscles are still of equal size. The ark shells can be distinguished from the majority of bivalves by the fact that their hinge is endowed with more or less similar teeth. The hinge structure is reminiscent of the comb teeth (ctenodont) of the protobranchs. In the ark shells this tooth arrangement is a secondary modification, as we know from examining their fossilized forms. The tooth arrangement originally was radial. This hinge tooth structure is therefore known as "pseudoctenodont" (from the Greek ψεῦδos—deception). The shells are elongate and broad and usually ribbed. The periostracum is hairy or scaly. One species, the BEARDED ARK SHELL (*Barbatia barbata*), was so named on the basis of its hairy periostracum.

Noah's Ark

The best-known European species of this family is probably NOAH'S ARK (*Arca noae*; L 6–9 cm; see Color plate, p. 139) from the Mediterranean. These shells are regularly sold in fish markets because they are both common and large in size. The local population eats the flesh raw. A similar species, the TURKEY WING (*Arca zebra*; 6–9 cm), is common in Florida and the West Indies. The significantly smaller MILKY WHITE ARK SHELL (*Striarca lactea*), distributed from the Mediterranean to the North Sea, is limited to areas with strong water currents. These shells are often found in groups. They are placed in the family Noetiidae. All ark shells have byssus strands which harden upon contact with water and anchor the animals to rocks, stones, or other solid objects.

In contrast, the COMMON BITTERSWEET SHELL (*Glycymeris glycymeris*; see Color plate, p. 139) as well as the TRUE BITTERSWEET SHELL (*Glycymeris pilosa*) and all other related forms, do not possess a byssus gland. Bittersweet shells, which measure from 6 to 8 cm and are almost circular, are frequently washed ashore, and therefore form part of the bivalve sand bottoms of European coast lines. As are the nut shells, the common bittersweet is often covered with hydroid colonies which seem to prefer to settle near the bivalves' respiratory apertures. They belong to the family Glycymerididae.

Bittersweet shells and ark shells possess eyes. As in most bivalves these are usually located along the margin of the mantle but in these animals the eyes are covered by the periostracum. Noah's ark possesses more than 100 eyes and each of these in turn is composed of approximately 250 individual visual units similar to the faceted groups in the compound eye of the arthropods (see Vols. I and II). In addition to these eyes, which are really a unique phenomenon, Noah's ark usually also possesses two simple eyes along the anterior mantle margin which are very similar to those of the limpets (*Patella*; see Chapter 5). Finally the ark shells, the bittersweet shells, and other representatives also have eyes on the gills. A light-sensitive organ is located on each side of the first gill filament. These are pigmented cups which are innervated by the cerebral ganglion. This is one reason why certain investigators regard these structures as "cephalic eye substitutes." All these eye structures are characteristic only in those species which inhabit regions where there is light. Species that occur in the deep

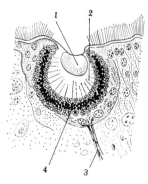

Fig. 6-8. Pallial eye of *Musculus marmoratus*: 1. Cornea, 2. Gelatinous material, 3. Nerve, 4. Retina.

sea are blind even though they may be closely related to forms that do possess eyes.

The almost symmetrical ark shells as well as the evenly rounded bittersweet shells are characterized by two well-developed adductor muscles, which is in concurrence with this symmetry. The muscle scars in the empty shells are evidence to this fact. The term "symmetrical" in reference to bivalves is used if the anterior half of the shell valve is the mirror image of the posterior half of the valve—independent of the mirror image of the right and left shell halves. The LIMOPSIDS (family Limopsidae), on the other hand, usually have small, obliquely oval shells. Since their internal organs are also shifted, the anterior adductor muscle has become greatly reduced or is absent. In *Limopsis aurita*, distributed in deep water in the Atlantic and Pacific, one can still clearly recognize the greatly reduced depression of the anterior muscle in the shells. Yet in the antarctic (and Californian) *Philobrya setosa* (family Philobryidae) this muscle scar is completely absent.

This discrepancy in the adductor muscles is a widespread phenomenon in the bivalves. In the following suborder, Leptodonta, this peculiarity has resulted in the much used term of unsymmetrical muscle (aniso-myaria). The hinge is usually toothless or is equipped with special structures. The shell valves are frequently dissimilar and usually nacreous. There are twelve families. Some of these are combined in superfamilies. There are approximately 1,000 species.

In contrast to the groups discussed up to now, the Leptodonta include a whole series of forms which are known at least in general even to the non-expert. The Leptodonta include such well-known animals as the mussels, the pearl oysters, edible oysters, and the scallops. Except for their usually well-developed inner pearly layers, these forms share few external similarities. In order to characterize this suborder we have to rely on similarities in morphological structures. The conspicuous lack of teeth along the hinge margin in most species and the reduction or even total degeneration of the anterior adductor muscle probably resulted from the fact that these bivalves are either periodically (in the juvenile state) or permanently attached to the substratum. "Many bivalves that are not closely related with each other possess unequal muscles," comments Sir C. M. Yonge. "In all cases, however, this condition is a result of a sessile way of life where the animal is attached by byssus threads. Although all the species concerned do not share a common origin, the evidence is clear that the degeneration of the anterior adductor muscle always proceeded in a relatively similar manner as a result of parallel evolution." It thus becomes evident that there are many unsolved questions that concern the phylogenetic relationship of the bivalves, which cannot clearly be answered on the basis of the structure and arrangement of the gills, for parallel evolution can also occur in the gills.

Even those well-known bivalves, the BLUE MUSSEL (*Mytilus edulis*;

Suborder: Leptodonta

Muscle degeneration

L 5–10 cm; see Color plate, p. 140) are characterized by unequally developed muscles. Mussels are often concentrated in masses along rocky shores, pilings, or harbor installations, where they are firmly attached by their brownish, bristlelike byssus threads. However, these animals are not always bound to the same location because they are able to tear each byssus thread with their powerful foot and reattach themselves at a new location. A mussel can even climb by employing this method. The mussel will attach new byssus threads at a higher spot, sever the old ones, and then pull its body up.

Mussel culture

Mussels enjoy a worldwide distribution and are extremely adaptable. Their meat is tasty, and contains a high percentage of vitamins, protein, minerals, and other important nutritive substances. Along the coastal regions of many countries mussels are regularly eaten and inland they are often considered tasty tidbits. In Europe over 100,000 t of mussels are eaten yearly. Holland alone delivers close to 60,000 t. In France the demand is heaviest, and approximately 50,000 t of mussels are consumed there annually. Certain countries cultivate edible mussels; in the Mediterranean region the MEDITERRANEAN MUSSEL (*Mytilus galloprovincialis*) is raised. Natural occurrence of these mussels in this tidal region greatly facilitates their cultivation. The related BEARDED MUSSEL (*Modiolus barbatus*; L 5 cm; see Color plate, p. 140) is also a valued food item in Mediterranean countries.

Musculus marmoratus (L 2 cm; see Color plate, p. 140) can only be distinguished from the mussels by its smaller size and heavily ribbed shell. *M. marmoratus* is widely distributed in the Mediterranean and the North Sea, where it is usually found on sedimentary bottoms. It is often also found on or in ascidians (Ascidiacea; see Chapter 18), to which the bivalves attach themselves during the juvenile stage.

Date mussel

The DATE SHELL (*Lithophaga lithophaga*; see Color plate, p. 140) is another edible shell of the Mediterranean. It does not possess byssus threads in the adult stage; byssus threads seem to have become unnecessary since the animals have adapted to a boring mode of life. In contrast to most boring bivalves, the date shell does not mechanically bore through calcareous rock, corals, and larger shells. The anterior mantle margin secretes an "acid" which dissolves calcium and thereby etches out a tunnel for the animal. This chemical boring process does not wear out the periostracum in any way. It remains entirely undamaged. The 12-m-high pillars of the Serapis Temple, near Pozzuoli along the Gulf of Naples, are perforated by date shell tunnels from the 3.6- to the 6.6-m level.

Hammerhead oyster

Just like the previously discussed forms of the Leptodonta, the PURSE SHELLS (family Isognomonidae) also possess pallial eyes. The purse shells and a few other families are united as the superfamily Pterioidea. The peculiar HAMMERHEAD OYSTER (*Malleus malleus*; see Color plate, p. 140), found in the Indian Ocean, is a member of the isognomonids. This form is characterized by a broad hinge margin and a short ventrally extended shell. In the hammerhead shell, as well as in the related species, *Vulsella rugosa*, the anterior adductor muscle is only present in the juvenile stage. *V. rugosa* parasitizes sponges, and is distributed in the Indian and Pacific Oceans.

Although the WING OYSTER (*Pteria hirundo*; L 6–8 cm) belongs to a different family (Pteriidae), its body form greatly resembles that of *M. malleus*. *P. hirundo* is found in the Mediterranean and is capable of climbing in the manner of the mussels. This form is very frequently found on gorgonians (*Gorgonaria*; see Vol. I) in the deeper coastal zones. The Floridian and West Indian representative is *Pteria colymbus* (L 6–8 cm).

The *Pteria* are closely related to the pearl oysters which play a great role in the history of civilization and commerce. The GREAT PEARL OYSTER (*Pinctada margaritifera*; L 25 cm; see Color plate, p. 140) belongs to this group, and is the most significant pearl producer. Pearl oysters occur in almost all warm oceans. They are characterized by a thick, roundish, flat shell with a scaly upper surface. The animals are anchored to the ocean floor by byssus threads. This is the reason why the ventral pallial eye is reduced.

The great pearl oyster

Pearl formation can be caused by a number of factors. Often foreign substances, parasites, or injuries can be the cause. At first, cells from the mantle surface, which usually secrete the shell, penetrate the connective tissue at the site of injury. It is at this spot that these cells form the pearl. The pearl is constructed in the same manner as the bivalve shell. It consists of a nacreous and calcareous layer and the periostracum which forms the outer layer in the shell. In contrast to the shell, however, the pearl is not secreted to outside but inside the mollusk. In the pearl the periostracum makes up the nucleus and the iridescent nacreous layers is on the outside.

The iridescence of the pearls lasts for 100 to 150 years before it goes "blind." When pearls are worn as jewelry they usually lose their shine long before this. The surface of the pearl is gradually worn off by exposure to the air and the chemical action of the human skin, as well as by frictional forces.

Aging of a pearl

Pearls are often gathered at great risk by special divers. The divers are equipped only with a clamp on the nose, and dive to depths of over forty meters to pry the bivalves from the reefs. In many regions, particularly along the coasts of Ceylon, considerably more pearl oysters are gathered than can be naturally replaced. Despite this wasteful exploitation, pearl fishing is only profitable if, in addition to the pearls, the mother-of-pearl shell itself is also manufactured into buttons, knife handles, ash trays, and similar objects. Artificial pearl culturing is carried out extensively in Japan.

Pearl formation can also occur in the PEN SHELL (*Pinna nobilis*; see Color plate, p. 140), the edible mussels, oysters, the fresh-water mussel, *Margaritifera margaritifera*, and several other species. The pen shell, endemic to the Mediterranean, has a shell which measures up to 80 cm. It is the

Pen shell

largest bivalve in the European region. These wedge-shaped shells are treasured trophy objects of sports divers. This has resulted in the near extinction of these shells in several of their natural habitats. Aside from being a trophy, this gigantic bivalve also serves as a human food source. In southern Italy the byssus threads of the pen shell were once used for woven and braided goods. The pen shells stick into the sea bottom mud or sand with the tip of their shells. The byssus threads serve to anchor the animal to the substrate. The animals filter food particles from the inhalent current which enters through a large opening at the upper margin of the shell. Younger specimens are endowed with numerous thorns and scale-like processes which become worn off during progressive growth. In the large shells only the coarse ribbing remains. Pliny the Elder (A.D. 23 to 79) was aware of the unusual relationship between a small crab, the oyster crab (*Pinnotheres pinnotheres*; compare Vol. I), and the pen shell. At the sign of a disturbance these crabs slip into the bivalve and cause it to close its valves.

Members of the next superfamily, PECTINACEA, differ from the PTERIACEA in their usually symmetrical round shells which as a rule are characterized by equal-sized "ears" (wings). *Amusium pleuronectes* (L 5–8 cm), distributed in the western Pacific, possesses such shell processes. This particular shell is further characterized by its differently colored shell valves. The left one is reddish-brown and the right one is yellowish-white. The conspicuous form of the hinge in the Pectinacea is particularly conspicuous in the SCALLOPS (family Pectinidae), of which there are approximately 360 living species. Like their closely related families, the scallops possess only one adductor muscle in the center. Yet a secondary symmetry was established in the scallops. The axis of the body is vertical to the toothless hinge margin. It is possible that this shift in the body axis is connected with the ability to swim, which is a characteristic feature of many scallops. If, for example, such a shellfish as the VARIED SCALLOP (*Chlamys varius*; L 5 cm; see Color plate, p. 141) is approached by a starfish, the resting animal quickly shuts its valves. The water which escapes from the mantle margin results in a backward push causing the bivalve to move from its location with its hinge facing frontward. Rapid repetition of this jerky movement enables the bivalve to escape from the dangerous enemy. Usually, however, the animals swim with the opening of the shell pointing anteriorly. The animal moves forward by closing and opening its valves and forcing water out dorsally from each side of the hinge. This type of swimming consists of jerky jet-propelled motions, whereby the animal constantly changes direction. There is a certain similarity to a fluttering butterfly.

The best-known scallop species is undoubtedly the PILGRIM'S SCALLOP (*Pecten jacobaeus*; L 5–15 cm; see Color plate, p. 141). In the past, many crusaders returning from the Mediterranean carried these shells on their hats or garments. Many pilgrims who had visited Santiago (St. Jacob) de

The scallops

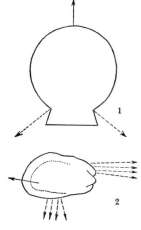

Fig. 6-9. Swimming of the scallop (1) and file shell (2), direction of movement (solid arrows), expelled water current (dotted arrows).

Compostela in Spain carried this scallop shell. The popular and scientific names for this bivalve were derived from this custom. Today, however, scallop shells are better known as service plates for "Ragoût fin en coquilles" or as a trade-mark for the Shell Oil Company.

Like many other scallops, the pilgrim's scallop is mobile. A series of prominent, highly developed, deep-blue eyes are situated among the numerous tentacles along the mantle margin. The location and degree of development of these visual organs can vary within a species. The eye or ocellus measures up to 1 mm, and consists of two successive retinal layers. As in the vertebrate eye, the retinal cells are inverse, directed away from the source of light. The pilgrim's scallop detect the difference between light and dark as well as motion with these eyes. The animal can thus perceive approaching starfishes or octopuses; in addition it can recognize its enemies by their scent. When endangered, the scallop escapes by its thrusting manner of swimming. The visual acuity of these bivalves is not sufficient to help the animal orient toward the source of light during swimming.

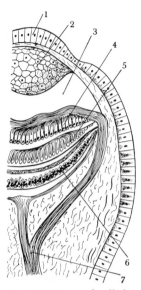

Fig. 6-10. Inverted pallial eye of the pilgrim's scallop: 1. Cornea, 2. Lens, 3. Gelatinous material, 4. Outer retina, 5. Inner retina, 6. Pigmented layer, 7. Nerve.

The right valve of the pilgrim's scallop is more convex than the left one. This feature influences the mode of locomotion, and the right valve is in a ventral position during swimming. Scallops with valves of the same size, on the other hand, nearly always swim in a vertical position.

The GIANT SCALLOP (*Pecten maximus*) is just as well known as the Mediterranean range of the pilgrim's scallop. The giant scallop is also found in the North Sea. The scallop species *Chlamys opercularis* is also distributed in the Mediterranean and up to the North Sea. This species occurs at greater depths and usually is anchored by byssus threads, although this bivalve is also able to swim.

The SPINY OYSTERS (Spondylidae) are regarded as a separate family. They are characterized by two interlocking teeth on each valve and by their occasionally very long spiny or scaly processes on the surface ribs. In contrast to *Chlamys*, which are only periodically attached to the substratum, the spiny oysters are always firmly anchored. However, these forms still possess pallial eyes, which is a clear indication that visual ability is not correlated with free mobility. The SPINY OYSTER (*Spondylus gaederopus*; see Color plate, p. 141) from the Mediterranean is a well-known species. It is found at greater depths and is usually covered by dense growth which hides the beautiful purple coloration of the shells. This bivalve is highly valued as a food item.

The spiny oysters

It is enchanting to observe the dancelike swimming motions of the FILE SHELLS (family Limidae). Their brilliant orange or reddish colored mantle margin with the numerous long, glandular tentacles give the appearance of a halo of light surrounding this animal. The bivalves jump about almost completely covered by these illuminating appendages. The water current that propels these animals is not only expelled alongside the hinge but also, quite sporadically, to the side of the ventral margin

Fig. 6-11. File shell within the nest that it built.

Fig. 6-12. Attached *Anomia* bivalve.

Fig. 6-13. Distribution of the window-pane oyster (*Placuna placenta*).

(see Fig. 6-9). The file shells can also obtain momentum by a frontward movement, and are even able to combine these two movements. This is one reason why these bivalves are also referred to as jumping shells. The FILE SHELL (*Lima lima*; L up to 4 cm), which occurs quite frequently in the Mediterranean and West Indies, is characterized by thick shells with prominent ribs, each with many scales. Most European species, however, such as the INFLATED FILE SHELL (*Lima tuberculata*; L 3–4 cm; see Color plate, p. 141) are petite animals with delicate, thin shells. The animals are frequently found attached by their byssus threads to other bottom-dwelling forms such as sponges or ascidians.

The GAPING FILE (*Lima hians*; L 2 cm) and a few other species have evolved an additional function for the byssus threads. This bivalve constructs a nest with these threads. With the aid of the byssus, which in this case serves as cementing material, the bivalve glues together stones, shells, and other solid objects, thereby constructing a cave which provides protection. The actual nest is lined with more delicate fibers. The structure measures approximately 12 cm. It is a type of fortress which not only hides its occupant from sight but also protects it against the attacks of predacious fish.

In another aspect file shells (and also the oysters) already indicate a transition to the eulamellibranchs (order EULAMELLIBRANCHIA). The filaments in the gills are cross-connected (although not completely, "pseudo-lamellibranch"). In this feature these "eulamellibranchs" do not comply with the characteristic properties of the remaining filibranchs but have achieved a higher degree of evolutionary development.

The FALSE SADDLE OYSTER (*Anomia ephippium*; L 3–4 cm; see Color plate, p. 141) is a rather unusual bivalve of the eastern Atlantic and the Mediterranean, which is regularly attached to rocks or other solid substrates. This species represents its own superfamily, ANOMIACEA. Its two shells differ greatly. The animal possesses only one adductor muscle and, in contrast to the primitive position, the internal organs are shifted to one side. The short byssus becomes calcified into a bone-like structure. When the young bivalve attached itself to stones or larger pieces of shell in the upper coastal regions, it always lies on its right side. The thin, almost transparent right valve, as a rule, contours itself around all irregularities of the substratum during subsequent growth, and a deep saddle-shaped bay forms around the calcareous byssus. The left upper valve is only slightly thicker but is more convex than the lower one. The upper valve only conforms to the substrate contours at its margin. Like many of the previously discussed bivalves, the saddle oyster is endowed with eyes, but only on the left side of the first gill filament. This unilateral development is a by-product of the sessile mode of life. In the false saddle oyster, as well as in the following species, the pericardium is absent. The heart is directly embedded in the tissue of the mantle cavity.

The WINDOW-PANE OYSTER (*Placuna placenta*; shell L 6-8 cm) of the

Indo-Pacific region, is characterized by very flat shell plates which are just as thin and transparent as those of the false saddle oyster. These delicate shells were used until recently as window panes in eastern Asia, particularly China, in a manner similar to our bull's-eye glass panes. They are used extensively in the manufacture of place-mats for tables.

The OYSTERS (superfamily OSTREACEA) have a totally different way of attaching to the substratum than all the bivalves discussed up to now. The newly metamorphosed juvenile oyster at first anchors itself to a suitable rocky substrate with the help of the byssus threads, and settles there for the rest of its life in the following manner: the animal sits on the ends of its shell margins and secretes a puttylike substance from its byssus gland. Then the juvenile oyster drops on this sticky material with its left, more convex shell plate. This shell half is thereby firmly cemented to the rock. From this period onward the byssus gland and foot atrophy. In the adult oyster only their rudiments remain visible.

In concurrence with the sedentary mode of life, the oyster possesses only a single adductor muscle, the posterior one, which has been relocated centrally. The lidlike upper shell plate is characterized by a less-well-developed umbo than in the lower one. The upper plate is controlled only by an inner ligament in a toothless hinge. Like all file shells, the oysters are characterized by eulamellibranch gills.

Just like several of their relatives, the oysters occasionally also form pearls although they are of little commercial value. The great popularity of these bivalves with the ungainly shells is mainly due to their great palatability. For centuries oysters have been a favorite food item. Oyster shells have been discovered in kitchen remains from prehistoric times, for example along the eastern coast of Jutland where oysters are extinct today.

The EUROPEAN or COMMON OYSTER (*Ostrea edulis*; approximately 5-10 cm; see Color plate, p. 141) is also known in Europe as the edible or table oyster. This species is endemic along the entire European coast from Norway to the Black Sea. Massive concentrations of oysters or oyster beds are limited to extremely rocky regions, and even there ruinous exploitation has taken its toll. In less favorable regions, on the other hand, such as the German North Sea coast, oysters are found in small colonies scattered over a wide area. Since oysters tend to settle at shallow depths, it is not too difficult to cultivate these bivalves in so-called "gardens." Large artificial oyster colonies have been laboriously established in England, Holland, Belgium, and France. Oyster culturing has been particularly successful in Bordeaux (Arcachon) and Sete (Lac de Thau) in France, where tens of thousands of tons are "harvested" annually.

The Roman writer Pliny described in his writings a man by the name of Sergious Orata, who was the first person to construct oyster basins on a great scale before the Marsian War, which was approximately 100 B.C. Today oyster farmers use large wooden frames, arranged in several tiers, covered with special shingles, cockle shells, or other objects suitable for the

▷
The giant scallop of European waters (*Pecten maximus*; compare Color plate, p. 141).
Above: Mantle margin (small segment) with eyes and tentacles.
Below: One valve side is more curved than the other.

▷ ▷
A file shell (compare Color plate, p. 141) with a halo of tentacles. Within are the gills, mantle and foot.

▷ ▷ ▷
Giant clam of the southwestern Pacific nestled in coral. (*Tridacna maxima*; compare Color plate, p. 167).

The edible oysters

Oyster culturing

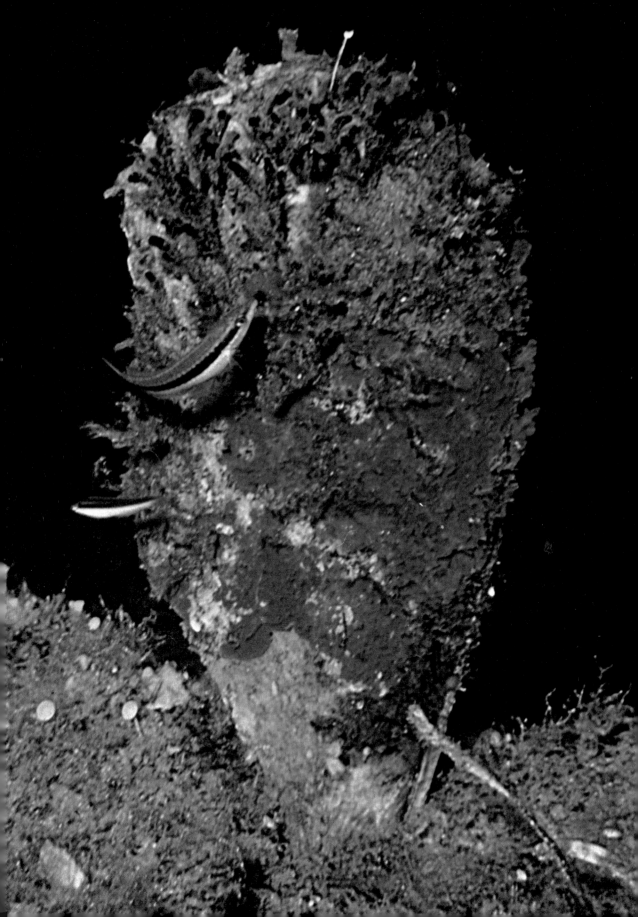

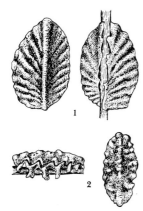

Fig. 6-14. The peculiar *Lopho folium* (1) and *Lopho frons* (2), variations of one oyster.

Control of spawning

◁
Pen shell of the Mediterranean (compare Color plate, p. 140) covered with sea growths.

attachment of the oyster larvae. As soon as the juvenile oysters have settled on this structure, it is transferred into gigantic brood basins protected from the open sea by a dam but still in contact with the ocean via sluices. The oysters grow in the shallow water of these "bivalve gardens." During the period of growth, the oysters are regularly raked, in order to prevent the animals from forming into clumps. Oysters can reach an age of twenty years, but become "edible" by the age of three or four. Before oysters are shipped off to their destinations they are put into a cleaning basin where they rid themselves of mud and excrements via the respiratory current. This then is the method whereby these living tidbits are made appetizing for the dinner table.

Aside from the European oyster and its subspecies which differ in the form and surface of the shell, the PORTUGUESE OYSTER (*Crassostrea angulata*) is also raised and eaten in Europe. Along the American seaboard the well-known AMERICAN OYSTER (*Crassostrea virginica*) is harvested commercially. Large species like the JAPANESE OYSTER (*Crassostrea gigas*; L 25 cm; see Color plate, p. 141), which are found in eastern Asia, usually inhabit shallow water. They have been introduced to the Pacific coast of the U.S.A. The small, reddish *Lopho folium* (L 5-7 cm), found in Indonesia, Australia, and the Indian Ocean, is characterized by fingerlike shell processes which hook around coral twigs or even mangrove roots and wooden branches. The smaller *Lopho frons* clings to the stems of gorgonians and is sometimes found on the beaches of the West Indies. All these species are distinguishable from the European oyster by their deeper shells. In addition they are dioecious in contrast to the European oysters, which are protandric. The European oyster also retains the eggs in its mantle cavity (brood case) until they hatch into larvae. The American oyster, on the other hand, releases its eggs into the surrounding water immediately.

The male and female gametes of the hermaphroditic European oyster do not, as a rule, mature simultaneously in the same animal. To a great extent this prevents self-fertilization. However, it is very important that the ova or sperm of the various individuals in an oyster bed mature at approximately the same time. Only in this manner can a satisfactory rate of reproduction be insured. As we now know, this synchronization in the oyster is controlled by the lunar periods. The Dutch zoologist P. Korringa noticed that swarms of oyster larvae of approximately eight days of age always appeared during specific periods in the year: "The peak period of these swarms can be expected between June 26th and July 10th of every year, and usually this event occurs approximately ten days after full moon or new moon." According to this the eggs have to be laid and fertilized within the mantle cavity two days after full moon or new moon. "Thus we understand," adds the Dutch ethologist Niko Tinbergen, "that the external factor influencing spawning is the tides. Spawning takes place during the spring tide which always follows a full or new moon. It is still unknown what specific factor triggers spawning in the oysters. There is a possibility that it could be the pressure of the water which is strongest during a spring tide. The same applies to light which at this time penetrates to the sea bottom." However, there is still another factor, as yet unknown, that influences spawning in oysters, because the animals do not lay eggs during every spring tide but only during the summer months, in the limited time span between the end of June to the beginning of July. This periodicity ensures that sufficient sperms enter the mantle cavity of spawning bivalves via the respiratory current and permit the fertilization of the eggs.

As has already been indicated, most bivalves go through a larval stage of which the trochophore type (compare Chapter 1; and Color plate, p. 26) is most frequently found. This developmental stage can be derived from the yolk larva type (see Chapter 2) with a simplified ciliated test. Almost all marine bivalves are characterized by this type of larva which is already endowed with hindgut as well as protonephridia. The veliger type larva, already described in the gastropods, is distinguishable from the trochophore, and the former occurs also in the oysters, for example. In the veliger larva, the ciliated band becomes enlarged into two lobes which are known as velum. Like all free-swimming (planktotrophic) larvae, the juvenile forms of the bivalves constitute an important part in the ecology of the ocean. They serve as food for numerous other animals. This becomes evident if one considers that the gonad of one single American oyster, for example, releases up to fifteen million eggs at one time. Other less fertile bivalve species still release several thousand eggs.

Before the larvae settle down to their sedentary mode of life, their ciliated parts degenerate. In some free-swimming larvae, for instance in the mussels, we can already detect the presence of the shells which measure barely 0.25 mm. F. Haas writes, "The developmental stage just described is the last step prior to loss of the planktonic existence and the attachment of the byssus threads. According to other observations, this stage is also characterized by the appearance of small gas and oil bubbles which greatly enhance the floating ability of the larvae and are a contributing factor to ensuring wide distribution of the species via the ocean currents."

The free-swimming bivalve *Planctomya henseni* is probably only such a transitional stage between the larva and the sedentary bivalve stage. This animal was first discovered in 1894 on only one occasion in the South Atlantic but recently this unique bivalve, with a length of only 1 cm, has been rediscovered. Its equivalve and uncalcified shell is endowed with two adductor muscles and encircles the body, which lacks siphons and a clearly defined foot. There are numerous oil droplets in the mantle which serve as a propulsive organ and flotation device. Sexual organs were not observed in this animal. It is still uncertain if in fact this is an amazingly large larval stage or the only planktotrophic bivalve.

Free-swimming adult bivalves

In the EULAMELLIBRANCHS (order Eulamellibranchia) the gill filaments are more completely united by vascular cross-connections than in the

Order: Eulamellibranchia

pseudolamellibranch gills of the file shells and oysters. The gill threads are fused (eulamellibranch) into true lamellae, which means that adjacent filaments are connected by interfilamentary tissue functions. As a rule, adductor muscles are present, although on rare occasions the anterior muscle is reduced. Often inhalent and exhalent siphons are present. There are great variations in the structure of the hinge. Four suborders are distinguished on the basis of the hinge structure and other characteristics: 1. Schizodonts (Schizodonta or Palaeoheterodonta), 2. Heterodonts (Heterodonta), 3. Adapedonts (Adapedonta), 4. Toothless bivalves (Anomalodesmacea); with an approximate total of 6,000 species.

This order encompasses the great majority of the bivalves. The SCHIZODONTS (suborder Schizodonta) include many fresh-water unionid mussels, as well as a few marine groups. The hinge consists of a central tooth on the right shell valve which interlocks with a tooth from the left shell which is of simple structure anteriorly but is forked posteriorly. One or two additional teeth may be present, sometimes in reduced form, in each valve. The nacreous layer in the shell is present. The byssus gland in the foot is absent. The mantle is devoid of siphons. There are five families with an approximate total of 1,200 species.

Only one small Australian family, represented by only a few species, is marine. These are the trigoniids (Trigoniidae), which includes the species *Neotrigonia margaritacea*. Its gill filaments are still incompletely interconnected.

All other schizondonts are strictly fresh-water forms. They are classified as one superfamily, the unionids or naiads (Unionoidea). This group is characterized by its unusual method of reproduction. Equally interesting is the method of pearl formation in these bivalves, which formerly were of economic and cultural significance.

One of the best-known species is the European PEARL MUSSEL (*Margaritifera margaritifera*; L 10–15 cm; see Color plate, p. 142). Unfortunately, exploitation and particularly pollution and channeling of the rivers have resulted in the extinction or great rarity of these bivalves. Great effort is being put forward today to save and maintain the last remains in Austria, the heath of Luneburg, and other regions in Europe. Originally the pearl mussel was widely distributed throughout central and northern Europe, Siberia, and North America, and was closely watched and managed as a pearl producer. Today, however, it has lost its great importance in pearl production. The artificial culturing of pearl oysters has taken over the market. Although pearl mussels can reach an age of sixty to eighty years, a pearl can be removed from one animal only once every six years if conditions are favorable. Despite this meager yield per animal, over 156,000 pearls were harvested and registered between the years 1814 and 1857 in the Bavarian Forest and Franconia alone. A truly good pearl is found in approximately every 2,700th bivalve. Curiously, this animal is found only in water with a very low percentage of calcium. If other conditions are

Fig. 6-15. Pearl mussel (*Margaritifera margaritifera*) with various shapes of fresh-water pearls.

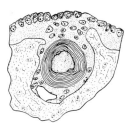

Fig. 6-16. Pearl formation within the mantle tissues.

also favorable these bivalves may settle there in great density. This is all the more astonishing since shell and pearl formation require a high concentration of calcium. Yet the bivalve possesses unusually solid shell valves which can reach a thickness of 1 cm along the anterior margin.

However, the chemical action of the surrounding water has a strong eroding effect on the shell if the protecting periostracum has been injured. The constant threat of calcium leaching away with the water current is particularly imminent in these bivalves. They protrude out of the river bottom and point their slightly gaping valves with the posterior mantle margin against the current. For this reason numerous individuals of these species possess greatly disfigured shell valves particularly along the umbo. The valves can only be protectively strengthened from the inside.

The well-known RIVER MUSSELS (family Unionidae) are not subjected to quite the same unfavorable conditions as the pearl mussels. They are found more frequently in European waters. Wherever fresh water has not been polluted or poisoned by industrial or household effluents one can still regularly find the PAINTER'S MUSSEL (*Unio pictorum*; L 10 cm; see Color plate, p. 142) in rivers, creeks, and lakes. The shells of these bivalves are used by painters as mixing pallets for water colors. Another species of this same genus, *Unio crassus*, only inhabits flowing water. Since each of these species can occur in discontiguous areas and in habitats which differ from each other, several geographically isolated subspecies have originated. This serves to demonstrate the great adaptive and transformative abilities of this bivalve.

Unio crassus is similar to the European SWAN MUSSEL (*Anodonta cygnaea*; L 7–20 cm; see Color plates, pp. 65/66, and 142). This species prefers more stagnant waters and accordingly does not possess teeth in the hinge. (*Anodonta* comes from the Greek ἀν—not, without, and ὀδοῦς—tooth; meaning "without teeth.") The small planktonic particles on which this bivalve feeds do not circulate past it with the water current. Instead the animal disturbs the mud on the bottom with trembling motions. The stirred-up bottom sediment is sucked up into the mantle cavity and the food particles, including bottom-dwelling microscopic organisms, are filtered out. Thus the bivalves plow up the substratum, occasionally leaving grooves up to one meter long. *Anodonta cygnea* shows an even greater tendency than *Unio pictorum* to form geographically isolated subspecies. In central Europe alone there were over eight different forms. It was shown, however, that all these variations could be condensed to two well-defined species. Interestingly, *Anodonta cygnea* and *Pseudoanodonta complanta* are able to perceive motion. More exactly, they are capable of detecting shadows that pass over them. If the edge of a shadow moves 3 to 10 mm per second and if the degree of light is only 7.7 percent, these two *Anodonta* species respond by folding their tentacles. The response increases as the shadow darkens. The ability to perceive these stimuli is related to the sensitivity of certain sensory cells which are evenly distrib-

▷
Heterodont Bivalves (one-half natural size): Tridacnidae: 1., 2. *Tridacna crocea*, (compare Color plate, p. 161), 3. *Hippopus hippopus*.

Painter's mussel

Detection of motion

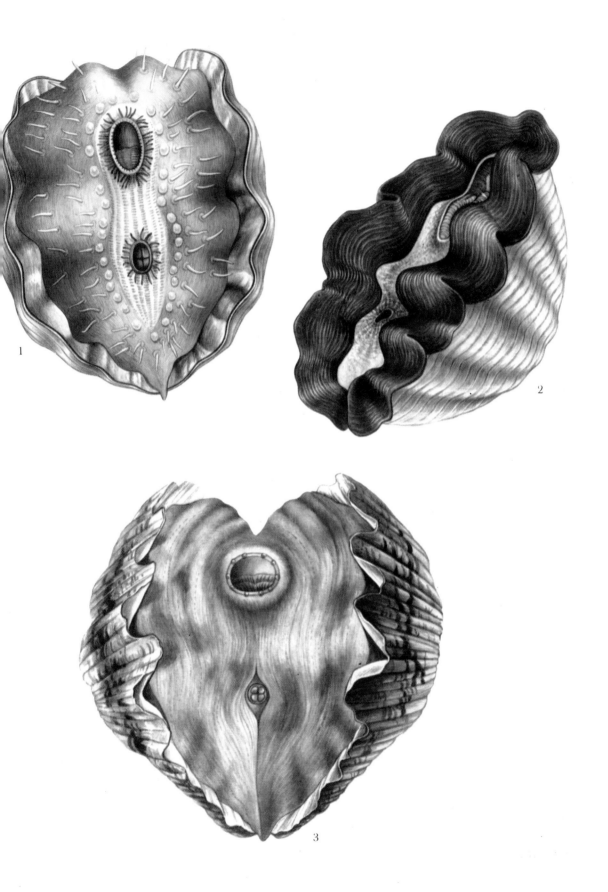

MILLA

◁
European Heterodont
Bivalves (natural size):
1. *Chama gryphoides*;
2. *Cardium fasciatum*;
3. Edible cockle (*Cerasto-derma edule*); 4. Warty
Venus (*Venus verrucosa*);
5. *Callista chione*; 6. Baltic
tellin (*Macoma balthica*);
7. Peppery furrow shell
(*Scrobicularia plana*).
Adapedont Bivalves:
8. Great trough shell
(*Mactra stultorum*).

Larval parasitism

uted over the surface of the mantle margin. This response is advanta-geous to the bivalve since a shadow often indicates the approach of an enemy. True "eye spots" are found on the posterior mantle lobes of the North American *Lampsilis* bivalves and their close relatives (subfamily Lampsilinae). In these animals, for example *Lampsilis ventricosa* and *Lampsilis fasciola*, the shells of the males and females differ. In addition the females are characterized by a unique aeration structure. The mantle flaps at the posterior end are enlarged and often are reminiscent of the outline of a fish. These flaps beat rhythmically and thereby fan fresh water over the brood which is sheltered in the mother's gills.

In the gastropods we already observed that the development of the fresh-water and terrestrial forms often followed different evolutionary lines than those of species that were strictly marine. A similar situation is found in the fresh-water bivalves. The development from the egg to the mature adult deviates greatly from the developmental process in the marine bivalves. The larvae of the NAIADS (Unionacea) parasitize fish. In their larval morphology the fresh-water forms have little in common with the marine forms. Free-swimming larvae would easily drift away and would be destroyed in the water current. For this reason fresh-water bivalves do not lay eggs or larvae into the surrounding water. During the summer the unfertilized eggs find their way into the spaces (ostia) between the gill filaments, which as a result often become swollen. The sperms enter the mantle cavity via the inhalent opening, and fertilization takes place in the gills. Here also the larvae—approximately 300,000 per female—undergo their development until they are released in the following spring. At a superficial glance the larvae resemble small bivalves, but closer examination reveals that their structure is different. Each of these larvae, known as glochidia (singular: glochidium), is characterized by a pair of triangular or oval shells. The exposed margin of the shells in the larvae of the *Anodonta* and *Unio pictorum* (but not those in their related forms) bears a toothed hook. The interior of the shells is characterized by a number of sensory bristles and a long, sticky, adhesive byssus thread. The early beginnings of the other organs of the adult bivalve are barely recognizable. When the glochidia larvae are expelled through the maternal exhalent opening, they form small matted clumps on the mud surface. At the approach of a fish, the snapping shells hook into the skin of the fish like a pair of pinchers, and thereby they become attached to their host. The glochidia of *Margaritifera margaritifera*, on the other hand, are swept into the fish's gills by the fish's respiratory current. The glochidia settle on the fish's gills, which they parasitize. The small wound caused by the hooks of the glochidia larvae quickly heals in the fish. The host's tissue grows around the larva and encapsules it. During this stage the glochidium grows into a true little bivalve. It grows a completely new pair of shells. After two to ten weeks the skin capsule bursts open. The fish, feeling an itching stimulus, rubs itself on this spot against stones or plants and thereby rids itself of the bivalve, which drops to the bottom and commences the adult phase of its life.

The Unionidae are closely related with the mutelids (family Mutelidae), which also includes the South American species *Anodontites patagonicus*. These bivalves have developed a different kind of larva. Their larval form is known as a lasidium. This larva is more reminiscent of a rotiferan (see Vol. I) than a mollusk, for its anterior body section is pear-shaped and ciliated; its central section is endowed with a thin shell. The tail is forked and bears stiff bristles. Lasidia also parasitize fish before they metamorphose and start the free-living bivalve stage.

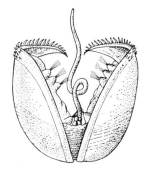

Fig. 6-17. Glochidium larva of *Anodonta*.

The fresh-water bivalves are of importance in American and European waterways. Usually the filter activity of these bivalves is greatly underestimated: each individual bivalve circulates over forty liters of water per hour through its filtering system. Destroying the bivalves in a body of water means upsetting the biological balance and destroying the ecosystem. Even human living conditions may be influenced since we are dependent on the cleanliness and quality of our water supply. The shells are also used to monitor the amount of radiation leakage into streams and rivers from atomic energy plants.

The HETERODONTS (Heterodonta) are the second suborder within the order Eulamellibranchia. Their hinge consists of a few alternating interlocking main teeth. Often up to four anterior and posterior ledgelike lateral teeth are also present. The shells are usually devoid of the nacreous layer. The byssus gland in the foot may be absent or present. Inhalent and exhalent siphons are frequently present. There are eighteen superfamilies with over forty families and approximately 3,000 species.

Suborder: Heterodonta

The heterodonts include the most species within the Bivalvia. It is rather difficult to describe the homogeneity of all these many forms on the basis of a few easily recognizable characteristics. The group's great diversity meant that members formerly were put in many different classifications. The morphology of the hinge alone is a relatively useful characteristic and this is the reason why more recently the group has been named Heterodonta. In the more pronounced case one shell valve, usually the left, bears two hinge teeth which either hook around one central tooth of the other valve or interlock with three teeth from the opposite side.

The equivalve, solid shells of the ASTARTE BIVALVES (family Astartidae) are characterized by a heavy periostracum. Representatives of this family usually occur in colder oceans. The GROOVED ASTARTE (*Astarte sulcata*) is found along European coast lines. This species has twenty-four to forty dark-brown concentric ridges, arranged around the umbo. The shell measures only a few centimeters and is found at water depths of ten meters. In contrast, the NORTHERN ASTARTE (*Astarte borealis*; see Color plate, p. 142), which occurs quite frequently in the North Sea and New England, is found at moderately deep muddy bottoms. The northern astarte and two species of tellins form the characteristic fauna of the soft bottoms of the Baltic Sea.

Some members of the carditids (family Carditidae) belong to the few marine bivalves that practice brood care. Out of this group the South African *Thecalia concamerata* and the Californian *Milneria minima* (L in each barely more than 1 cm) have even developed special brood chambers. In the female of both of these species a fold or depression develops in the shell along its lower margin. This space is sealed off by the mantle and the progeny develops within it.

The brood chambers of the fresh-water PEA MUSSELS (family Pisidiidae) fulfill a similar function but here, as in most bivalves that practice brood care, the chambers are double-walled sacs situated within the gill spaces. The Pisidiidae are fresh-water forms, and, just as for the Unionidae, the habitat of flowing water necessitated protection for the young brood. Only a very small number of young bivalves develop within the tiny membranous brood pouch where they are protected and nourished by the cells of the maternal body wall. In this case the larval stage is omitted and the eggs develop directly into complete bivalves. Only when the juvenile bivalves are fully developed do they leave the brood chamber. In all Pisidiidae the inhalent and exhalent openings, or at least one, are extended into siphons.

The European species RIVER ORB MUSSEL (*Sphaerium rivicola*; L 2 cm) and the more frequently occurring COMMON ORB MUSSEL (*Sphaerium corneum*; L just over 1 cm; see Color plate, p. 142), as well as all their immediate related forms from the genus *Sphaerium*, produce only ten to sixteen fully developed juveniles. These, however, may be quite large when they "hatch." *Sphaerium corneum* is widely distributed throughout Eurasia. This bivalve not only inhabits creek bottoms but also likes to climb about on various aquatic plants. *S. corneum* slowly creeps along with its foot extended very much in the manner of a looper caterpillar. The bivalve anchors its posterior end with the sticky byssus, then extends the tip of the foot widely and again attaches itself with the byssus substance. The somewhat smaller *Sphaerium lacustre*, which inhabits ponds, pools, and other stagnant waters, moves in a similar manner.

Fingernail clams

The smallest bivalves known are found in the genus of FINGERNAIL CLAMS (*Pisidium*). Some of these forms can hang suspended from the water surface and creep along it. In Europe and North America alone close to forty species have been identified. Some of these are difficult to distinguish from one another. One of the most common species is the COMMON FINGERNAIL CLAM (*Pisidium casertanum*; L 4 mm) which prefers slow-flowing waters. *Pisidium* bivalves like to settle in unusual habitats where other mollusks are very rare. They have been found at 40 m under the water surface of lakes, or at heights of over 2,500 m in Alpine ponds which remain frozen for most of the year. They also inhabit prairie lakes up to the Arctic Circle. It is highly probable that not all species and their ecological niches have been discovered. The PEA MUSSEL (*Pisidium amnicum*; see Color plate, p. 142) lives on gravel in creeks and rivers, while the remaining species prefer muddy bottoms. Since *P. amnicum* is "gigantic" among the gem clams—measuring 1 cm—it is relatively easy to find. The smallest known species is *Pisidium torquatum*, which barely measures 2 mm. These two species differ greatly not only in size but also in their respiratory apparatus. *Pisidium torquatum* possesses only one gill and only the exhalent opening is extended as a siphon. *Pisidium amnicum*, on the other hand, is characterized by both an inhalent and an exhalent siphon, and two gill leaflets on each side. Because of these differences these two species are classified as two subgenera.

Fig. 6-18. Distribution of the heart cockle (*Glossus humanus*).

Among the KELLIELLIDS (family Kelliellidae), *Kelliella militaris* and a few of their relatives also measure only a few millimeters. *K. militaris* is not a fresh-water dweller but is found at great depths in the Atlantic and Mediterranean. Its small, roundish shells are thin and fragile, as are most deep-sea bivalves. However, this statement does not apply to the HEART COCKLE (*Glossus humanus*; L 6–8 cm) which belongs to the same related group. Its shell valves are solid and are particularly conspicuous because of the firm, coiled umbo which gives each valve the appearance of a weakly curled gastropod shell. When observed from the frontal view, the animal has the appearance of a heart. This bivalve is not common. It is found on muddy sea bottoms at great depths. Its range of distribution extends from the Mediterranean and the adjoining coastal region of the Atlantic to Iceland and Norway. The heavy periostracum gives the usually yellow shells a brown coloration. The edge of the mantle is not tubular. The edge of the foot is hatchet-shaped and bears a byssus gland.

The OCEAN QUAHOG (*Arctica islandica*; see Color plate, p. 142) is very similar to *Glossus humanus* although its umbos are not curled. This species is also characterized by a heavy periostracum, and lacks an inhalent and exhalent siphon. The ocean quahog is about the same size as the *G. humanus*, but the two species differ considerably in their internal morphology. *A. islandica* is classified in its own family. This bivalve is found only in the North Atlantic, the Baltic Sea, and the northeastern U.S.A., in which it forms a major part of the animal community in the upper regions of the soft bottom. The calcareous boring TRAPEZIID CLAMS (family Trapeziidae) belong to the same superfamily as the ocean quahog. Trapeziid clams are distributed along cliffs and reefs in the warmer oceans. Like the date shells (see Chapter 5), the trapeziid bivalves are capable of secreting "acids" which aid them in chemically boring through calcareous rocks. *Corolliophaga lithophagella*, as well as other forms, also belongs to this family. This bivalve occurs in the Mediterranean, where it is found in the cracks of cliffs and in shells which have been vacated by their original occupants.

The ocean quahog

The boring Trapeziidae

The GAIMARDIIDS (family Gaimardiidae) are an unusual family. They are represented by only a few forms which are found in eastern Asia and

Australia, the Pacific, and the antarctic region. The gaimardiids are among the few marine bivalves that practice brood care. Their eggs develop within the gills, as exemplified by the antarctic species *Gaimardia trapezina*. Peculiarly, the mantle margin of this bivalve is fused except for three openings. The foot moves very little since the animals are anchored by the byssus to drifting algae or crabs. Formerly, before the anatomy of the animal's body was actually known, the Juliidae (see Chapter 5) which belong to the sacoglossans (Sacoglossa; see Chapter 5) were classified as immediate relations to this bivalve family.

After discussing the various aspects of bivalves living in a fresh-water environment and the brood care method of reproduction, it may seem surprising that there are fresh-water forms that do not practice brood care. These bivalves belong to the family ZEBRA MUSSELS (Dreissenidae). The history of the ZEBRA MUSSEL (*Dreissena polymorpha*; see Color plate, p. 142) may serve to exemplify this conspicuous exception to a rule which can usually be applied. These dreissenids are actually marine inhabitants which have invaded inland waters only during recent times. The rapid distribution of this bivalve was favored by the fact that the animal maintained the veliger larva state (veliger; see Color plate, p. 26) in the fresh-water habit although this is a characteristic feature for so many marine bivalves. The dreissenids often attach themselves to boats with their byssus threads, which often results in the transfer of these bivalves over vast distances. Finally, the dreissenids have a far higher tolerance to different degrees of salinity than do other bivalves. By 1824 they were found in London and at the mouth of the Rhine River. Subsequently these bivalves have spread throughout all central European river systems. Today, this animal with its two short siphons is found almost everywhere in the river systems of the Danube, Rhine, Weser, and Elbe. The bivalve adheres to rocks, shells, woodwork, and even to snails and crabs.

In Europe the LUCINES (Lucinidae) are represented by species that inhabit soft substrates. They lack well-developed siphons. Members of this family lack the outer gill filament. They possess only one gill filament on each side (anatomically only one half of a gill). *Divaricella divaricata*, which measures only 1 cm, and the larger *Loripes lacteus* possess a long, wormlike foot which protrudes from the round to triangular shells while the animal is burrowing. *D. divaricata* is distributed along the northern coasts while *L. lacteus* is more frequently found in the Mediterranean. The still larger species *Lucina borealis* (L 3 cm) is limited to the northern oceans.

The MONTACUTIDS (family Montacutidae) are also characterized by a single gill filament on each side of the body. Numerous members of this family have formed feeding associations (commensalism) with other animals. *Mysella bidentata*, which occurs in the Atlantic and the Mediterranean, occasionally forms a feeding association with various animals. *Montacuta substriata* and *Montacuta ferruginosa*, which occur frequently in the North Atlantic (L of both 4–5 cm), on the other hand, regularly

Fig. 6-19. Present distribution of *Dreissena polymorpha*.

Commensalism

attach themselves to the echinoid species *Spatangus purpureus* and *Echino-cardium cordatum*. The montacutids *Mysella donacina* and *Devonia perrieri*, which inhabit the French and English coasts, live on the sea cucumber species *Leptosynapta inhaerens*. The Japanese species *Devonia semperi* also attaches to a sea cucumber species (*Protankyra bidentata*, a species which is not mentioned in the sea cucumber chapter, Chapter 12). The large, oval, and flat foot of this bivalve is separated from the animal's body by a stem. The semi-parasitic life habits of this bivalve have resulted in a decrease of their very fragile shells. The broad margins of the mantle entirely cover the shell plates. This developmental trend is most advanced in *Entovalva mirabilis*, found along Africa's eastern coast, and closely related forms. In these forms the shells are not visible externally. This unusual structure is an adaptation to the animal's life habits, for *Entovalva* penetrates into the anterior portion of the digestive tract of the sea cucumber (*Patinapta crosslandi, Synapta coplax*), where it lives a semi-parasitic existence.

Fig. 6-20. *Devonia semperi;* the foot protrudes out front.

To ensure propagation of the species, the offspring are retained in a brood chamber up to the trochophore stage. The brood chamber originated from a fusion of the posterior mantle margins. In addition to sea cucumbers, many other animals also serve as hosts for these bivalves. In this context the chiton *Ischnochiton* (see Chapter 2) and the bivalve *Montacuta oblonga* have already been mentioned. It lives in a pallial groove or on the shell plate of this primitive mollusk. In contrast, another species of this same family, *Litigella glabra*, from the French coast, attaches to the sipunculid (*Sipunculus nudus*; see Vol. I). One could go on citing examples of such feeding associations. The LEPTONIDS (family Leptonidae), which are closely related to the montaculids, are characterized by similar commensal relationships, although they tend to prefer crabs. One species of this group, *Galeomma turtoni*, distributed along the west and south European coasts, is found on sandy substrates or on sponges (see Vol. I) and tunicates (see Chapter 18). The leptonids as well as the montacutids possess an anterior inhalent opening. In some species, for example *Kellia suborbicularis*, a short siphon has developed at this location.

Family:
Leptonidae

Because of their sessile mode of life, the JEWEL BOX CLAMS (family Chamidae) are reminiscent of oysters. Most species of this family have become extinct. The shell valves of this bivalve are differently shaped and are not joined by a hinge. As with the Spondylidae the upper, usually the right, valve of the jewel box is sculptured with numerous scalelike foliations and spines. The whole animal is irregularly shaped and has little resemblance to an ordinary bivalve. Because of their highly irregular shell surfaces, it is almost impossible to differentiate the jewel box shells from the rocky substratum to which they adhere.

Jewel box clams

The characteristics of this family are well exemplified by *Chama gryphoides* (see Color plate, p. 168) which is distributed from the Mediterranean to the Azores. Their thick, long shell, which measures approximately 2 cm, is structured with an irregular, rough exterior which seems to blend the animal into its background. The animals are difficult to find although they occur quite regularly along coastal cliffs in moderately shallow water. The left but deeper valve adheres to the substratum like a bowl. The flat right valve fits on top of the left one like a lid. The upper valve is endowed with thorns and grooves which are concentrically arranged around the umbo. *Chama gryphina*, which is very similar to the *C. gryphoides* but is slightly larger, is easily distinguished from *C. gryphoides* by its more scaly exterior surface. *C. gryphina* is also found in the Mediterranean.

The COCKLES (family Cardiidae) are worldwide in distribution and contain about 200 species. The COMMON EDIBLE COCKLE (*Cardium edule*, also known as *Cerastoderma edule*; L 4–5 cm; see Color plate, p. 168) is one of the most common and best-known bivalves in Europe. Shells of this form have often been found in superabundance swept up on sandy beaches from Iceland to western Africa. The animals are characterized by whitish-yellow shells with regularly spaced ribs. The shells are overlapped by the tentacles protruding from the mantle margin. There is an exhalent and inhalent siphon. The long foot is bent and tapers to a point. It enables this bivalve to move in a peculiar way. The animal extends the foot as far out of the shell as possible, up to over 5 cm, and gropes for resistance from any object. When the bent portion of the foot has located a stone or another resisting object, the entire foot is suddenly jerked into a straight position which results in a push which flings the bivalve a distance of over 50 cm. Dugout animals use similar jerky foot motions to burrow themselves rapidly into the sand. Many other related species are also capable of using these motions. Some cockles possess eyes. The OBLONG COCKLE (*Laevicardium oblongum*) and *Cardium tuberculatum* are only capable of perceiving motion, just as in the *Anodonta* shells, because they lack true visual organs. *Cardium muticum* has eyes which are around the siphonal openings. The eyes are closed and have a sunken retina and have inverted sensory cells located at the end of the pallial tentacles. In *Cardium edule* the visual organs, on the other hand, are only slightly developed.

Cardium aculeatum (L 6–7 cm or more) is found along the western and southern European coasts. This bivalve is easily distinguished from related forms by its size and significantly heavier shell spines. The two smaller European species, *Cardium fasciatum* (see Color plate, p. 168) and the LITTLE COCKLE (*Parvicardium exiguum*), can easily be differentiated. The former species is characterized by beautiful dark brown bands; the latter possesses a triangular shell. Both species measure only a little over 1 cm. Of all the *Cardium* species found in the Mediterranean, *Sphaerocardium paucicostatum* is most difficult to recognize. In size it is similar to the *Cardium edule*, but its shell shape is more reminiscent of the *Cardium tuberculatum*. *Sphaerocardium paucicostatum* is more frequently found on muddy substratum while all other species prefer sandy substrates. The beautiful and tasty cockles are popular for good reason. They are very easy to gather. The natives in

Fig. 6-21. The Pacific heart cockle, *Corculum cardissa*, has greatly compressed valves.

the South Seas make necklaces from these shells. Among the South Sea species, the HEART COCKLE (*Corculum cardissa*; L 6–7 cm) is noteworthy because its shell valves are strongly compressed from front to back. A sharp keel is located along its midline. This gives the impression that the bivalves are "upside down." This bivalve lives exposed on the surface of shallow, flat reefs.

The TRIDACNIDS (family Tridacnidae; see Color plate, p. 161) show little external resemblance to the cockles. However, the internal morphology of both groups indicates a close relationship. Extinct forms verify the hypothesis of a close relationship. The giant bivalves inhabit the Indian and Pacific Oceans. One of these is the GIANT CLAM (*Tridacna gigas*) which has become popular through a variety of fictitious stories. The morphology of this gigantic bivalve is rather unusual. Like many other bivalves, they sit on the substratum with their reduced foot, gills, and mouth opening turned towards the substrate. However, the entire mantle and shell are turned about by 180° so that the dorsal side of the animal is facing the exposed shell opening and the ventral side is facing the hinge. This shift occurred as a result of the animal's permanent sessile mode of life. In the large forms this is caused by the weight of the shells alone. For example, the giant clam, which often has a shell that measures more than 1 m in length, can reach a weight of over 200 kg and is therefore unable to move. These super-large shell valves were formely used frequently as wash or baptismal basins. Smaller species, however, such as the *Tridacna crocea* (see Color plate, p. 167), which measures approximately 10 cm, are capable of boring, umbo facing frontward, into the porous calcium substrate by shell motions. Most species live embedded among corals. Yet not all giant bivalves are reef dwellers. *Hippopus hippopus* (L 25 cm; see Color plate, p. 167), for example, is usually found on sandy substrates.

In science fiction the giant bivalves are often described as "sinister animals." In reality, however, they are totally harmless filter feeders which live on microscopic organisms. The giant bivalves have developed a special feature in this connection. The exposed mantle margin, often brilliantly colored, houses a number of small unicellular algae (*Zooxanthellae*). These organisms and the bivalve form a feeding association which is of mutual benefit. The algae obtain their food substances from the tridacnid clam, and they, in return, provide an additional supply of oxygen to the bivalve. Like all green plants, these algae produce oxygen as a by-product when they turn carbon dioxide and water into sugar by trapping the energy of the sun. This symbiotic relationship explains the fact that tridacnid clams, as well as corals that also form symbiotic associations with algae, are almost always found in the upper levels of water that receive the most sunlight. When the algae reproduce excessively, their cells enter the digestive system of the bivalve and their numbers are thereby decreased.

The danger of the notorious giant clam is not related in any way to

Giant *Tridacna* clam

▷
Adapedont Bivalves:
1. Pod razor (*Ensis siliqua*; L 12 cm); 2. *Solen pellucidus* (L 8 cm); 3. Soft-shelled clam (*Mya arenaria*; L 8 cm); 4. Basket clam (*Corbula gibba*; L 15 mm); 5. Red-nose clam (*Hiatella arctica*; L 4 cm); 6. Flask shell (*Gastrochaena dubia*; L 3 cm); 7. Norwegian shipworm (*Teredo norvegica*; L 15 cm); 8. Frilled piddock (*Zirfaea crispata*; L 6 cm); 9. White piddock (*Barnea candida*; L 7 cm); 10. Shipworm (*Teredo navalis*; L 20 cm).

Symbiosis with algae

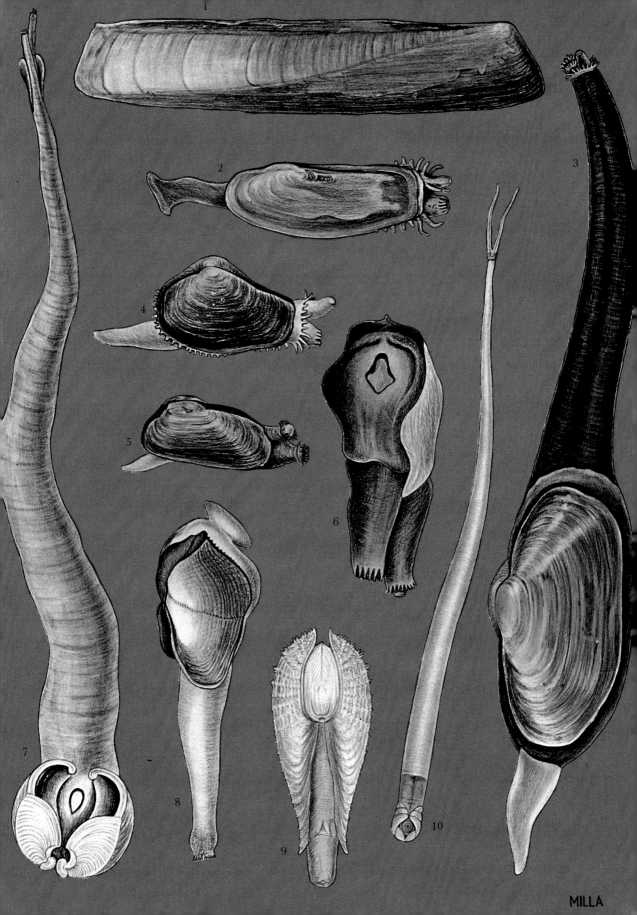

MILLA

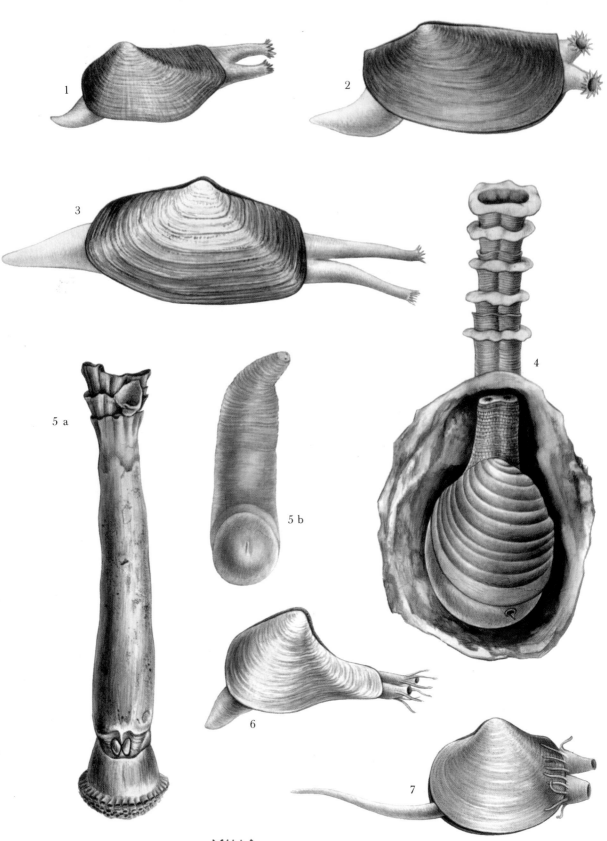

MILLA

aggressive behavior but rather is a by-product of the innate protective response of the animal. At the sign of disturbance the shell valves close up. The giant clam is potentially dangerous to man because the bivalve is extremely well camouflaged, and because of the tremendous strength of the single adductor muscle (which is considered a delicacy). Once an object is caught between the two valves it can only be released again by either severing the adductor muscle or breaking the shell. Hans Hass was able to demonstrate these properties of the giant clam on his exploration trip with the *Xarifa*, although no authentic cases are known of people having been trapped and drowned by a giant clam.

The largest pearl

In the early 1930's a crude pearl weighing 7 kg and measuring 23 by 15 by 14 cm was extracted from a tridacnid clam in the Philippines.

The venerid clams

The VENUS CLAMS (family Veneriidae) are represented by some noteworthy forms in Europe and the Americas. They are quite frequently encountered on sandy and muddy ground. The Venus clams are able to jump a short distance, with the aid of their foot. In the Venus clams the siphons are much longer and are partially fused, and externally these shells are also quite distinct from the cockles. Their shells are almost always weakly ribbed and lack spines, and they are often decorated with dark brown bands. The WARTY VENUS (*Venus verrucosa*; L 4 cm; see Color plate, p. 168) is one of the few ribbed species. In certain locations this species is extremely abundant, and is offered for sale in fish markets. In contrast to the cockles, the European Venus clams are almost always characterized by distinct lamellar ribs which are arranged in concentric rings. The light brown coloration of the shell valves is often complimented by three or four interrupted bands which cross over the concentric rings from the inside to the outside. *Venus gallina* is somewhat smaller. It is found in great numbers in the North Sea, where it is the characteristic species of an entire animal ecosystem on the sandy ground at moderate depths. The edible *Callista chione* (L 8 cm; see Color plate, p. 168) is equally abundant in the Mediterranean.

In contrast to the true Venus clams, the CARPET SHELL (*Venerupis perforans*) is clearly more elongated and therefore its form is reminiscent of a blunt arcid bivalve.

The American QUAHOG (*Mercenaria mercenaria*) is another member of the Venus clams. It was introduced to northwestern France from the eastern coast of North America. The Algonquin tribes, which formerly inhabited what is now the eastern U.S.A., used to make beads out of the shells, which they pierced in the middle and strung together into necklaces. These strings of shell beads also served as money which was known by the Indian name of *wampumpeag*. The Indians also strung these shell beads into very intricately designed belts known as wampums, which were used as tokens to bind treaties.

◁

Anomalodesmacea: 1. Glassy lyonsia (*Lyonsia hyalina*); 2. Pandora clam (*Pandora inaequivalvis*); 3. *Thracia pubescens* (L 3 cm); 4. *Clavagella aperta*; 5. Watering-pot shell (*Penicillus vaginiferus*; L 12 cm), (a) Calcareous tube, (b) Soft body; 6. *Cuspidaria cuspidata* (L 18 mm); 7. *Poromya granulata*.

At the turn of the 19th Century the FALSE ANGEL WING (*Petricola pholadiformis*) was also introduced into Europe from the western part of the Atlantic. From the human point of view this bivalve is a nuisance because it uses its anterior toothed shell margins to drill through wood, peat, and clay (without the aid of chemical secretions). This borer also infests small reefs composed of the many sandy tubes of the polychaete (*Sabellaria spinulosa*; see Vol. I) which often occur in masses. The European *Petricola lithophaga* is not even half as large as the American false angel wing. This borer prefers calcareous rock and various calcareous shells. In the Mediterranean it frequently inhabits spiny oysters and the bittersweet bivalves.

Just as the quahog was used as a monetary token by the Algonquins on the American East Coast, the PISMO CLAM (*Tivela stultorum*) was used by the Indians on the West Coast.

The TELLINS (family Tellinidae) are represented in Europe by the THIN TELLIN (*Tellina tenuis*; L 2 cm). Their shells are characterized by a beautiful reddish coloration which is broken by concentrically arranged rings. The BALTIC TELLIN (*Macoma balthica*; see Color plate, p. 168) is a related species found in Europe and the eastern U.S.A. This species and the edible cockle form the major components of a life community inhabiting sandy flats at less than 15 m depth. *Gastrana fragilis* is found less frequently. It is characterized by a white to faintly pink coloration.

Just like the cockles and Venus clams, the tellins are also represented by a great variety of species. Individuals of one species are often highly concentrated in certain locations. Unlike most other bivalves, tellins are not filter feeders. They possess extremely long siphons which they use to suck up food particles from the surface of the sea bottom. This "vacuum cleaner" mode of feeding is well exemplified in the roundish PEPPERY FURROW SHELL (*Scrobicularia plana*; L 4–5 cm; see Color plate, p. 168). This tellin is found quite frequently on all sandy and muddy sea bottoms along the European coasts. In southern Europe they are sold in the fish markets. *Donax trunculus* (family Donacidae; L 3 cm) occurs in great masses on alluvial land, for instance in the nature sanctuary at Camargue in southern France. Here the animals fill up irrigation ditches and lagoons. These beautiful brownish-purple-striped bivalves are not only found in the Mediterranean region but enjoy a range of distribution extending from northern Europe to western Africa. The ventral margin of the *D. trunculus* is serrated. Despite the bivalve's frequent presence in muddy substrate, it is frequently eaten raw just like oysters and other bivalves by Africans. In Florida the COQUINA (*Donax variabilis*; L 1 cm) is abundant on certain beaches, and is sometimes used to make clam broth.

The ADAPEDONTA is the third suborder within the order Eulamelli-branchia. The hinge is weakly developed. Central teeth may be absent or present. The lateral teeth are missing in the Mactridae. They have no nacreous layer. Their foot may or may not have a byssus gland. The inhalent and exhalent siphons are always present. In some families the siphons are very long; in the Solenidae they are short and fused at the base,

Suborder: Adapedonta

and are partly surrounded by a calcareous layer which also encloses the shell. There are eleven families within four superfamilies, and approximately 1,000 species.

Razor clams

Despite their barely visible siphons, the RAZOR SHELLS (family Solenidae; see Color plate, p. 177) can easily be distinguished from all other bivalves. Their characteristic feature is a long, narrow shell which is reminiscent of the sheath of a knife or a sword. In Europe *Solen vagina* (L 12 cm), with its drawn-out shell, is the most familiar species of the razor shells. Its range of distribution extends from England to the Black Sea. The slightly curved *Ensis ensis* and the POD RAZOR (*Ensis siliqua*; L up to 20 cm; see Color plate, p. 177) also enjoy a wide range. All three species are regularly found on sandy flats, where they inhabit deep tubes. The bivalve's anterior section—not the ventral side—with the extended foot is actually quite similar to a knife or sword handle with a short blade. The shells are usually of a whitish to pinkish coloration. *E. siliqua* in particular is characterized by numerous violet stripes on its ventral side. *Solen vagina* is able to move well on solid substratum. When the bivalve, with its siphonal openings closed, jerks back its foot, a strong pressure builds up in the water-filled mantle cavity. As a result, the water is suddenly expelled and the animal is propelled a distance of from 30 to 60 cm by force of repulsion.

Mactrid clams

The GREAT TROUGH SHELL (*Mactra stultorum*; L 6 cm; see Color plate, p. 168) is mainly found on sandy ground as is the razor shell, but it is sometimes also present in muddy subsoil. The shell of the great trough shell is characterized by a grayish-yellow valve with light stripes which radiate out from the umbo. Therefore, this bivalve is also known as the rayed basket. The roundish, triangular shells with the delicate purple coloration on the inside are not uncommonly washed up on the coasts of Europe. Unlike all other Adapedonta, the TROUGH SHELLS (family Mactridae) are endowed with lamellarlike lateral teeth in the hinge. The genus *Spisula* is characterized by a triangular groove which contains the hinge band which joins the two shell valves. Unlike the great trough shell, *Spisula subtruncata* (L 2–3 cm) inhabits the deeper sand flats, where it reaches high concentrations in certain localities. Particularly in the North Sea these mass concentrations of trough shells (as well as tellins) serve as abundant food supplies for certain species of bottom-dwelling fish, for example the flounders. Indirectly, these bivalves are very important for the fish industry. In the eastern U.S.A., the SURF CLAM (*Spisula solidissima*) is the major commercial source of canned clams.

Red-nose clam

The RED-NOSE CLAM (*Hiatella arctica*; L 1 cm; see Color plate, p. 177) is widely distributed in the North Atlantic, the Mediterranean, and along Africa's western coast. Unlike other borer bivalves, this species only rarely drills its own place of residence with the mechanical action of its shells, preferring to utilize already existing cracks and holes. This particular bivalve also penetrates into shelled crustaceans which belong to the

Cirripedian (see Vol. I), roots, sponges (*Geodia*; see Vol. I), and empty bivalve shells. The red-nose clam's hinge teeth are only weakly developed. These, however, are totally missing in the closely related *Hiatella rugosa*. This species is somewhat larger than the red-nose clam and some zoologists even consider it as a variety of the red-nose clam. *Hiatella rugosa* is distributed over the entire world. It only inhabits rocks. Its long and fused inhalent and exhalent siphons are fused and covered by a cubicle which protects it during its burrowing activities. The gills extend into the lower portions of the siphons.

The SAND GAPER or SOFT-SHELLED CLAM (*Mya arenaria*; L 5–12 cm; see Color plate, p. 177), frequently found in New England, the North Atlantic, the North Sea, and Baltic Sea, is externally very similar to the red-nose clam, although it lacks the extended gills. The soft-shelled clam is found burrowed into sandy flats of moderate depth. Only the fused inhalent and exhalent siphons are in contact with the upper surface. These siphons can be extended to several times the length of the shell. Consequently, this bivalve can be buried in the sand at a depth of more than 30 cm, and thereby be safely hidden without suffering a food or oxygen shortage. They are a major source of clams in the eastern U.S.A.

Soft-shelled clam

The BASKET CLAMS (genus *Corbula*, family Corbulidae; L 1 cm) are closely related to the soft-shelled clam. The former genus includes the COMMON BASKET CLAM (*Corbula gbiba*; see Color plate, p. 177), found along the European coasts, and the MEDITERRANEAN BASKET CLAM (*Corbula mediterranea*). Unlike the soft-shelled clams, the basket shells possess only short siphons which lack the protective outer layer. The right valve wraps around the smaller left one. The byssus gland is well developed in the basket shells, and the animals anchor themselves with the byssus threads.

The FLASK SHELL (*Gastrochaena dubia*; L 1 cm; see Color plate, p. 177) is a totally different ecological type. Just like the date shell and *Coralliophaga*, the flask shells utilize acid-gland secretions to drill through rocks, stones, or thick-walled shells. The animal surrounds itself with its own layer of calcium which means that the shell and inhalent and exhalent siphons, which are longer than the body, are covered by a calcareous layer. This may take on the shape of a pear or a vial. The animal often incorporates sand or other foreign objects into this "secondary shell." For this reason the secondary shell of a *Gastrochaena* may vary greatly, but the double opening of the inhalent and exhalent siphon is always recognizable as a characteristic figure eight. The outer casing of dead *Gastrochaena* often contains the loose shell valves.

On several occasions we discussed the mechanically and chemically drilling bivalves. The last superfamily of the Adapedonta is the Adesmacea, which comprise the "classical" piddocks. The drilling mode of life of these bivalves requires that their shells lack hinge teeth and a hinge ligament (Adesmacea means "without ligament," from the Greek à—

Fig. 6-22. Distribution of the flask shell (*Gastrochaena dubia*).

without, and δέδμος—ligament). The two valves are held together only by muscles. Thus the two shell halves can move freely against each other and act together as a drill. Additional transformation of the shell shape and the shift of several organs are further adaptations to this mode of life.

The COMMON PIDDOCK (*Pholas dactylus*; L 10–15 cm) developed a drilling mechanism which consists of protruding the leverlike shell portions on the hinge margin of each shell half. The hinge margin serves as an articulating joint in this action. The posterior adductor muscle, on the other hand, has retained its original function: closing the shell valves. These morphological features provide the animal with powerful muscular action in opening and closing its valves, which are covered with well-developed spikes, ribs, and teeth. Boring proceeds in rasplike motions. The common piddock is widely distributed along the European coast and is found in soft stone. The WHITE PIDDOCK (*Barnea candida*; L 7 cm; see Color plate, p. 177) and the FRILLED PIDDOCK (*Zirfaea crispata*; see Color plate, p. 177) prefer soft materials such as peat, clay, or wood. The frilled piddock is limited to the Atlantic coasts of Europe and North America, but the white piddock is found along all European coast lines and causes considerable damage to harbor installations. The globular WOOD PIDDOCK (*Xylophaga dorsalis*), which is less well known, is distributed from Norway to the western Mediterranean. It can be found in driftwood and also in deep-sea cables. There is sexual dimorphism in this bivalve. All species of this family (Pholadidae) are endowed with additional calcareous plates in the umbo region. These structures serve as protection for the anterior adductor muscle which opens the shells and is situated outside the shell proper. Depending on the species, one to four of these plates may be present. Although these piddocks are so completely unlike other bivalves in their body structure, they are "normal" filter feeders. The two extended inhalent and exhalent siphons are totally fused and serve as the link with the external environment. The piddock is characterized by five glandular areas along the opening of the siphons and the shell margin, which secrete a greenish luminescent mucus when irritated. This luminescent substance, expelled through the exhalent siphon, obviously serves as an attractant for microscopic organisms on which the bivalve feeds. As far as is known, the piddocks are the only bivalves that demonstrate luminescence.

The TEREDINIDS (family Teredinidae) are still more specialized to the boring mode of life. Their foot is reduced and the entire shell is shifted to the most anterior portion of the body, where it functions as a boring apparatus. The body proper can frequently reach considerable length. This is exemplified by the NORWEGIAN SHIPWORM (*Teredo norvegica*; see Color plate, p. 177) which can measure from 50 to 100 cm. The soft body of these bivalves is always naked and, together with the long inhalent and exhalent siphons, as with the *Gastrochaena*, is surrounded by a thin calcareous layer which lines the bore tunnel like wallpaper. The anterior end

Wood piddock

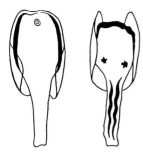

Fig. 6-23. Luminescent areas (in black) of the frilled piddock *Zirphaea*.

of the shell is covered by filelike ribs which encircle the body like a ring. The shell valves touch only at a point on the dorsal and ventral side. This serves as the cardan joint and allows the shell halves to pivot around a vertical axis. Two calcareous structures known as pallets (see Fig. 6-24) are found at the terminal end of the siphons. These structures differ greatly in the various species. Some are reminiscent of palettes, ears, or similar shapes. When the siphons are withdrawn, the pallets serve to seal off the drill hole.

All the species of this family drill in wood. The destructive activity of these bivalves can easily be compared to termite damage. The SHIPWORM (*Teredo navalis*; L 10–45 cm; see Color plate, p. 177) in particular causes great damage to harbor installations and the bottom of ships. These animals, which measure about 20 cm, are often found in such high concentrations that they can damage and destroy entire harbor constructions. In the year 1731/32 the high concentration of shipworms nearly caused an incalculable disaster in Holland. The wooden construction of the dams was heavily infested with shipworms and total destruction seemed inevitable. These mass occurrences are curbed only minimally through natural enemies, for example polychaetes and the gribble (*Limnoria*). Large concentrations of these shipworms are due to a high reproductive rate and effective brood care. A female produces 1 to 5 million eggs, three or four times a year. The young are retained in the gill brood pouch of the female up to the last larval stage. The shipworm is already sexually mature before it is three months old. At this age they measure not more than approximately 5 cm. Shipworms have a life expectancy of approximately three years. Because of the former popularity of wooden ships, to which they would adhere, the shipworms became widely distributed throughout the world. In the Black Sea, for example, this species is predominant. In the Mediterranean, however, *Teredo utricula* is more numerous. This species often infests ropes and deep-sea cable. *Teredo pedicellata* is also frequently found in the North Atlantic, the Mediterranean, and on Africa's western coast. *Bankia carinata*, on the other hand, is limited to the Mediterranean. This species can reach a length of over 10 cm, and can be distinguished from other shipworms by its earlike pallets. This animal generally is rather flexible in its habitat selection. It has been found in solid wood, driftwood, and bark. *Teredo megotara* of the North Atlantic behaves similarly.

Unlike the previously mentioned piddocks of the family Pholadidae, the shipworm and its relatives feed on the substance into which they drill. These animals are well adapted to this rather unusual bivalve mode of feeding. The shipworms are able to digest up to eighty percent of the ingested celluloses and up to fifty percent of the hemi-celluloses. As with the "wood worms" (larvae of the wood beetle; see Vol. II), these bivalves literally eat themselves through the wood. This fact in conjunction with their particular body shape has resulted in the misnomer of this animal as

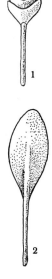

Fig. 6-24. Pallets of *Bankia carinata* (1) and *Teredo utricula* (2).

"shipworm" or "drill shell." Despite these rather unappetizing names, the larger species of both families are occasionally sold in fish markets.

Suborder: Anomalodesmacea

The TOOTHLESS BIVALVES (Anomalodesmacea) make up the last suborder of the eulamellibranchs. The hinge either lacks teeth or has weakly developed ones, as in the genus *Cleidothaerus*. The nacreous shell layer is usually present. As a rule there is no byssus gland in the foot. The mantle margins are almost entirely fused, allowing only a slit for the foot and the inhalent and exhalent siphons. The hinge ligament is often associated with a piece of calcium known as "lithodesma." There are nine families in two superfamilies. Altogether there are less than 200 species.

Some species of the Anomalodesmacea are also found along the European coasts, but they are little known. For example, *Lyonsia striata* (L 4 cm) is attached by its byssus to the sea substrate, and as a result of this sessile mode of life it is usually hidden. Yet the species is endemic from the Lofoten to the Azores and the Mediterranean. *Lyonsia striata*, with its almost rectangular shell, shows certain resemblances to the ark shells, but the two species will not be mistaken for each other because *Lyonsia striata* has short inhalent and exhalent siphons. The somewhat smaller North American GLASSY LYONSIA (*Lyonsia hyalina*; see Color plate, p. 178 is similar to the European species but is characterized by a whitish transparent shell.

Pandora clam

All toothless bivalves are characterized by a shell that is generally inequivalve. In the PANDORA CLAMS (*Pandora inaequivalis*; L 2–3 cm; see Color plate, p. 178), this characteristic is particularly pronounced. The left shell valve is curved in the usual manner at the anterior section but the right valve is completely flat. The entire shell is slightly bent out posteriorly in the form of a bill. This feature is characteristic for this species. The Pandora clam is distributed from Spitzbergen in the western Atlantic to western Africa and the Mediterranean. Like all toothless bivalves with the exception of the *Lyonsia* species, they do not secrete permanent byssus threads.

The equivalve, thin shell of *Pholadomya candida*, with its well-developed ribs radiating from the umbo, is somewhat exceptional among all the species discussed so far. This bivalve is a true deep-sea form that inhabits the Atlantic in the vicinity of the West Indies. Its small foot is endowed with a posteriorly directed process which is known as an opisthopodium. We are already familiar with the great variety of foot enlargements, usually lateral expansions found in the gastropods (see Chapter 5). In the bivalves, however, this feature is a rare exception.

Family: Thraciidae

The only Anomalodesmacea which are regularly represented by several species along the European coast lines are the THRACIIDS (family Thraciidae). Yet because of their inconspicuous body form and their presence in a muddy habitat, these forms are known only to a few experts. *Thracia pubescens* (L 3–4 cm; see Color plate, p. 178) may be an exception to this generalization because it is the most frequently occurring species. Like all thraciids, *Ixartia distorta* does not possess a nacreous layer. Their long inhalent and exhalent siphons are not fused and are totally evertible. *Ixartia distorta* also differs from its relatives in its highly irregular shell. This form is distributed from Scandinavia to the eastern Mediterranean.

Someone holding a representative of the CLAVAGELLIDS (family Clavagellidae) for the first time will hardly believe that he is looking at a bivalve. The greatly reduced, minute shells can, at best, be described as appendages. The valves make up only a small section of the animal. The mantle margins are completely fused except for the openings of the siphons and the small tip of the foot. The mantle and the long siphons give the animal a cylindrical appearance. As in *Gastrochaena*, the mantle forms a calcareous casing which covers the animal proper.

In the genus *Clavagella*, of which we already mentioned *Clavagella aperta* (see Color plate, p. 178), only the left shell valve is incorporated into the "secondary shell." The right valve remains free. Almost all *Clavagella* species drill through their substratum with the help of calcium-dissolving glandular secretions. As the animal grows, the calcareous rings around the siphons are newly deposited, resulting in a peculiar "smoke signal" appearance.

The WATERING-POT SHELLS (genus *Penicillus*) are equally unusual. The species *Penicillus vaginiferus* (see Color plate, p. 178) had already been mentioned. These burrowing forms inhabit sand or mud. The anterior portion of these bivalves is additionally covered by a sievelike calcareous plate. In members of this genus the two tiny shell valves are completely incorporated into the "secondary shell." Calcareous ruffles are also deposited around the siphons as the animal grows. None of the *Clavagella* species and watering-pot shells occur in Europe. Some species of this genus are found in the Red Sea, for example the illustrated 12-cm-large *Penicillus vaginiferus* (see Color plate, p. 178). Despite their aberrant body form, these animals possess true eulamellibranch gills, which, however, are long and thin and extend into the inhalent siphon. The minute foot has lost its functional significance in these forms.

The last order of the bivalves is comprised of the SEPTIBRANCHS (Septibranchia). They lack true gills. The mantle cavity is divided by a perforated muscular horizontal septum which has taken over the function of the gills. The hinge teeth are absent or are only weakly developed. Both adductor muscles are present. Inhalent and exhalent siphons are present. There are three families with a total of approximately 300 species.

It is still not quite clear to what extent the perforated and highly vascular horizontal septum of the septibranchs corresponds to the gills of the remaining bivalves and mollusks. Yet this specialization seems to be sufficiently important as to warrant a separate order for this group. According to more recent investigations, the separate status of the septibranchs is also substantiated by the morphology and physiology of the intestinal tract and in particular the stomach. In this aspect the animals seem to resemble the eulamellibranchs. The septibranchs are frequently

▷
Nautiluses: 1. New Caledonian nautilus (*Nautilus macromphalus*); 2. Ram's horn squid (*Spirula spirula*); 3. Rondlet's sepiolid (*Sepiola rondeleti*); 4. *Rossia mastigophora* (compare Color plate, p. 224); 5. Common cuttlefish (*Sepia officinalis*; compare Color plate, p. 224).

Order: Septibranchia

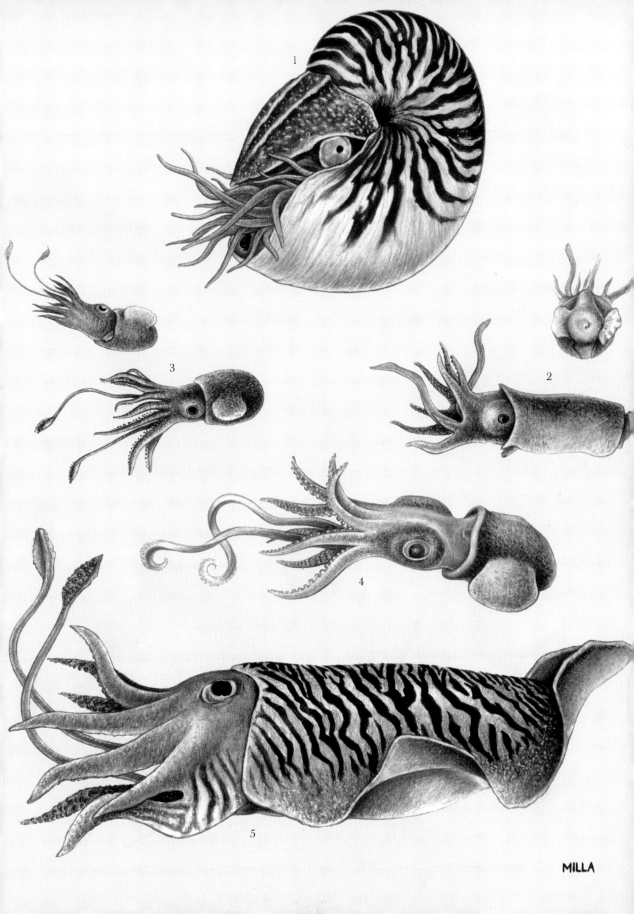

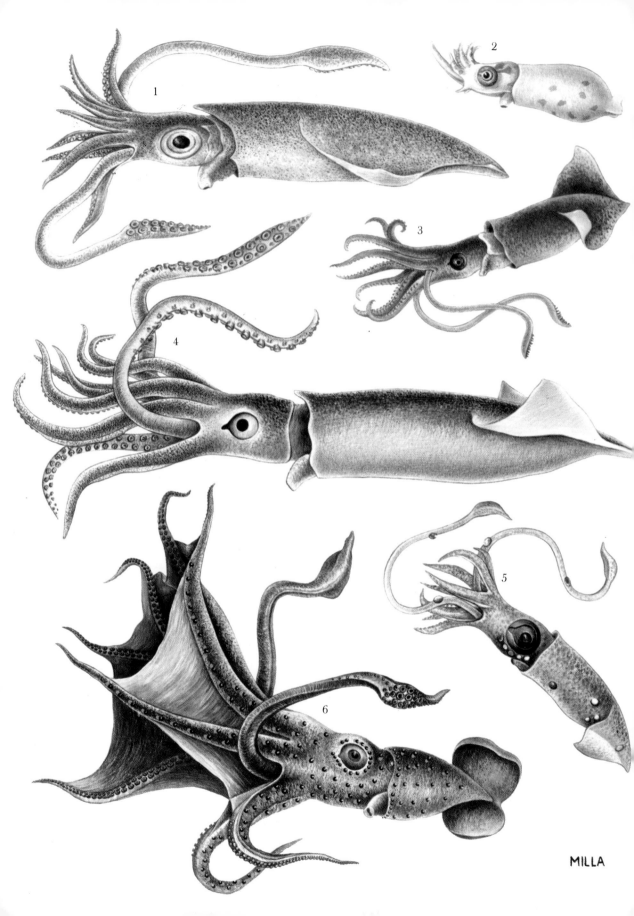

MILLA

classified with the Anomalodesmacea because both groups are similar in their body morphology. In the septibranchs as well as the Anomalodesmacea the mantle margins are almost totally fused, the hinge teeth are absent or weakly developed, and the hinge ligament usually contains a calcareous concretion (lithodesma). Just these few remarks should serve to indicate the difficulties and insufficiences that are encountered in the systematics of the bivalves.

Carnivores of the deep sea

All septibranchs are deep-sea inhabitants. Most species are not filter feeders but are carnivores. There is one species, *Lyonsella abyssicola* (from the Greek ἄβυδδος—abyss, and the Latin *colare*—inhabit: "dweller of the abyss"). Members of the genus *Lyonsella* and related forms deviate from other bivalves only in their body morphology. Representatives of the other two families, the CUSPIDARIIDS (Cuspidariidae) and the POROMYIDS (Poromyidae), on the other hand, are also characterized by a special adaptation. In these bivalves the muscular horizontal septum has the additional function of a pump. The septum fluctuates like a diaphragm and the water is transported from the eversible inhalent siphon to the mantle cavity by this action. The body position of these animals is remarkable. The dorsal side and front end are embedded beneath the mud. Food particles thus glide along the septum towards the mouth opening, which is situated ventrally. Here the prey is grasped by two labial palps surrounding the noticeably large mouth and is then transported along the digestive tract. The crystalline style in these forms is very small and therefore only minute amounts of microscopic particles are digested at one time. The shells of the deep-sea forms *Lyonsella*, *Poromya granulata*, and related species possess a nacreous layer on the shell's interior. In the cuspidariids, on the other hand, this layer is absent.

Poromya granulata (see Color plate, p. 178) is one of the few species found in Europe. It is distributed from the North Atlantic to Central America. The almost globular shell measures only approximately 1.5 cm. *Cuspidaria cuspidata* (see Color plate, p. 178), distributed from the North Atlantic to the Mediterranean, is characterized by a somewhat conspicuous shape. Like all members of this family, this bivalve's shell tapers off posteriorly, giving it a club-shaped form. This extended "beak" covers and to a great extent protects the separate siphons, which are fringed by three or four tentacles each. This shell extension is particularly evident in *Cuspidaria rostrata*, which is limited to the abyssal depths in the North Sea section of the North Atlantic.

◁
Teuthoidea: 1. Common squid (*Loligo vulgaris*); 2. Larval form of *Todarodes* (formerly classified as the genus *Rhynchoteuthis*); 3. Hooked squid (*Onychoteuthis banksi*); 4. *Todarodes sagittatus*; 5. Jeweled squid (*Lycoteuthis diadema*); 6. *Histioteuthis bonellii*.

It was only possible to cite a few examples from nearly 10,000 bivalve species. Yet these few examples serve to illustrate that even the most inconspicuous animals possess an unforetold wealth of fascinating and informative properties of life. Granted, the life of a bivalve is not "sensational," but on the other hand it is a highly specialized and amazingly diverse creature. Only a superficial observer would regard these animals as "dull."

7 The Cephalopods

Animal life on our planet has achieved three impressive highlights: first, the vertebrates with the mammals including man; second, the arthropods with the insects; and third, the mollusks with the cephalopods. Highly specialized sensory organs, particularly the eyes, significant brain evolution, and an excellent capacity to swim allow the cephalopods (class Cephalopoda) to compete with vertebrates for food and territory in the same habitat.

Cephalopods range from small to very large; total length of adults (including the tentacles) may be from 1 cm to 20 m. The visceral sac is shifted by almost 45 degrees to a posterior position from the original mouth-anus body axis. The mantle surrounds all internal organs. The shell is reduced, modified, and covered by the mantle; only the nautiloids have a well-developed external shell. The molluscan head-foot is distinctly set apart from the body, and comprises the head, arm apparatus, and funnel. The head has noticeably large, highly developed eyes; the mouth, surrounded by at least eight muscular arms, consists of circular double lips, a radula, and parrotlike jaws. The tubelike funnel at the base of the head is a narrow, transverse opening from the mantle cavity. There are one or two "salivary" glands, some with poisonous secretions. The stomach consists of three parts and is associated with a frequently paired digestive gland. Usually the kidneys are paired. The ventricle is unpaired and there are one or two auricles, depending on the number of gills. The well-developed anterior aorta and thin posterior aorta transport the blood, which is pumped by the ventricle to the entire body. The circulatory system is almost closed, and in some cases true capillaries are present. The sexes are separate. The gonads are fused and unpaired and are located at the posterior end. Frequently the gonads discharge only into one gonoduct. In females the nidamental gland, which provides substances for the egg coverings, opens into the mantle cavity. The usually yolky eggs (telolecithal) undergo, so far as is known, direct development without a true larval stage. Growth is unlimited. The life

Class: Cephalopoda, by L. v. Salvini-Plawen

Distinguishing characteristics

expectancy is from one to several years. Cephalopods are strictly marine and predacious.

The cephalopods are divided into two subclasses: 1. Nautiloids (tetrabranchs), with only one family, 2. Coleoids (dibranchs), with three orders. There are approximately 750 living species. Another 10,500 species are extinct.

The cephalopods (Cephalopoda; from the Greek χεφαλη—head and πούς—foot) are easily distinguished from other mollusks on the basis of a distinct head region which is surrounded by eight or more powerful, and often rather long, arms which, among other activities, aid in locomotion (in octopeds especially), and the saclike to torpedo-shaped body. Only the nautiloids still have a visible true shell. In the coleoids the vestigial shell has been relocated in the interior of the mantle tissue as a result of the fusion of the two mantle margins. The large, highly developed eyes are a prominent feature of the cephalopods. The structure of the lens and the iris of most cephalopod eyes is reminiscent of vertebrate eyes. However, unlike those of vertebrates, the cephalopod eyes do not originate as brain bulges (see Vol. IV) but rather as epidermal invagination. As a consequence the photoreceptors are not directed away from the lens as in the vertebrates but are turned towards it, undoubtedly to make better utilization of the light source. The visual process is quite similar to that of the batrachians, reptiles, and insects. A "photograph" of the recorded image is not traced on the retina as in man; instead cephalopods record and interpret as stimuli (pattern recognition) only the light and color variations of a moving object.

The jaw apparatus consists of a horny substance that is secreted by the buccal walls. The jaws resemble a parrot's beak. However, unlike the parrot's beak, the upper jaw does not fit over the lower jaw, but, rather, the sharp tip of the upper jaw is pressed deeply between the laterally overlapping cutting edges of the lower jaw via powerful muscular action. This jaw can bite through the exoskeleton of crustaceans and fish skulls. The radula is spoon-shaped and is made up of transverse rows of usually seven small, well-developed teeth, which because of the uniformity of the diet are usually also of about equal size. In forms which feed on microscopic organisms or plankton, for example *Spirula* or the cirroteuthids, the radula is reduced or completely absent.

Fig. 7-1. The parrotlike jaw of the northern squid (*Loligo forbesi*).

The kidneys are associated with the blood vessels in a characteristic manner. The veins which surround the kidney sacs form numerous grapelike sacculi (renal appendages) at the point of contact. The renal clusters penetrate into the kidney tissue, increasing the surface area. Excretory products are discharged into renal cavities. The hemolymph consists of blood cells which are constantly replenished from a blood gland near the eye. The plasma also contains hemocyanin, a copper-containing compound with oxygen-carrying capacities. Oxygenated hemocyanin looks bluish but is colorless in the deoxygenated state.

All male cephalopods produce spermatophores. These capsules, which measure 3–10 mm, are characterized by well-developed walls and a tubelike shape. They are produced in a special gland. Only the posterior end of the spermatophore contains the sperms. The anterior section contains a gelatinous substance which discharges explosively upon contact with the female glandular structure, to which the posterior section, the sperm reservoir, adheres. During copulation the male utilizes one or several arms to transfer the spermatophores into the female.

Numerous cephalopods have a remarkable capacity to regenerate lost body parts. This particular feature was already described in pre-Christian times, and the legends about Medusa's head and Hydra, which grew two new heads for every one chopped off, were probably based on the ability of the octopus to replace a lost arm with one or several new ones. Damaged shells, suction cups, and fins are also regenerated in numerous species. In the species *Octopus defilippii*, self-mutilation of the arms seems to be a regular occurrence (see Fig. 7-3).

The great diversity in life habits of individual cephalopod forms also finds its expression in their locomotory habits. Although all species are able to apply high pressure to force the respiratory current out through the funnel and channel this propulsive force for swimming, other unusual modes of locomotion have also evolved. Depending on the cephalopod group concerned, crawling with the aid of arms may predominate, or hovering via waving motions of the fin margins, or moving by adjusting buoyancy by exchanging gas, or moving by regular jet propulsion. Some species are even able to jet themselves clear of the water by a backward thrust, gliding through the air a distance of up to 15 m.

The well-developed predatory nature of the cephalopod necessitates a solitary way of life which is only interrupted during the breeding season. However, many species congregate in schools which then migrate over great distances. Consequently, some species have become important commercially, for example the NORTH AMERICAN COMMON SQUIDS (*Loligo pealei* along the eastern coast, *L. opalescens* along the western coast of North America). Usually these mass concentrations are spawning migrations. In other instances cephalopods congregate to follow schools of fish, in food migrations. This results in the regular recurrence of cephalopod concentrations along certain definite migratory paths. Vertical migrations are known to occur in several deep-sea forms. The nocturnal movement of these animals toward the surface of the water possibly is associated with their reproductive cycle of copulation and spawning migrations, or, more likely, with feeding activity. Cephalopods feed mainly on crustaceans, fishes, and mollusks. Even conspecific juveniles may be eaten. Most cephalopod juveniles, and possibly some of the specialized gelatinous adult Cranchiidae, feed on the minute organisms, zooplankton and phytoplankton.

Cephalopods fall prey predominately to toothed whales such as the

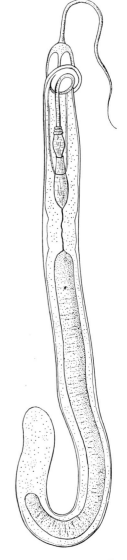

Fig. 7-2. A spermatophore of the common cuttlefish (*Sepia*).

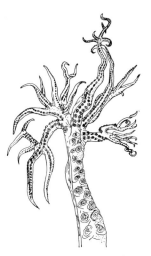

Fig. 7-3. Regeneration of an arm tip with many branches in the common octopus.

sperm whale and bottle-nosed whales (see Vol. XI) as well as fishes including sharks, rays, and cods (see Vol. IV). Many cephalopods are consumed by seals, petrel (see Vol. VII), and penguins. Carl Chun reported that the stomach contents of an emperor penguin included sixty squid jaws.

The cephalopods prefer habitats with a normal oceanic salinity content (3.2 to 3.75%). Most cephalopod species live in the warm oceans, but the number of individuals is higher in the colder regions. A few highly specialized deep-sea cephalopods have been found at depths below the 5000-m mark. Approximately 130 out of a total of 750 species occur in European waters. Approximately thirty-five species inhabit northern European waters, forty-five species, western European, and fifty, southern European oceans. Approximately 150 species of cephalopods occur in the waters contiguous with the North American continent. Cephalopods have not been recorded in the Black Sea and the Baltic Sea because the salinity content is too low.

Our knowledge of cephalopods goes far back into history. They have been known from "sailor yarns" and legends, and even Aristotle and Pliny the Elder described a few species. Some forms were illustrated on Minoan and Mycenaean records from approximately 4000 years ago. Cephalopod drawings were used for decorative purposes. During the late Minoan period on Crete (around 1500 to 1470 B.C.), a unique ornamental "Oceanic style" developed. In addition to dolphins, starfishes, and purpura snails (banded or purple murex), cephalopods (*Octopus* and *Argonauta*) also were predominately represented. Even today primitive tribes still utilize shell segments and the horny rings of the suction cups to make necklaces and rings. Additionally, extraordinary and unexplained discoveries and events have always played a significant role in folklore and superstitions. In former times fossil cephalopods provided sufficient material to be used in religious teachings and medicinal uses.

Phylogeny of the cephalopods

There is still considerable uncertainty about the phylogeny of the cephalopods, the nautiloids, and the coleoids. More recent discoveries are significant because they indicate that the fossil bactritids and goniatids, which are closely related to the ancestral ammonites and coleoids (compare Chapter 1), did not have more than ten arms, and had a seven-toothed radula (at least in the goniatids) and an ink sac. If these two evolutionary lines of the present-day coleoids and extinct ammonoids possessed ten tentacles, an ink sac, and a seven-toothed radula, then a similar development probably also occurred in the "archeo-cephalopod." In the light of this evidence it would appear that the third evolutionary line, the nautiloid cephalopods, which favored a reduction of the ink sac, multiplication of the tentacles, and the presence of four gills, followed a separate course early in their development. Therefore it does not represent a primitive condition as had previously been assumed.

However, the controversy about the six-to-ten-pointed buccal membrane in the living coleoids still remains unsettled. Some zoologists consider it as a vestigial ring to which the tentacles were attached.

If the last assumption were true it would mean that originally there were sixteen to twenty tentacles. The number of tentacles would have been reduced to eight or ten in the coleoids and ammonoids, and in the tetrabranchs it would have been increased to eighty or ninety.

Notwithstanding the phylogenetic controversy, today's living nautiluses (tetrabranchs) are classified as a separate subclass from all remaining orders since their anatomy shows marked differences from those of the coleoids (dibranchs).

Length varies from 10 to 27 cm. The external shell is well developed and symmetrical. The shell is planospirally coiled and is partitioned into chambers. The siphuncle or tubule extension of the visceral hump extends through the center of the septa. The modified foot consists of eighty-two to ninety tentacles which lack suction cups. The funnel consists of two lobes which are not fused. Each transverse row of the radula contains thirteen teeth. There is no ink sac. There are two pairs each of kidneys, gills, osphradia, and auricles. The eyes are simple and faceted, without a lens and aqueous chamber. The embryology is unknown. There is only one living family: Nautiloids (Nautilidae) with a single genus *Nautilus* and six species: 1. PEARLY NAUTILUS (*Nautilus pompilius*; L up to 20 cm); 2. SOLOMON'S NAUTILUS (*Nautilus scrobiculatus*; L up to 25 cm); 3. NEW CALEDONIAN NAUTILUS (*Nautilus macromphalus*; L up to 20 cm; see Color plate, p. 187). Three other species are found along the Australian coast.

The uniqueness of the few living nautiloids has earned them a variety of names in each of several languages. Since the animals sometimes drift slowly along near the surface of the water and because they have an irridescent naucreouslike inner layer, they are sometimes called "pearly boats" (compare Chapter 6). The description "chamber snail" refers to the nautiloid's shell structure. A longitudinal section will reveal that the shell consists of several successive chambers. These are filled with gas and are connected by a tubelike siphuncle with the "animal proper." The animal's body is located in the outermost chamber of the shell. As the animal grows, new chambers are added, one at a time, to its shell. The new growth ring is seen as a suture line on the outside. Just as in the primitive gastropod larvae, the shell is coiled towards the head region. This type of coiling is known as exogastric (compare Chapter 3). The *Nautilus* species are also known as Tetrabranchiata (four-gilled) as opposed to the two-gilled Dibranchiata-coleoids.

As in all mollusks, the mantle encircles the dorsal surface of the visceral sac, although in the nautiloids a dorsal flap is also developed which folds back and lines the shell. The many tentacles, up to ninety, are particularly conspicuous. The individual tentacles are short and consist of a thick shaft and a thin, coiled tentacle which can be retracted into the shaft. The shafts of the four dorsalmost tentacles are fused into a hood. This

Subclass: Nautiloidea

Distinguishing characteristics

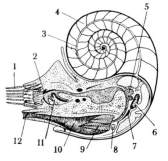

Fig. 7-4. Anatomy of a chambered or pearly nautilus: 1. Arms, 2. Jaw, 3. Siphuncle, 4. Shell, 5. Stomach, 6. Gonad, 7. Heart, 8. Kidney, 9. Gills, 10. Mantle cavity, 11. Radula, 12. Pallial funnel.

beautifully designed "shield" for the animal is rather conspicuous (see Color plate, p. 187). The other tentacles surround the mouth opening in two incomplete circles, the inner having from forty-four to fifty-two, the outer, thirty-eight. Some tentacles are modified as copulatory organs. The ring of tentacles is interrupted by the funnel on the "ventral side" (posterior to the mouth). Unlike the dibranchs, the nautiloids lack suction cups on the tentacles. Instead, each tentacle is equipped with an adhesive pad facing the mouth opening, which secretes a sticky, glandular substance. Depending on their function, tentacles can be differentiated into "chemosensory tentacles" and "grasping tentacles."

Sensory organs

The prominent stalked eyes are uncomplicated. A comparison with a simple pinhole camera comes to mind. These eyes lack a lens and an aqueous chamber but possess a well-developed retina which receives the light rays that fall through the visual hole. Nevertheless, this compound eye is weak visually. The eyes and the brain are partially located on an H-shaped cartilage which supports the bipartite funnel and also serves as muscle attachment. A fingerlike process known as a rhinophore is situated below each eye. This is the center for chemoreception. Additional sensory organs are found on each of the four gills. These are pleatlike organs that also function as olfactory structures. In function they correspond to the osphradia of other mollusks (osphradium—sense organ which consists of a ciliated epidermal swelling at the entrance of the mantle cavity; compare Chapter 1).

The pearly nautilus is a nocturnal bottom predator. During the day the animals usually remain hidden, occasionally drifting like boats near the water's surface if they have been disturbed. The shell chambers of the nautiloids contain a fluctuating amount of fluid in addition to a constant amount of gas. The amount of fluid can be regulated by the siphuncle which extends into the last chamber. The amount of fluid influences the gas pressure and thereby its buoyancy potential. The animal can sink or float by regulating the gas/fluid ratio in the shell chambers. The nautiloids can adjust to the external pressure of the water at various depths. After having observed the *Nautilus macromphalus*, the French scientist René Catala was led to assume "that it seems to be the lack of external pressure rather than the proper choice of food which has resulted in the failure of keeping these animals in captivity for longer periods of time." At depths of from 50 to 650 m, pearly nautiluses feed on crabs and carrion they find on the ocean bottom. One can feed this same fare to captive animals without hesitation. When devouring a prey, the thirteen-toothed radula probably serves only as a "shoveling device." Often digestion takes an unusually long time. Thirty hours or more after feeding, pieces of food could still be detected in the crop.

Arms

In 1705 the Dutchman Georg E. Rumpf, known as the "Pliny of the East Indies," and in 1831 the British researcher George A. Bennett, had already described how these animals swim with jerky jet-propelled motions wherein their tentacles are spread out widely. On the ocean bottom, however, these creatures crawl or attach themselves via the tentacles. Their use of the tentacles for crawling still has not been observed in the aquarium, a point emphasized by the British researcher Anna M. Bidder. When the animals are at rest or are swimming rhythmically, all tentacles are withdrawn into their shafts. Occasionally their tips may protrude. As soon as one of the external lateral (tasting) tentacles touches food, however, all other (grasping) tentacles are extended. The prey is then entangled and pulled underneath the hood. The interior circle of tentacles holds the prey while the jaws tear off piece after piece. The nautilus species do not utilize the entire mantle cavity as a pressure-producing apparatus; only the strongly modified halves of the funnel are contracted and expanded. According to some lay observers, the nautilus can move just as quickly as a fish. By approximately one year of age the nautilus has reached maturity. At this stage its shell consists of twenty-three to twenty-seven chambers. Nothing is known about the embryology. On the basis of the yolky eggs one can assume that there are close similarities to the dibranchs.

The nautiluses are limited in their distribution to the Indo-Pacific region of the tropics. *Nautilus pompilius* is found from the eastern Indian Ocean to the Fiji Islands. Other species do not enjoy such a wide range of distribution. *N. scrobiculatus*, for instance, is limited to the region around the Solomon Sea near New Guinea, and *N. macromphalus* is found only near the island of New Caledonia. The few existing species are distinguishable only by insignificant features. Classification, for example, is based on differences in body size and the number and arrangement of the brownish-red shell bands. The juvenile nautilus is characterized by numerous jagged color strips which cover the entire shell. *N. scrobiculatus*, on the other hand, has a rough shell surface. The shell aperture lacks color bands.

Often masses of nautiluses and their empty shells are swept onto the beaches following storms. The local people collect these to use as food and to make the shells into jewelry.

All other cephalopods are classified in the subclass Coleoidea (L, including the tentacles, from 1 cm to about 20 m). The shell is completely enclosed by the mantle. The shell is usually strongly reduced. Pigment cells (chromatophores) are present in the dermis. Chromatophore expansion and contraction is controlled by muscular action. The modified head-foot consists of eight to ten appendages equipped with suction cups. These are rarely degenerated. The funnel is fused and unpaired. The radula which is rarely reduced, consists of transverse rows usually with seven teeth each. The diverticulum of the hindgut forms the ink sac. The dibranchs have one pair each of kidneys, gills, auricles, and branchial hearts. They have no osphradia. The nervous system is greatly concentrated and is surrounded by a cartilaginous capsule. The eyes, which

▷
Teuthoidea: 1. *Chiroteuthis veranyi*, 2. *Taonidium suhmi*, 3. *Desmoteuthis pellucida*, 4. *Sandalops melancholicus*, 5. Larva of *Chiroteuthis veranyi* (formerly regarded as its own species, *Doratopsis vermicularis*); Vampyromorpha: 6. Adult vampire squid (*Vampyroteuthis infernalis*), 7. Juvenile vampire squid (*Vampyroteuthis infernalis*).

Subclass: Coleoidea

Distinguishing characteristics

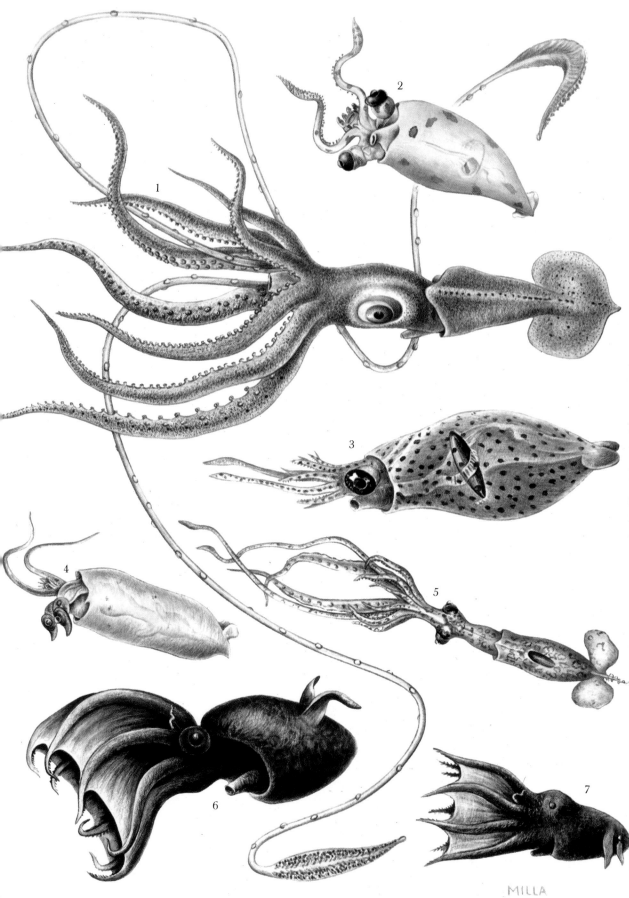

MILLA

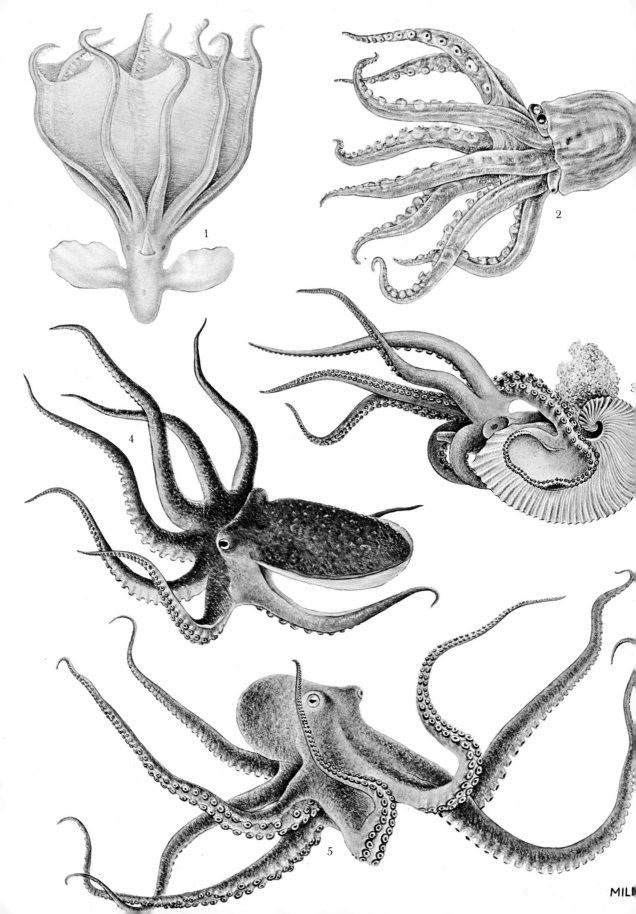

1

2

4

3

5

MIL

contain a lens, are highly developed. The male transfers spermatophores to the female with the aid of a modified arm, the hectocotylus. The eggs are quite yolky and undergo direct development. There are three orders: 1. Teuthoidea, with two suborders, 2. VAMPIRE SQUIDS (Vampyromorpha), and 3. Octopoda, with two suborders.

Only the dibranchs (approximately 745 species) have an ink sac. At first glance the large head-foot of the dibranchs seems to be the most noticeable feature. The rest of the body appears like a saclike appendage. The eight or ten arms, which bear many suckers, function as weapons and as tools and in some forms aid significantly in locomotion. Furthermore, the jaws and gigantic eyes indicate a mode of life totally different from that of other mollusks, including the nautilus. The greatly degenerated shell, completely covered by the mantle, is another feature which deviates still further from the usual image of a mollusk. Nevertheless, all molluscan characteristics are still present in the dibranchs, although at a higher level of specialization and somewhat modified. Although a shell is not visible in any of the species, it is almost intact in one form. In another species, similar to the nautilus species, the shell is still present as a chambered structure including a siphuncle. Highly specialized dibranch forms possess only a vestigial shell which is embedded in the mantle (compare legend to Color plate, p. 64, and Figs. 7-6, 7-7).

The reduction of the shell signifies a loss of external protection, but the loss is largely compensated for by special protective measures. Beneath the highly glandular single-layered and transparent epidermis of the dibranchs lies a multi-layered leathery or gelatinous dermis which contains a pigment cell (chromatophore). A chromatophore consists of pigment cells surrounded by a stellate circle of from four to twenty-four fine muscle fibers (fibrils). The rapid contraction and expansion of the chromatophores results in changing color patterns. The localized change of the chromatophores is influenced by sensory stimuli. Its effect is increased by the multicolored cell content. The deeper layers of the dermis also contain iridocytes which are immobile but which can produce varying degrees of iridescence by breaking down and reflecting incoming light. The highly developed eyes are capable not only of distinguishing between light and dark but also between blue and yellow. These forms also have light-sensitive skin, just like the bivalves and other mollusks. The "shape-color sense" via tactile stimuli is well developed. This unique feature is caused by tactile stimuli conducted from the suction cups of the arms. These stimuli are transmitted to specific regions in the brain and are correspondingly expressed in a suitable color display by the chromatophores. Even an octopus which has lost or been deprived of its eyesight will display a fine granulation in its skin when sitting on a sandy background. On gravel or other deposits the skin will look coarsely granular. Many bottom-dwelling forms dig themselves into the soft substrate and accentuate their camouflage by a similar coloration of their

Degeneration of the shell

◁
Octopoda:
Cirrata:
1. *Cirrothauma murrayi*;
Incirrata:
2. *Vitreledonella alberti*,
3. Paper nautilus (*Argonauta argo*, female) with egg case,
4. and 5. Common octopus (*Octopus vulgaris*; compare Color plate, p. 223).

skin. For similar reasons, deep-sea forms are either glassy-colorless or black, brown, and brownish-violet. Because red light is invisible below depths of approximately 20 m, it is not uncommon that some of these forms are also purplish-red, although these red animals appear black, and thus invisible, at their normal living depths.

The release of a dark brown fluid, or ink, is a well-known protective mechanism of the dibranchs. This secretion is produced in the unpaired diverticulum, or ink sac, located ventrally to the hindgut, and is discharged to the outside through the funnel via the anus and mantle cavity. According to more recent investigations, the ink secretion not only seems to camouflage the animal but also temporarily immobilizes the predator's olfactory sense. This of course facilitates a successful escape by the pursued cephalopod. In certain families inhabiting the greater ocean depths, the ink sac has become degenerated; this has also occurred in the *Nautilus* species. Yet other deep-sea species utilize the ink sac to great advantage. These forms possess external pouches which contain luminescent bacteria coupled with the ink sac. In conjunction with the ink secretion, they produce a glaring "light cloud" which deceives the opponent for a moment.

Luminescent organs have reached a high degree of specialization and bewildering degree of variety in the cephalopods. Over four-tenths of all dibranchs possess such structures, which can utilize up to ninety percent of the energy of the rays of the so-called cold light (as contrasted to only four percent of our artificial light!) Often these light-producing organs are highly complex, but there are two basic forms which are usually embedded in the skin. Some shallow-water species have so-called open organs. Deep-sea species more often have enclosed luminescent organs which produce the luminescent material themselves. The open organs, on the other hand, achieve luminescence on the basis of a symbiotic relationship with luminescent bacteria that live in these organs. The mechanism of light production is, however, the same in both cases. Hans E. Gruner described the process: "Living light-producing cells manufacture a special substance known as luciferin with the aid of an enzyme, photogenase, a catalyst. Luciferin itself is not luminescent, and has to undergo additional chemical reaction. A second enzyme, luciferase, enters into the reaction. However, the mere presence of luciferin and luciferase does not produce light. Molecular oxygen, even in minute quantities, is necessary to complete the reaction. Luciferin is oxidized or rather dehydrogenated by the oxygen. In this chemical reaction, energy is liberated in the form of light rays. This of course is the light we perceive." Light, therefore, can be produced either by the action of luminescent bacteria or by the organism itself. The open light-producing organs containing the luminescent bacteria are situated in the mantle cavity. There are tangled tubes of various bacteria within these organs.

In contrast to this true symbiotic relationship between the luminescent

Luminescent organs

bacteria and the dibranch, there are forms that do not have to "steal" light from another source but produce their own light. The natural bluish-green color of the cold light is transformed in tone by the squid by various structures such as a color filter and a diffraction mirror into red, blue, green, and white light. These structures are highly complex and are equipped with light-scattering lenses and color-transforming reflectors. These luminescent structures may play an important role in bringing together members of the opposite sexes. Additionally, the patterns and colors of light are species-specific; they also serve in the capture of prey.

The high degree of sensory and expressive capacity in the cephalopods is practically equal in complexity to the behavior of higher vertebrates and certain arthropods. Cephalopods orient via sight and olfaction, a sense the receptor of which is located underneath the eyes in the form of an olfactory pit, and by a sense of direction and territory. They also

Learning and training ability

demonstrate a certain degree of learning and training ability, as well as behavior based on insight. Experience gained after a single or repeated applications of a suitable behavior pattern and the appropriate utilization of a tool in response to a certain set of circumstances are feats which involve the entire cephalopod nervous system and should not be underestimated. The previously mentioned optic process is also an example of a high degree of nervous integration. The eye can adjust to focusing at various distances (accommodation). Just as in the batracian, the lens is squeezed forward by muscular contraction when focusing at close objects—in the fishes the lens is squeezed for looking at distant objects; in reptiles, birds, and mammals the lens is not moved but the lens shape is altered. It is particularly amazing that numerous dibranchs have the ability to perceive polarized light. They share this ability with most insects (bees, see Vol. II) and certain spiders, and are able to orient with the help of polarized light. We still do not understand this sensory ability. The perceived light intensity reaches high values. The squid, with its 160,000 retinal elements, possesses the same number of units per square millimeter as does man. The common *Sepia* (cuttlefish), with its 70 million cells per mm², even surpasses man (50 million cells per mm²). Visual acuity is only slightly less well developed than the chemical sense. In poorly lit zones dibranchs use their olfactory sense to locate prey. Olfactory and taste stimuli also play a large role during the mating season.

Brain capsule

Essential body organs such as the brain, the eyes, and the chemo-receptors need protection and so in these animals they are surrounded by a cartilaginous brain capsule. The sephalic cartilage shelters the highly specialized superior buccal ganglion, which coordinates all parts of the body, and associates present and past sensory impressions, as well as the inferior buccal ganglion, which controls the statocysts present in all dibranchs to regulate body orientation. The cephalic capsule (skull) also serves as an attachment area for various muscles including one pair of shell muscles. In dibranchs one can almost speak of a true internal skeleton because there are several other cartilagenous components which give support to the animal's body. For example, there are also the locking cartilages (knobs that fit into sockets) which close the mantle. The dibranchs are predators that have to respond quickly to sensory stimuli. Therefore, unlike other mollusks, they have evolved striated muscle fibers. Only striated musculature can respond rapidly to the great demands on the nervous system.

Another specialized feature is the circulatory system, which is mostly of the closed type and is associated with two gills (Dibranchiata—two-gilled). The system consists of an additional organ: a brachial heart of spongy muscular consistency is found between the enlarged brachial veins near the kidneys and the afferent branchial vessels. It functions to overcome pressure resistance and to increase blood circulation through the gills, and concomitantly to facilitate oxygen exchange. At this point we may mention the group of parasites occasionally found in some dibranchs. These parasites, represented by the order Dicyemida and belonging to the mesozoans (Mesozoa, see Vol. I), have been found only in the renal sacs of bottom-dwelling cephalopods. The parasites measure only a few millimeters in length and swim about in the renal fluid. They lack a digestive system, and absorb food substances directly through their body cells. Despite this, their relationship with the dibranchs seems to be symbiotic, since there are several indications that the mesozoan and cephalopod each benefit from the presence of the other (symbiosis).

Branchial hearts

With the exception of a few forms, particularly Vampyromorpha and Cirrata, the males possess one or two arms that have been modified (hectocotylus), for the transference of spermatophores to the female. Spawning follows soon after copulation, but, unlike almost all other mollusks, dibranchs do not have a true larval stage. Dibranch eggs contain a great deal of yolk, and cleavage is therefore incomplete (discoidal cleavage) and is restricted to one end of the egg. This causes the embryo's ectoderm to spread over the yolk mass. The body grows upward and the head-foot plus the ever-shrinking yolk sac extend downward into the water (see Fig. 7-12, and Color plate, p. 224). Phylogeny recapitulates the rearrangement of the body axis in the development of the individual (ontogeny): the entire body of the dibranch represents the actual dorsal side of the animal while the arms and tentacles and the funnel alone correspond to the ventral side, as is shown by the nerve arrangement in these body parts. The anteroposterior (vertical) body axis shortens and the dorsoventral (horizontal) one lengthens only at a later stage.

Today the majority of cephalopods belong to the squids and cuttlefishes. There are two orders: 1. Sepioidea, the cuttlefishes, with five families, 2. Teuthoidea, squids, with two suborders and twenty-five families. There are approximately 525 species. Length, including the tentacles, from 1 cm to about 20 m. There are ten appendages bearing adhesive suction disks (suckers). Two of these, the tentacles, are much longer than the others. They are characterized by long shafts and short flattened ends which bear the suckers located on short stalks, cup-shaped, and usually reinforced by serrated horny rings. The funnel can be closed by a valve. The mouth is surrounded by an epidermal fold consisting of six to ten flaps that bear small suction cups in some species. Since the animals are usually free-swimming, the body shape is almost always elongate, not saclike. The mantle is equipped with lateral fins.

Viewed from above, the fourth pair of appendages, the tentacles, is usually greatly elongated, extending beyond the remaining arms. In cuttlefishes, these tentacles can be coiled and retracted into a skin pocket, giving the appearance of an animal with only eight arms. A movable pad located on the floor of the stalked suction cup (Fig. 7-14) acts like a muscular "suction piston." The vacuum created when this pad is pulled back causes an adhesion to the substratum or to the prey.

Order: Sepioidea

The body of members of the order Sepioidea is usually bulky. The shell is in various degrees of reduction. The suction cups are lined with a simple horny rim without hooks. There are five families with approximately 150 species.

The Sepioidea includes the well-known ram's horn (family Spirulidae) with only one species, *Spirula spirula* (BL 4–6 cm; see Color plate, p. 187). This form still possesses a shell which is characteristically coiled. The shell is chambered and a siphuncle extends throughout it. The coiled shell is located internally in the posterior end of the body, so it is not visible externally. *Spirula* was, until recently, only rarely found, because it inhabits tropical and subtropical oceans at depths of 200 to 600 m. The depth at which the animals are found seems to be temperature dependant. According to extensive studies by A. Bruun, *Spirula* prefers temperatures between 10° and 20° C. One specimen was caught off western Africa at a depth of 100 m at 18.6° C. Reports of *Spirula* found at 1,750 m are questionable.

Fig. 7-5. Distribution of *Spirula spirula* in the Atlantic Ocean.

Like the *Nautilus* species, *Spirula* is pelagic and is found near mainland coast lines, and just like the *Nautilus* and the common cuttlefish *Sepia*, "the buoyancy is altered by adjusting the fluid to gas ratio in the chambers of the shell. The regulation of the shell fluid is not under the influence of hydrostatic pressure but rather it is an osmotic process," according to E. Denton, J. Gilpen-Brown, and J. Howarth. The regulation of the gas volume inside the chambers is directly proportional to the amount of fluid present. This exchange of shell fluid is regulated by the animal's body and takes place through small perforations in the middle of the septa. The corresponding amount of gas is simultaneously fixed or released by the blood.

Fig. 7-6. *Spirula spirula* with the shell. An exposed section shows the chambers within the shell. (below at left).

A. Bruun's experiments with *Spirula* shells demonstrate that there is a maximum of 50 to 75 atmospheric pressures which corresponds to a water depth of 500 to 750 m. This coincides with the data on the distribution of these animals. The shell itself is "somewhat lighter than the surrounding water, and when the animal dies the shell rises to the water surface," as A. Bruun observed, "for this is the only possible explanation for the unusual phenomenon that a deep-water form represents such a significant portion of coastal debris." *Spirula* possess a unique circular light-producing organ on its posterior end which is reminiscent of a "rear reflector." The Englishman Frank W. Lane explained, "*Spirula* possesses two small posterior fins which by constant fluttering motions during swimming support the water current expelled from the funnel. The disk, with a small, central, buttonlike organ, which emits an even ray of yellowish-green light, is situated between these end fins. The light can 'burn' for hours. Since *Spirula* usually is found in a vertical position with arms dangling down, the light appears like a taillight. The function of this organ probably is to keep a number of these animals concentrated in 'schools.' Groups of *Spirula* migrate mainly in a vertical direction for several meters, unlike most animals, which migrate in a horizontal direction. The buttonlike light organ is also equipped with a diaphragm which obviously serves to cover up or bring on the luminescence." If *Spirula* is disturbed in its vertical rest position—the body position corresponds to the original orientation of the actual body axis—it pulls its head and all the arms inside the mantle cavity. The mantle margins can close over the opening. This protective measure makes it extremely difficult for smaller predators to take hold of such a withdrawn animal, since the mantle is very leathery, tough, and slippery.

Let us turn to the true cuttlefish. The family Sepiidae encompasses approximately eighty species, several indigenous to European oceans. The majority of species, however, occur in warm waters of the eastern Atlantic and throughout the Indo-Pacific. Cuttlefish do not occur in the western hemisphere. The cuttlefish is represented by several species in the eastern Atlantic ranging from the Faroe Islands to South Africa. The COMMON CUTTLEFISH (*Sepia officinalis*; see Color plate, p. 187) is found in the Mediterranean and the adjoining Atlantic. Large specimens (up to 65 cm, the arms measuring approximately 30 cm) have been recorded from this region. Along the coasts of Yugoslavia, Italy, and Spain, fish markets offer cuttlefish in addition to small fish, crayfish, tunicates, and various mollusks. Cuttlefish are characterized by a body shape that is dorsoventrally compressed and oval; fins extend the length of the mantle. Even the layman can recognize the conspicuous zebralike brown to violet cross-patterning on the upper surface and the spotted bluish-green belly. One can often search in vain for the two elongated tentacles, which are usually retracted within a pouch beneath the eyes. Thus only eight arms, bent slightly downward, are visible. The protruding black eyes with the W-shaped pupils appear fixed. During the day the animals are usually buried in the sandy ocean bottom, where they assume a mottled pattern to match the surrounding environment. If a prey, for example a crab,

The cuttlefish

Capture of prey

should approach such a buried cuttlefish, a wave of color spreads over the animal's back and arms. If the crab is out of reach the cuttlefish will "sneak" almost imperceptibly out of the sand to pursue it. Slow undulations of the fin margins and the directing water ejection from the posteriorly pointing funnel permit the hunter to come within close range of its prey. The tentacles, extended in the direction of the crab, undergo rippling color changes. Finally the almost colorless long tentacles with the two club-shaped ends are shot out to seize the prey. Armed animals are always attacked from their least dangerous side. Crayfishes and crabs that possess powerful pincers are seized from behind. Nevertheless, many prey animals often escape the cuttlefish by quickly moving off prior to the attack. The seized prey is pulled toward the mouth and is held by the arms while the sharp jaws tear into the victim. Fishes, crayfishes, and similar animals are devoured almost completely.

Despite its cryptic mode of life, its color change, and its defensiveness, the cuttlefish does not enjoy a protected existence. Small sharks and rays are its major enemies. Along the ocean floor the voracious ocean pike (*Merluccius merluccius*; see Vol. IV) and the dangerous eel (*Conger conger*; see Vol. IV), cruising along the rocky coastal zones, are two powerful enemies which prey upon cuttlefish. Dolphins, seals, marine birds, and numerous fishes also prey on cuttlefish. The cuttlefish can usually escape by disappearing in a cloud of "ink." Last but not least, man can also be counted as the cuttlefish's enemy. In countries around the Mediterranean this sea food is highly valued and attracts commercial fishermen and sports fishermen. The refined ink secretion has long been used as sepia, a painter's color.

Bird lovers are familiar with cuttlebone, the cuttlefish's shell which provides the essential calcium requirements for their canaries and other pet birds and gives them a chance to sharpen their beaks. Craftsmen use the cuttlebone as a fine whetstone for their instruments. The cuttlefish used to be significant in medicine. Its meat was believed to be an aphrodisiac, its eggs a cure for bladder catarrhs, and the ground-up cuttlebone was used in healing eye infections, asthma, and similar ailments. The cuttlebone (Ossa sepia) represents the entire shell of the mollusk. Despite its greatly reduced state, it shows the original chambered effect. The siphuncle is still present but the septae are only developed in the upper section.

We already mentioned color changes in relationship to cryptic behavior and excitation during prey capture. Color changes play an even greater role in cuttlefish courtship. Since dibranchiate cephalopods are predominately "visual animals" because of their large, highly evolved eyes, courtship displays solicit responses primarily with visual stimuli. The fourth arm on the left side of the male is modified into a copulatory organ (hectocotylus) which transfers spermatophores to the female in the latter part of the mating ceremony. The particularly broad and beautifully patterned arm is prominently displayed by the male in the

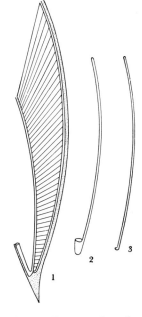

Fig. 7-7. Degeneration of the shell in 1. *Sepia*, 2. *Todarodes sagittatus*, 3. Teuthoid.

direction of the "chosen one." The two sexes respond differently to this signal. Another breeding male will return the display. In this manner the two rivals will try to displace each other. A receptive female, on the other hand, does not have a sexual color to display, and remains quiet. During subsequent mating, the circular cleft or bursa copulatrix near the female's funnel becomes significant. This organ is the receptacle for spermatophores, which the males places there with the hectocotylus. While all the other arms of the cuttlefish are characterized by four rows of suction cups, the hectocotylus possesses suction cups only at the outer edge of the tip. The remaining arm surface is covered with numerous transverse ridges and atrophied suction cups. Copulation takes place in a characteristic posture and has been described by Richard Bott in Naples. The two partners face each other. "The arms are intertwined like the fingers of praying hands." In this position the hectocotylized arm of the male transfers bundles of spermatophores into the vicinity of the bursa copulatrix of the female.

Eggs are usually laid after an interval of several hours. The partners frequently stay together, sometimes even after the female has died. The female has a tendency to fasten the eggs on corals or plants where already densely packed "spawn clusters" are present. Eggs, one at a time, are blown out of the funnel and are transferred along the club-shaped tapering space between the front end of the arms which are pressed together. The egg glides past the bursa copulatrix and becomes coated by a secretion from the nidamental gland and ink sac. Finally it reaches the tips of the tentacles and becomes fastened to a designated spot. There the cuttlefish female anchors the eggs by two adhesive strings (which are processes of the black egg shell) to the back of the stalk (or branch) with two of her lateral arms. In this fashion each female may attach over 500 fertilized, lemon-shaped eggs to various locations. These deep-brown to black clusters are known in countries around the Mediterranean by phrases which mean "ocean grapes."

Fig. 7-8. Egg cluster of *Sepia*.

The spawn of *Sepia orbignyana*, which inhabits the Mediterranean and eastern Atlantic, is similar, and is produced in the same manner. A close relative, the *Sepia elegans* (L 10 cm), however, deposits its oval, bluish eggs on sponges. This species enjoys the same range of distribution as the gray-and-brownish-speckled *S. orbignyana*, but *S. elegans* occupies the deeper layers of water and possess a glandular light-producing organ. *S. elegans* is easily recognized by its smaller and more slender body shape and the yellowish-red color.

Only a few of the numerous *Sepia* species occurring outside Europe are mentioned here: *Hemisepius typicus* (L 3–4 cm) is the smallest member of this family. The tiniest representative of all the cephalopods is *Idiosepius pygmaeus* (L 1–1.5 cm), which belongs to the family Idiosepiidae. This form is found along the coasts of the Indian Ocean and Japan. The females are always larger than the males (sexual dimorphism). It lives hidden

Family: Sepiolidae

among algal growth, where it preys on small crabs and fishes. Within the family Sepiolidae (bob-tailed squid) one can trace the progressive degeneration of the shell. In the various genera the shell shrinks from a distinct remnant to an insignificant structure. *Rossia macrosoma* possesses a shell similar to that of the cuttlefish but without calcification. *R. macrosoma* is distributed in the western North Atlantic and parts of the Mediterranean. Like all species of this family, this form is characterized by a short saclike body with two rounded fins near the middle of the body. The arms have four rows of suckers. At the base the arms are joined by a velar fold or interbrachial web. The various species are mainly distinguishable on the basis of their body form and coloration: the northern form (BL up to 25 cm) is brownish and its elongated tentacles are relatively short. The deep-water species *Rossia mastigophora* (see Color plate, p. 187) is found in the Indian Ocean. The European lesser *Rossia* is equipped with a light-producing organ in the immediate vicinity of the funnel. The smaller *Allorossia glaucopis*, from the North Atlantic, has no luminescent organs, and has only two rows of suckers on its arms.

Heteroteuthis dispar is a regular deep-water form. It is found at depths of from 1,200 to 1,500 m in the Mediterranean, particularly around Naples. Instead of ink, *Heteroteuthis* discharges a cloud of luminescent bacteria in order to confuse its pursuer. Many other species of this family have luminescent organs, for example the closely related Mediterranean *Sepiola rondeleti* (BL 4 cm; see Color plate, p. 187), the sexes of which are of equal size, and *Sepiola atlantica* (BL 4 cm), of the Atlantic. These are the smallest dibranchiate cephalopods found in the European region. The Mediterranean *Sepiola* is also known as the dwarf *Sepia*, and is conspicuous because of its characteristic swimming motions. All representatives of this family swim in a birdlike manner by beating their fins up and down. The well-defined, uncalcified shell of the Atlantic *Sepiola* is more highly reduced in the other species. In *Sepietta oweniana* the shell is only represented by a long, horny structure. Copulation in *Rossia*, *Sepiola*, and *Sepietta* is unlike the process in the cuttlefish. Instead, the male grasps the female from the ventral side with the long tentacles so that both heads face in the same direction. Then the male's left first arm, the hectocotylus, is inserted into the female's mantle cavity, where it deposits the spermatophores. These smaller-sized Sepiolids are exposed to far more danger than the large cuttlefish. Since the sepiolids inhabit the open sea close to the coast line, they are not only preyed upon by sharks and rays but also by large codfishes and other predacious fish (cod, haddock, pike; see Vol. IV). Sepiolids employ two escape mechanisms when pursued. They either quickly dig themselves into the sea bottom with the aid of the jet current from the funnel, in the manner of cuttlefishes, or they discharge an ink cloud that simulates their body shape. This maneuver is aimed at fooling the attacker long enough to allow escape under the protection of the "double."

Fig. 7-9. Copulation in *Sepiola atlantica.*

Fig. 7-10. Egg cluster of *Sepietta oweniana.* Below: Eggs (magnified) with embryos.

As a rule the SQUIDS (order Teuthoidea) are torpedo-shaped long-distance swimmers with large posterior fins. The shell has been reduced into a sword-shaped chitinous lamella. The suckers are frequently equipped with hooks (Fig. 7-14). The rows of the radula consist of sharp triangular or sickle-shaped teeth; occasionally some rows have double-pointed teeth. There are two suborders with twenty-five families and approximately 375 species.

In the suborder Myopsida the eye is covered with a transparent membrane punctured by a tiny anterior pore. The so-called COMMON SQUID of North America and Europe (*Loligo pealei*, western Atlantic; *L. opalescens*, eastern Pacific; *L. vulgaris*, eastern Atlantic; L 20–50 cm; see Color plate, p. 188) have long, tapering, cigar-shaped bodies characterized by two large, elongate, rounded fins; they are very dissimilar to the squat, rounded bodies of *Sepia* and *Rossia*. The species of the COMMON SQUID (*Loligo* species) inhabit the nearshore regions of the warm to cold waters of the world but they are excluded from the very cold waters. *Loligo* attaches its spawn to solid objects on the ocean floor. Many forms congregate in schools, for example the NORTH AMERICAN SQUID (*Loligo pealei*), which regularly follow large fish schools. These "associations" are not merely determined by food requirements but also by the time of spawning. Schools of young squids are composed of a larger number of individuals than are schools of mature animals. Juveniles fall prey to fish predators and to adult squid, so their schools are reduced during migration. It is amazing to observe how an entire school of squids will suddenly change direction in its arrowlike procession without causing the slightest confusion or collision in the many animals. Despite numerous enemies which prey on the adult squid, or plankton-feeders which prey on the young, the survival of the species is ensured by the very large number of eggs produced. The onset of the breeding migrations is brought about by the amount of daylight and the water temperature. The annual northern migrations begin later in a colder year. Under these circumstances there are fewer individuals in a school.

The eggs are laid in colorless, transparent, gelatinous capsules containing up to several hundred eggs. The eggs generally are laid at night, and are attached in large clusters to the hard substratum. The swollen capsules of from 10 to 20 cm shrink as the embryos become increasingly more developed, causing irregular bulges in the egg string. At the time of hatching, following an embryonic development of about thirty days, the young resemble the adult form even though they measure only 5–7 mm. For a short time these newly hatched forms still maintain a vertical body position and drift passively in the water. During the summer the young grow relatively quickly. Sexually mature individuals may be found from June to October. Most *Loligos* probably die after 1–1½ years, but some may reach an age of two, and sometimes even three, years.

While it is quite easy to keep cuttlefish and octopus in captivity, squids

Order: Teuthoidea

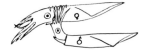

Fig. 7-11. Copulation in *Loligo pealei*.

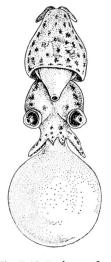

Fig. 7-12. Embryo of a squid (*Loligo*).

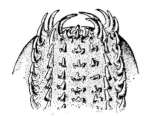

Fig. 7-13. Radula of *Loligo forbesi.*

present difficulties. The size limitation of aquaria and the absence of natural hunting possibilities have caused the failure to keep squids alive for an extended period of time. The torpedo-shaped body is adapted for fast swimming, and the body weight is greatly reduced because of the advanced shell reduction. The shell, or gladius, a mere lancelike horny (chitinous) structure, is enclosed by the mantle and serves as a supporting rod for the body. The gladius is also an adaptation characteristic of the swimming mode of life and is a feature common to all Myopsida. Of the Myopsida only species of *Sepioteuthis* have a dorsoventrally compressed body and long, narrow fins, similar to those of the cuttlefish *Sepia*; this resemblance is indicated by the name: *Sepioteuthis*, the "cuttlefish squid." *S. sepioidea* occurs in the coral reef area of the western Atlantic and Caribbean.

In addition to the North American squid, numerous other loliginids also have commercial significance as human food, fish bait, and even in fish-meal production, especially in the Mediterranean. The NORTHERN EUROPEAN SQUID (*Loligo forbesi*; L up to 75 cm) can cause considerable "damage." Found along European coast lines, these animals congregate into huge schools and consume large quantities of herring. Only man, who has grossly *overconsumed* the herring, could consider the natural prey-predator relationship of squid and herring as "damaging." The DWARF SQUID (genus *Alloteuthis*), which are a few centimeters shorter than the northern squid, also prey on smaller fish. *Alloteuthis subulata* has even

penetrated into the western part of the Baltic Sea. *Alloteuthis media* is found in the Sea of Marmara up to the limit of the tolerable salinity level.

The suborder Oegopsida has twenty-three families. The eye is not covered with a membrane but is open and in constant contact with the sea water (Oegopsida). This group is made up of forms that follow highly diverse modes of life. The Oegopsida include not only the common squids of the high sea but also the fabled giant squid and the highly specialized species occupying the ocean's great depths. In this context special attention should be focused on the spawn: as with all squids, the eggs of the oegopsids are surrounded by gelatinous capsules; however, as an adaptive feature to a particular habitat, the spawn, which often consists of many very small eggs, is not attached to the substratum but instead floats freely in the ocean. The spawn appears ribbonlike or sausage-like and may measure up to 1 m in length. There is a direct relationship between the number of eggs laid and the survival rate of the progeny.

The JEWELED SQUID (*Lycoteuthis diadema*; BL 12 cm; see Color plate, p. 188) is one of the most beautiful representatives of this group. It is found in the open waters of the Atlantic. The mantle is transparent, and the body glitters like a diadem covered with colored jewels. This spectacular glitter is caused by twenty-two luminescent organs which are represented by ten or more different morphological structures. This demonstrates the "ingenuity" and diverse possibilities of nature, making this animal a fascinating creature not only to a casual observer but also to the scientist. Other species are equally diverse, including *Pterygioteuthis giardi*, in which one can also distinguish between at least seven different types of luminescent organs. The cephalopod specialist Carl Chun, who contributed greatly to our knowledge of light-producing organs, was able to differentiate between four types of luminescent organs, generally around the eyes, the anal organs, the gills, and the abdomen, and around the tentacles of *Lycoteuthis diadema*. These organs differ not only in their histology but also in light intensity and color.

An additional specialization was discovered in several other deep-sea forms. *Octopoteuthis sicula*, which measures several centimeters and is distributed throughout the Mediterranean Sea and warm Atlantic waters, is characterized by the presence of only eight arms. The two elongated tentacles are missing on the adults. Once the juvenile development of *Octopoteuthis* was known, the apparent eight-armed phenomenon no longer was a puzzle, as newly hatched animals have ten appendages, and the two tentacles degenerate only during the course of further development until they disappear in the adult and sexually mature individuals.

The family Onychoteuthidae are active predators of the high seas. The horny rings around some of the suckers on the tentacles have been modified into asymmetrically arranged hooks of various lengths. *Onychoteuthis banksi* (BL 15–20 cm; see Color plate, p. 188) is characterized by two rows of these hooks on the terminal end of each tentacle. This structural arrangement is a very effective grasping organ. Similar structures are evident in the North Atlantic *Gonatus fabricii* (family Gonatidae), which is somewhat larger and inhabits colder regions of the ocean. Hooks and claws which are less highly developed and less conspicuous than in *Onychoteuthis banksii* are also present on the entire surface of the arms.

The GIANT SQUID (family Architeuthidae) are forms of totally different appearance. We are familiar with them from the frequently exaggerated, fear-inducing "sailor yarns." Although we should not be carried away by legends from earlier centuries, there remains a bit of truth in these stories. Tales of giant squids attacking sailboats and dragging the entire crew to abyssal depths belong in the realm of fables, of course, but not everything can be dismissed as tall exaggerations. Frank W. Lane described several "reliable" reports from recent times where giant squids actually attacked boats on occasion, perhaps mistaking the vessel for their enemy, the sperm whale. Giant squids which have washed up on beaches, and examples from the stomach contents of several sperm whales, demonstrate the large body size reached by some individuals; some, including their extended tentacles, measure nearly 20 m. A North Atlantic species, *Architeuthis clarkei*, which was washed up on the eastern coast of England, had a total length of over 6 m. Similar and larger measurements have been recorded for *Architeuthis harveyi*. Several forms, including *Architeuthis princeps*, reach the unbelievable body length of 6.5 m, with tentacles that measure

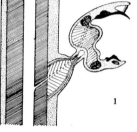

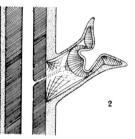

Fig. 7-14. A longitudinal section through the suction disk of 1. a squid and 2. an octopus.

Giant squids

10 m, and a body weight of 3 t. The current record is held by a specimen caught in New Zealand in 1933. The animal's body length was 8 m and the tentacles measured 14 m: a total of 22 m! Unfortunately, storms and rigor mortis greatly contribute to the decomposition of the soft body of the giant squids before they reach the upper water surface or the shore. Often one can determine the species of the washed-up individual only on the basis of the form of the jaw. Sometimes even this is impossible.

Aside from these single discoveries of giant squid, many astonishing facts about these animals have been obtained from the stomach analysis of sperm whales (see Vol. XI, Chapter 16). For example, the eyes have a diameter of 40 cm, the largest eyes in the entire animal kingdom. Some observers suggest that giant squid may achieve a body length of 10 m, a length of over 25 m, and a body weight of several tons! However, such large forms have never been found; possibly they are extraordinary individuals of species that usually have an average length of several meters. The data suggest that these animals have a longer life span than just one or two years. According to data based on the stomach analysis of sperm whales it can be deduced that giant squids occur relatively frequently at intermediate depths in all oceans.

Histioteuthis bonellii

Histioteuthis bonellii (L approximately 50 cm; see Color plate, p. 188), found at great depths, in the Mediterranean and Atlantic, is not less remarkable than the giant squids. The mantle and arms have a conspicuous purplish-red coloration and are endowed with numerous luminescent organs. The long tentacles are connected by a broad interbrachial web. The right and left eye are of different size, giving the animal a striking appearance. In the closely related form *H. dofleini*, which occupies the same habitat, the size difference of the eyes is particularly pronounced. The somewhat smaller, orange-red body lacks the interbrachial web but is also covered by luminescent organs. The left eye is characterized by an enlarged "diseased-looking" eyeball (bulbus) which is larger than the right one. This enlarged eye can reach a size of one-third of the animal's head. The development of the ocular light-producing organs is also influenced by the unequal development of the eyes. The right eye is surrounded by a luminescent ring; the left eye, enlarged four times, is only associated with an occasional light-producing organ. Many are also partially degenerated. The significance of the development of eyes of different sizes is not known.

Illex illecebrosus

Like its closely related forms, the SHORT-FINNED SQUID (*Illex illece-brosus*; L 20 cm) is well adapted to swimming in high seas. The animals congregate into schools and pursue their prey, particularly herring and shrimps. The Mediterranean species, *Illex coindeti*, is common along European and tropical western African coasts, while the western Atlantic species, *Illex illecebrosus*, occurs along North America's eastern coast from Labrador and Newfoundland to the Gulf of Mexico. The muscular body and the fused, heart-shaped fins characterize these forms as excellent swimmers. The powerful expulsion of water through the funnel is the only mode of propulsion. During sudden backward darts the large terminal fin is closely rolled around the body to decrease water resistance. In addition to the male's hectocotylus, the *Illex* species also show sexual dimorphism in the body size.

Todarodes sagittatus (see Color plate, p. 188), one of the larger species of this group, is an avid predator. As inhabitants of the high seas, schools of them follow the migrations of their prey fish. The species can be distinguished from a similar *Illex* species by two conspicuous characteristics. The fused fin is not heart-shaped but rhombic, and the elongated tentacles are nearly entirely covered by suction cups rather than only on the club-shaped ends. The juvenile form of *T. sagittatus* has a different appearance than the adult's, and for this reason they had been described as a separate genus (*Rhynchoteuthis*; see Color plate, p. 188). In the juvenile stage the underdeveloped tentacles are still fused and form a trunklike appendage.

Todarodes sagittatus is one of the squid species which are able to dart out of the water with a backward motion when escaping a predator. *Ornithoteuthis volatilis*, from around Japan, is well known for this backward escape motion, as is *Ommastrephes bartrami* (L approximately 1 m), from the Atlantic and other warm oceans. Its slender body is bluish-red. Like several other squids, this species has a very broad skin fold on the third pair of arms. In the species *O. caroli* this protective membrane has developed into a three-cornered "sail."

Among the cephalopods *Chiroteuthis veranyi* (L approximately 15 cm; see Color plate, p. 197) is an "oddity." This bluish, transparent, pelagic animal occurs in the Mediterranean among jellyfishes (see Vol. I) and Salpida (see Chapter 18). It is characterized by conspicuous large eyes and, on the posterior end of the body, a heart-shaped fin which is sharply set off from the rest of the body. The two elongated tentacles seem to be totally out of proportion. The small body appears like an appendage when compared to the large head and threadlike tentacles which measure up to 1 m and are covered by small, red, wartlike pads (light organs). The long tentacles serve as fishing lines for acquiring food. The juvenile forms of *Chiroteuthis veranyi*, which were formerly classified under a separate name (*Doratopsis vermicularis*; see Color plate, p. 197), are characterized by relatively normal arms and a slender body form. They have a posterior rodlike elongation often as long as the body. This long spine is part of the gladius that is not encircled by the mantle.

Other extraordinary and even bizarre forms are found in the family Cranchiidae (L usually 3–10 cm). The transparent gelatinous mantle is fused with the head. Parts of the internal organs and photophores shimmer eerily through the transparent mantle. It seems that barrel-shaped or streamlined animals are adapted to living conditions in zones with little light. The reduction of the arms (but not tentacles; compare Color plate,

Todarodes sagittatus

Chiroteuthis veranyi

Deep-water adaptations

p. 197) in juveniles of numerous species seems to indicate a specialized mode of acquiring food. The cranchiids are characterized by their extraordinary eye structure, which Siegfried H. Jaeckel described: "The eyes are frequently enlarged, and in a number of cases are stalked. Stalked eyes were particularly common in the juvenile forms of deep-water, freeswimming species which have normal eyes in the adult stage. *Bathothauma lyrroma* [from the central Atlantic], and *Sandalops melancholicus* [of the southern Atlantic; see Color plate, p. 197] have permanent stalked eyes. The unique ventral location of the eyes in *Sandalops* is probably associated with its upward-slanted swimming posture, which has also been observed in *Cranchia scabra*."

Cranchia scabra is widely distributed and is one of the few cranchids with a deep, dark brown coloration, which is, however, characteristic for numerous other deep-living forms. In addition to the unique structure of the eyes, many species have a great diversity of photophores. In *Desmoteuthis pellucida* (see Color plate, p. 197) the light-producing organs can illuminate up or down in a spotlightlike fashion. Cranchid fins usually consist of two small, almost terminal, nearly circular, weakly muscled structures. Most forms control their bouyancy by concentrating light-weight ions in the body; they float without self-induced movements and drift vertically or horizontally depending on the distribution of the bouyancy fluid in the large coelomic sac in the mantle cavity. These fluids are isotonic in relation to the surrounding sea water but their specific density is less. The specific weight can be reduced drastically and thereby cranchids can effectively regulate their passive movement.

Order: Vampyromorphya

The VAMPIRE SQUIDS (order Vampyromorpha) are dibranchiate cephalopods characterized by eight long arms and two (the second pair) small, filamentous tendrils. True tentacles are absent. The suction cups are not stalked, but are slightly raised and are hardened internally, although without a recognizable horny ring. The mantle has reduced fins. The funnel is equipped with a valve. The ink sac is degenerate, and there are no egg-shell glands. The oral epidermis does not consist of flaps. There is only one species, the VAMPIRE SQUID (*Vampyroteuthis infernalis*; L 10–28 cm, BL 6–10 cm, arm L 6–8 cm; see Color plate, p. 197).

"One of the most legendary figures among the pelagic [living in the open sea] deep-water forms," is how Carl Chun described the vampire squid, which he introduced to the world for the first time after the German *Valdivia* Deep-Sea Expedition in 1903. Despite the species' unique appearance, and the discovery of several other specimens during the course of the following thirty years (which led to the mistaken description of eleven different "species"), a truly sensational fact about this animal remained a secret from the public until the North American scientist Grace E. Pickford revealed for the first time in 1936 that the vampire squid (or the literal translation of the scientific name, the "infernal vampire squid") was in reality a relict of long extinct forms which link the squids and octopuses. The external features and the plump body shape of the vampire squid are reminiscent of the Octopoda, to which *Vampyroteuthis* had been classified up to that time, but the internal anatomy and a few specializations are indicative of the Teuthoidea. Pickford was able to demonstrate in individual representations that all of the captured animals of this group really belonged to a single species. The mistaken classification of vampire squids into eight different genera was primarily a result of faulty preservation techniques and the various degrees of decomposition of the individual specimens. The vampire squid enjoys a worldwide distribution in the colder deep-water zones of tropical and subtropical oceans.

The vampire squid is characterized by a plump, saclike body enclosed by the mantle, and two shining over-sized, deep-red eyes. The eight arms are connected by an interbrachial web. The large umbrella thus formed is responsible for the name "vampire." The impression of this sinister apparition is emphasized by the purplish-red to dark coloration which hardly contrasts to the dark black interior of the interbrachial web. Two paddle-shaped fins are present on the dorsal surface. The arms are equipped with protruding papillae (cirri; see Color plate, p. 197). The papillae are found near the mouth in groups of five to eight, starting in front of the first suckers (primary cirri). There are up to 110 cirri on each side. The exterior suction cups are larger than are the interior. Toward the tips the suckers become wartlike papillae. Up to now the largest vampire squid (L 28 cm) was caught on the southeastern coast of Africa. This specimen did not show any deviation in the number of primary cirri but unfortunately the arms, which showed over thirty suckers, were greatly decomposed.

A closer observation of a vampire squid reveals that the animal possesses a pair of small, usually coiled structures aside from the eight large arms which are connected by the interbrachial web. These filaments can be retracted into pouches between the uppermost and second pair of arms. Originally these filaments represented a fifth pair of arms (the second pair in the arm row). The arms were greatly reduced and were obviously modified into tactile organs. Two wartlike organs in the vicinity of the mantle opening seem to have an olfactory function. The vampire squid also has light-producing organs, a characteristic for deep-water cephalopods. There is a pair of large organs on the dorsal surface at eye level consisting of thirty to seventy-five individual units. Two luminescent organs, equipped with a cornea, are located posterior to the fins. Each of these luminescent organs can be closed by a lid. Numerous smaller photophores are situated in the mantle covering, particularly in the regions of the funnel and the bases of the arms.

Like most cephalopods, the vampire squid does not possess a true mollusk shell, but there is an uncalcified internal shell remnant, which is significant in relationship to the body size. Viewed from the dorsal side,

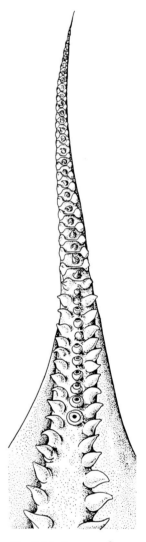

Fig. 7-15. An arm of *Vampyroteuthis infernalis*, showing the cirri and suckers.

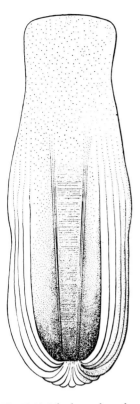

Fig. 7-16. The boat-shaped gladius of *Vampyroteuthis infernalis.*

Habitat

the boat-shaped, transparent gladius is three times as long as it is broad, usually measuring from 3 to 6 cm. The gladius of the specimen from southeastern Africa measured nearly 10 cm. This unique gladius shows a slight downward coil at its posterior end, as is seen in *Spirula spirula*. The gladius is blunt at the head end, and has no recognizable similarity with the gladius of any living form. Pickford suggested that the structure could only be understood meaningfully if compared with specific extinct ancestral forms.

An examination of the embryology of the vampire squid serves to explain this animal's uniqueness. The female, usually larger than the male, seems to discharge the eggs singly into the water. The eggs are globular, measuring 3 to 4 mm. The male probably transfers the spermatophores to the female from his funnel, since he lacks a hectocotylus. The fins undergo an unusual development during the course of juvenile development. A freshly hatched vampire squid is characterized by a head with eyes, stumpy arms, and a finless body with two posterior photophores. Length is approximately 7 or 8 mm. Two small flaps, the larval fins, soon begin to grow just behind the photophores. When the juvenile has reached a length of 3 to 3.5 cm, a second pair of fins develops in front of the posterior photophores. Finally we have a juvenile form of 4.5 cm, sporting four equal-sized fins on the dorsal posterior end of the mantle (see Color plate, p. 197). Surprisingly, the posterior larval fins degenerate during the course of further development. The adults have only one pair of fins, which are paddle-shaped, situated in front of the photophores. These observations partly explain why individual specimens at different stages of development were classified as different species.

According to present-day data, the habitat of these "living fossils" is limited to areas with specific temperature, salinity, and oxygen-concentration ranges. The vampire squid is a free-living form occurring at depths of 300 to 3,000 m, usually in the zone between 1,500 to 2,500 m. It prefers a temperature of 2 to 6° C and a layer of water where the oxygen content is barely above the minimum requirement. It may seem that the eyes (as in other deep-sea forms) are superfluous in this dark abyss; however, these zones are inhabited by numerous organisms with luminescent organs which, like the vampire squid, are independent of sunlight. The geographical distribution is worldwide, but seems limited north-south by the fortieth degree north and south latitudes. The VAMPIRE SQUID (*Vampyroteuthis infernalis*) is found from the latitudes of the Azores to South Africa across the Atlantic, and from the near-shore deep waters of Oregon, California, and Mexico in the Pacific.

Up to now about 120 vampire squids in all developmental stages have been captured. It is probable that deep-sea research will uncover more animals of this ancient group of dibranchiate cephalopods and perhaps also discover some other species, thereby greatly enhancing our knowledge about vampire squids.

The last order of cephalopods and also the last mollusk group to be discussed is the Octopoda. Length including the arms is from 1.5 cm to over 3 m. There are usually eight equally developed arms, with no filamentous or tentacular arms. The labial membrane is missing. The suckers are not stalked but have a broad base without horny rings and hooks (Fig. 7-14). The mantle is fused with the head on the dorsal side. Fins are usually absent. The funnel lacks a valve. The nidamental glands are absent. The pericardium is greatly reduced. There are two suborders: 1. Cirrata, with three families. 2. Incirrata, octopuses (in the broad sense), with nine families.

Order: Octopoda

The unusual shape of the octopus is well known even outside zoological circles, due in no small degree to the numerous illustrations and caricatures. The body of the animal is often mistaken for the "head" because the octopus swims "backward" when making a rapid escape. There are some species that have well-developed interbrachial webs between the arms, reminiscent of *Histioteuthis bonellii* and the vampire squid.

The arms of members of the suborder Cirrata are equipped with suckers and two rows of papillae (cirri). The arms are always interconnected by an interbrachial web. No hectocotylus is known. There are two fins. A shell vestige is present. The radula, posterior salivary gland, and ink sac are reduced or absent. There are three families with approximately thirty species, all very poorly known.

Suborder: Cirrata

The Cirrata are deep-sea forms. They feed on plankton, minute crabs, and other small organisms. The type of food explains the absence of the radula. Possibly the well-developed interbrachial web and cirri on the arms have a special function in capturing small prey.

The Cirrata still possess well-recognizable uncalcified internal shells, somewhat saddle-shaped, probably because of their free-swimming mode of life and the development of fins.

Cirroteuthis muelleri (L 25 cm) was first found around Greenland at depths of 400 to 3000 m. This form has two paddlelike fins which are supported by a saddlelike gladius, an arrangement which facilitates regular fin swimming. *Cirrothauma murrayi* (L 13 cm; see Color plate, p. 198) is similar to *C. muelleri* but lacks large eyes; it is the only presumably blind cephalopod. It lives at depths of 2000 to 3000 m in the North Atlantic. It has a noticeably small funnel which could hardly aid in the animal's locomotion. Probably the large interbrachial web, as in the jellyfish, has taken over this function. To Carl Chun, *C. murrayi*'s locomotion and appearance resemble those of a jellyfish. *C. murrayi* is also characterized by a gelatinous body consistency, an adaptation to the abyssal habitat. Like jellyfishes, it appears semitransparent and gives the impression of fragility.

The STAUROTEUTHIDS (family Stauroteuthidae) represent the Cirrata family with the greatest variety of species, approximately twenty. This family enjoys a worldwide distribution. *Grimpoteuthis umbellata* (L 15 cm)

Fig. 7-17. *Opisthoteuthis*—lateral and dorsal views.

Suborder: Incirrata

belongs to this group. This plump, purple, deep-sea species occurs in the Atlantic from the Azores to the Cape Verde Islands. It does not possess a large interbrachial web. The two protruding fins of all stauroteuthids are supported by a V-shaped or U-shaped internal shell (gladius).

In the OPISTHOTEUTHIDS (family Opisthoteuthidae; L up to 30 cm) the vestigial shell has been reduced to an insignificant rodlike structure. The animal can do without this support since it is completely adapted to living on the bottom of the deep sea. Its brownish-purple body seems to consist only of the interbrachial web, which is molded to the ocean floor, and the protruding semi-globular eyes. The flattened, greatly reduced body, with the vestigial fins and small funnel, hardly seems to stand out at all. There are six known species. *Opisthoteuthis agassizi* is distributed in the Atlantic from Ireland and Greenland southward to the Gulf of Mexico and western Africa. *O. californiana* occurs along the western coast of the U.S.A. Their unusual body structure indicates a crawling mode of life or a jellyfishlike mode of swimming. Hectocotylus and spermatophores are present.

Octopuses in the more common sense (suborder Incirrata) possess arms without cirri. The interbrachial web is only weakly developed or is absent. The mantle lacks fins. There are nine families, with approximately 170 species.

The gelatinous, often transparent body of members of the family Bolitaendae is somewhat reminiscent of the cranchids which live in similar habitats. The particular mode of life of the bolitaenids has resulted in the reduction of cartilagenous support, and the two jaws are less strongly developed than in benthonic octopods. It seems probable that the bolitaenids feed on soft substances. The group is transitional in appearance from the Cirrata, since they possess a well-defined but not large interbrachial web.

The genus *Bolitaena*, which only measures a few centimeters in length, inhabits the mid-depths (600–1500 m). The Indo-Pacific *Amphitretus pelagicus* (L 5–6 cm) is also pelagic and occurs at depths of about 250 m. This form is characterized by the dorsally directed telescopic eyes which presumably perceive incoming light from above. Siegfried H. Jaeckel mentioned, however, that "it must be emphasized that this type of eye does not function like a telescope. The term merely describes a degree of similarity between the two." Both eyes are synchronized and perceive the same image. The eyes are neither far-sighted nor near-sighted, but the distance between lens and retina has been increased. The ability for accommodation is very limited. Probably they can recognize motion and the distance to close objects. The animals are covered with a gelatinous coating which also extends over the interbrachial web except for the arm tips. *Vitreledonella alberti* (see Color plate, p. 198) is reminiscent of a patterned glass ball. Its organs appear extremely small. According to Louis Joubin, the species *Vitreledonella richardi* "bears" live young, but, in fact, the eggs and hatchlings are brooded in the chamber formed by the arms and web.

Unlike the previously discussed octopod cephalopods, the octopuses (family Octopodidae) are generally well-known animals.

The COMMON OCTOPUS (*Octopus vulgaris*; L up to 3 m; see Color plates, pp. 198 and 223) is distributed nearly around the world in warm waters. The octopus, the most familiar of all cephalopods, lives on the bottom in shallow water along coasts. Its saclike body lacks fins. Two rodlike vestigial remnants of the shell are present only in two genera. The arms, each equipped with two rows of suckers, are long and very flexible. The large, lidded eyes appear fixed. Various subspecies enjoy a worldwide distribution. Individuals show great variations in color and size. An octopus in the North Sea barely reaches a length of over 70 cm while specimens caught in the Mediterranean have measured 3 m and weighed 25 kg. When the animal is excited, the usual yellowish-gray to dark brown marbled skin coloration may become red and flushed. Camouflage coloration is well developed.

The octopus is a bottom dweller, not found in the open waters. When threatened, it retreats by swimming backward. The octopus uses its extremely powerful arms for locomotion; it either crawls or stalks, preferably on a rocky substrate, to find a natural cave from which it can prey on other animals. Octopuses respond to every tactile stimulus by moving (thigmotaxis). Under unfavorable conditions an octopus always has an escape route. If no suitable niche is present, it will construct a gravel "nest" or erect a stone wall with its arms. The female practices brood care, watching over the spawn, grapelike clusters of small eggs, in a small cave or a stone nest. Occasionally a female may look after a brood of up to 150,000 eggs. She circulates the water around the eggs, touches and cleans them, and does not leave them for the entire brooding time of one month. Some octopuses inhabit empty bivalve shells during the brooding period, closing them for protection when disturbed. In *Octopus aegina* brood care even includes aid during hatching. After the brood period, of different duration in the various species, the females usually die, because they do not feed during the brooding period.

The octopus hunts everything that it is able to overpower, its main sources of food being crayfish, crabs, and various bivalves. Like a true waylayer, the animal ambushes its prey from the background. An octopus often stores several captured animals in its web to be eaten later. The octopus keeps a refuse pile of food remains outside its hiding place. The octopus captures prey with its arms, pulls it toward its mouth, and cracks it open with the jaws. Some prey may defend itself, for example the lobster (*Homarus vulgaris*; see Vol. I). The lobster is protected by a heavy armor and is equipped with powerful pincers. Ulrich K. Schulz described a fight between an octopus and a lobster which "approached the octopus with the extended pincers and attempted to grasp the predator

The octopuses

Brood care

Predator and "ambusher"

in the center of the trunk. The chips seemed to be stacked in favor of the lobster. The large octopus discharged the entire content of its ink sac in response to the pain inflicted by the lobster. The octopus took advantage of the lobster's resultant confusion and prepared for an all-out offensive. Under the protection of the 'smoke screen,' the octopus sinisterly approached its foe from all sides with its eight undulating arms. Arm after arm adhered itself to the lobster with hundreds of suction cups. This time the entanglement spelled doom for the lobster, its dangerous pincers held in a tight embrace. The jawed mouth of the octopus lowered itself over the opponent and crushed the chitinous shell of the lobster's head region, also injecting a poisonous secretion from the salivary glands. A final quiver and jerk from the lobster and the octopus was the undisputed victor."

As in all social interactions with conspecifics, the territorial animal always has the "psychological advantage" when defending his own territory. Competitors and rivals are attacked in the same fashion as are representatives of other octopod species. In order to give an indication **Brain performance** of the high intellectual capacities of the octopus, here are some observations made by Anton Dohrn. He observed fights between a lobster and several octopuses at the zoological station in Naples. In order to avoid the constant conflicts between them, Dohrn placed the lobster in an adjoining basin. "The lobster was separated from the two preceding basins by a wall which contained an open door, and a cement wall which protruded two centimeters over the water level. Yet this attempt to protect the lobster from the aggressive octopuses was in vain. During the course of the day one of them crawled over the wall and attacked the unsuspecting lobster. After a brief tussle the lobster was literally pulled in half. In the time span of barely forty seconds the victor had not only completed the fight but had also started to eat the victim! This act shows behavior that goes far beyond an innate behavior pattern. It demonstrates intelligence. It is probable that the octopus had observed the keeper placing the lobster into the next basin, or it became aware of the close proximity of the prey via the circulating water supply. Whatever the reason, it remains a fact that the octopus sensed the prey even though the latter was not visible, and it crawled out of the water to the lobster.

Similar chance observations by zoologists stimulated a series of planned experiments to test the cognition and retention of the octopus. Alfred Kaestner conducted such an experimental series: a home for an octopus was constructed in a water tank. Three parallel channels led off from one side of the tank. The ducts were separated by opaque plates. The central duct was open to the home but the two lateral ones were closed by a glass plate. The three ducts were interconnected outside the living area. "A crab inside a deep beaker was placed in the left duct. The glass beaker prevented chemical stimuli leaking from the crab into the water; however, the octopus was able to view the crab through the glass.

From twenty-nine animals, eight (experiment was repeated because of the high number of test animals) ventured into the central duct at the first try and always turned correctly towards the left although the test animal was not able to see the crab all the time it was in the central chamber." The other octopuses soon learned to perform the same feat. One test animal was shown a crab within a glass in the left duct for thirty consecutive days, and it always turned off to the left. However, on the "thirty-first day the prey was put in the right side. The octopus took the correct right turn. The animal could only have performed this test because of its reasoning abilities." Another octopus was even able to uncork a glass jar and extract the prey. These capabilities far exceed any performance observed in other invertebrates; rather they approach the brain performances of higher vertebrates.

The octopus is not only an agile fighter and hunter in the water but occasionally it may venture from its own habitat to pursue a crab on the dry land. It seems that Anton Dohrn's observation of the octopus climbing over a wall protruding out of the water was not a unique occurrence. Furthermore, the octopus is often very skillful in avoiding its enemies. Some of the octopuses' predators are the Mediterranean moray eel (*Muraena helena*; see Vol. IV, Chapter 9), the eel (*Conger conger*; see Vol. IV, Chapter 9), as well as sharks and dolphins (*Grampus griseus*; see Vol. XI, Chapter 16). Irenäus Eibl-Eibesfeldt described the meeting of an octopus and morays: "At dusk the morays came out of their hiding places to search out their prey under the protective darkness. A hunting moray orients itself primarily by its olfactory sense. The defensive mechanisms of the octopods which are some of their victims is adapted to this ability, for during escape the octopuses discharge a fluid that temporarily anaesthetizes the predator's olfactory sense. In this case the ink secretion does not serve as superfluous smoke-screen. During visible daylight hours both the ink cloud and the olfactory immobilizer come into effect. By the time the enemy has recovered his full sensory capacities the octopus has long since escaped."

In contrast to many squids, the octopus is solitary except for a brief interval during the mating season. Bitter fights between rivals often center around a female. The victor claims the female and inserts the hectocotylus into her mantle cavity. Both mates sit apart and only the hectocotylus join the two. The spermatophores move along the hectocotylus in wavelike contractions until they finally reach the female's genital opening. Copulation lasts for over an hour.

Similar modes of behavior are found in most other related species, for example in *Octopus macropus*, which enjoys a worldwide distribution, and *Eledone cirrosa*, which occurs along Europe's coast from the Adriatic to the Lofoten Islands. *E. cirrosa*, frequently gray to brownish-marbled, can easily be distinguished from the octopus by its smaller size and well-developed interbrachial web. Furthermore, *E. cirrosa* is equipped with

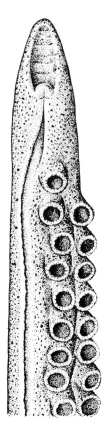

Fig. 7-18. The hectocotylus of the common octopus (*O. vulgaris*).

Fig. 7-19. Distribution of *Eledone cirrosa*.

only one row of suction cups, which become reduced into transverse ridges and cirri at the arm tips.

Eledone moschata (L up to 40 cm) is also characterized by an interbrachial web and a single row of suction cups. This animal possesses a distinct musk odor. *E. moschata* is limited to the Mediterranean, where it inhabits mud and sandy flats. Like *E. cirrosa*, this form feeds primarily on small organisms, and because of its life habits is rarely seen by man, although numerous specimens have been collected by bottom dredging. Like other octopuses, this form fights vigorously when captured. Under normal circumstances it avoids man by either retreating in its home niche or escaping. These octopuses (*Octopus, Eledone*), which measure from 10 to 20 cm, can inflict biting wounds when captured, but the wounds are only the size of an ant's or mosquito's bite and cause limited swelling of the skin.

The various octopod species differ only slightly in their behavior. In *E. moschata* the interpretation of sensory input is highly developed. The manner of spawning is similar in all octopuses. The spawn is attached to solid objects just as in *Sepia*. *E. moschata* and *E. cirrosa* do not practice brood care.

Graneledone verrucosa and *Bathypolypus arcticus* (L rarely over 25 cm), both distributed in the North Atlantic, lack the ink sac. This feature is directly related to the deep-sea habitat of both species. The smaller-sized *B. arcticus* is equipped with a unique hectocotylized arm. The third right arm terminates in a spoonlike structure characterized by nine to seventeen (usually eleven) transverse ridges on the inner surface, without suction cups or similar structures. Unlike the other mid-water octopuses, *B. arcticus* is exceptional in its manner of movement; occasionally it swims by pushing against the water by closing and opening the interbrachial web. As in the decapods the funnel is bent posteriorly during swimming, particularly when capturing prey.

Argonautidae

The hectocotylus reaches its peak of specialization in members of the family Argonautidae. In the various species of this group the females may have one or several detached hectocotylized arms with suction cups in their mantle cavities. Aristotle discovered these structures and until the 19th Century it was assumed that these were parasitic worms. They had been described as *Trichocephalus acetabularis* and *Hectocotylus octopodis*. In 1852 Heinrich Müller was able to explain the true nature of these mysterious structures. The "worms" were in fact the hectocotylized arms of the males. During copulation or prior to the process, the arms become detached from the males and penetrate into the female mantle cavity. This type of spermatophore transfer ensures fertilization. Argonautid males are dwarfs. The unique hectocotylus is coiled within a bulbous sac. When the hectocotylus is fully developed, this sac bursts open and the whiplike arm is released.

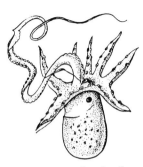

Fig. 7-20. Dwarf male of the argonaut, with freed hectocotylus.

The North Atlantic *Alloposus mollis* (L approximately 10 cm) is another member of this peculiar octopod group. This form is characterized by a broad, well-developed interbrachial web and short arms. It is extremely soft and gelatinous. In *Alloposus mollis* the hectocotylus does not detach, but in other aspects it is similar to other forms.

Tremoctopus violaceus occasionally is found in schools. In this species not only does the male's hectocotylus become detached, but in the female (L approximately 15 cm) sections of two highly broadened upper arms reportedly are shed during the breeding season (segmented autotomy). The brown to purple animals are distributed in warm waters of the world. In addition to ring lines on the arms, caused by "self-mutilation," this species is characterized by another oddity. The mantle is connected with the head on its dorsal side by a skin fold which is perforated by large holes. These openings penetrate into the underlying tissue which contains sensory organs for the detection of water currents (compare lateral lines on the bony fishes, Vol. IV, Chapter 3).

In *Ocythoe tuberculata* (L, ♂♂ 3–4 cm, ♀♀ 25–30 cm) the reproductive process is different again. The species is distributed throughout the Mediterranean, Pacific, and Atlantic. The male's hectocotylus also becomes detached and the male itself occupies a pelagic tunicate (thalicid) shell (see Chapter 18) from which the contents have been eaten. The female practices internal brood care. The female broods the eggs in the oviducts and "bears" living young.

The PAPER NAUTILUS (*Argonauta argo*; see Color plate, p. 198) is one of the best-known species of this group and also is one of the most remarkable cephalopods. It inhabits the Mediterranean and other warm and temperate oceans. The first time one sees a female paper nautilus float on the water in a shell measuring about 20 cm, one could assume that the animal has retained the true mollusk shell. Closer examination reveals that this "brood shell" is not a structure of pallial origin. The boat-shaped shell serves primarily to protect the minute eggs. In the juvenile stages the female develops a flaplike enlargement at the tips of the upper arms. The inner surface of these flaps secretes an eggcase, or secondary shell, where the halves fuse at the lower margins (secondary because this shell has nothing in common with the original mollusk shell, which is present in most octopuses as an internal vestigial shell). The boat-shaped shell is supported by the arm flaps, since it is not attached to the body.

The female usually swims with its mouth pointed slightly upward, and the well-developed funnel which propels the animal lies in the axis of swimming. During this process the shell-supporting arms encircle the shell.

The dwarf male (BL 1–1.5 cm; hectocotylus 2–3 cm; Fig. 7-20) lacks the shell and flaps on the dorsal arms. His "purpose in life" is copulation, whereby he loses his large hectocotylus.

We have now come to the end of our journey through the world of the mollusks and have concluded with the extremely highly evolved

Paper nautilus

▷
Above: A gliding squid (*Sepioteuthis*). Below: The common octopus (*Octopus vulgaris*; compare Color plate, p. 198).

cephalopods which demonstrated special adaptations both to their external environment and in their mode of reproduction. We became acquainted with an animal phylum whose basic body plan followed many different evolutionary patterns. Ever new and more elaborate adaptations were developed and perfected in response to evironmental conditions. This survey reveals much more than the shear diversity of nature: it demonstrates the full wonder of life evolving through millions of years.

◁
Above: The common cuttlefish (*Sepia officinalis*; compare Color plate, p. 187) showing various colorations. From left to right: during courtship, camouflage, and fright. Center: First and second figures from the left: Embryological development and hatching of *Rossia macrosoma*. Third figure from the left: *Rossia caroli* (compare Color plate, p. 187). Below: Egg clusters of the common octopus (*Octopus vulgaris*; compare Color plates, pp. 198 and 223).

8 Lophophorates: The Phyla

Lophophorates, by
E. Popp

The lophophorates are the last major group of the Protostomia. They are characterized by a crown of ciliated tentacles surrounding the mouth. The cilia produce currents which sweep food particles to these exclusively aquatic animals. Similar tentaculate structures are evident in the rotifers (see Vol. I), but those are adaptations to a specific mode of feeding and do not indicate a phylogenetic relationship. The lophophorates are also interrelated by various other features, including the original segmentation of the body into three sections, and the trend toward formation of an internal skeleton. The phoronids and bryozoans have similar larval forms, the actinotrocha larva. The first two characteristics, however, also indicate a phylogenetic relationship to the Deuterostomia, whose formation is the same as that of the lophophorates.

The lophophorates have little in common with the other protostomian phyla. These groups apparently branched off from the common root to the deuterostomes and protostomes early in evolutionary history.

The lophophorates are diverse and inconspicuous creatures. The phoronids are reminiscent of the rotifers (see Vol. 1), the bryozoan of the colonial coelenterates (see Vol. I), and the bivalved brachiopods of mollusks, in particular the bivalves (see Chapter 6). At one time the brachiopods were even named Molluscoidea (mollusklike animals) because of their resemblance to the bivalves. However, all lophophorates share one common characteristic, the arch or ring of ciliated tentacles surrounding the oral region. The tentacles arise from a lophophore supported by skeletal elements. The tentacles supply the animal with food (plankton) and oxygenated water. They serve as a type of "cephalic gill." All lophophorates, except the fresh-water bryozoans, inhabit all depths and zones of the ocean, where they are either permanently or periodically sessile. All forms show a trend toward colonization, which is particularly evident in the bryozoans.

With few exceptions, the lophophorates are diminutive animals measuring only a few millimeters or centimeters. Only the brachiopods

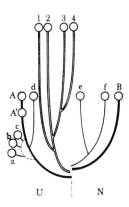

Fig. 8–1. The major phyla of the coelomate animals. Double lines = Lophophorates; 1. Brachipods, 2. Phoronids, 3. Freshwater Bryozoans, 4. Marine Bryozoans. Thick lines = (U) Protostomia and (N) Deuterostomia: A. Arthropods, A'. Annelids, B. Vertebrates. Thin lines = Branching phyla: a. Flatworms, b. Nematodes, c. Nemertines, d. Mollusks, e. Acorn worms, f. Echinoderms.

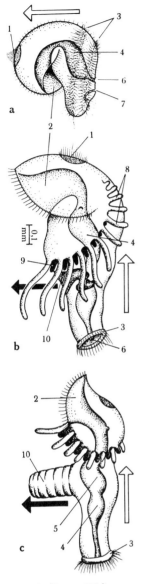

Fig. 8-2. (See p. 228.)

Phylum: Phoronida

reach the sizes of small bivalves. The bryozoans include dwarf forms. Usually the body consists of three segments. The most anterior part of the body is the tentacles, which, however, arise from the central body segment. The fore body is reduced except for a few vestiges. The hind body hangs downward like a sac. It is surrounded by a tube consisting of chitin and calcareous substances. In the brachiopods this part of the body is encased in a pair of bivalvelike valves. As a special adaptation to constant enclosure in a tube, the animal's anus is located outside the tentaculate crown. The intestine is U-shaped and hangs suspended in the coelom, where it is constantly bathed in coelomic fluid. Metabolic substances are exchanged in this fluid and in the more or less closed vascular channel system, except in the Bryozoa. Under favorable feeding conditions the gametes mature and are released into the water via the nephridial ducts. The fertilized egg develops into a free-swimming larva with ciliary bands (actinotrocha larva). Many species practice brood care. All lophophorates possess a stalk on their hind bodies which attaches to the substratum. The animals can withdraw the lophophore inside the protective tube by contracting a muscular trunk, and expose it again by increasing coelomic pressure. There are three phyla: 1. Phoronids, 2. Bryozoans, 3. Brachiopods. There are approximately 5,000 species.

In paleontology the lophophorates and in particular the brachiopods are the most important indicator fossils of the early Paleozoic era. These forms have been represented by a wealth of species since the Lower Cambrian (approximately 600 million years ago). Since then the brachiopods, along with the inarticulate forms, have maintained their original body structure and mode of life almost without modification.

The PHORONIDS (phylum Phoronida) look like polychaetes at first glance, but their affinities are shown by the embryological development of the larva. Following a three-week planktonic existence, the actinotrocha larva, which is similar to a trochophore (see Vol I), sinks to the bottom. Within a few minutes a long tube emerges from the hind body in the direction of the mouth, like a tube out of a burst tire. Subsequently the intestine, suspended from the body wall by mesenteries, is drawn out in a U-shaped loop. Since it is customary to refer to the mouth-anus line as the longitudinal body axis in bilaterally symmetrical animals and to call the body side with the mouth the ventral surface and the one with the anus the dorsal surface, one has to regard the entire hind body tube of the phoronids as the ventral visceral sac and the shortest distance between mouth and anus, between the two ends of the U-shaped intestine, as the longitudinal body axis. Therefore the tentacles are located on the dorsal side of the animal, on the anterior surface of the longitudinal axis. Keeping this "T" symbol in mind, where the left arm indicates anterior and the right arm indicates posterior, we should have no trouble understanding the anatomy of the phoronids and the other lophophorates.

Of the three body segments which characterize the lophophorates, the fore body in the phoronids is evident as the epistome. It is the highly reduced and modified upper lid (epistome) which lines the crescent-shaped oral cavity. The central body (mesosoma) is separated from the posterior body segment (metasoma) by a semi-circular groove. The mesosoma consists of the lophophore, projecting out of the tube. The metasoma protrudes out of the middle segment like an ampulla. The epidermis covers the metasomal tube, which usually measures only 1 mm. The epidermis consists of circular folds, and, in conjunction with the well-developed underlying muscular trunk, it facilitates a great degree of contraction and expansion of the phoronid body within its tube. The thin-walled saccule containing the U-shaped digestive tract lacks muscular elements. It serves as a pressure equalizer for the animal's telescopic motions. The epidermal layer of the saccule contains numerous glandular cells which secrete a sticky substance that hardens upon contact with water. The tube formed from this secretion is transparent and delicate, yet resilient because of the chitinous components. The tube elongates downward. Sand granules, bits of mollusk shell, and similar objects tend to adhere to the tube's outer wall.

Phoronids that are transferred to an aquarium with a sandy bottom will vacate their tubes and burrow into the sand with the intestinal saccule. The animals will construct new tubes, but these will not acquire the same consistency and circumference as the original tubes for two or three weeks. This demonstrates that the animal's body wall is in no way attached to the outside tube. Usually the tubes are much longer than the animal. Phoronids inhabiting muddy flats have tubes that go straight down. In a substrate consisting of shell fragments and sand, the tubes follow the contours of the habitat, usually horizontally. On a rocky substrate, certain phoronid species intertwine their tubes in nestlike colonies, suggesting that entire "schools" of similar-aged larvae settled on a crevice in the coastal cliffs and finished their development there. Many phoronids are capable of burrowing, probably with the aid of acidic secretions, into the shells of oysters, scallops, calcareous rock, and even concrete harbor installations. The burrows are smooth and of uniform diameter. The phoronid then lines the burrow with an integument of glandular secretions (Fig. 8-3).

When viewing the dorsal surface of a phoronid with its tentacles extended one notices a double spiral of tentacular rows that are open posteriorly. Their appearance resembles a horseshoe, or a ram's horns, or a handlebar moustache. Phoronids have also been called "bush worms." The tentacles arise from an epidermal ridge. They are hollow, tubular extensions of the coelomic cavity located in front of the mouth. The original single-row outer coil of tentacles is joined by another parallel inner row which develops during continued growth, starting from the tips of the horseshoe shape and meeting behind the margin of the mouth and the epistome. Growth takes place only at this small breaking point

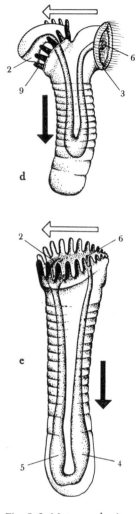

Fig. 8–2. Metamorphosis of the phoronid from the larva to the sessile stage: a) Primary larva. 1. Apical sensory organ, 2. Pre-oral cavity lined with cilia, 3. Ciliary bands, 4. Intestine, 5. Stomach, 6. Anus, 7. Protonephridia; b) Actino-trocha larva, 8. Larval tentacles, 9. Budding tentacle of an adult animal, 10. Inverted intestinal tube; c) Later stage of the actino-trocha larva. Intestinal tube is everted (black arrow); d) The intestine is pulled into the tube, larval tentacles have degenerated; e) Juvenile phoronid (white

Fig. 8-3. A subspecies of *Phoronis hippocrepia* burrows to form tubes in a scallop (*Pecten*) shell. These tubes are lined by a secreted integument. The section of the phoronid tube that protrudes above the bivalve shell is covered by sand grains. Egg deposits (black) are located in the lophophore. Two sections of the tube are shown. Magnification 2x.

of the double spiral, and new tentacles, at first very small, are added. The tentacles are supported by a cartilaginous supporting mass secreted within the ciliated epidermal cells of the tentacles. This rigid mass corresponds to an internal skeleton, and acts as an opposing force to orally directed tentacle muscles. In a way these supporting rods act like the spokes of an umbrella when the tentacles extend out of the tube in their normal spread-out posture. The rhythmically beating cilia lining the tentacles suck in a water current from above and between the two rows of tentacles in the direction of the oral cavity. By lightly sprinkling powdered charcoal on the water surface, one can clearly trace the conical water path (Fig. 8-5).

Food particles (organic detritus, unicellular organisms, diatoms) are entangled on the tentacles and groove by mucus, and are carried to the mouth by cilia. At intervals the stomach swallows small amounts of food. The phoronids avoid indigestible particles by closing the epistome over the mouth and by spreading the tentacles widely part around the mouth, deflecting the ciliary stream. The tentacles and the epistome test the incoming stream of food particles, rejecting unfavorable particles. The number and length of the tentacles increase with a concomitant increase in the phoronid's body size. Correspondingly, more coils are added to the double spiral of the lophophore (Fig. 8-6).

The digestive tract consists of a short esophagus extending into a globular stomach, and a small intestine which terminates at the anal papilla. Circular muscles cause continuous wavelike contractions of the digestive tract. Digestion is intracellular. At the lower end of the intestinal loop, well-developed cilia carry fecal particles on a constantly turning mucous string to the anus, where sphincter muscles pinch off small balls of fecal material. The shells of digested diatoms are retained in the intestinal cells of the phoronid; their empty shells are often found in the empty tubes of phoronids.

Broken-down food particles are distributed to various parts of the body by a quite simple closed circulatory system. A descending artery is connected to the ascending vein by a hemal plexus or capillary net on the stomach wall. The artery and the vein both break through the transverse septum between the mesosome and the metasome and unite to form a circumesophageal ring which gives off blindly ending vessels into each of the arms of the lophophore. The inner cavity of the lophophore artery is comparable to the incomplete ventricular septum in the frog heart. The cavity of this artery is subdivided into two layers by laterally located folds, each of which is connected with a blindly ending blood vessel via a three-way connection. This arrangement is highly advantageous because when the direction of the blood flow in the ascending abdominal vein changes every thirty minutes, the upper layer of the lophophore artery receives the venous blood from the body, transporting it to the tentacular vessels for oxygenation. As soon as the flow direction

arrow indicates the longitudinal body axis and swimming direction).

changes due to a change in the contraction wave of the circular muscles, the oxygenated blood is redirected to the lower layer of the lophophore vessel, probably by Y-shaped valves, and then to the abdominal artery which ascends to the head region anterior to the stomach.

Numerous vascular villi surround a specialized fatty tissue, in addition supplying the gonads that originated from the external wall of the artery. Finally, there is a system of vascular tubes within the saccule containing the intestine which performs undulating movements and serves to equalize the great fluctuations in the body content of the animal by an appropriate blood supply. The colorless blood plasma contains disc-shaped nucleated red corpuscles 5 to 20 microns (1 micron = one millionth of one meter) in diameter. The corpuscles contain a hemoglobinlike pigment which carries oxygen.

Within the primary coelomic cavity of larva only a few hours old one finds unattached mesenchymal cells of mesodermal origin which flaked off the embryonic archenteron. These cells aggregate into two little pouches situated on each side of the small intestine. The pouches cover the inner surface of all the organs, and thereby a paired (secondary) coelom has originated. During the course of the lophophore's development, a pair of coelomic cavities arises in a similar manner in the remainder of the larval body. At the point of contact these coeloms give rise to septa along the logitudinal axis of the body, and to a transverse wall subdividing the body into two segments (metasome and mesosome). The body segment anterior to the mesocoel always lacks a secondary coelomic cavity. The coelom is filled with a colorless, albuminous fluid. During the reproductive period, sperms or eggs are shed into this fluid, and the gametes complete their maturation there. The coelomic fluid circulates constantly, supplying oxygen and food substances to all those organs neglected by the vascular system. This pertains mainly to the posterior section of the intestine and the epidermal muscular trunk. The coelomic fluid also plays an important role as an opposing force to the contractile epidermal muscle trunk which maintains the shape of the vermiform body, and in sustaining hydrostatic pressure, particularly when the lophophore is being erected.

A pair of tubular metanephridia to the left and right sides of the hindgut carries out the excretion of metabolic by-products produced outside the digestive system. A ciliated funnel-shaped opening at the end of the coiled duct of the metanephridium protrudes into the coelomic fluid in the posterior body segment, collecting excreta including the "spindle-shaped bodies" of the fat organ. The metanephridial duct extends between the epidermis and the coelomic lining, and, like the intestine, is U-shaped. The ducts open laterally to the anus on the anal papillae. The phoronid metanephridia also serve as gonoducts for the gametes.

At the slightest touch of a tentacular tip the entire lophophore rapidly

Fig. 8–4. *Phoronis hippocrepia* forms intertwining, nestlike colonies via transverse division. The tubes are covered with a coffee-brown, finely granular coating (natural size).

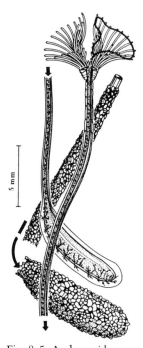

Fig. 8–5. A phoronid removed from its tube. The black line on top of the tentacles represents the direction of the lophophore. The transparent body reveals a lateral view of the digestive tract (dotted and short lines) and the two blood vessels (black).

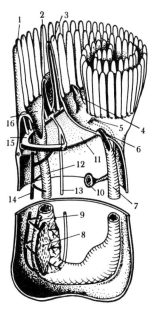

Fig. 8–6. Anatomy of a phoronid (anterior and posterior segments; the uniform, long central part has been cut out): 1. Outer, 3. Inner row of tentacles; the most anterior tentacle (sectioned) shows the course of the blood vessel. 2. Epistome representing the vestigial protosome, 4. Lophophore organ to which the egg deposits are attached, 15. Double-chambered vascular vessel with lateral artery (14, going to the body) and central vein (13, in direction of lophophore); both blood channels are united by a circumesophageal ring (9), which also supplies the testes and ovaries (8), 16. Oral cavity which leads into the esophagus (12) which penetrates through the mesosoma and metasoma into stomach, 7. Hindgut, 6. Anal papilla, 10. Nephrostome, 5. Nephridiopore.

withdraws into the tube. This response suggests that the phoronid perceives tactile stimuli and responds by muscular contractions, although sensory organs have never been demonstrated. The phoronid has a septum of nerve fibers with a nerve ring situated in the coelomic lining at the dorsal side directly in front of the anal papilla. The nervous mass has the appearance of a curved clasp. Lateral nerves branch off from here to the inner and outer tentacles. A giant fiber extends along the left body side to the abdomen, innervating the retractor muscles. Slender neurosensory cells are found among the epidermal cells, most frequently in the tentacles, as well as on the body surface. The nervous net covers the entire body. Mechanical impulses are conducted, with a noticeable drop in excitation, to the neighboring muscle fibers. The degree of sensitivity can be tested by carefully applied needle pricks. Beyond the threshhold that is adapted to the tactile stimulus of touching planktonic particles, the phoronid retracts the erected lophophore back into the protective tube.

In sexually mature phoronids, which except for *Phoronopsis viridis* seem to be hermaphroditic, the testes and ovaries develop along the numerous blindly ending lateral branches of the descending blood vessel. Gametes reach maturity within the coelomic fluid but never are fertilized there; they are expelled through the metanephridia. Like other marine animals, in particular the sessile forms, the phoronids release their gametes at specific synchronized periods. Strings of sperms are sucked into the lophophores of neighboring animals, and most probably are collected and stored in the ciliated groove of a conical structure (lophophore organ) at the base of the tentacular coil. Some species produce large yolky eggs with a diameter of 0.1 mm. They practice brood care when a small number, usually a pair, of mucus-covered eggs become deposited in the lophophore depression between the anus and the regenerating zone of the tentacles, where they are fertilized. In the extensively studied species *Phoronis hippocrepia*, the larvae remain within the protection of the parental body for approximately five days. Under the benefit of the adult's respiratory current, the larvae eat the yolk supply until they have developed two pairs of tentacles. The phenomenon of brood care has proven adaptive (Fig. 8–7) for those phoronid species inhabiting moving waters along the coastal zones.

In the smallest species, *Phoronis ovalis*, which measures only 6 mm and which possesses eggs with a diameter of 120 microns, the ova are too large to slip through the narrow metanephridia. In this case the parent casts off the lophophore just prior to "egg laying" in order to make the birth canal sufficiently large. During egg laying the maternal animal withdraws deeper into its tube because the upper two-thirds become filled with successively deposited eggs. After one week the young larvae leave the parental tube. They have a rich supply of yolk. At this point the adult stretches out again and develops a new lophophore.

It is easy to perceive the advantages of a good yolk supply in the phoronid larva. It decreases the length of the dangerous planktonic phase, and the food requirements of the larva are fulfilled. This kind of larva is also termed lecithotrophic. The free-swimming larval stage lasts approximately four days. The entire surface of the larva is covered by short cilia which facilitate some measure of mobility. After the planktonic existence, the larva sinks to the bottom and crawls around on its mucoid ventral body surface, which bends upward. Three days later the larva attaches itself to the substratum and rolls up into a semicircle. During this entire period the larva has maintained itself with the yolk supply in its intestine. Subsequent developmental stages remain to be investigated.

Large yolky eggs cannot be produced in the same numbers as small eggs with little yolk content. Small eggs are not protected by the parental body, but are released into the surrounding water via the metanephridia. Up to 500 eggs may be laid one by one. The water current disperses the eggs. Those that manage to survive develop into young larvae after a few days. Each pear-shaped larva develops a bill-like upper lip. Two ciliary bands propel the larva through the water while a third band sweeps food particles to the mouth region. During the course of further development the larval body becomes cylindrical along the longitudinal axis. The oral ciliary band changes into tentaculate processes. The well-developed paddling bristles in the posterior ciliary band propel the transformed actinotrocha larva through the water like a turning screw, with the preoral hood facing forward. After an average period of three weeks the larva sinks to the bottom and undergoes rapid metamorphosis.

Actinotrocha larva

The actinotrocha larva depends on outside food sources (planktotrophic), so its free-swimming phase lasts approximately four times that of the lecithotrophic larva. The higher number of eggs compensates for the heavy losses during the planktonic existence. In the long run many "cheap" eggs ensure the survival of a species just as well as a few "precious" ones. Many marine animal groups, including several crustaceans, gastropods, and polychaetes, practice brood care.

During the period from December to March the tubes of *Phoronis hippocrepia* (L 35 mm) in Naples' harbor contain diminutive individuals or even single body segments measuring from 2 to 3 mm. In response to a drop in the water temperature, certain phoronids constrict their circular muscles just below the transverse septum between the metasome and mesosome. This constriction results in the total amputation of lophophore and body trunk. If by chance a fish or crab should bite off the lophophore before the phoronid could withdraw it into the tube, the remaining trunk is sealed off in a similar manner. The regeneration process of replacing body parts lost due to autotomy or injury starts almost immediately. The two fork ends of the exposed circulatory system close off in both severed body segments just like the two intestinal sections in the upper portion of the animal. Esophagus and hindgut are regenerated in the lower body by invagination of the body wall. New body orifices also develop. It is

▷
The marine bryozoan *Retepora beaniana* forms part of a living community on a coralline substratum (calcareous red algae) which also includes a salmon-colored net coral and *Hippodiplosia foliacea*, which branches like a moose's antler. This species belongs to the Cheilostomata. The carmine-red horny coral *Paramuricea chamaeleon* is also present. Magnification 2x.

obvious that this body segment is at a disadvantage because it has to regenerate a new lophophore, but after only a few hours small, wartlike structures emerge at the amputation site. The tentacles develop from these structures.

Phoronids, for example *Phoronis ovalis* (which can reach population densities of up to 150 individuals per cm² of oyster shell), can quickly occupy a favorable habitat because they are able to reproduce asexually via transverse division and regeneration. Only the original animals of such a colony developed from planktonic larvae which in turn originated via sexual reproduction. However, the phoronids do not have stalk formations as is found in the bryozoans, where all the zooids are interconnected.

Understandably most phoronids have been found in the vicinities of marine biological stations. For this reason our knowledge of the distribution of these animals is extremely fragmentary. Numerous actinotrocha larvae occur in plankton, but these specimens usually cannot be identified to species level, nor can their point of origin be determined. Additionally, many phoronids are characterized by several actinotrocha larval forms. Presently one can distinguish a considerable number of "Actinotrocha-species" as with adult phoronids. It is assumed that phoronids are found along all coast lines as well as the continental-shelf regions, because, due to the relatively long planktonic larval phase, the progeny can drift over long distances. Up to now phoronids have been recorded from the tidal zones at the low water mark down to depths of around 50 m.

The famous scientist Johannes Müller spent the fall of 1845 with his students on the island of Heligoland in the North Sea. He was primarily interested in the developmental history of the marine animals, and his work strongly affected zoological thinking. During their three months on Heligoland, this research group noticed minute animals, measuring only 3 mm, on the surface of sea-water samples. Müller described this phenomenon in the following manner: "Once we had become aware of these organisms, we could identify them in the collecting vessels on the basis of their rhythmic movements through the water with the sole aid of a wheellike (teletroch) organ." This, in fact, is the first description of a phoronid larva. Müller coined the term actinotrocha larva (see Color plate, p. 253) because of the ring of tentacles (from the Greek ἀχτος = ray, τροχός—wheel). Müller, however, did not classify this newly discovered creature with any of the existing animal groups. He only said, "At first glance one could arrive at the assumption that the actinotrocha larva is the larval stage of some mollusk." Because he found that these larvae possessed organs in addition to the alimentary canal, he rejected this as a larval phase, thinking the organs were sexual organs. The presence of the teletroch suggested a similarity to the rotifers (see Vol. I) and a possible taxonomic affinity to the Turbellaria (see Vol. I).

◁
The gelatinous bryozoan *Alcyonidium gelatinosum* forms fleshy branching zoaria. The zooids, characterized by delicate tentacles, are arranged in dense rows within a spongy ground substance of the colony. Natural size.

In 1856 the scientist T. Stretill Wright discovered the adults of two phoronid species. These were *Phoronis hippocrepia*, which grows on the algae-encrusted stalks of the coral *Caryophyllia smithi* in the channel of Bristol, and *Phoronis ovalis*, which is attached to oyster shells along England's western coast. Finally in 1867 Alexander Kowalevsky demonstrated that the actinotrocha larva was a developmental phase of *Phoronis*. He believed that the phoronids were phylogenetically related to the mollusks.

In the following decades, attention was focused on the juvenile development of the phoronids. It became known how the eggs matured and were fertilized, and how organogenesis proceeded in the larva and the adult. Once these essential facts were known, a satisfactory classification of the phoronids within the zoological system became possible. The phoronids, bryozoans, and brachiopods were united under "superphylum" Lophophorata. All three phyla are characterized by a common type of ovum, which in the course of cell division and cleavage does not differentiate into distinct regions, which would indicate potential organogenesis. All also possess the actinotrocha larva which eventually metamorphoses into an adult animal. Finally, there is great similarity in the anatomy of these phyla in addition to their common life habits and behavior. It therefore seems unlikely that this is a mere example of convergent evolution. The high degree of homology in these groups probably goes back to a common ancestral type which arose before the advent of the annelids. The greatly similar trochophore larvae of the annelids are already indicative of a higher evolutionary stage where the mesoderm originates from the primary cells.

Even today very little is known about the physiology of the phoronids, their chemical processes during digestion, their respiration, or their excretion of metabolic by-products. The behavior of phoronids in their natural environment is still an open question. It requires much painstaking work to maintain these animals in an aquarium.

Although the phoronids represent an evolutionarily very ancient group, there are surprisingly few phoronid species. Carl Cori, who has worked extensively with this animal group, has expressed the opinion that because the phoronids are attached to the substratum, they are not greatly influenced by external forces. Up to now approximately twenty species have been identified, and some of these are probably only subspecies. There are two genera: A. *Phoronis*, with the lophophore separated from the body trunk by a circular groove, and B. *Phoronopsis*, characterized by a conspicuous circular fold. In many cases not all of the developmental stages from larva to adult have been discovered. Because of this lack of information, phoronids have been classified according to their mode of settling:

1. Phoronids that are horizontally embedded in a scattered manner, in a bivalve-sand substratum along the coastal zones: *Phoronis psammophila*

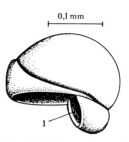

0,1 mm

Fig. 8–7. Incomplete swimming larvae develop from the yolky eggs of certain *Phoronis* species. These forms live from the yolk supply stored in the intestine. After a few days these larval forms sink to the bottom, where they crawl about with the aid of cilia and mucus for three days; then they metamorphose into phoronids.
1. Anus.

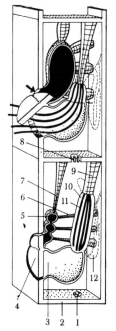

Fig. 8–8. A rigidly calcified zooecium (2) requires some arrangement that produces an internal pressure that would force out the lophophore. The polypide emerges via a vestibule (3) through an aperture which is closed by an operculum (4). The operculum opens when transverse and longitudinal muscles (6, 7) stretch the compensation sac (5). The resultant suction opens the operculum to the inside (arrow). When the pressure equalizes, the lophophore (11) is pressed out of the lophophore sheath (10) into the vestibule. 9. Polypide retractor muscles (which bring about the reverse process), perforated rosettes (interzooidal pores; 1, sectioned in 8) facilitate food exchange. 12. Stomach and intestine of polypide.

Phylum: Bryozoa

(see Color plate, p. 253; L 5 cm), flesh-colored; lophophore pale red; up to 124 tentacles; tube covered with sand granules and bivalve fragments. *Phoronis architecta* (L 5 cm), flesh-colored to yellowish-red; up to 100 tentacles; tube up to 15 cm long, covered with a regular pattern of apparently selected granules of sand.

2. Vertically embedded phoronids scattered in muddy substrate along the coastal zones: *Phoronis muelleri* (L 8 cm); larva of this species was described by J. Müller as *Actinotrocha branchiata* (see Color plate, p. 253) in 1846. Flesh-colored with up to 60 reddish-tinted tentacles; tube up to 8 cm long, covered by a hardened mud crust containing tiny sand granules. *Phoronis pacifica* (L 9 cm), with up to 200 tentacles; tube covered with sand granules.

3. Intertwining phoronid colonies which resulted from transverse asexual division and which are always found on hard substrate (cliffs, wharves, breakwaters). Mediterranean race *Phoronis hippocrepia* (L 3.5 cm), light gray; up to 130 tentacles; tube up to 4 cm long, covered by a finely granular, coffee-colored coating; *Phoronis vancouverensis* (L 4 cm), transparent with white splotches; up to 100 tentacles; brown tube.

4. Phoronid colonies resultant from transverse asexual division. They burrow in limestone or bivalve shells. North Sea race *Phoronis hippocrepia*; yellowish tube consists of delicate chitin; the protruding tube end is fastened with shell fragments. *Phoronis gracilis* (L 1 cm), grayish-yellow; up to 80 tentacles; tube membraneous and transparent.

5. Phoronids that live in the oral regions of the mucous skins of *Cerianthus*: *Phoronis australis* (L 12 cm), lophophore purplish-red; up to 300 tentacles. It is also found in sandy substrate.

There are no data available about the habitat of the second genus, *Phoronopsis*. The largest species yet discovered is *Phoronopsis viridis* (L 20 cm), with over 300 tentacles, each of which measures approximately 3.5 mm. The lophophore and trunk are greenish; the tube, covered with sand granules, is 20 cm long and 3 mm in diameter. The species is found in Morro Bay, California.

Anyone who has paid close attention to objects such as bits of plants, wood fragments pulled out of a pond, seaweed, driftwood, or bivalve shells washed up on the ocean beach, may have noticed a brownish, slimy coating or gelatinous clumps on these objects. This seemingly insignificant substance is in reality a highly organized aquatically adapted group of animals known as colonial bryozoans.

Although the encrustations or clumps formed by the stalks of bryozoans (class Bryozoa) may reach dimensions of up to 1 m, the individual animal (zooid) has a maximum length of only 4 mm, as does, for example, the species *Nolella alta*. Essentially, the body structures of the phoronids and the bryozoans are similar. One can regard bryozoans as dwarf phoronids. As a result of the small body size, the circulatory system and special excretory organs (kidneys) are absent. The digestive system consists of an esophagus, a stomach, and an intestine which is suspended in the coelomic cavity as a V-shaped loop. Within the intestinal loop a gelatinous rod turns constantly, approximately 100 times a minute. The anus terminates in the vicinity of the mouth but outside the lophophore. This feature is responsible for the term Ectoprocta (from the Greek ἐχτός = outside, πρωχτός = anus).

By looking at the lophophore of a bryozoan and the method of eversion of the lophophore, one can differentiate between three types of arrangements which divide the phylum into three classes: 1. Class GYMNOLAEMATA, with a circular lophophore. The episome is absent. The zooids are cylindrical to flat, and can be everted only by muscular deformation of the body wall. Two orders with a total of 650 genera, all marine. 2. Class STENOLAEMATA, with a circular lophophore. The episome is absent. The zooids are cylindrical, and the mechanisms for everting the lophophore is not dependent upon muscular deformation of the body wall. Four orders, all but one (Cyclostomata; Paleozoic-Recent; about 250 genera) extinct. 3. Class PHYLACTOLAEMATA, with a horseshoe-shaped lophophore. The episome is present. All fresh-water forms, comprising about 12 genera.

Even just a strong magnifying glass reveals the fanning of the numerous delicately feathered tentacles of a bryozoan colony. This part of the bryozoan is soft-skinned and can be evaginated. It is therefore, very appropriately, known as a polypide, because it is reminiscent of the tentaculate crown of the polyps (see Vol. I). The polypide is able to stretch its tentacles to ten times their normal length, with ciliary activity throughout this length. The resultant current sweeps in food such as rotifers, diatoms, ciliates, flagellates, algae, and rhizopods. Just as in the phoronids, the increased surface area of the tentacles functions as a region where gaseous exchange takes place. At times of danger the polypides are completely withdrawn into the zooecium, with the aid of well-developed muscles. The zooecium is made up of rigid chitin or is often reinforced by calcareous material. Extinct bryozoan colonies still maintain the regular pattern of calcified zooecia. Fossils of this animal class can be traced back to the Lower Ordovician period (500 million years ago). In the Cretaceous period (135 to 63 million years ago) the bryozoans were represented by a greater wealth of species than today. Presently we can identify approximately 15,000 species from the Cretaceous period.

Even though the polypide can be withdrawn into the neck region which surrounds the tentacles, there remains a very vulnerable orifice at the upper margin of the zooecium. Foreign substances or enemies may penetrate at this location, or the open exposure may hasten dehydration of the colony during low tide. Various adaptive structures have evolved to lessen or avoid such dangers. The gymnolaemates are subdivided into two orders on the basis of these structures: Ctenostomata, with a circular (or nearly so) terminal orifice, often closed by a pleated collar surrounding

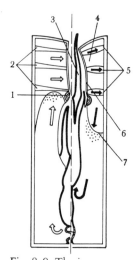

Fig. 8-9. The inner space of a stiff-walled cheilostome is divided into two sections. Muscles (5) of the upper compartment (4) compress the wall of the vestibule (6), the elastic separating membrane (7) of the two compartments functions as a "pressure equalizer" which presses downward, the polypide (3) along with the lophophore are pushed out like an elevator (black arrows). When the vestibule muscles (2) relax, the vestibule wall resumes its original position, and the separating membrane (diaphragm) is pulled upward (white arrows). Decrease of the pressure in the lower compartment causes retraction and recovery of the lophophore below the sphincter muscles (1). This shift of the fluid within the interior necessitates a closed system. This schematic drawing demonstrates these two processes in the two opposing half sections.

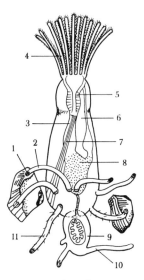

Fig. 8–10. The solitary bryozoan species, *Monobryozoon ambulans*: 4. Lophophore, 5. Pharynx, 6. Hindgut, 3. Esophagus, 8. Stomach, 7. Polypide retractor muscle, 2. Adhesive processes with sticky glands (1), 11. Locomotory processes, 9. Bud with stolon (10).

the neck region of the extended polypide. Cheilostomata, with a frontal orifice, closed by a hinged operculum.

An increase in the internal pressure of the coelomic fluid pushes the polypide out again, and the tentacles become unfolded. This excess pressure can be produced by various methods. The fresh-water bryozoans and gymnolaemates, with their elastic zooecium walls, decrease the animal's entire volume by simply constricting the circular musculature. In this manner the anterior body is pressed out. Many cheilostomes possess a heavily clacified zooecium with a soft, elastic operculum which can be indented by the action of muscular bundles. This maneuver causes a decrease in the internal cavity of the zooecium, and the polypide becomes extended. Others are characterized by a rigid circular zooecium which is totally calcified. This adaptive feature necessitated a flexible compensating sac for lophophore movement. The walls of this sac are equipped with transverse muscles which permit the entry of sea water when they contract. Since this sac is suspended on the inside of the animal, the internal pressure is consequently increased. Suction causes a tightly closing operculum to open to the inside when the water flows into the sac. Concomitantly, a hinge in the center of the operculum is turned to the outside in order to expose the extending polypide (Fig. 8-8).

The stenolaemates which are rigid-walled are able to protrude the lophophore without the application of sea water but by shifting the coelomic fluid within the coelom. Special flexor muscles in the atrium of the chitinous zooecium press the coelomic fluid posteriorly, thereby causing a constriction of the upper zooecium. The withdrawn polypide is extended as if carried on a rising platform. A tendon (funiculus) maintains the polypide at a specific elevation. The funiculus consists of vestigial longitudinal coelomic septa which anchor the intestine at the most posterior curvature with the basal plate of the zooecium.

Almost without exception, the bryozoans form cohesive colonies by repeated budding. Only the species *Monobryozoon ambulans*, which belongs to the ctenostomes, is solitary. This species, discovered by A. Remane, caused a scientific sensation. He introduced it in the *Zoologischer Anzeiger* in 1936, with the following words: "In September 1934, when I was examining *Amphioxus*-sand [sand inhabited by the lancelet] on the island of Heligoland, I suddenly noticed among the sand granules a lophophore which obviously belonged to a gymnolaemate bryzoan [bryozoan without an epistome]. This discovery was disconcerting, since marine sand is generally avoided by colonial and sessile animals, and bryozoans also were, up to this point, never recorded in this habitat. This animal's body was solidly surrounded by sand granules, which could only be removed with great difficulty because they were firmly anchored to the animal. Careful exposure of the specimen revealed a bryozoan which was unusual in more than one aspect. I named this new species *Monobryozoon ambulans* [new genus and species]. The first unusual characteristic was the solitary nature of this bryozoan. It was the first example of an ectoproct bryozoan that was not colonial. This solitary form is relatively simple in body structure. Its mouth is surrounded by thirteen or fourteen ciliated tentacles. The polypide and zooecium are subdivided by indistinct circular folds. The hind body is blunt and unattached. Two types of tubular processes arise from the hind body. The first type surrounds the hind body in a slightly irregular circle, made up of approximately ten to fifteen processes which are strongly reminiscent of the ambulacral tube feet of the echinoderms (see Chapter 10). These structures are hollow tubes that actively grope around and attach to sand granules with a secretion. Fine, hairlike threads are scattered over the outer surface of these processes. They are probably tactile hairs. These tentaculate processes function not only as attachment organs but also aid in the locomotion of *Monobryozoon*. This bryozoan is not only solitary but also free-moving, albeit slowly. The second type of process is represented only by a single structure which arises from the hind body. As far as I could observe, this process was not actively mobile, but was dragged along. The base of this structure is strongly enlarged and bears recognizable tentaculate rudiments arranged in a circle, a typical budding phenomenon of bryozoans. This discovery clearly demonstrates that *Monobryozoon* is not a colonial form. Slow development (and some detachment) of the individual buds suppresses colony formation."

The second process described by Remane corresponds to the long, repeatedly forking stolon of many ctenostomes. Zooids arise from this process at regular intervals. Most bryozoan zooecia, however, are clustered in dense patches. These form flat crusts that adhere to a variety of substrates such as stones, bivalve shells, and seaweed, or they may form clump-shaped colonies when growing around a stalk of reed. Bryozoans inhabiting calm areas of ocean water may not need the support of the substratum and may form self-supporting colonies that are either fan-shaped or brushlike. The shapes of the colonies of representatives of the genus *Hornera* are reminiscent of the stony corals. In these forms the wall of the zooecium that faces the water current is double-layered and heavily calcified.

The colonies of marine gymnolaemates are made up of individual animals (zooids). For example, a colony of the genus *Flustra*, which resembles *Fucus* (brown algae), measures 10 cm in height with a surface area of 80 cm², and is composed of nearly one million zooids. Zooids of fresh-water bryozoans are, on the average, three times the size of marine bryozoan zooids. Consequently, a fresh-water colony consists of fewer zooids. When the bun-shaped, transparent stalks of *Cristatella mucedo* reach a certain size (a width of 1 cm and a L of 20 cm), additional growth is contrictive and separates the colony into two parts. The younger section of the colony glides away from the older and larger section on a mucous base measuring 3 mm, located between the creeping sole and the sub-

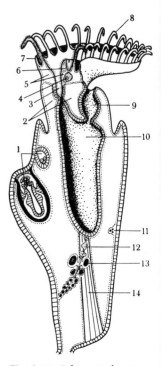

Fig. 8–11. Schematic longitudinal section of a fresh-water bryozoan: 8. Lophophore with circular diaphragm groove (7), 6. Epistome (first body segment, the protosome), 5. Second body segment (mesosome), which surrounds the lophophore and contains the central nerve mass (4), 3. Oral aperture which continues into the metasome (2) via the digestive organs, the stomach (10) and intestine (9). 1. Embryonic buds are asexual outpocketing in the metasomal wall. 14. Polypide retractor muscle. Funiculus with statoblasts (13), testes (12). Eggs (11) originate along the coelomic lining.

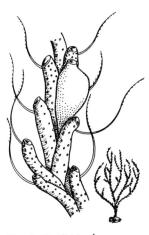

Fig. 8–12. *Crisia eburnea* attaches to the substratum via modified zooids (keno- zooids). Other zooids are modified into gonozooids which house embryos that develop at the expense of the body mass of the poly- pide. A colony (below right) grows to a height of approximately 15 mm.

strate (e.g., a leaf of a sea lily) at a speed of a few millimeters per hour until it is several centimeters from the former site. Other species of fresh-water bryozoans also are able to crawl on a substrate. This is parti- cularly true for their younger colonies.

In the gymnolaemates the walls of the individual zooecia within a stalk are perforated by holes (interzooidal pores), or are reduced to single bandlike ridges, or are totally absent, as in the phylactolaemates. Because of this arrangement, an exchange of metabolites can take place via the coelomic fluid in the lower body segment (metasome), which is particu- larly important for the heterozooids. Many marine gymnolaemates display a great deal of polymorphism, in which several types of zooids exist. This phenomenon offers many advantages to the entire animal colony.

Zooids arise from buds which usually form at specific spots on the bryozoan colony. Although the exact nature of this process has not been investigated, it is known that these zooids usually do not possess a polypide and are not able to produce gametes. These are the kenozooids (Greek χενδς = empty, δῶον = creature). These zooids have the function of anchoring the colony to the substrate. The former can be compared to rhizoids or tendrils which are equipped with runners and branches. The kenozooids consist of a zooecium that is closed from all sides.

The stalks of stenolaemates are characterized by barrel-shaped en- largements that are much larger than the normal zooids. These structures are either original complete zooids or zooids that were devoid of the lophophore from the beginning and which were modified into gono- zooids (brood zooecium) for the self-fertilized ovum. In the gonozooids the intestine is reduced to a clump of cells which, together with the egg, is surrounded by a membrane that provides the embryo with the necessary food substances. Judging by the relatively small number of gonozooids on a colony, one would assume that there is only a small number of progeny; but in fact several embryos are produced from one zygote by separation of the cells in early cleavage (see Vol. I). By this process (polyembryony), up to 100 larvae can be produced, finally breaking out of the membranous sac and being discharged out of the brood chamber (ooecium or ovicell).

In the rigid-walled cheilostomes an operculum covers the orifice of the water sac and the passage of the lophophore. With the advent of continued growth, the operculum undergoes amazing adaptive modifi- cations so that the actual zooid on the colony is obscured. In the so-called vibracula the opercula are modified into long rods up to ten times the length of the zooids. The vibraculum is equipped with a socket joint and powerful muscles, which allow for a sweeping radius of ninety degrees, and in some species even 270 degrees. This bristle functions like a whip that sweeps over the water surface of adjacent zooids once every minute. This sweeping activity is highly adaptive, for it keeps the colony clean and free of mud and sand granules, and prevents the settling of adhering organisms. The arrangement of the vibracula on the colony is irregular. The structure is present in only a few species of cheilostomes (Fig. 8-13).

The avicularium is another type of modified zooid, found more frequently than the vibraculum. It occurs in a great variety of shapes within the same colony. The operculum in this modified zooid forms a mandible that can open and close by muscles. This musculature, several times more highly developed than in its original form when it was functional in closing the operculum, performs the "bite." The opening of the mandibles is facilitated by an epidermal muscular trunk which has become modified into a muscular strand. The intestine and lophophore of this modified zooid has become reduced to an insignificant cellular clump. Adjacent zooids, therefore, supply the avicularium with food and oxygen via pores in the body wall. The avicularia which occupy a place in the colony normally occupied by an autozooid are termed "vicarious." Those which are reminiscent of a bird's skull and are attached via a movable stalk to the ventral side of the usual autozooid are termed "adventitious" (Fig. 8-14).

"Bird head"

If a diminutive larva of some marine animal touches one of the clumps of sensory bristles which line the "palate" of the immovable upper mandible, the avicularium shuts closed immediately. Even the proteinaceous juice of a bivalve elicits a state of "irritability" in the avicularium. This zooid swings back and forth approximately every six seconds. The avicularia seem to patrol the immediate area surrounding the colony. They help keep the colony free of undesirable adhesive organisms. More recently it has been discovered that avicularia of the genus *Bugala* trap and hold the amphipods *Corophium insidosum* and *Jassa falcata*, which measure from 1 to 5 mm. These amphipods attempt to build tubes between the spaces of the colony, which would restrict the field of activity for the lophophores. The avicularium does not release its victim until it has decayed. This organic debris, along with the associated organisms, is swirled to the adjacent feeding individuals (autozooids). Charles Darwin even observed that detached avicularia continue to snap for some period.

Almost all ectoprocts (bryozoans) are hermaphroditic. The testes develop in the coelomic cavity. The ovaries are usually located along the abdominal lining. Only in the stenolaemates do the gametes develop along the margin of the colony, the most recently developed area of the colony. Sexual reproduction takes place primarily in young bryozoans or juvenile sections of a colony. In those fresh-water bryozoans whose breeding stock dies during the winter, sexual reproduction occurs in the spring.

The eggs are freed from the ovaries by the swinging motions of the intestinal loop. The eggs swim in the coelomic fluid and are fertilized there, probably by sperms from the same individual. Only a few species among the ctenostomes and cheilostomes release masses of microscopic eggs, with little yolk content, into the surrounding water throughout the

▷
The fresh-water bryozoan *Plumatella repens* (compare Color plate, p. 254) with extended lophophore. Numerous peritrich protozoans (*Vorticella*) adhere to the zooecium. Magnification 130x.

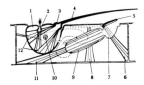

Fig. 8–13. In many cheilo-stomes a vibraculum (4), equipped with muscles (12) and a socket joint (3), moves in a semicircle to keep the upper surface of the colony free of particles. This cleaning apparatus (vibraculum), and its associated atrophied poly-pide (2) and interzooidal pores (10) for metabolic exchange, are attached (as an autozooid) to another normal zooid, which shows the operculum (5) and associated retractor muscles (6) and extensor muscles (8) for the vestibule (7) and tentacle sheath (9) into which the lophophore is pulled by a muscle (11).

◁
Membranipora pilosa.

entire year. The eggs are discharged through a pore at the posterior side of the lophophore. Otherwise bryozoans practice a great variety of brood care.

Several species of the ctenostomes (e.g., genus *Alcyonidium*) retain groups of six to eight eggs attached by membranous stalks to the epi-dermis of the neck of the lophophore. In many ctenostomes (e.g., genus *Flustrellidra*) the eggs reach the vestibule between the zooecium and lopho-phore neck through a spontaneously developing rift in the coelomic wall. Within the protection of this location the eggs undergo additional de-velopment. The polypide of the parental zooid becomes contracted and finally atrophies. The majority of bryozoans possess special brood cham-bers to house the fertilized eggs. In the fresh-water bryozoans the lateral wall of the zooid evaginates, is constricted, and separates from the external environment (internal brood chamber). It is still not known how the egg reaches this sac. The embryo is supplied with nutrition from the inner evagination, and is suspended by a girdle-shaped area of fused tissue from the outer evagination; at a later stage potential polypide buds become noticeable on the evaginated walls as little depressions. Two flaps start to grow from the center of the embryo's body over the section with the polypide buds. These mantlelike flaps are covered with swimming bristles on the outside (Fig. 8-11).

At this phase of development the embryo, which measures approxi-mately 1 mm, breaks through the parental zooecium wall at the site of the former evagination. This form, known as a primary zooid, swims with the aid of the swimming bristles. The free-swimming phase may last for minutes or hours, until the larva becomes fastened to the substratum with the end opposite to the polypide. A plate of nervous cells guides the larva to the most suitable substratum. Within a few minutes the pre-formed polypides are everted and the two protective flaps are flipped back and used as a stiffening for the zooecium sides and as a pad for the lower segment. Continual budding (asexual reproduction) of the primary zooid finally gives rise to a new bryozoan colony (Fig. 8-16).

Many species of cheilostomes have an elongation at the upper margin of the zooecium, which grows into a caplike bladder with a slit that permits the entry of sea water (external brood chamber [ovicell]). Sub-sequently, one egg at a time will be rolled back and forth in the narrow coelomic cavity, by motions of the intestinal loop, until it has elongated and can be pressed through the small pore near the lophophore, with the aid of the polypide muscles. During this procedure the birth canal is directed toward the entrance of the brood chamber. The egg turns into a larva within two days, on the average. During this period another egg is maturing in the coelomic cavity, but it is transferred to the brood chamber only when the previous larva has vacated it. It is assumed that this amazing example of synchronization is controlled by hormones (Fig. 8-15).

Those bryozoans that do not practice brood care produce large numbers of eggs with a small amount of yolk. These eggs measure only a few hundredths of a millimeter. The surplus of eggs compensates for the great losses that occur. The presence of a protective brood chamber, however, greatly increases the rate of egg survival. In this case, fewer eggs with a greater amount of yolk and a size up to ten times that of the smaller variety suffice to ensure the survival of the species. In the fresh-water bryozoans, only one egg per zooid enters the brood chamber, undergoing additional development there. The appearance of the embryo can reveal whether the reproductive rate is increased via asexual budding or, as in the surviving stenolaemates, via constriction and division of the egg (polyembryony). The fact that larger bryozoan colonies are surrounded, nevertheless, by dense clouds of young larvae can be explained by the tremendous numbers of fertile zooids.

When examining a sample of plankton from the coastal zones a ciliated organism reminiscent of a rimless hat frequently appears. This creature is barely discernible with a magnifying glass. In 1830 the scientist Christian Gottfried Ehrenberg (1795–1876) described it as "*Cyphonautes compressus*" (see Color plate, p. 253) because he assumed that this organism was an adult protozoan (see Vol. I). In 1869 Schneider discovered that *Mempranipora pilosa* developed from "*Cyphonautes compressus*." The name "cyphonautes" (Greek $\chi\upsilon\phi\delta\varsigma$ = humped, $\nu\alpha\upsilon\tau\eta\varsigma$ mariner) has been retained until the present for this true bryosoan larva. A typical *Cyphonautes* larva (see Color plate, p. 253) develops only from those eggs that have a low yolk content. During its planktonic existence, which may last from one to two months, the larva filter-feeds on microscopic organisms. Larvae which develop from yolky eggs, on the other hand, enjoy brood care. Their swimming existence lasts for only a few hours, until they find the appropriate habitat. These larvae lack a mouth and intestine for independent feeding. These forms also usually lack the two fused chitinous valves on the front margin, which offer some protection to the flanks of the *Cyphonautes* larvae. However, both these larval types possess three structures that perform specific functions.

The cilia, which line the lower margin of the *Cyphonautes* larva and which beat successively but discontinuously, propel the larva through the water in a spiral fashion. A retractile knotty structure on top of the larva is characterized by numerous sensory bristles and a red pigment. Microscopic studies of this structure have revealed nerve fibers. This probably is a tactile organ that is also light-sensitive. This crown organ or apical sense organ is regarded as the "brain" of the bryozoan larva.

A second sensory organ is located underneath the anterior larval margin. It consists of columnar grandular cells that give rise to a tuft of long tactile hairs. This sensory pit is found in the immediate vicinity of the oral depression. This "pear-shaped organ" has a dual function: it serves as a grasping and food-capturing organ for the free-swimming *Cyphonautes* larva and, during the process of attaching to the substratum, this

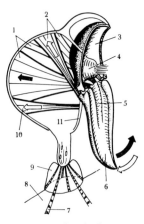

Fig. 8–14. The cheilostomates possess defense structures that look like bird skulls (avicularia). The avicularium is attached to an autozooid (8) by a peduncle (9) which is movable by the action of muscles (7). This snapping organ is reminiscent of a bird's beak. Mobile mandible (6), stationary mandible (3), penduncle (tactile sensory organ; 4). When the mandibles close, the muscles (1) pull a ligament (5; black arrows). When the muscle depresses the elastic wall of the zooecium the attachment of the mandible is shifted toward the inside, and the two muscles (2) open the capture apparatus (white arrows).

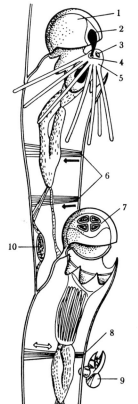

Fig. 8–15. The interior of the helmet-shaped brood chamber (1) is in contact with the surrounding sea water. The ovum (2) is pressed into the inside of the brood chamber through a narrow opening (4) near the mouth (3) from the coelomic cavity. During this process the lophophore (5) is tipped backward. In the lower brood chamber a multicellular embryo (7) is shown. Muscular action (6) causes a decrease in the internal pressure, and an evagination of the lophophore (black arrows = pulling direction of the muscles, white double arrow = relaxed muscle; 8). An avicularium (9) is attached to the lower zooid. The "brown body" (10) consists of aggregated toxic waste products, because the bryozoan does not possess kidneys.

ciliary tuft tests for a suitable area of settlement. When it is ready, the larva everts the adhesive pad situated between the two valves, and uses suction to attach itself to the substratum. The larva anchors itself with a secreted cement. During a succession of muscular convulsions, all soft body parts, such as the circular bulges that bore the locomotory cilia, the ciliary depressions in the digestive tract, and the intestine, are retracted underneath the valves. Lateral pressure exerted by the soft body tears the adductor muscles of the valves, and the *Cyphonautes* larva collapses. In the final stages the partially stacked valves cover this rather depressed sacculus, which is filled with the cellular debris of the dissolved larval organs.

The continuation of this process is described in rapturous words by Carl Cori: "The wonder of this process lies in the fact that the image of [an adult] stelmatopod [= gymnolaemate] is organized from this heap of ruins. This is not the result of embryological forces but is due to the almightly power of the budding process. It is this power that completes the total blueprint of the bryozoan organization." Just as in the asexual reproductive process, the first complete bryozoan individual buds from this cellular clump. This first bryozoan (ancestrula) forms the breeding stock for a new bryozoan colony. The phenomenon is unique in the course of biology: the process of sexual reproduction via egg and embryo is intercepted shortly before metamorphosis to the "adult" bryozoan by the asexual budding process. The metamorphosis of the uncomplicated yolky larvae follows a similar course. In the genus *Mempranipora* the reconstruction from the point of the disintegration of the larval organs to the emergence of the ancestrula lasts five days.

The bryozoan colony is formed by continuous budding. The coelomic wall of a potential autozooid is developed in the hindgut region of the ancestrula; then the intestinal loop of this zooid grows. The ancestrula seems to have special developing properties, for concomitantly several buds and specific attachment structures, such as an adhesive disc or rhizoid processes, are produced.

The structural arrangement of a colony is determined by the number of buds and the species-specific type of organization. In a creeping bryozoan colony, each zooid gives rise to two buds which branch off in a forklike manner. In certain species where the colonies freely protrude into the water in leaf shapes, the surface of the colony resembles a meshwork. The pores help decrease the resistance to the water current. A flat colony is arranged like a mosaic. Zooids that are adjoined along their longitudinal surfaces fuse together within the chain. Adjacent rows are frequently out of line with a parallel chain by half a zooecium (quincunx arrangement). Just as in a wall where the bricks are shifted by half a length, the bryozoan arrangement is adaptive, because it prevents tearing. In addition this alternating arrangement prevents the lophophores from interfering with each other.

The manner of budding reveals the eventual destiny of the zooid.

If the zooecium is calcified, it is even possible to identify the bryozoan from past geological eras.

If the ancestrula produces a higher number of buds (e.g., six) there will be a ring-shaped arrangement of bryozoan generation around the first animal. Many crusty and clump-shaped forms originate in this manner. The vast extent of asexual reproduction via budding becomes evident if one considers that *Schizoporella sanguinea*, starting from one ancestrula, formed a colony consisting of 38,000 zooids, with a diameter of 12 cm, in the harbor of Rovinj, Yugoslavia, within five months. Disturbing factors, such as a change in the salinity or temperature of the water, or in food availability, will influence the characteristic growth form of the colony. Bryozoans kept in an aquarium usually atrophy because of the unnatural living conditions. At periodic intervals the polypides of marine bryozoans degenerate in conjunction with the sexual reproductive phase. During this time the lophophore remains withdrawn and distintegrates along with the foregut. The resultant reddish-brown end-products (brown body) collect in the stomach and become surrounded by a cellular lining. This is the manner in which the bryozoan rids itself of toxic excretions which it cannot discharge in any other way. A new polypide buds from the wall of the zooecium. It grows around the brown body and often forces it to the outside. In *Membranipora membranacea* this transformation process takes five days. Fresh-water bryozoans also rely on the components of the polypide when the colony requires an increase in growth products for budding, for the formation of ovicells, or for the regeneration of injured sections.

Marine bryozoans are not seriously affected by changes in the seasons; fresh-water forms, however, do not survive the winter in the cooler latitudes, or the dry periods in the hot countries. In these cases the life span of a colony is only one year.

In order to survive this dilemma, these species have developed resistant bodies (a permanent stage known as a statoblast) that give rise to new colonies under favorable conditions. During the early summer, epidermal cells of the peritoneum start to grow along the funiculus in the direction of the intestinal loop. At the apex of this cell migration, gobletlike crescents develop which become filled with yolky cells from the coelomic cavity and which finally close into lentillike, hollow balls. The outer cells secrete a dark brown, solid chitinous shell. The fully developed statoblast maintains the location of the former colony or sinks to the bottom as soon as the zooecium in which it originated has distintegrated. The outer part of the chitinous shell may be modified into a pneumatic ring (annulus). After the colony has died, statoblasts equipped with annuli rise to the surface of the water and become attached to pieces of driftwood, floating leaves, or the feathers of an aquatic bird. Some fresh-water species are characterized by statoblasts that bear hooks around their margins (spinoblasts).

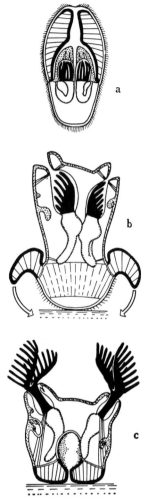

Fig. 8–16. Fresh-water bryozoan: the primary zooid escapes through a rupture in the ventral body wall of the parental body, and starts to swim freely for a few minutes or up to 24 hours (a) via the mantle cilia. Then the larva attaches to a substratum and the mantle lobes (long lines) are flipped downward (6) and are tucked under the animal like a cushion (c). The first buds are evident in the lateral walls (compare Fig. 8–11).

Fig. 8-17. The position of the tentacles is indicative of the degree of stimulation caused by various concentrations of a solution: moderate stimulation elicits the "disgust response" (b) which can be differentiated from the normal position (a) by the kinked tentacles. Saline solution twice the concentration of sea water relaxes the tentacles (c).

Many fresh-water species, particularly those inhabiting lakes and ponds, have achieved extensive ranges of distribution by the dissemination of resistant bodies; but no other species have exceeded the adaptability of the bryozoan statoblast. Colonies of the genus *Plumatella* which were found in Lake Michigan produced approximately 800,000 statoblasts per m² of inhabited leaf-covered areas. Along the shore of this lake a 50-cm-deep and 1-km-long border of washed-up statoblast shells could be found.

Statoblasts are able to survive periods of desiccation and cold but do require a dormant period of several months before they are able to germinate. At a water temperature of approximately 10°C a polypide bud develops at a specific location on the stratoblast wall within five days. This structure develops from epidermal cells and cells from the funiculus, and feeds on the stored yolk. The polypide finally emerges through a split groove. A statoblast originating in the spring can give rise to a new colony within the summer of the same year.

In contrast to the free-swimming larvae, the sessile bryozoans lack all sensory organs. Individual subepidermal sensory cells are receptive to external stimuli. These cells are concentrated in the lophophore region. The central nervous center (ganglion) is found on the dorsal side of the pharynx. Stimuli of different intensities elicit the "disgust response" (compare Fig. 8-17) in the tentacles, which means that they are kinked to the outside. The genus *Electra*, which is adapted to a water temperature of 19°C, kinks the tentacles when the temperature increases by six degrees or decreases by eight degrees. The genus *Farrella* demonstrates the "disgust response" if one part of citric acid is added to 7,200 parts water (compare Fig. 8-17). Man, by comparison can only taste citric acid in salt water at a concentration which is ten times that strong. The lophophore is quickly withdrawn into the protective zooecium in response to more intense stimuli. When irritated, a cheilostome from eastern Asia, *Acanthodesia serrata*, emits a pale blue light from a pair of globular, glandular cell masses located at the lophophore neck.

Bryozoan colonies are most abundant in shady locations in the littoral zone of the continental shelf. The *Cyphonautes* larvae have red pigment spots which are probably light-sensitive. However, it has also been observed that the lophophores of colonial forms turn away from sunlight. Only few species have rhizoids that anchor them to sandy or muddy substrates. Usually these animals adhere to solid substrates such as stones, driftwood, empty or living gastropods or bivalve shells, or the smooth surfaces of broad-leafed algae and water plants. Except for needing an adequate source of food, the bryozoans in general do not seem to be too specific in their choice of habitat. They prefer water depths of 0.5 to 200 m. Exceptional species, like *Menipea normani* from Iceland, penetrate to depths of 1000 m, where the temperature is only 1°C above the freezing point. Bryozoans have even been recovered from depths of 6000 m.

Varying degrees of salinity do not seem to influence the distribution of the individual species, if one does not consider the fact that the Gymnolaemata and Stenolaemata are marine, and the Phylactolaemata are fresh-water forms.

However, even this general rule is broken by approximately twenty species of Gymnolaemata which have invaded the mainland fresh-water sources via the river mouths. The genera *Paludicella* (see Color plate, p. 254) and *Plumatella* (see Color plates, pp. 243 and 254) may even cause damage to water pipes in Europe. In 1885 in Hamburg, a water pipe was dug up which was covered with a layer of bryozoans which measured 15 cm. It must be considered that this pipe was kept in total darkness, and was subjected to an atmospheric pressure of five atm. Similar annoying incidences were reported from Rotterdam in 1877, and from Paris and many English towns. Bryozoans plug these water pipes. The disintegration of these organisms in turn brings about an increase in bacterial growth. Modern water mains are protected from these organisms by filters.

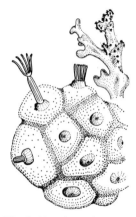

Fig. 8–18. *Alcyonidum gelatinosum* forms clump-shaped colonies which are attached to supporting objects. These colonies grow to a height of 10 cm. One magnified terminal lobe of the colony is shown.

Bryozoans constitute a significant part of the sessile animal communities found in the oceans. They are found in association with hydroids, actinarians, polychaetes (see Vol. I), *Zostera* meadows, and oyster and bivalve beds. Other creatures such as flatworms, threadworms, annelids, crustaceans, gastropods, and asteroids which are not sessile are often also found in these animal communities. They either ignore the bryozoans or feed on them, as do certain fishes. The habitat of the fresh-water bryozoans is far less varied. Many minute crustaceans, earthworms, and the larvae of various water insects seek the protective shelter of bryozoan colonies, which often reach the size of a child's head. Often fresh-water sponges and bryozoans fuse together into a clump and compete in their growth. The large blood-red midge larvae nibble long channels into all directions of the bryozoan colony. These channels are lined with finely spun secretions and probably hasten the distintegration of the colony during the fall. Some marine bryozoans burrow into the shells of bivalves and barnacles and use the ground-up calcium in reinforcing their zooecia. The species *Hypophorella expansa* forms a symbiotic relationship with certain tube-dwelling polychaete worms (e.g., *Lanice conchylega*; see Vol. I). The bryozoan colony, however, is in constant danger of being coated by the secretion which the worm discharges in order to reinforce its own tube. The bryozoan has adapted to this situation in that each zooid chisels a hole into the newly secreted layer with a crescent-shaped rasping plate located on the neck of the polypide. *Harmeriella terebrans* destroys the polypides of *Tubiporella* and then occupies the latter's zooecia, activities reminiscent of "cannibalism."

The class Gymnolaemata is represented by bryozoans in all oceans. Since the animals are sessile, they are dependent on the ocean currents for dispersal. The effect of these currents is more intense along the coastal

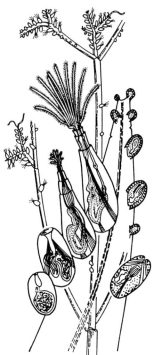

Fig. 8-19. The branched stalks of *Zoobotryon verticillatum* are easily mistaken for algae. The zooids are attached to slender branches measuring 2 mm in opposing rows. A terminal section of such a branch, magnified about 50x, is superimposed on a section of the colony. The right side shows buds in various stages of development; the left shows a fully and a partially extended polypide; below shows two old autozooids which have supplied the colony with structural components (dissolved viscera) via the funiculus (small lines).

zones, since there is a greater availability of driftwood and strings of algal growth to which the bryozoans are attached and thus transferred. With the exception of the true *Cyphonautes* larvae which live for months in the open ocean, the free-swimming yolky larvae with their brief period of swarming are comparable to ephemeral insects, and capable of only limited extension of their range of distribution. However, because of the extreme age of this animal class, they have nevertheless become cosmopolitan in distribution. Favorable local living conditions can be fully exploited by the bryozoans because of almost unlimited budding. Marine gymnolaemates are classified into two orders on the basis of the means by which the polypide opening is closed and protected. These orders are the Ctenostomata and Cheilostomata.

The CTENOSTOMES (order Ctenostomata), characterized by their uncalcified zooecia and orifices with a pleated collar for a closing apparatus, are differentiated into two different developmental lines on the basis of their external appearance.

A. Suborder Camosa: true stolans are absent. There are two superfamilies.

1. Superfamily Halcyonelloidea: the colony is fleshy or leathery. The oval or boxlike cylindrical zooecia are embedded in the surface of the colony along the ventral side. *Alcyonidium gelatinosum* (see Color plate, p. 234) is greenish-yellow, with a freely suspended lobular colony measuring up to 90 cm in height. Each polypide possesses fifteen to seventeen tentacles. It is distributed along the eastern coast of the North Atlantic. *Flustrellidra hispida* has reddish-brown, densely prickly colonies which form spongy linings on kelp. There are thirty or thirty-five tentacles per polypide. The species is found along the eastern coast of the North Atlantic.

2. Superfamily Paludicelloidea: the colony is chainlike and horny. The bottle-shaped zooecia branch off the main trunk at an upward angle. The family Paludicellidae includes *Monobryozoon*, and the water-pipe-clogging bryozoan species *Paludicella articulata* (see Color plate, p. 254). These reddish-brown, netlike colonies adhere to the stalks and leaves of water plants, roots of shore brushes, bivalve shells, and stones, as well as the insides of water pipes. There are sixteen tentacles per polypide. This form is easily overlooked because of its delicacy. This fresh-water species is cosmopolitan in distribution. The winter-resistant phase is known as a hibernaculum, which is an oval zooid containing a yolk supply, and which is surrounded by a calcified shell reminiscent of the statoblasts of the true fresh-water bryozoans. The species *Victorella parida*, of the family Nolellidae, was discovered in the brackish water of the docks in London. This species forms fuzzy colonial covers on woods and rocks in brackish water with a salinity of one percent. There are eight tentacles per polypide. It is found in harbors and in the Baltic Sea.

B. Suborder Stolonifera; zooecia are tubular or cylindrical and open at the top. The zooecia arise singly or in netlike or branching groups from distinct stolons composed of modified zooids which are divided by nodes. There are two superfamilies.

1. Superfamily Vesicularioidea includes *Amathia lendigera*, characterized by beige zooid groups which may be arranged with four to six pairs in a row borne on forking stolons. Each polypide has eight tentacles. The species is found in association with algae and laminar bryozoan of the calcareous red algae (see Vol. I) in the Mediterranean. The same superfamily includes *Zoobotryon verticillatum*, which is characterized by milky-white tubular stolons which measure several meters and are forked. The diameter is 2 mm. The zooids protrude from two opposite rows at certain intervals. The zoaria are reminiscent of filamentous algae, and for this reason are easily overlooked. The colonies disintegrate in the fall, and remain dormant on the ocean floor during the winter. In the spring new side branches emerge from the old colony. New buds form on these branches (spring colonies). Segments of these colonies break off and are dispersed by the water current. At an optimal location a segment will anchor and produce larvae via sexual reproduction. These larvae give rise to the summer colonies. The stomach of these particular polypides is equipped with serrated horny linings and special chewing muscles, which enables it to crush and digest armored dinoflagellates and radiolarians. Each polypide has eight tentacles. This form is frequently found in large concentrations on buoys, drifting material, and resting boats in the Mediterranean (Fig. 8-19).

The genus *Penetrantia* also belongs to the same superfamily. In 1945 Lars Silen discovered a series of bryozoan species which occurred on both empty and occupied bivalve shells along Sweden's western coast. Subsequently he found these same species in mollusk collections. These bryozoan species grow in the upper calcareous layer of the bivalves, where they form a network of flat stolons that run parallel to the upper surface, with a diameter of 0.02 to 0.08 mm. Connective channels, approximately 0.5 mm long, procede from the stolons at regular intervals, forming a vertical tube to the outside. A burrowing zooid corrodes these channels with phosphoric acid, and subsequently a budding autozooid penetrates into this space. The presence of these zooids is revealed by the two fitted zooecia.

2. Superfamily Walkenoidea includes *Hypophorella*, the bryozoan commensal of tube-dwelling polychaetes.

The order Cheilostomata, represented by the greatest number of species, is characterized by calcified zooecia which have orifices which can be shut by hydraulic opercula. Four suborders are currently recognized:

1. Suborder Anasca, characterized by a membranous anterior wall which can be pushed in by muscular action, thereby producing the internal pressure necessary to extend the polypide. It includes the species *Aetea anguina*, which is represented by hairlike, transparent zooids in a single

▷

Brachiopods:
1. Three articulate brachiopods (*Terebratula*) which are attached with their short peduncles to the substratum; natural size.
2. *Lingula unguis* buried in the sand. The tubes have been cut longitudinally. The length of the shells is 4 cm, and the tubes measure 16 cm.
3. Two sessile articulate brachiopods (*Terebratulina*) with long peduncles; natural size.
Phoronids:
4 to 6. *Phoronis psammophila*:
4. Top view of the extended lophophere.
5. Four animals (BL 30 mm) on a razor shell.
6. Dorsal view of an extended lophophore.
Tentaculate larvae:
7. Swimming larva of an articulate brachiopod (BL 0.2 mm).
8. Swimming larva (*Cyphonautes*; width 0.5 mm) of *Mempranipora membranacea*.
9. Actinotrocha larva of a phoronid (BL 0.6 mm).

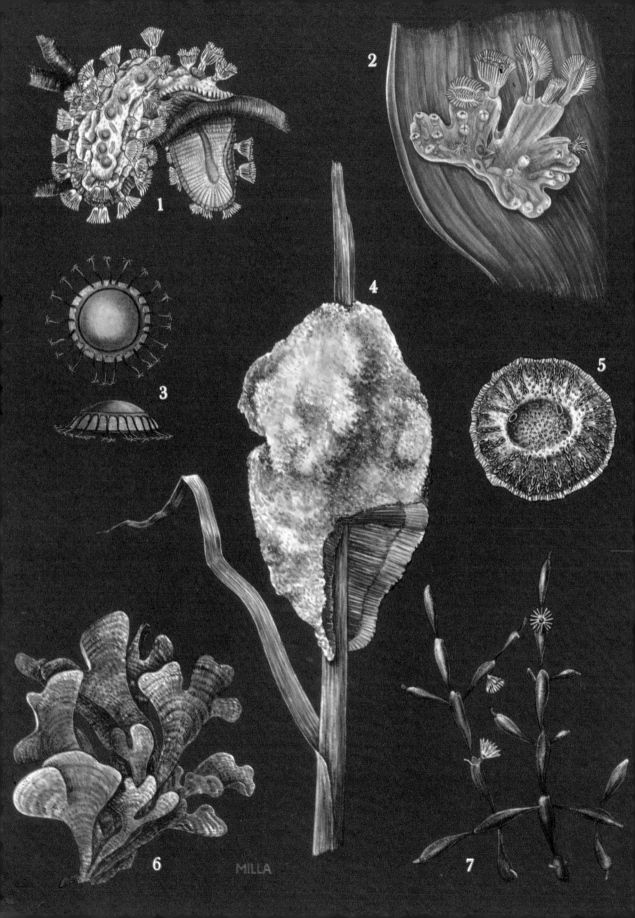

row with polypides that measure 1 mm in length and emerge vertically from the substratum, spreading in a spoonlike fashion. The species is found on kelp and stones in shallow coastal zones. *Bugula neritina* is found in high concentrations on buoys and even on fast-moving boats. This species is characterized by grayish-brown, horny, shrublike colonies (L up to 3 cm). *Bugula* is characterized by a biserial attachment of zooids. *Flustra foliacea* (see Color plate, p. 254) is characterized by greenish-brown colonies that measure up to 18 cm in height, and are branched in the forklike lobular manner of the brown alga *Dictyota*. The two sides of the lobes consist of autozooids interspersed with scattered avicularia. The species has a worldwide distribution, at depths greater than 25 m. The attachment disc of the colony adheres to broken bivalve shells and coral moss. *Membranipora membranacea* forms a fanlike, whitish, transparent cover on algae (for example on *Laminaria* and *Fucus*). It is found in the shallow coastal zones of all oceans. The zooecia are out of phase by one half length, and are characterized by tubular attachments (tower zooids).

2. Suborder Cribrimorpha, with a membranous anterior wall, with spines which fuse in the midline, forming a protective sievelike shield. Includes *Cribrilina*, a characteristic intertidal species.

3. Suborder Gymnocystidea, with a membranous anterior wall and an outgrowth of the anterior wall growing over the membrane. Includes *Umbonula*, which is common on undersides of intertidal rocks.

4. Suborder Ascophora has a heavily calcified outer covering; therefore ascophorans have evolved a compensation sac which aids in increasing the internal pressure needed to expel the polypide. *Retepora beaniana* (see Color plate, p. 233) is characterized by salmon-colored zoaria that measure up to 10 cm in height. The colony resembles a filled funnel that is rigid and fenestrated. The zooids oppose one another like shingles on a roof. The polypide orifice is located on the inner surface of the funnel. This species is found in heavily shaded caves and along the northern slopes of the rocky coastlines of European oceans. *Schizoporella sanguinea* is characterized by clumplike colonies that are orange-red because of the carotene they contain. This species is found on the trunks of algae, and also in high concentrations on buoys and their anchors, or the hulls of ships, where it is considered an important fouling organism. *Cellepora pumicosa* is characterized by colonies which hang from gulfweed (*Sargassum*) like a swarm of bees. The light-red, urn-shaped zooids are closely packed together. Each polypide orifice is associated with an avicularium. The species is found in the slowly moving waters of the North Atlantic.

In the class Stenolaemata the terminal orifice of the polypides is closed by a circular muscle. The zooecium is made of a chitinous cuticle with calcareous deposits. Of four orders, only the Cyclostomata has living representatives. Four of six living cyclostome superfamilies are mentioned here. The superfamily Articuloidea is characterized by shrublike colonies which are anchored by rhizoids. The branches of the colony

◁
Phylactolaemata, freshwater bryozoans:
1. A colony of *Cristatella mucedo* (BL 1.5 cm) crawls along a plant stalk.
2. A young colony of *Lophopus cristallinus* which is attached to a leaf of a reed.
3. Statoblasts of *Cristatella mucedo* seen from above and from the sides.
4. *Plumatella fungosa* (compare Color plate, p. 243) grows around a reed; the exposed section shows the annual growth layer (colony L 20 cm).
Gymnolaemata, marine bryozoans:
5. Calcareous colony of *Lichenopora radiata* with retracted lophophores.
6. *Flustra foliacea* colony (height 80 mm).
7. *Paludicella articulata* (L of autozooid 0.6 mm); one of the few species that inhabits fresh water.

are interrupted by flexible articulating joints. The species *Crisia eburnea*, a representative of the family Crisiidae, is characterized by an ivory-colored colony which reaches a maximum size of 3 cm. It adheres to algae found at shallow depths in all oceans (Fig. 8-12).

The superfamily Tubuliporoidea is characterized by unsegmented zoaria in the form of laminar discs or lobes. The species *Diastopora patina* consists of white discs which measure up to 0.5 cm and grow on algae and granular substrates in the Mediterranean.

The superfamily Canceloidea forms vertical colonies which are branched in a forklike manner much like the stony corals. The external surface of the cylindrical zooecium consists of two layers connected by heavily calcified transverse and longitudinal ridges. *Hornera lichenoides* is characterized by a dirty-yellow colony which grows up to 20 cm and resembles a coral stalk. This species grows on underwater cliffs at depths greater than 50 m, in arctic oceans.

The superfamily Rectanguloidea can be identified by its disc-shaped, crusty colonies. The ends of the zooecia are arranged like rays in the calcareous mass of the colony. *Lichenopora radiata* (see Color plate, p. 254) is characterized by a yellowish-white disc measuring up to 3 cm in diameter. This species grows on corals, bivalve shells, and algae in the continental-shelf area of the North Atlantic and the Mediterranean. The margin of this colony is enlarged by a thin marginal lamella.

When compared with the great diversity of the marine gymnolaemates and stenolaemates, the class Phylactolaemata seems primitive and simplified. It is represented by only a few genera with a total of slightly more than fifty fresh-water species, which are cosmopolitan in distribution. The small number of species is compensated for by a well-developed degree of polymorphism, and a rapid and extensive rate of dispersal. The extension of the range of distribution is greatly aided by a free-swimming juvenile colony of primary polypides (Fig. 8-6), enclosed in a ciliated cover, which take the place of true larvae. The accelerated rate of development and the tendency for early colony formation in primary polypipes are adaptations to the seasonally unfavorable conditions of the fresh water. In addition, the Phylactolaemata have evolved a very effective resistant form (statoblasts; see Color plate, p. 254) which remains dormant during the winter or dry period, ensuring the survival of the species. The development of these special adaptations makes the Phylactolaemata just as effective (as living organisms) as the more diverse marine bryozoans. Unlike the bewildering modifications encountered in the gymnolaemate zooids, the anatomy of the phylactolaemate zooid is simplified in favor of the common structure of the whole colony. The innermost regions of the zooecia are not separated by septa. The coelomic fluid, which functions as an excretory and vascular organ and also as a "buffer" for changes in fluid caused by the eversion of the polypides, is continuous in all the zooids. The terminal orifice is closed by a simple diaphragm. The phylac-

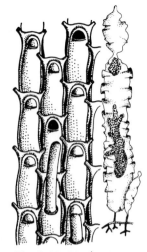

Fig. 8–20. *Membranipora membranacea* forms a laminar whitish coating on leaf kelp. Each zooecium of this bryozoan colony measures 0.5 to 1 mm in length and is out of phase with the adjacent zooecium by half a length. Some zooecia have been emptied by parasitic worms (compare black polypide opening). Two zooecia give rise to the so-called tower zooids. Right, two colonies on the kelp *Laminaria rodriguezi*.

tolaemate bryozoans are devoid of any "luxury items." Calcified zooecia, avicularia, and vibracularia are absent; but the essential features are well emphasized. The two lophophore ridges permit the growth of a larger number of tentacles, which in turn provides a greater surface area for respiration—an adaptation to a habitat of low oxygen content—and also provides a more effective organ for capturing prey, which is scarcer in fresh water than in the ocean. The phylactolaemate bryozoans are classified into two families on the basis of the means by which the zooids are arranged in the colony:

The family Plumatellidae is composed of fresh-water bryozoans characterized by chitinous, dark brown zooecia either joined into antler-like chains or with the lateral walls adhering to each other to form massive clumps. The statoblasts are round or oval. The main genus, *Plumatella*, is represented by species that can be differentiated on the basis of their growth characteristics. *Plumatella repens* (see Color plate, p. 243), the most frequently encountered fresh-water bryozoan, adheres to the underside of leaves and rocks. In the spring the free-swimming juvenile colonies adhere to the substrate and fuse together. In the fall *P. repens* forms a thick, walnut-brown lining containing masses of statoblasts. The species *Plumatella fungosa* (see Color plate, p. 254) is characterized by colonies that are clumplike and may grow to the size of a child's head, weighing up to 1 kg. The zooecia may form short or broad chains. This species grows around the stems of reeds and the roots of aquatic plants, in the form of a spindle-shaped colony. At times in the larger rivers the shells of the gastropod *Viviparus* are completely covered by this bryozoan species, making the snails look like chocolate balls with only the aperture remaining free of the bryozoan. The snails do not seem to be harmed by this coating. *Fredericella sultana* forms streamerlike colonies suspended from stones and water plants in small, stagnant pools. Ripples caused by autumn winds break up these streamers and disperse them to new habitats.

The few species representing the family Cristatellidae are characterized by colonies forming the shape of a loaf of bread. The zooids are surrounded by a gelatinous cover. Their boat-shaped statoblasts are always surrounded by an annulus, and are often embedded with spinoblasts. *Lophopus cristallinus* (see Color plate, p. 254) has the distinction of being the most beautiful bryozoan. Its filigree fragility is enhanced by the bluish cover, a lemon-yellowish lophophore consisting of sixty tentacles, and a shimmering, reddish-brown intestinal loop with yellow stripes. This species forms pea-sized colonies adhering to aquatic plants, preferably the roots of the water lentil (*Lemna*), in ditches and pools. When the colony has reached a diameter of 2 to 4 cm its margin divides into lobes which become detached. These fragments crawl away from the parent colony on the gelatinous mucus at a speed of 1 cm per thirty minutes. The anterior region of the budding zone of this fragment pulls the entire colony in the manner of the pseudopodia in the amoebae (see Vol. I). The species

Cristatella mucedo (see Color plate, p. 254) is characterized by a colony with the shape of a shell-less gastropod. Longitudinal rows of polypides, with ninety tentacles each, protrude from the colony's dorsal surface. The youngest and most functional polypides are located on the colony's margin, while the older ones with the now-superfluous lophophores are situated more to the center, where they are fused into the colony. During sexual reproduction the larvae emerge out of these empty zooecia. When the colony becomes too long, it divides and the two halves creep in opposite direction on their "creeping soles." Budding ceases in the fall. The colonies lack the polypides which have been destroyed by waves caused by autumn winds. This physical force also releases the statoblasts (see Color plate, p. 254), which adhere to aquatic plants and birds by their spinoblasts. This bryozoan species is even found in highly acidic swamp waters, where it grows on the underside of floating leaves.

The souvenir shops of coastal towns often offer as high-priced rarities peculiar-looking shells resembling bivalve shells. Even during the 19th Century many zoologists still assumed that the animals associated with these shells were related to the mollusks, although they differentiated these animals from the familiar gastropods (see Chapter 5) and cephalopods (see Chapter 7) because of the obvious differences. These forms were called Brachiopoda because it was assumed that the spiral lophophore arms were coiled capture apparatus. This unfortunate description of this animal class has persisted although the brachiopod's relationship to the bryozoans, or even more convincingly, to the phoronids, has been recognized.

Present-day brachiopods or lamp shells (phylum Brachiopoda) are the last representatives of an animal phylum that was formerly greatly abundant. Although the approximately 30,000 forms do not give a direct indication of their evolutionary origin, they nevertheless are witness of the great geological age of this animal group and its great diversity. The geologically oldest remains of brachiopods have been dated from the Lower Cambrian era (approximately 600 million years ago). A wealth of inarticulate and articulate brachiopods was already recorded from the Lower Paleozoic era, reaching a peak number of species during the Devonian period (405 to 345 million years ago). Since the Carboniferous (345 to 280 million years ago), a decline set in, with a brief flourishing period again during the Jurassic (181 to 135 million years ago).

Phylum: Brachiopoda

It is interesting that some inarticulate brachiopid genera belong to the oldest known animal forms. The shell fragments of the genus *Lingula* from the Paleozoic can hardly be differentiated from those of today's forms. *Lingula unguis* is a true "living fossil" that has remained practically unchanged for millions of years. Only a few conspicuous representatives from the great diversity of extinct brachiopods shall be mentioned. In contrast to most of the living forms, these fossil forms inhabited shallow waters. The superfamily Productacea, which enjoyed a worldwide distribution during the Upper Paleozoic, evolved gigantic forms with the shell measuring 30 cm wide, e.g., *Gigantoproductus giganteus* from the Carboniferous. Members of this group attached to the substratum via spicules.

The spirifers (order Spiriferida; genera *Spirifer, Anthracospirifer, Imbrexia*, and others), which flourished during the Devonian era and are considered as indicator fossils for that period, were characterized by ribbed and strongly, transversely elongated shells of the helicopigmate type. The superfamily Richthofeniacea (including genera *Richthofenia, Prorichthofenia*, and *Coscinaria*), from the Permian period, were "reef"-building brachiopods (hippurite type). They were characterized by a tubular ventral valve, which attached to the substratum, and a lidlike dorsal valve. In the suborder Oldhaminidina (including genera *Oldhamina* and *Leptodus*), from the Carboniferous to the Triassic, one dorsal valve is serrated along the margin. In some of the terebratulids (family Pygopidae; genus *Pygope*), from the Jurassic and Cretaceous, holes occur in the center of the shell. This phenomenon is due to the increased growth of the shell margin.

During the Mesozoic era, when the marine bivalves and gastropods flourished in the number of species and diversity, the brachiopods diminished in numbers or survived in the greater depths of the oceans, like other animal phyla, where they occur to the present.

In order to classify the brachiopods correctly, one has to start by pointing out the differences between the brachiopods and the mollusks among which they were incorrectly included. The bivalves are characterized by a hinge on the dorsal surface which serves as a connection between the right and left shell valves. The only possible plane of symmetry through the body of the brachiopod would make the two conspicuous valves a ventral and dorsal shell. This would apply at least to the subclass Testicardines (or Articulata), which is characterized by possessing an interlocking teeth-and-socket arrangement. In addition, the ventral valve in the brachiopods is almost always more convex and tapers to a peak posteriorly. This process projects beyond the posterior margin of the dorsal valve and is perforated for the passage of the fleshy pedicel.

As has already been mentioned in the introduction to the lophophorates, all representatives of this group have a common body plan. In the brachiopods the unique shell valves seem to interfere with this body plan. In reality, however, these valves are comparable to the structured tubes of the phoronids or the protective zooecia of the bryozoans which have been split open along the sides. This body-covering evolved into a tightly closing, calcareous, protective capsule which proved to be highly adaptive to the brachiopods which are freely suspended into the water. This anatomical arrangement proved to be successful without any changes for a long geological time span of at least 500 million years.

The two shell valves are produced by mantle lobes of the hind body

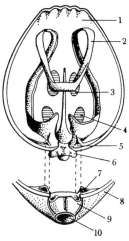

Fig. 8–21. The exposed calcareous components of an articulate brachiopod: The looped, transversely connected pair of lophophore skeletal arms (2) arise at the posterior end of the dorsal valve (1). The teeth (5) articulate with the sockets (7) in the partially illustrated ventral valve (8). The extensor muscles insert at the hinge (6). The aperture of the peduncle (10) is constricted by the two plates (9). The inner surfaces of the valves reveal the muscle scars of the valve extensor and flexor muscles. The pattern of the scars is species-specific.

(metasome). The germinal layer for the mantle and shell is located in a groove surrounding this lobe. Long bristles frequently originate at this location. They arise out of deep saccules and are movable by muscles. The shell is covered by a periostracum consisting of nearly equal amounts of chitin and calcarious substance. The upper shell surface is usually smooth, and only rarely ribbed like a scallop, where the ribs radiate from the hinge. Fossilized shells are often covered by spicules. The tightly closing shell margins may be curved or indented in a variety of ways, which helps identify the various species. The shells of the brachiopods may be all shades of brown. Several species are porcelain-white or have a greenish tint because of algal growth. Like all sessile marine animals, the brachiopods are also plagued by adhering plant and animal growth.

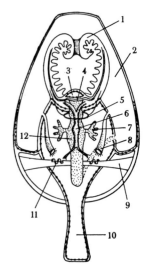

The major portion of the space between the shell valves is occupied by the lophophore. This organ arises from a thin, transparent septum which separates the posterior region of the inner shell space (coelomic cavity) in a dorso-ventral direction. Small brachiopods, measuring 0.5 cm, possess a transverse membranous disc which bears the tentacles in its margin (trocholophe type). In the larger species this tentacular disc is slit or lobed (schizolophe type). Most brachiopods, however, possess a calcareous loop from the dorsal valve which supports the lophophore. This skeletal support may run parallel to the symmetry of the body (plectolophe type) or maybe vertically coiled (spirolophe type). The outer margin of the lophophore support bears thousands of tentacles. In order to produce a water current that provides the animal with a sufficient amount of suspended food particles and oxygen, it is highly advantageous that the lateral arms (brachia) of the lophophore are rigid. Even in the primitive species of the inarticulate brachiopods, a well-developed strand of cartilagenous, elastic supporting tissue penetrates through the lophophorate arms. The articulate brachiopods have either snowflakelike crystals of calcite that are embedded into the supporting tissue, or, as in the majority of species, a calcareous loop. This structure serves as an important feature in the classification of fossil brachiopods because it does not decompose like the soft parts of the body. The calcareous loop is produced by two cellular tubes which merge into the two lateral arms from the hinge section of the dorsal valve. When the lateral arms are not supported by skeletal elements, they are moved by flexor and extensor muscles.

A nerve strand extends to the tips of the lateral arms just below the dermis. This strand arises from an esophageal nerve ring with a small ganglion on its dorsal side and a larger one ventrally. The most recent examination of the brachiopod nervous system, carried out in 1883, revealed an additional pair of nerves supplying the arms, as well as a number of nerves going to the valve musculature.

The transverse mouth opening lies within a groove located at the origin of the two lateral lophophore arms. The brachial groove is formed by a lower lip consisting of the basal part of the tentacular fringe and an

Fig. 8–22. The anatomy (schematic) of a brachiopod with the ventral valve removed: the body is divided into three segments as is indicated by dotted lines. There is the anterior section (prosome), the upper lip (3; epistome), a central section (mesosome) with the double spiral of the lophophore arms (1) and the mantle cavity (2) which is exposed to the outside water. The hind section (metasome) is composed of the viscera and peduncle (10). The brachial groove is located at the base of the tentacles. It widens near the mouth (4) which leads into the blindly ending digestive tract (12) or a tract that ends in an anus near the peduncle aperture. The digestive tract is located in the right mantle cavity (interrupted lines). Digestive pouches (7) in the central gut secrete digestive juices. Tubular, forking (on both ends) circulatory system (6) with contractile vessel, the heart. Paired metanephridia (8) transport the gametes (11) to the outside. The gametes grow on a transverse strand of connective tissue (9). Nerve ring (5).

upper-liplike transverse fold (corresponding to the epistome of the phoronids and fresh-water bryozoans). In contrast to the bivalves, the brachiopods open the valves with a set of well-developed muscles extending from the ventral valve to the hinge process of the dorsal valve. The brachiopod valves may gape 5 mm, for example, in *Gryphus* (which measures 4 cm), or up to an angle of forty-five degrees, as in certain smaller species. Two water currents produced by the ciliary action of the tentacles enter the anterior gape, move past the lophophore into the cavity between the valves, passing out along the anterior margin of the shells. Food particles, particularly diatoms, are filtered from the water currents and are carried by the cilia which line the outer margin of each tentacle to the brachial groove. In addition to diatoms, the intestinal contents of *Lingula* may include dinoflagellates, foraminiferans, radiolarians, various larvae of mollusks and sea urchins, polychaete bristles, and sponge spicules. A mucous string, constantly moving towards the mouth, collects food particles like a conveyor belt. The mucous flow is amazingly rapid: in *Lingula* a particle traveled 1.4 cm, from the tip of the arm to the mouth, in only seventy-five seconds. Inedible particles are immediately rejected by an outflowing ciliary current running over the tips of the tentacles.

The food-laden mucous strand is conducted into the mouth. In *Lingula* the mucus is fragmented within the esophagus and then reaches the globular stomach. The ciliated stomach lining turns the food string until four glandular outpocketings pump the stomach contents into these pouches at a rate of approximately five to ten times per minute over a time span of several hours. Digestion is intracellular within the digestive glands. Inedible particles are rejected from the glands and are conducted to the intestine via ciliary action. Feces are voided by the anus, located to the right of the right lateral arm. In the similarly inarticulate *Crania* the anus terminates at the posterior end of the body. The articulate brachiopods which lack an anus excrete fecal segments through the mouth. In order to avoid pollution of the water surrounding the lophophore, metabolic by-products and wastes are fused into a gelatinous strand which is constantly churned toward the mouth by the blindly ending stomach. Periodically segments of this strand are ejected from the stomach by a reverse beat of the cilia.

It is rather difficult to define the three body segments which characterize the lophophorates in the brachiopods. The upper-liplike transverse fold (epistomal lamella), which makes up the dorsal wall of the brachial groove, originated from the fore body (prosome). The lophophore and visceral organs are components of the central body (mesosome) which is usually indistinctly separated from the shell-secreting hind body (metasome) by perforated septa. The coelomic cavities of the two posterior body segments contain coelomic fluid. This fluid in turn contains food-absorbing amebocytes and reddish-brown granulocytes. The granulocytes are composed of a colorless, iron-containing hemerythrin, comparable to the vertebrates' blood pigment, hemoglobin, and which, like hemoglobin, takes up oxygen and becomes reddish-brown. The coelomic fluid is constantly kept in circulation via ciliary action. The coelomocytes supply the body organs with oxygen and food elements. The coelomic fluid reaches the tentacles and shell-forming mantle lobes via canals. A simple, open circulatory system is found in the brachiopods. The system extends through the mesenchyme of the entire body with a longitudinal vessel in the dorsal mesentery. Part of this collecting vessel is enlarged into a contractile vessel or heart, which beats every thirty to forty seconds.

A metanephridium extends along each of the two lateral mesenteries supporting the stomach. Each metanephridium has a fringed nephrostome that opens into the coelomic cavity, a tubule extending anteriorly to the mantle cavity, and a nephridiopore that opens on each side of the mouth. If one injects a solution of carmine into a living brachiopod, the nephrostomes pick up this substance within thirty minutes and conduct it along the tubule, where it is coated by mucus from the flagellated cells. After a maximum of five hours, a red string of mucus emerges from the nephridiopores.

Except for a few species, all brachiopods are dioecious. The metanephridia provide the gametes their only possible passage to the outside. The eggs or sperms are produced in gonads found in the peritoneal bands (in the lingulids) or in the coelomic pouches of the mantle lobes. The mature gametes are discharged into the metacoel, where they are sucked up by the nephrostomes. If the males and females are in close proximity, as in *Lingula* for example, the gametes are expelled into the water with great force through the narrow valve slit via the middle bristle fringe. In *Terebratella inconspicua* the gametes are retained in the protective inner cavity of the lophophore until the early larval stage. Many species practice brood care until the larvae are free swimming. *Notasaria nigricans*, which produces about 8000 tiny eggs, each of which has a diameter of 0.16 mm, houses this brood in the two vertical lophophore spirals. In *Argyrotheca* the metanephridial tubules seem to be enlarged as brood pouches from the beginning. In this species the eggs are fused to food-storage cells. *Lacazella* is characterized by a different brood chamber. The two oldest tentacles, which are farthest removed from the mouth, become fused with the club ends. A mufflike ring of eggs is pushed onto this structure and is maintained within this depression of the ventral mantle.

According to many reports, eggs are discharged during the summer months. Several species reproduce twice a year. A tiny brachiopod with a diameter of 0.35 mm will hatch from a *Lingula* egg after six days. This tiny creature possesses wax-paperlike valves which protect the transparent and delicate body. Six tentacles protrude from the valve gap like a funnel. The *Lingula* larvae close the valves at the slightest disturbance. A pair of statocysts permits the animals to perceive vibrations as well as changes in

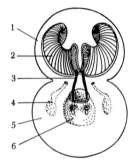

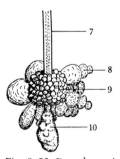

Fig. 8–23. Brood care in an articulate brachiopod (*Lacazella*): 1. Dorsal valve, 2. Lophophore, 3. Brood-care tentacles, 4. Metanephridia and ovaries, 5. Ventral valve, 6. Brood chamber, 7. Terminal section of brood tentacle with rings of eggs (9) and developing embryos (8), 10. Club-shaped enlargement.

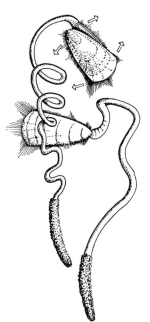

Fig. 8–24. *Lingula,* an inarticulate brachiopod, makes undulating motions with its 30-cm-long muscular peduncle, for wormlike, progressive movement. The valves, which have a slight ability to shift in opposing directions (white arrows), function as burrowing structures.

their normally vertical body position. The larvae sink to greater depths during even moderately rough seas.

In the adult phase, *Lingula unguis* burrows tubes up to 35 cm into the sand and mud of the intertidal zone, with the aid of the dorsal valve and the wormlike peduncle. The walls of the tube are soaked and fastened with mucus secreted by a glandular bulge along the margin of the mantle. The peduncle, which is equipped with longitudinal and transverse muscles, is amazingly agile. It is approximately six times as long as the valves, which are 4 cm long. The basal part of the peduncle is enlarged, and is extraordinarily sticky, causing a clump of sand granules to adhere to it. This added weight anchors the animal to the substratum. The anterior, transversely worn-off valve margin is at the same level as the upper surface of the sand. The three rows of bristles which protrude out of the narrow shell split the inhalent opening to both sides and also protect the central exhalent opening against "cave-ins" if the animal should have to retreat suddenly into the tube because of a vibration in the ground or a passing shadow. The brown pigment spots at the corners of the mantle are apparently light-sensitive. The peduncle is able to contract, stretch, or coil spirally. If the *Lingula* is exposed by a storm or receding tide, the worm-like motions of the peduncle quickly move the animal back to a safe place.

The development of the articulate brachiopods differs from that of *Lingula* and its hingeless relatives which attain the brachiopod body form within the egg. In the articulates one finds a mouthless true larva which maintains itself from a yolk supply. The larva swims about with the aid of a ciliary tuft for a few hours or days, and then attaches to a substratum by means of a secretion. At this point metamorphosis begins. Prior to this process the secondary coelom of the potential hind body had already originated within the egg. Although classified as Protostomia, the articulate brachiopods undergo a development similar in many ways to that of the Deuterostomia (see Chapter 9). A transverse fold constricts the archenteron of the gastrula into an upper section, the future digestive tract, and a lower tube, the coelum. The primitive mouth remains closed until the intestine has again penetrated to the lower section and thereby has separated the coelomic tube into a left and a right half. This feature is indicative of the deuterostomes. The mesocoel is formed much later, in the same manner as in the phoronids. The mesocoel arises as a cavity formation of migrating single cells (mesenchymal cells) from the central germinal layer (entero-coelous method). This method is familiar from the *Lingula* species (Fig. 8-25).

During its planktonic existence, the larva has divided into three body segments. The central portion forms a ring-shaped bulge around the hind body. Within a few hours after the larva has become attached, this bulge rolls over the fore body. During the following months the larva becomes progressively more flattened, thimble-shaped tentacles become evident on the fore body, and the two shell valves are secreted by the bulge and hind body. The final, species-specific shape of the lophophore arms is obtained by a series of intermediate stages. Smaller species with a lower food requirement never progress beyond the intermediate developmental stage. The mouth opening ruptures as soon as the tentacles are able to filter-feed on plankton. The nephridiopores are formed very late in the development. In *Terebratella* they develop only after two years, when the animal is sexually mature. This series of events seems to indicate that the true function of the metanephridium is that of a gonoduct.

Brachiopods are benthonic. They are found from the lowest tide levels to abyssal depths, although most are restricted to the continental-shelf region. Only the lingulids are able to leave the original location of settlement. All other species are firmly attached to the substratum by the peduncle or the ventral shell. *Chlidonophora chuni* roots in *Globigerina* deposits (a deep-sea deposit deficient in minerals and gravel but rich in calcium; it is named after the foraminiferan species *Globigerina*; see Vol. I), with delicate fibers arising from the long peduncle. The species is found near the Antilles at depths of 2000 to 3000 m. The stalked species maintain a horizontal body position, with a slight angle. The ventral valve points upward. The dispersal range of the brachiopod larva is limited because of its brief free-swimming period. This phenomenon has resulted in the concentration of brachiopods of the same or different species at certain favorable locations. Then individuals of the species *Glottidia albida* were counted on an area of 0.1 of 1 m². This species occurs along both coasts of the Americas. In the northern Arctic Ocean a clump of up to 100 individuals of *Hemithiris psittacea* was found.

In the aquarium the brachiopods have proven to be resistant to temperature fluctuations, bright sunlight, changes in salinity, and vibrations during transport. *Crania anomala*, which firmly attaches to hard substrates or stones with its frequently reduced platelike ventral valve, was reported to have lived a seemingly normal life in a laboratory for a period of fourteen months.

It is rather difficult to describe the exact geographical distribution of the brachiopods because of the lack of a sufficiently large number of recovery sites, the uncertainty of the species taxonomy, and the variations of individual species due to environmental influences. Living brachiopods total some 260 species classified within sixty-nine genera, as compared with approximately 30,000 extinct forms. The worldwide distribution and frequency of these fossil brachiopods helped in the delineation of certain geological time periods. The emergence of *Gigantoproductus giganteus* approximately 345 million years ago introduces the Carboniferous period. This largest brachiopod, which had a valve width of 30 cm, originated at a time period when the brachiopods shared the dominance of Paleozoic oceans with the trilobites (see Vol. I). This species was far larger than present-day representatives of this animal group, which have a body size of from 1 mm to 8 cm.

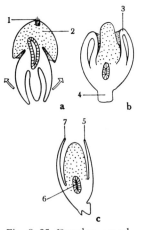

Fig. 8–25. Developmental stages from a free-swimming brachiopod larva to an adult sessile form: a. Free-swimming larva with a tuft of sensory bristles (1) after it has left the parental body. The sensory tuft faces in the direction of movement of the larva, which measures 0.2 mm. b. The larva has attached itself to the substratum with the peduncle (4). The circular bulge (3) is rolled upward. c. The mantle bulge splits at the lateral sides of the body, and assumes the shape and firmness of the future dorsal (5) and ventral (7) valves. The mushroomlike protosome (2; dotted) becomes the visceral cavity, where only the midgut (6) remains unmodified.

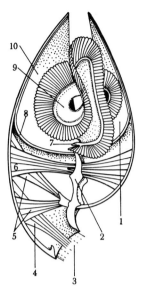

Fig. 8–26. Schematic longitudinal section of an articulate brachiopod: 10. Ventral valve, 1. Dorsal valve, 2. Calcareous loop which supports the coils of tentacles (9), 3. Peduncle, with tilt and twist muscles (4), Valve extensor muscles (5), 6. Valve flexor muscles with a double branch extending to the dorsal valve, 7. Epistome, the "vestige" representing the protosome; the mouth is located between the epistome and tentacular fringe. 8. Coelomic lining which surrounds the open mantle cavity housing the visceral organs (not shown here).

The brachiopods are divided into two phylogenetically parallel classes, the Inarticulates and the Articulates.

The INARTICULATE BRACHIOPODS (class Inarticulata) include approximately fifty species which seem primitive, but actually are not ancient. The two valves, consisting of a crust of calcium carbonate, can be closed by transverse muscles. An increase in the internal pressure of the body is supposed to open the valves. The lophophore is not supported by a skeletal axis. The intestine terminates in an anus. The order Lingulida is represented here by the species *Lingula unguis* (see Color plate, p. 253). The valves are spatulalike and greenish. Shell length (SL) is 3 to 5 cm; the peduncle is muscular, and measures 6 cm in the contracted state, 30 cm in the extended state. The genus *Lingula*, except for minor changes, has been distributed in the world's oceans since the Paleozoic period. Fifteen present-day species of this genus are the brachiopods found most frequently along the coasts of Japan. In Japan and in various regions of the South Seas the lingulids are eaten, particularly the peduncle. *Crania anomala*, from the order Acrotretida, is cemented to the substratum by by ventral valve, and therefore is often asymmetrical. The dorsal valve is dark brown, with a diameter of 1.5 cm. This species is distributed along the rocky coasts of European oceans at depths of 20 to 30 m.

The ARTICULATE BRACHIOPODS (class Articulata) are represented by approximately 200 species. Both valves are constructed of parallel layers of chitin and calcium phosphate. The shells articulate with each other along a hinge line. Muscles from the ventral valve insert on the dorsal valve, in front of the hinge for closing, and behind the hinge for opening. The intestine ends blindly. The lateral arms of the lophophore are supported by a calcareous loop. *Hemithyris psittacea*, from the order Rhynchonellida, is characterized by highly convex brownish-purple shells; SL 3.5 cm. The ventral valve tapers to a long, highly curved peak. There are two spirolophe lophophore spirals, and two pairs of metanephridea. The species is circumpolar, and occurs at depths of from 10 to 1300 m. *Terebratulina retusa*, from the order Terebratulida (SL 3 cm; see Color plate, p. 253), is characterized by oval yellowish to rust-colored shells which bear longitudinal ribs and are perforated by papillae. The lophophore loops are plectolophe. The peduncle is short. The species occurs on rocks along the northern coasts of Europe. *Argyrotheca cistellula*, from the suborder Terebratellidina (shell width 1.5 cm; see Color plate, p. 253), has a yellowish-brown, four-cornered ventral valve and a five-cornered dorsal valve. The lophophore loops are schizolophe. The species is hermaphroditic. The peduncle is minute. The animals are found on the underside of rocks along the European coast at depths of from 2 to 60 m.

9   The Arrowworms

The arrowworms or chaetognaths are small, slender, essentially pelagic animals which have an extremely uniform body structure, as a result of adaptation to life in the open-sea plankton. The phylum shows very little diversity in itself, and it stands in an isolated position with respect to the other coelomate phyla. As the blastopore of the chaetognath embyro becomes the site of the anus of the developing adult, some affinities with the deuterostomes are suggested. However, in some features of their development the chaetognaths show resemblances to the brachiopods (see lophophorates, Chapter 8). The bipartite secondary coelomic cavity further indicates affinities with the lophophorates. On the other hand, the paired nervous system, with abdominal extensions of longitudinal strands and the yoke-shaped section above the esophagus, resembles the situation in some lower coelomates (protostomes; see Vol. I).

Phylum: Chaetognatha, by L. von Salvini-Plawen

The arrowworms are bilaterally symmetrical coelomates characterized by a transparent arrow-shaped body that is clearly subdivided into a head, trunk, and tail. Length ranges from 3 to 100 mm. The head bears chitinous hooks and eyes; the trunk section is equipped with lateral fins, and the tail section with caudal fins. There is an oval ciliary loop which consists of modified epidermis. The thin cuticle covering this structure is made up of several layers in certain areas, for instance in the neck region. In the juvenile the secondary coelomic cavity consists of two parts, but in the adult it is dipartite with five cavities. The body musculature is striated and consists only of longitudinal strands. Specialized circulatory, nephridial, and respiratory organs are missing. The alimentary canal is straight and uncomplicated. The anus is located on the ventral side at the posterior end of the trunk.

Distinguishing characteristics

Nervous system: unpaired cerebral ganglion with paired nervous strands branching off to the eyes and ciliated loop (corona ciliata), and descending to the subesophageal ganglion and ventral ganglion within the trunk region. The subesophageal ganglion is made of a lateral and an abdominal strand which fuse into an unpaired strand behind the head, which then extends along the foregut. The central ganglion consists of paired strands and a dense network of nerves (plexus) which supplies the posterior trunk region and the tail section.

Arrowworms are hermaphroditic. The gonads are paired. The ovaries in the trunk are found just anterior to the trunk-tail septum, and the testes are just behind it. This septum (genital) develops along with the growth of the gonads. There is a seminal vesicle for foreign sperm and a seminal recepticle for its own sperm. Following reciprocal fertilization, the eggs are released into the sea, where they develop into juveniles without undergoing a larval stage.

The chaetognaths are strictly marine, and mainly planktonic. There are seven genera and about fifty species.

Arrowworms are relatively uniform animals characterized by a more or less well-defined head and a tapering tail section. Outstanding features are the well-developed grasping hooks which serve as capture appratus for these carnivorous animals (Fig. 9-2), and the lateral and caudal fins located on the sides of the trunk and tail respectively. Along with the ciliated loop along the nape and the sensory organs, these features help to classify the various species.

The fin structures are supported and held horizontally by rays of epidermal cuticle. These structures function as excellent stabilizers for the body. Therefore the resting animal sinks only gradually. This adaptive feature is assisted by the high proportion of water in the animal's body and its rigid posture.

The unpigmented, glassy-transparent consistency of most arrowworms is another adaptive feature to their planktonic existence. Several species from the deeper zones, for example *Eukrohnia fowleri* or *Sagitta macrocephala* (see Color plate, p. 271), found in the tropics, are characterized by red color hues, as are other deep-sea animals. Consequently, these forms also blend into the light-absorbing deep-sea darkness.

The arrowworms show a great amount of uniformity even in their life habits. With the exception of the *Spadella* species, the arrowworms are planktonic. During their lifetime (one to two years), they either swim or float without ever touching the sea floor. The distribution range of a particular species is primarily influenced by temperature and salinity.

Most species are concentrated in the warmer zones of the Indian and Pacific Oceans, where they may have a wide distribution. They include *Sagitta enflata* (see Color plate, p. 271) and *Sagitta bipunctata* (see Color plate, p. 271). *Sagitta setosa* and *Sagitta elegans* are distributed throughout the North Atlantic. Arrowworms are found primarily in those water layers offering sufficient amounts of natural food, which means they are not limited to the surface layer of the ocean. Like *Eukrohnia fowleri* and *Sagitta macrocephala*, *Eukrohnia hamata* (see Color plate, p. 271), from almost every ocean, lives primarily at greater depths, down to 3000 m, where it congregates in regular swarms. *E. hamata*, however, is a cold-adapted

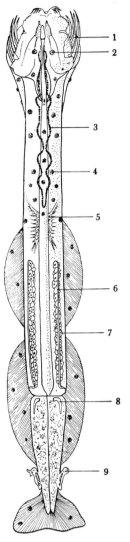

Fig. 9–1. Anatomy of an arrowworm (*Sagitta bipunctata*): 1. Capture hooks, 2. Eye, 3. Ciliary loop, 4. Tactile organ, 5. Ventral ganglion, 6. Seminal receptacle (Receptaculum seminis), 7. Ovary, 8. Testes, 9. Seminal vesicle.

form which also inhabits the upper surface of the polar oceans. *Ptero-sagitta draco* (see Color plate, p. 271), for example, is restricted to the region beyond the southern Adriatic, because of the degree of salinity.

Aside from such geographically differing habitats which are tempera-ture- and salinity-dependent, vertical migrations have been observed. For instance, the juveniles of *Sagitta gazellae*, from around Antarctica, inhabit the 50–100-m zone. During the course of development, these forms migrate deeper. Thus sexually mature individuals are found only from the 750 m mark downward. Large adults, up to 7 cm, are found only below 1000 to 3000 m. Similar conditions are present in the widespread *Sagitta decipiens*.

Arrowworms may undertake vertical migrations during the day to escape the warmer surface layers, returning at night when it is cooler. However, these diurnal rhythms comprise differences of only 100 to 150 m.

Aside from these characteristic planktonic forms, there are a few species inhabitating the sea floor. *Spadella cephaloptera*, from the Mediter-ranean and the northwestern Atlantic, is one of the best-known examples of this mode of life. This species can also be kept in the aquarium. *S. cephaloptera* (L barely 1 cm) has a dull brownish-yellow color, and at-taches itself to algae, stones, or similar objects in seaweed meadows at quite shallow depths. The animals attach with the aid of special adhesive papillae located along the ventral side of the tail section. In the juveniles these adhesive discs are developed along the entire ventral side. The animals seem to move freely only when they are changing location, for even when capturing food they seem to adhere to the substratum. These species feed on very minute crustaceans; the free-swimming arrowworms also consume other planktonic creatures. Even "cannibalism" is not an uncommon occurrence. Food is captured by a sudden thrust of the slightly upwardly bent body. Concomitantly, the large capture hooks enter into action. These hooks, which are almost completely covered by a hood, are exposed within seconds by pulling back this skin fold. When the hooks close inward, it is as if the prey were hemmed in by numerous daggers. The hooks perform true mouth-directed stuffing motions. The loss or back-sliding of the prey is prevented not only by the action of the hooks but also by the teeth which are pointed posteriorly and which are ar-ranged in a funnel-shaped manner close to the oral cavity. Additionally, the prey seems to become immobilized by substances secreted by the walls of the oral cavity. The arrowworms in turn make up an important com-ponent of the plankton and serve as an important food source for whales, jellyfishes, and other plankton-feeders of the high seas.

Under certain circumstances there seem to be great differences in the life rhythms of development, maturation, and spawning in the arrow-worms. Temperature seems to be the influencing factor. *Sagitta gazellae*, a cold-adapted form, reproduces only once every year, while *Sagitta*

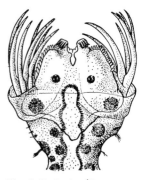

Fig. 9–2. *Sagitta bipunctata*: Head with the capture hooks partially exposed and with hood pulled back half-way.

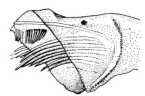

Fig. 9–3. *Sagitta setosa*: Capture hooks in relaxed position, partially covered by the hood.

Reproduction

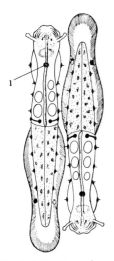

Fig. 9–4. Reciprocal copulation position in *Spadella cephaloptera*, 1. Spermatophore.

enflata, distributed worldwide in the warmer seas, is characterized by two or more annual reproductive periods. Members of the genus *Spadella* also reproduce repeatedly in the same year, but in this case other circumstances also play a role. E. Ghirardelli reports: "One can say with certainty that in *Spadella*, which inhabits *Posidonia* growth, there is a definite relationship between growth, number of eggs, and environmental conditions. During the spring, when these conditions improve, an increase in the growth of the population takes place, with the result that a larger number of eggs is produced, resulting in turn in a rapid resettling of the *Posidonia* growth which was unoccupied in the winter. Moreover, animals born during the fall or late summer seem to survive over the winter."

As has already been mentioned, all arrowworms are hermaphroditic. The male gonads mature before the female (protandry). Self-fertilization is not an uncommon occurrence. Reciprocal copulation has only been observed in a few species, even though this process may predominate. E. Ghirardelli has described the copulation in *Spadella cephaloptera*: "Prior to mating, the two animals approach each other and repeatedly touch each other's heads, neck, and perhaps the ciliary loops. These motions occur very rapidly, often making it impossible to record their exact sequence. Subsequently the two partners position themselves with the heads in opposite directions and with the bodies touching at the sides. Both immediately perform a scissorslike motion, which is repeated several times. Then the neck region and the anterior trunk of one animal are crossed over the other's tail section. After this rapid sequence of motions, which may last less than one second, each animal has discharged it seminal receptacle on the side of the partner. The contents of the seminal vesicle appear at the neck region of each partner. The size of the discharged spermatophore ball corresponds to the size of the seminal vesicle. The sperm balls are always placed along the midline of the dorsal side, directly behind the ciliary loop. Both animals stay in contact until the second clump of sperms is deposited on the head of the partner. A few seconds after the disposal of the spermatophores, the posterior section of the sperm clumps seems to disintegrate in the direction of the tail, resulting in a stream of sperms flowing toward the tail segment. At a specific point near the tail, the stream of sperms forks, each branch leading to the opening of a seminal receptacle into which the sperms penetrate. It takes no longer than two to four minutes before the flow of sperms begins, but the penetration of all sperms into the opening of the seminal receptacle requires a somewhat longer time. When mating takes place between two animals which already have more or less filled receptacles, sperm penetration may take up to twenty minutes."

If one experimentally relocates the sperm clumps to other parts of the animal's body, a normal sperm flow, whereby a certain amount of sperm reaches the openings of the seminal receptacle, is very rare. The majority of sperms usually flow to the margin of the body, from where they usually glide off. Similar events also take place in injured or sick animals. Therefore, this highly specialized "sperm migration" really functions as a protection for the species as a whole. The fertilized eggs are released into the surrounding water. The gelatinous eggs float near the surface of the water, either singly or in clumps. Only the bottom-dwelling *Spadella* species attach their egg clusters to objects within their environment.

Aside from the previously mentioned unusual life habits, *Spadella cephaloptera* also possesses several unique anatomical features. One of these oddities is the structure of the ciliary loop (Corona ciliata). This feature, very characteristic for the arrowworms in general, consists of differently arranged loops of ciliary cells which fringe a specific region of the animal's neck. The outline of the ciliary loop varies from species to species but occasionally there may be great variations within the same species, as is exemplified by *Sagitta enflata*. The development of this epidermal organ is directly connected with the prominent ventral ganglion, and correspondingly it serves as a sensory organ. In the bottom-dwelling *Spadella* species, however, there is a second ring of secretory cells within this transversely oval ciliary loop. The function of these cells is secretory. This second ring has not yet been demonstrated for any other chaetognath species.

Finally, *Spadella* is the only arrowworm that also has a pair of unsegmented tentacles on its head, and only one pair of large lateral fins. Both these features are strongly reminiscent of *Amiskwia sagittiformis*, a Canadian fossil species from the Middle Cambrian (approximately 500 million years ago), which possessed these same characteristics. However, *Amiskwia* possessed an alimentary canal which extended to the end of the body. Because of this feature, certain experts have classified this species as a bathyplanktonic nemertine (order Nemertini; see Vol. I). On the other hand, this extended intestine may prove that the development of the tail represents a phylogenetically relatively late specialization which resulted from the anterior relocation of the anus and the development of the "genital septum" (which grows with the gonads between the trunk and tail). However, this question cannot be settled until more data are available.

The metazoan animals discussed up to now have been classified as protostomes (Protostomia; see Vol. I) on the basis of certain embyryological characters, or as gastroneuralia or zygoneura in reference to the structure of the main nervous system. All animal phyla discussed in the following chapters belong to the deuterostomes (Deuterostomia; see Vol. I), because of the manner in which the mouth is formed. Aside from the mouth development, the origin of the coelom, which is produced as a mesodermal outpocketing from the primitive gut, is another feature shared by almost all deuterosomes. But are we justified in classifying such diverse creatures as an acorn worm, as starfish, an ascidian, and a vertebrate in one common category on the basis of these embyrological

▷
Arrowworms:
1. *Sagitta bipunctata*
2. *Sagitta enflata*
3. *Pterosagitta draco*
4. *Sagitta macrocephala*
5. *Sagitta planctonis*
6. *Eukrohnia hamata*

Fossil arrowworms

Deuterostomia

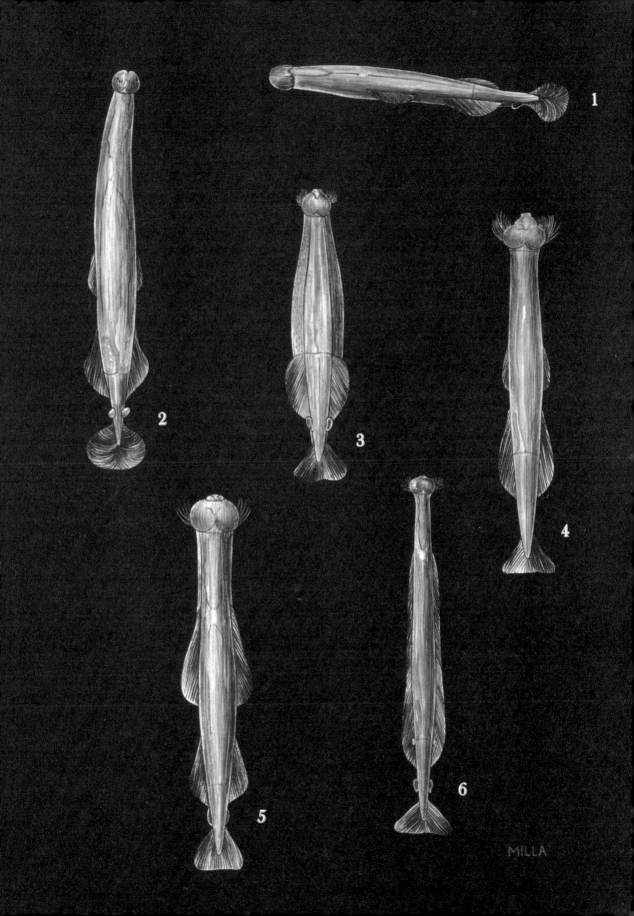

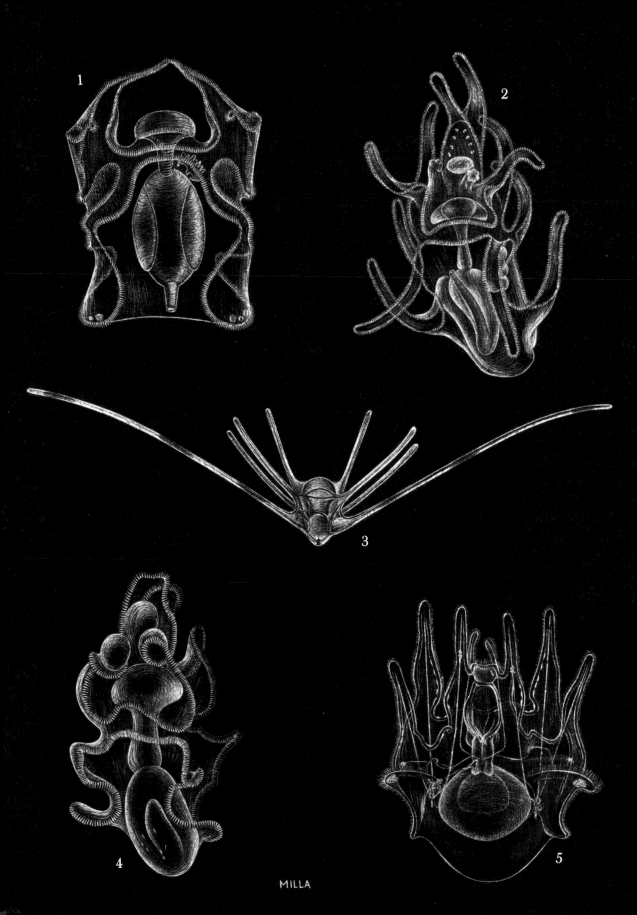

MILLA

characteristics? The new techniques of protein analysis have substantiated the hypotheses of the embyrologists, showing that the proteins of all deuterostomes yet investigated have demonstrated a certain degree of similarity which each other, a similarity not evident between the protostomes and deuterostomes. Despite the great external diversity found in the deuterostomes, this group probably represents a true phylogenetic unit.

◁
Larval stages in the echinoderms:
1. Auricularia larva of a sea cucumber;
2. Brachiolaria larva of a starfish;
3. Ophiopluteus larva of a brittle-star;
4. Bipinnaria larva of a starfish;
5. Echinopluteus larva of a sea urchin.

10   The Echinoderms

Most people are familiar with at least one representative of the echinoderms (phylum Echinodermata), although this recognition may only
come from memories of holidays spent on a sunny ocean beach—and
the starfish. The vacationer may have become acquainted with another
echinoderm along the rocky coast. This encounter with the sea urchin
may have been painful, because of the animal's many spines.

These two animals introduce us to two different echinoderm body
types. There is the star form, evident in the feather-stars and brittle-stars
as well as in the starfish, and the closed spherical or cylindrical shape,
characteristic of the echinoids and sea cucumbers. These five different
animal groups comprise the phylum Echinodermata.

In comparison with other animals, the echinoderms appear highly
unusual and distinctive. When looking at a starfish or sea urchin, we can
readily say what is up and down, but not what is front or back—a question
which would hardly have bothered us when looking at most other
animals. "Front" usually connotes that body part which points forward
when the animal moves; but this concept will not help us out in this case,
since we will soon find that a starfish or sea urchin can move "forward"
with any side of its body. We also know that freely moving animals
usually have the mouth pointing anteriorly, and the anus pointing
posteriorly, with the connecting line between these two points representing the main axis of the body. The body is arranged symmetrically on
either side of this axis. If one is searching for the mouth and anus in the
echinoderms in order to obtain some points of orientation, one will
discover that the mouth is located in the center of the ventral side, and
the anus in the center of the dorsal side. The connecting line between
these two points, i.e., the main axis of body, is in vertical relationship to
the substrate on which the starfish or sea urchin is resting, and is usually
very short. The sea cucumbers are an exception, however, because their
long main body axis is parallel to the substrate. In the sea cucumbers the
oral surface is not simultaneously the ventral surface as in the starfishes,

Phylum: Echinoderms,
by H. Fechter

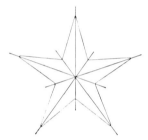

Fig. 10–1. Symmetry planes
in a starfish.

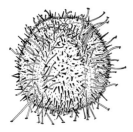

Fig. 10–2. From top to bottom: sea urchin, brittle-star, feather-star, starfish, sea cucumber.

sea urchins, and brittle-stars. Sea cucumbers have made one of their lateral sides into the ventral surface. When moving forward, the oral end of the sea cucumber points in the direction of movement. In this aspect and in their long, vermiform body shape, the holothurians seem to resemble many other invertebrates, and therefore do not appear to belong with the echinoderms. The symmetrical arrangement of the sea cucumber, however, clearly reveals its affiliation with the echinoderms.

When the body of almost all animals is cut through, the main axis usually divides the body into two halves that are mirror images of each other (bilateral symmetry). When applying this same procedure to a starfish, it is soon evident that one has to make five cuts, creating five sections, each of which has its own symmetry. The starfish is radially symmetrical, that is, it is divisible into equal symmetrical portions by any of three or more planes passing through the axis. Radial symmetry occurs in one other member of the animal kingdom, the coelenterates.

The anatomy of an echinoderm, as was shown in the example of the starfish, reveals five radii. Echinoderms are thus pentamerous and radially symmetrical. The five-rayed condition is not only applicable to the external appearance, as in the starfish, but also to the arrangement of the internal organs. All important organs occur in sets of five. Many echinoderms, for example the sea cucumbers, do not appear pentamerous at first glance, but their internal anatomy is radially symmetrical.

In examining the anatomy of an echinoderm from the outside to the inside we first encounter the body wall, consisting of outer, central, and inner cell layers. The external layer (epidermis or "skin") is very delicate, consisting of only one layer of cells, which covers the entire external surface including all processes and appendages. The central layer (dermis) consists of connective tissue and is relatively thick. This layer contains the endoskeleton that gives support and shape to the entire body. The inner layer is also thin, and like the outer layer consists of only a single row of ciliated columnar epithelium. The inner epithelium lines the entire coelomic cavity and all internal organs. It is known as "coelomic lining."

The endoskeleton in the central layer may consist of closely joining plates that form a shell or test as in the feather-stars and sea urchins, or it may also consist of separate articulating ossicles, as in the starfishes, brittle-stars, and the arm-joints of the crinoids. The skeleton is made up of calcite. This substance is extracted from the surrounding ocean water and is deposited in the connective tissue layer.

The structure of the skeletal elements is reminiscent of foam rubber. It is a reticulate network of tiny calcareous struts and spaces filled with connective tissue cells. Along the surfaces and margins of the supporting elements the cells continue to secrete skeletal substances, resulting in continuous growth. In the sea cucumbers, the skeleton is reduced to form microscopic calcareous ossicles embedded in the dermis, which is frequently leathery and tough.

The larger appendages of the body wall are also usually supported by skeletal elements or even consist, to a large degree, of skeletal material. These parts would be the spines, warts, and tubercles, which, however, are almost always covered with a thin layer of skin. The subepidermal region contains muscle fibers which have developed into powerful muscle strands in the holothurians and certain sea urchins. Almost all the cells of the external body wall, skin, and the inner layers of the coelom are ciliated. Ciliary action keeps the body surfaces clean and keeps the coelomic fluid in motion.

Aside from the spines and tubercles, each echinoderm also possesses five double rows of tubelike epidermal processes on its external body wall. In the sea urchins, most starfishes, and many sea cucumbers these tubelike processes are tipped by an adhesion cup. In the brittle-stars the tubelike process is slender, and in the feather-stars it is often branched. In the star-shaped echinoderms these skin tubules are located on the oral side of the arms (radii) in a groove of varying depth. In the globular sea urchins and cylindical sea cucumbers the tubules are arranged as meridians between the anal and aboral poles. Echinoderms move with the aid of these appendages, which are called tube-feet or "podia." The tube-foot consists of epithilium and longitudinal muscles. Lining the lumen of the tube-foot is the coelomic epithilium, continuous with the rest of the water-vascular system. All tube-feet are connected with a system of fluid-filled canals known as the ambulacral system.

This water-vascular system distinguishes the echinoderms from all other animals, and is their most characteristic feature. The arrangement of this system probably best serves to demonstrate the five-rayed radial symmetry of this animal group. A ring canal surrounding the pharynx gives off a radial canal into each of the five rays. The radial water canal extends from the oral region to the arm tips of the star-shaped echinoderms, while in the globose and cylindrical forms it extends to the anal region. In the feather-stars and starfishes the radial canals are located outside the endo-skeleton, and in the other echinoderms they are on the inside. Each of the radial water canals gives off a double row of lateral canals to each of the tube-feet. From the ring canal a stone canal extends to the aboral surface between the radii. A madreporite plate connects the stone canal, which has calcified walls, to the outside. In the brittle-stars the stone canal terminates on the oral side, and in feather-stars and most sea cucumbers it terminates within the coelomic cavity.

The interior of the echinoderm body is often characterized by a voluminous body cavity which, depending on the body shape, may be star-shaped, globose, or cylindrical. This cavity, which is filled with coelomic fluid containing coelomocytes, houses the intestine and gonads, which are suspended from the body wall. The feather-stars are an exception, because their gonads are located in the lateral branches of the arms. To be precise, we should recognize three coelomic cavities, which can be

Body appendages

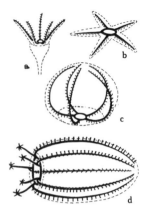

Fig. 10–3. The arrangement of the water-vascular system in the sea lilies and feather-stars (a), the starfish (b), sea urchin (c), and sea cucumber (d).

The water-vascular system

Coelomic cavities

of different shapes and sizes in the echinoderms. These cavities have their origin in the three successive coeloms within the larval forms. The first cavity, the protocoel, is lost in the feather-stars and sea cucumbers. In the sea urchins, starfishes, and brittle-stars it surrounds the stone canal. The second cavity, mesocoel, develops into the water-vascular system, and the third cavity, the metacoel, becomes the general, large body cavity. A part of the metacoel also gives rise to the canal system that lies underneath the water-vascular system. This canal system parallels the water-vascular system, but is characterized by a ring canal on the coelomic wall opposite the mouth. The genital coelomic sinus gives rise to genital coelomic sacs in each of the arms. These sacs contain the gonads. The ring canal contains a cellular tube made up of gametes (genital strand).

Digestive system

A jaw apparatus in association with the mouth is developed only in the sea urchins, starfishes, and brittle-stars. The digestive tract is suspended by mesenteries, and extends transversely through the coelomic cavity to the anus. In the starfishes and sea urchins the anus is on the aboral surface, in the feather-stars on the oral side, and in the sea cucumbers on the posterior end of the body. Certain starfishes and brittle-stars do not have an anus. Their intestine ends blindly and all indigestible remains are discharged through the mouth. In the irregular sea urchins (see Chapter 13) the anus has moved from the center of the aboral surface posteriorly, usually to the margin or to the oral surface. Often this modification also results in the movement of the mouth toward the margin. This may establish a secondary bilateral symmetry. Feather-stars, sea cucumbers, and sea urchins possess a relatively long, coiled, tubular intestine, which is hardly enlarged in any section. In the starfishes, however, the intestine is short, with a relatively strongly enlarged stomach connected by ducts to a pair of large pyloric caeca in each arm. Actual digestion of the food takes place in these highly lobed caeca.

Nervous system

The arrangement of the nervous system also follows the pentamerous symmetry. Generally the echinoderms possess a nervous system that functions at three different levels. These systems are the ectoneural, hyponeural, and aboral. All systems follow the scheme of the water-vascular system, or rather the canal system of the coelomic (metacoel) cavity. In the starfishes the ectoneural system is just beneath the epidermis, and in the sea urchins, sea cucumbers, and brittle-stars it is underneath the endoskeletal ossicles. With the exception of the brittle-stars, the ectoneural system is well developed in each of the echinoderm groups. The system consists of an extensive subepidermal plexus which innervates the sensory cells of the body surface. The hyponeural system is found in the lateral oral wall of the hyponeural sinus beneath the coelomic epithelium. This system innervates the body musculature. In the sea urchins, where the skeletal plates are fused and do not move, this system is greatly reduced. The aboral system extends along the outer margins of the ambulacral groove in each arm. This system is connected to the sex organs. The aboral system is particularly well developed in the feather-stars, because it is of the utmost importance in controlling their arm movements. Sea cucumbers and starfishes lack the aboral system.

The echinoderms lack a single, central nervous center. Each radius has its own autonomous center located at the junction of the radial nerve and the pentagonal circumoral nerve ring. The nerve ring coordinates the activity of the various centers and sends out the appropriate impulse. It can be regarded as a "switchboard" of the individual centers.

Gaseous exchange takes place at all thin surfaces of the body wall, particularly on the numerous tube-feet. The tube-feet are connected via the water-vascular system to the large coelomic cavity. Here the diffused gases pass directly into the body and are transported out again. Often the starfishes also have numerous protrusible, thin-walled papillae where gas exchange takes place. The sea urchins have a few papillae that have a respiratory function; these structures are referred to as gills. They are found in the vicinity of the mouth. Brittle-stars have special organs for respiration, called bursae. These are saclike invaginations of the thin oral wall of the disc near the arms. Each bursa is located between two stomach pouches. The ciliated epithelium within the bursae is constantly bringing in freshly oxygenated water, and expelling deoxygenated water. In the thin-walled sea cucumbers, gas exchange takes place over the entire body surface, and in the thick-walled species the cloaca regularly pumps water to and from the paired, branching respiratory tubes which join to the hindgut. Gas exchange takes place in these respiratory tubes.

The respiratory system

Hemal vessels are developed in all echinoderms. The hemal system is particularly well developed in the sea cucumbers and sea urchins. However, the hemal vessels within which the hemal fluid flows do not have definite linings like true blood vessels. For this reason these vessels are frequently referred to as lacunae. These lacunae are enclosed in canallike coelomic spaces between organs and tissues. The larger hemal lacunae have boundaries of connective tissue and muscular fibers, which pump the hemal fluid by contraction. There is no central pumping mechanism which could be called a "heart." The hemal fluid does not transport oxygen-carrying substances. The fluid consists of the same cells, the coelomocytes, as those present in the coelomic fluid.

The hemal (blood) system

The hemal lacunae can be differentiated as three major systems. First, there is an area close to the intestine that functions in adsorption of the digestible products; second, an oral distributing system; and third, an aboral distributing system. The oral hemal system extends along the canals of the water-vascular system and the large coelomic cavity. The aboral system supplies the genital hemal strands and gonads. All three systems are interconnected with the highly vascular axial gland, which extends along the stone canal.

No echinoderm has excretory organs in the form of kidneylike organs.

Soluble waste products are probably excreted via the respiratory body surfaces or via the intestine.

Reproduction and embryology

Echinoderms are usually dioecious, although a few hermaphroditic forms are found among the sea cucumbers and brittle-stars. It is almost always impossible to differentiate externally between the sexes. Eggs and sperms are usually discharged into the open sea. Species that practice brood care are found in all five echinoderm groups. Usually the young are protected in special brood pouches until they have reached a certain size. Typically, the eggs develop into characteristic free-swimming larvae which are bilaterally symmetrical. The five-rayed radial symmetry of the adult form becomes evident during the course of metamorphosis.

In summary, the echinoderms can be characterized in the following manner: in the larval stage, bilaterally symmetrical; in the adult stage pentamerous-radially symmetrical, with a calcareous endoskeleton of separate (reticulate) plates which frequently bear external appendages such as spines and tubercles. Radial symmetry is also present externally in the arrangement of the skeletal plates, and internally, pentamery is evident in the water-vascular system and nervous system. The water-vascular system, characteristic for this phylum, gives off numerous projections to the entire surface of the body wall. These hollow projections may be the tube-feet (tentacles) which function as food-catching devices, or the podia which have locomotive functions or often perform both functions. There are five existing classes: 1. Sea lilies and feather-stars (Crinoidea; see Chapter 11), with four orders and nineteen families, 2. Sea cucumbers (Holothuroidea; see Chapter 12), with three subclasses, five orders, and twenty-four families, 3. Sea urchins (Echinoidea; see Chapter 13), with two subclasses, fifteen orders, and forty-nine families, 4. Starfishes (Asteroidea; see Chapter 14), with two subclasses, seven orders, and twenty-nine families, 5. Brittle-stars (Ophiuroidea; see Chapter 15), with three orders and sixteen families.

Phylogeny of the echinoderms, by E. Thenius

Although all existing echinoderms, despite their different forms, are characterized by radial symmetry, a water-vascular system, and a similar basic anatomy, this does not apply to the various extinct groups of this phylum. One assumes that the echinoderms, like other invertebrates, already evolved into individual structural patterns during the Pre-Cambrian (at least 600 million years ago). However, Pre-Cambrian fossil remains are not available, and therefore the phylogeny of the echinoderms cannot be explained on the basis of fossils. Nevertheless, fossils from more recent times have revealed information concerning the origin and phylogenetic development of the various groups, and also the existence of totally extinct echinoderm classes.

The phylogenetic origin of the echinoderms has presented various problems which have become controversial again with the discovery of new fossils. The main issues of the controversy are concerned with the symmetrical arrangement of the archi-echinoderm, its mode of life, and whether it was sessile or free-living.

The geologically oldest echinoderms found date back to the Cambrian (540 to 450 million years ago). Some of these forms, for example Eocrinoidea and Edrioasteroidea, were sessile. Free-living species were found among the Homalozoa and Helicoplacoidea. These groups represent some of the few important extinct echinoderm forms, along with the Cystoidea and Blastoidea. However, all extinct classes have not yet been enumerated.

The sybphylum HOMALOZOA, which lived during the Paleozoic era (from the Cambrian to the Devonian, 540 to 350 million years ago), deviate from all other echinoderms because of their bilateral symmetry. This group also represented a unique type of echinoderm because of the manner in which its skeletal and stalk plates were arranged and structured, and its lack of arms or armlike outgrowths with brachioles. The functional significance of the various body apertures and the mode of life of many homalozoans still remain to be clarified. Jefferies has recently suggested classification of the Carpoidea as a separate chordate subphylum (Calcichordata).

The subphylum Crinozoa includes several classes of sessile echinoderms, of which only one, the Crinoidea, survives today.

The CYSTOIDEA, which lived from the Ordovician to the Devonian (approximately 450 to 350 million years ago), often were not pentamerously symmetrical; forms with two to five water-vascular areas have been found. Instead of arms, the Cystoidea usually had well-developed brachioles. This animal consisted of a globular, pear-shaped, or flask-shaped theca located on a stem. The theca was made up of a varying number of many-sided plates. Some of these always carried a characteristic arrangement of pores.

In the BLASTOIDEA, however, which lived from the Ordovician to the Permian period (450 to 240 million years ago), pentamerous symmetry was completely developed. This blastoid form was characterized by a theca consisting of a specific number of plates which had five ambulacra with pleatlike grooves known as hydrospires, and brachioles. The blastoids reached the height of their evolutionary development during the Carboniferous (310 to 240 million years ago), as was exemplified by a form such as Pentremites.

The only crinozoan group which has survived into recent times is the SEA LILIES (class Crinoidea). This group passed its phylogenetic high point long ago, apparently in the Silurian (400 to 350 million years ago). Of such groups as the Camerata, Flexibilia, and Inadunata, which were represented by many species and were widely distributed during the Paleozoic, only the Inadunata managed to survive the Permian-Triassic period (200 million years ago). This group gave rise to the Articulata, which have survived up to today as "living fossils" in deep-water zones, for example Metacrinus, or the (secondarily) free-swimming feather-

▷
A feather-star of the genus Comanthus.

stars like the representatives of the order Comatulida (see Chapter 11). The feather-stars have evolved since the Jurassic (175 to 140 million years ago). The ancestral forms of recent deep-sea forms were still shallow-water inhabitants during the Mesozoic era.

Of the subphylum ECHINOZOA, the classes Helicoplacoidea, Camptostromatoida, and Edrioasteroidea are the oldest forms (Cambrian), while the ophiocistioids, cyclocystoids, and edrioblastoids, all small groups, arose during the Ordovician.

The HELICOPLACOIDEA (genus *Helicoplacus*, from the Lower Cambrian) were free-living echinoderms which had a flexible oval to cigar-shaped body with spiral plating and a single ambulacrum. The oral and anal openings were apparently at opposite poles of the body. In contrast, the EDRIOASTEROIDEA of the Cambrian to Carboniferous period were usually sessile (but stem-less) echinoderms characterized by a bag-shaped or disc-shaped body wall consisting of numerous irregular plates. The central mouth opening was surrounded by five usually curved ambulacra. The anus and hydropore (exit of the water-vascular system) were also located on the dorsal side, but in the interambulacral areas. Brachioles are not developed. The Edrioasteroidea and the sea urchins (class Echinoidea) can be derived from common ancestral forms.

Numerous fossils of the SEA URCHINS (Echinoidea), which also first appeared in the Ordovician, have been discovered. These fossils showed that the test (corona) of most sea urchins from the Paleozoic era (for example, the genera *Bothriocidaris*, *Eothuria*, *Lepidocentrus*, and *Melonechinus*) was not regular at all, but carried a variable number of columns of plates in the ambulacra and interambulacra. The plates were arranged in scalelike fashion, which resulted in a flexible test. The spinal arrangement and structure varied, and reached bizarre heights of development in the Mesozoic cidaroids, which possessed club-shaped spines. In these "regular" (symmetrical) sea urchins the number of thecal-plate rows was already constant. The first representatives of these forms, the family Miocidaridae, became apparent during the Carboniferous period. The first "irregular" (secondarily bilaterally symmetrical) sea urchins appeared in the Jurassic period. These forms demonstrate a number of modifications which evolved in response to a different mode of life. The most conspicuous adaptive features in the irregular forms are the degeneration of the jaw apparatus, the diminution of the spines, the relocation of the anus and mouth, and the development of secondary bilateral symmetry. This newly evolved type triggered a phylogenetic blossoming within the sea urchins, which is still in progress.

Unfortunately, little can be said about the phylogenetic development of SEA CUCUMBERS (class Holothuroidea), because of the lack of fossil evidence. Fossil data consists almost entirely of the microscopic endoskeletal ossicle. There is some indication that early forms were heavily plated.

◁
Mediterranean feather-star (*Antedon mediterranea*; see Chapter 11 and Color plate, p. 291) in capture position on top of a horny coral (*Paramuricea chamaeleon*).

The subphylum ASTEROZOA includes a single class, Stelleroidea, with three subclasses, Somasteroidea, Asteroidea (starfishes), and Ophiuroidea (brittle-stars).

The somasteroids are believed to exhibit some characters intermediate between crinoids and other asterozoans. As fossils they range from the Ordovician to the Devonian, although several scientists believe that the genus *Platasterias*, from the eastern Pacific, is a living somasteroid, a surviving member of this ancient group.

11 Sea Lilies and Feather-stars

Sea lilies and feather-stars, by H. Fechter

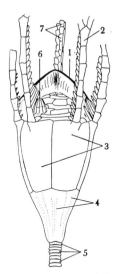

Fig. 11–1. Structure of the calyx, stalk, and arms of the sea lily *Hyocrinus bethellianus*: 1. Ambulacral groove, 2. Skeletal plate of arm, 3. Radial plates, 4. Basal plates, 5. Stalk segments, 6. Anus, 7. Lappets that can cover the ambulacral grooves of the arms.

SEA LILIES and FEATHER-STARS (class Crinoidea; see Color plates, pp. 291 and 292) represent a group of echinoderms that are either stalked and sessile or have become detached from the stalk and are free-living. The thimble-shaped calyx (theca) is very small in relation to the entire body mass. It bears five branched arms that are usually forked several times. The oral side is always turned away from the substrate. Sea lilies are sessile. The feather-stars are those crinoids that are free-living in the adult stage. Approximately 88% of present-day crinoids are feather-star species, and only 12% are sea lily species. The animals vary greatly in size. The diameter of the calyx of the largest living sea lily (*Metacrinus superbus*) is 1.2 m, the arm length is 19 cm, and the stalk is over 2 m long. The arms of the largest feather-star (*Heliometra glacialis*) measure up to 35 cm. In the sea lilies and feather-stars, pigment cells are located in and between the cells of the connective-tissue layer, which is beneath the extremely thin skin (epithelium). The connective-tissue layer also contains the supporting elements of the endoskeleton. The pigment produces the orange, reddish, violet, brown, green, or black shading in the sea lilies and feather-stars. Aside from monochromatic animals, there are many that are variegated.

The stalk is a very characteristic feature of this class, although in the feather-stars it is present only in the juvenile stage. The stalk attaches to the aboral surface of the calyx. In the sea lilies as well as in the young feather-stars the stalk is made up of a single row of superimposed disc-shaped ossicles called columnals, which are held together by elastic ligaments. The terminal columnal is usually modified into an adhesive disc that attaches to the substratum. Primitive sea lilies also have tentaclelike offshoots which arise at regular intervals along the stem. These offshoots, or cirri, which occur in groups of five, are made up of individual columnals. These cirri are able to grasp any convenient projection. The feather-stars also develop cirri, but only on the basal plate, usually called the centrodorsal plate, which supports the calyx. In the feather-stars, therefore, the top of the larval stalk (centrodorsal) with the cirri is retained.

This single plate serves as a mount for the main structure of the animal's body, while the projecting curved cirri which possess claw-shaped terminal joints enable the adult animal to attach itself firmly to the substrate (Fig. 11-2).

The cirri-bearing basal plate, usually referred to as the centrodorsal plate, bears the thimble-shaped calyx. In those primitive sea lilies which still exist (i.e., genus *Hyocrinus* from the order Cyrtocrinida), the skeletal wall still consists of two overlapping circlets consisting of five tall plates each. The five lower plates are known as basals, and the upper ones are known as radials. The radials make up the skeletal elements at the base of the arms. However, in most of the somewhat more evolved sea lilies and almost all feather-stars, the basals and radials are quite small. The basals might even be present within the calyx, where they are fused together, forming a rosette which acts as a cover to the depressed centrodorsal. Instead of the large skeletal plates, the lower arm sections which are interconnected by a leathery skin now form a more or less solid body wall which protects the viscera.

The small radial plates give rise to the arms. The arms in turn are made up of cylindrical brachial ossicles (segments). The arm segments are joined by elastic ligaments which enable the arms to extend, and by flexor muscles which enable the arms to bend inward. The arms subdivide by forking, often at the second arm segment, resulting in ten arms which branch out from the first five arm sections. Frequently the number of arms remains constant at ten, although it is not uncommon to find repeated forking in the arms of feather-stars. In extreme cases there may be up to 200 arms. Almost every arm segment supports a short, delicately jointed lateral branch, the pinnule. The pinnules branch off alternately at each segment, one left, one right, etc., resulting in the "feathered" appearance.

The flat oral surface of the calyx is covered by the same leathery skin as are the lateral walls. The mouth is in the center of the oral surface. Five ambulacral grooves run from the mouth to the bases of the arms. The grooves parallel the many branchings of the arms along their oral surfaces (Fig. 11-4). They are ciliated, and function like conveyor belts. Food particles caught by the arms flow to the mouth within these grooves. These ambulacral grooves have raised edges that are sculptured into lappets, and occasionally also skeletal elements which can seal the grooves. The inner side of the base of each lappet bears podia which seem to function as food collectors. The arms, with the pinnules, podia, and ambulacral grooves, form an extensive food-collecting apparatus.

The radial canals of the water-vascular system are located on the interior of the arms. Underneath each groove, even into the finest branches, lies a canal of the water-vascular system. The radial canals connect to the ring canal which surrounds the esophagus. Usually there are several stone canals, although there is no direct connection to the exterior as with a

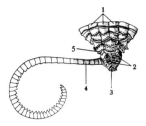

Fig. 11-2. Calyx, arm bases, and cirri buds of the feather-star *Thalassometra marginata*: 1. Arms, 2. Cirri bases, 3. Centro-dorsal plate, 4. Cirri, 5. First arm segment.

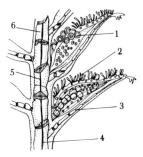

Fig. 11-3. Part of the arm of *Antedon bifida*: 1. Ovary, 2. Podium, 3. Skeletal element of a pinnule, 4. Nervous strand, 5. Muscles between arm segments, 6. Arm segment (Brachial ossicle).

madreporite. Instead, the water passes first into the perivisceral coelom, and then to the outside via numerous funnels which pass out through the oral plate (tegmen). The perivisceral coelom gives off three canals which penetrate into the arms and parallel the branching water-vascular system but are underneath it. In the sea lilies the perivisceral coelom penetrates throughout the stalk. A vestige of this coelom is still present in the depression of the centrodorsal plate in the feather-stars. This aboral coelomic structure is radially divided by septa into five chambers called the chambered organ. The chambered organ gives off a paired lateral canal to each cirrus. These canals penetrate through the center of the cirri segments.

The mouth, which obtains the food particles from the ambulacral grooves, is surrounded by oral tentacles which probably function in testing the palatability of the food. A short, muscular esophagus leads into a somewhat wider intestine, which makes a circular loop and consists of several diverticula. The intestinal tract of the feather-star family Comasteridae, on the other hand, lacks diverticula, instead being coiled into four spirals. A short, highly muscular hindgut finally terminates in the anus, which is located at the margin of the tegmen at the tip of an anal cone.

Of the three nervous systems usually found in the echinoderms (see Chapter 10), the aboral system is particularly well developed in the feather-stars. The center of the aboral (entoneural) system consists of a cup-shaped nervous mass, which surrounds the chambered organ and sends out nerves to all body appendages. The center may also send a nervous sheath out into the coelomic canal that runs through the cirri, and the stem canal of the sea lilies. Five brachial nerves pass from the same central mass through a pentagonal nerve ring in the radial plate of the calyx and then branch off into the arms and all the forks. The main function of this nervous system is concerned with the movement of the arms. The ectoneural system runs directly underneath the food grooves, and the hyponeural system flanks the two sides of the groove. Both systems complement the aboral system. The entire surface of the body is covered by sensory cells, which are particularly numerous around the ambulacral grooves. The nervous cells are concentrated on the podia as papillae which probably function as taste and tactile organs.

The gaseous exchange of oxygen and carbon dioxide takes place over the entire body surface, which is covered by only a thin epidermis. Special respiratory organs are superfluous, since the hemal system is in very close proximity. The hemal system consists of a large number of interconnecting hemal lacunae. Branches from the dense plexus around the intestinal tract extend to the axial organ (see Chapter 10). This organ extends along the main axis of the perivisceral coelom and consists of a glandular nucleus enmeshed by many hemal lacunae. This network is connected to a ring lacuna around the esophagus. This esophageal hemal plexus in turn sends out hemal vessels into each of the arms and their branches.

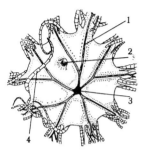

Fig. 11-4. Tegmen of a feather-star: 1. Ambulacral groove, 2. Anus, 3. Mouth, 4. Oral pinnules.

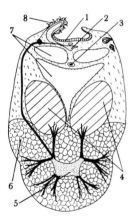

Fig. 11-5. Cross section of a feather-star arm: 1. Ambulacral groove, 2. Radial canal of water-vascular system, 3. Strand of sexual cells, 4. Brachial muscles, 5. Brachial nerve (aboral), 6. Brachial (ossicle) skeleton, 7. Aboral coelomic canal, 8. Podium with sensory papillae.

The feather-stars represent a single systematic and phylogenetic unit (order Comatulida), while the surviving sea lilies fall into three different orders. The class thus comprises the following living orders: 1. Cirri-bearing sea lilies (Isocrinida), 2. Feather-stars (Comatulida), 3. Sea lilies without cirri (Millericrinida), 4. Cyrtocrinida.

Order: Isocrinida

The Isocrinidae are the only family of the CIRRI-BEARING SEA LILIES (order Isocrinida). These are long-stemmed sea lilies. The columnals are usually pentamous in cross-section. Circlets of cirri are found at specific intervals, every fifth to fifteenth segment. This stem bears a crown with many branched arms. The lower segments of the arms make up part of the calyx. The ambulacral grooves are characterized by raised edges which can close over the former. The most common genus, *Metacrinus*, is mainly known from Japan, Malaysia, and Australia. The individual species have arms that are forked five times. The arm length is 16–20 cm. The diameter of the calyx is up to 1.2 cm, and that of the stem is up to 0.8 cm. The coloration may be light yellow, orange-red, reddish-brown, or green, and on occasion the animals may be striped or spotted. The yellowish-brown *Cenocrinus asteria* (crown diameter up to 1.7 cm, arm L 10 cm; see Color plate, p. 292) is found in waters around the West Indies, at depths from 200 to 600 m, and is frequently the subject of illustrations.

Order: Feather-stars

In contrast to the sea lilies, which are sessile, the FEATHER-STARS (order Comatulida) are free-swimming in the adult stage. The feather-stars are only attached by a stem during the fixed crinoid stage, the "sea lily phase." They break free when mature. This entire order is made up of approximately 550 species out of an overall total of 620 living sea lily and feather-star species. This large number of species is classified into four suborders: Comasterina, Mariametrina, Thalassometrina, and Macrophreatina. The first three suborders are characterized by a small, shallow depression in the centrodorsal plate. Because of this distinguishing characteristic, the first three groups are frequently combined as one unit, the Oligophreata, as opposed to the Macrophreata which makes up the fourth suborder. The individual families are primarily distinguished on the basis of the number, shape, and method of joining of the arm segments, and also on the shape and number of the cirri and pinnules.

Members of the suborder COMASTERINA are characterized by a disc-shaped centrodorsal plate. Cirri are found only in the young forms. They are absent or reduced in the adults. These feather-stars attach to the substratum by one or several arms, with the aid of the movable, claw-shaped terminal ends of the pinnules close to the mouth. Arms used for attachment are not used for acquiring food and are often shorter than the others. Frequently ambulacral grooves are not developed in these attachment arms, although they still contain the gonads. The mouth has been relocated toward the margin of oral disc (tegmen). The anus is located near the center of the tegmen. The arms are frequently branched so that there are often more than fifty arms. There is only one family,

Comasteridae, with the genera *Comatula, Comaster, Comissia, Comanthus* (see Color plate, p. 281), *Comatella, Capillaster,* and *Comantheria.* There are over 100 species, including *Comantheria grandicalyx.*

The suborder MARIAMETRINA also has a disc-shaped centrodorsal plate with a narrow, shallow depression. The cirri are always well developed, and are rarely found in more than two alternating rows. There are no terminal claws on the anal pinnules. In cross section the pinnules are triangular, at least at the base. The ambulacral groove is devoid of lateral lappets and other structural covers. The mouth is central and the anus is located along the margin of the tegmen. There are from five to one hundred arms. Usually there are from fifteen to forty, and not uncommonly only ten. These forms are primarily found in the coastal zones of the Indian and Pacific Oceans and the West Indies. There are six families.

The family ZYGOMETRIDAE includes the very colorful genera *Captometra* and *Zygometra,* and the only five-armed genus, *Eudiocrinus.* The family HIMEROMETRIDAE includes more than fifty species, each of which has ten or more arms. The genus *Heterometra* is represented by the most species. It includes the better-known species *Heterometra savignyi* (see Color plate, p. 292), which is frequently seen in underwater photos of the Red Sea. This species usually has twenty arms which may measure up to 15 cm. Of the approximately thirty usually long-armed species of the family MARIAMETRIDAE, only the uniformly blackish-brown *Lamprometra klunzingeri* will be mentioned here. This species is frequently seen on diving excursions in the Red Sea. *L. klunzingeri* has up to thirty-one smooth, slender arms which measure approximately 11 cm. The family STEPHANOMETRIDAE, represented by only a few species, is characterized by one or more oral pinnules that are enlarged and rigid like a spine. The ten-armed TROPIOMETRIDAE possess relatively short and stout pinnules that are triangular in cross section. The cirri are also short. The single genus *Tropiometra* is one of the most common feather-stars along the tropical coastal zones. *Tropiometra afra macrodiscus* is one of the largest known feather-stars. It arms measure up to 26.5 cm. In the approximately fifty species of the COLOBOMETRIDAE the cirri segments bear one or two transverse combs with two or three thorny projections that are curved toward the inside. Of the eighteen genera, *Decametra, Oligometra, Colobometra,* and *Iconometra* have the most species.

The centrodorsal plate varies greatly in the suborder THALASSOMETRINA. In this group the centrodorsal is club-shaped to disc-shaped but with a somewhat deeper depression than in the previous groups. Cirri are arranged in several rows. There are ten to twenty arms which are usually triangular in cross section. The ends of the arms bear five to seven segments with curved pinnules. In cross section, pinnules appear rectangular to triangular. The ambulacral grooves are equipped with lappets and cover plates. There are five families.

The family THALASSOMETRIDAE has approximately sixty species which possess long, slender cirri made up of more than twenty-five segments with ends curved to the inside. The curved inside has thorny or keellike projections. Of the fourteen genera, *Thalassometra*, *Cosmiometra*, and *Stiremetra* have the most species. The family CHARITOMETRIDAE, with thirty species, is characterized by short, stout, strongly curved cirri usually consisting of fewer than twenty-five segments without distinct keels or bulges on the inwardly curved surface. The genus *Glyptometra* is represented by the most species. The family CALOMETRIDAE has approximately sixteen species. In these forms the first pinnules are extremely delicate, weak, and pliable. The first two segments of the pinnules are larger than all others. The second and some of the subsequent pinnules are enlarged and rigid. The tegmen is dome-shaped and completely covered by skeletal plates. *Neometra* and *Pectinometra* are the most significant genera. The family ASTEROMETRIDAE has ten species. The ventral surface of the pinnules is protected by side-plates and covering-plates. Includes the genus *Asterometra*. The family NOTOCRINIDAE is unique. It is characterized by weakly developed ambulacral plates. In this group the segments of the arms and pinnules are curved on the aboral side and the cirri are arranged in ten regular rows. The tegmen is covered with skeletal plates. The gonads are located at the bases of pinnules along the arms. The single genus *Notocrinus* is found in the Antarctic. It practices brood care.

The centrodorsal of the suborder MACROPHREATA is club-shaped or semiglobular, but seldom disc-shaped. The depression is more extensive and deeper. There are usually ten arms, rarely more. The pinnules are never prismatic. The pinnules near the crown differ in shape from all the others. The lappets and ambulacral plates are reduced. There are three families.

ANTEDONIDAE, with over 130 species, is by far the most diverse group of all feather-star families. This group is particularly widespread and common outside of tropical waters. Antedonidae also includes the feather-stars off the coasts of Europe. The MEDITERRANEAN FEATHER-STAR (*Antedon mediterranea*; see Color plate, p. 282) belongs to this group. This species is characterized by more than twenty cirri segments and an extremely varied coloration. This feather-star is found in deep algal growth and seaweed, clinging to cliffs, and on coral substratum, from the ocean surface to a depth of 220 m. *Antedon bifida* is pink to orange and has fewer than twenty cirri segments. This form inhabits the Atlantic coasts from the English Channel to Portugal from the ocean surface to depths of 500 m. *Antedon petasus* usually has more than forty cirri. It is light red to brownish-red and is frequently characterized by lighter stripes. This form is found from Iceland and the Faeroes to Norway. *Leptometra phalangium*, much larger and stouter than *Antedon mediterranea*, is found only in the Mediterranean. This greenish feather-star possesses slender cirri up to 8 cm long. It is found at depths from 50 to 1,300 m on coral, gravel, or sandy or muddy substrates. Several other members of this family shall be mentioned, such

▷
Color variations in *Antedon mediterranea* from the Mediterranean (see Color plate, p. 282).

as *Leptometra celtica* whose cirri measure from 3.5 to 4 cm. This species inhabits the Atlantic coast from the Faeroes to Maderia to depths of 500 m. *Heliometra glacialis* also belongs to this group. It is the largest known feather-star, with arms measuring up to 35 cm. This light yellow and occasionally purplish-tinged crinoid is distributed around the North Pole (circumpolar). It is found on sand, mud, and gravel bottom at depths from 40 to 1,150 m.

The last two families of this suborder are less well known. The ATELECRINIDAE are characterized by parallel externally transparent basal plates. They are primarily deep-water forms. The PENTAMETROCRINIDAE have only five arms.

Order: Millericrinida

The SEA LILIES WITHOUT CIRRI (order Millericrinida) are delicately stalked forms with greatly reduced cirri or none at all. Cirri might be present at the lower end of the stalk, where they appear as rootlike projections. Basal plates are distinctly visible. There are two families. The family BATHYCRINIDAE includes some species that are distributed in the North Sea, the best-known North Sea species being the grayish to yellowish *Rhizocrinus lofotensis* (see Color plate, p. 292). The stalk of this species measures 7 cm, and the arms approximately 1 cm. *R. lofotensis* is found at depths of 140 to 4,800 m. *Bathycrinus australis* has a height of 5 cm and has ten arms. It penetrates to depths of 8,300 m. The family PHRYHOCRINIDAE has two species, found at great depths in the Indian and Pacific Ocean.

The last order of this echinoderm class is CYRTOCRINIDA. These are sea lilies without cirri. These forms are attached to the substrate either by the stalk or directly by the calyx. The calyx is composed of basal and radial plates. The tegmen is covered by skeletal plates. There are two families. The HOLOPODIDAE are directly attached to the substratum by the stalkless, thick, and short crown. The ten well-developed arms can be coiled spirally above the tegmen and folded into each other. Only one species is known, *Holopus rangi* (see Color plate, p. 292), from depths between 200 and 300 m in the Caribbean. The HYOCRINIDAE, on the other hand, are long-stemmed and multi-segmented, with a very slender crown. The three genera, *Hyocrinus*, *Calamocrinus*, and *Ptilocrinus*, are primarily found in the antarctic oceans.

◁
Sea lilies:
1. Cirri-bearing sea lily (*Cenocrinus asteria*);
2. Sea lily without cirri (*Rhizocrinus lofotensis*).
Feather-stars:
3. & 4. *Heterometra savignyi*:
3. Capture position,
4. At rest;
5. *Holopus rangi*.

The stalked sea lilies of the orders Isocrinida, Millericrinida, and Cyrtocrinida are firmly anchored to the substratum, as has been mentioned already. These forms cannot change their location without breaking off from the stalk. However, sea lilies of this type are able to sway back and forth on their stalks. They are able to swing around in a semicircle whose radius corresponds to the length of the stalk. This ability should by no means be underestimated when one considers that some extinct crinoids possessed stalks that were approximately 20 m long. Even today there are still some sea lilies with stalks which probably reach a length of several meters. The longest known stalk fragments have a length of 1.5 m.

Some sea lilies are able to break off their stalks at certain places, thereby freeing themselves from the ocean floor, to move freely by beating the arms up and down, dragging the stalk behind them. At a new location the sea lily will anchor itself to the substratum with the circles of cirri which are present at specific intervals along the stalk.

However, we are much better informed about the locomotive abilities of the feather-stars. Aside from those forms that are very mobile, there are others that stay anchored with their cirri to one location for weeks at a time, provided living conditions are optimal. Slight disturbances that do not pose an immediate threat, such as the touch of another animal, cause the feather-star to release the clawlike grip of its cirri and crawl away.

In the feather-star this activity proceeds in the following manner. The arms opposite the source of irritation stretch out widely, parallel to the substratum, and attach themselves with the terminal pinnules. Subsequently, these anchored arms are curved greatly, pulling the body away **Crawling** from the source of aggravation. At the same time the arms that are turned toward the disturbance become active, curving widely underneath the crown and also hooking their terminal pinnules into the substratum. Then the arms stretch and push the body away from the disturbing factor. Thus the animal moves by pull and push. The arm that points in the direction of movement frequently does not participate in the crawling motions, instead seeming to serve as a direction indicator during this locomotive activity.

Most feather-stars do not have to turn or twist. They can move in any desired direction, and any arm can serve as a direction indicator. However, there are certain species with several arms that are longer than the others in the adult stage. These feather-stars have lost the ability to start out in every direction. Instead, they can only move in the direction of the long arms, and therefore they have to turn in the desired direction before they start to move away. During crawling, the long arms always pull the animal forward, and the short arms push. The crawling speed is approximately 40 m in one hour. Nevertheless, there are some species that can crawl for several hours.

The measured crawling pace can develop into regular running through **Running** an increase in the speed of the individual steps. With increasing speed, the feather-star tends to lift the steeply bent arms higher and higher until the animal seems to stride along the substrate. However, the principle of the driving force remains the same in this process, i.e., pulling by the leading arms and pushing by the following ones.

Many feather-stars, although not all, can switch from a rapid walk **Swimming** into a swimming motion by forcefully, yet very gracefully, beating their arms up and down. A swimming response can also be elicited by pulling away the feather-star's supporting substrate. During the upsweep beat the pinnules on the arms are rigid, creating a broad paddle surface and producing a more or less forceful lift for the animal. During the lift the

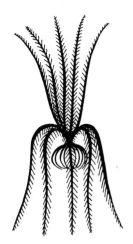

Fig. 11–6. Arm motions of a swimming feather-star.

Turning

pinnules are pressed against the arm surface. This action reduces the opposing frictional force and permits a new arm beat. Not all arms are synchronized in their motion. In adjacent arms the swimming motions are opposed. This means that in a ten-armed feather-star, arms 1, 3, 5, 7, and 9 would be lifted up. In feather-stars with more than ten arms this whole sequence, of course, becomes much more complicated and confusing, However, in these multi-armed feather-stars it seems that there are always five arms that are in the same place of movement but which are slightly out of phase with the motions of the remaining arms.

In the beginning the swimming motions are quite forceful. At various occasions 100 beats per minute were counted. However, this speed is maintained for only a few seconds and then switches into a more measured and slower pace. Feather-stars never swim for long durations, rarely longer than one minute, and they never swim for long distances. Each beat moves the animal approximately one arm's length. The longest distance known to have been produced by one arm-beat was approximately 3 m. The family Comasteridae, composed of many species and found primarily along the coasts of tropical oceans, is incapable of swimming; it can only crawl.

It is noteworthy that the feather-stars, in contrast to most other echinoderms, never turn the oral side toward the substrate when crawling. If one places a feather-star oral-side-down, it will turn over within a short time. The feather-star will bend several adjacent arms beneath the oral disc (tegmen) and push with the opposite arms until the crown is finally high enough that it can tip itself over its bent arms onto the proper side.

An examination of the contents of the feather-star's digestive tract will reveal remains of vegetable and animal matter. The composition of the intestinal contents varies with the location from which the animal has been taken. Aside from various unicellular organisms such as diatoms, dinoflagellates, foraminiferans, and radiolarians, remains of various crustacean larvae, small amphipods, copepods, and ostracodes, and fragments of algae, sponges, hydrozoans, and byrozoans may be found. Also there may be a substantial amount of detritus. How do feather-stars and sea lilies capture their food? In contrast to the majority of echinoderms, the feather-stars and sea lilies have maintained a manner of food-gathering that is regarded as the most primitive in the entire animal kingdom. The crinoids feed on plankton and microscopic organisms stirred up from the ocean floor.

Feather-stars inhabiting the sunlit regions of the ocean usually wait until dusk or night to capture food. At this time the animals will leave the crevices and holes where they have been hiding during the day. They will unfold their arms like an opened fan. The outspread pinnules with the closely spaced podia form a dense capturing net. This fan is always positioned diagonally to the water current, thereby providing the greatest possible surface for trapping flowing food particles. The feather-star extends the arms in such a manner that the aboral surface faces the current. In the quest for food, the feather-stars enjoy an advantage over the sessile sea lilies, since they can move about and choose a favorable location. Frequently feather-stars climb up on elevations in the ocean floor, such as rocks and cliffs, blades of seaweed, sponges, horny and stony corals, or other suitable places, to carry out their feeding activities.

It is quite impressive to observe the daily nocturnal departure for these feeding excursions along the many coral reefs. During the day one will find relatively few feather-stars, mostly in coral stalks where the animals are resting with coiled arms. At night, however, the reef seems to blossom and a multitude of feather-stars sways on the tips of the coral stalks. Food particles swept past by the water current become trapped on bits of mucus secreted by the glandular cells of the innumerable podia. The mucus is expelled from these cells by muscular contraction. Food collected by the feathery arms and podia is tossed into the ambulacral groove, which is also filled with mucus. The mucous-entangled food particles are swept toward the mouth by ciliary action along the extensive ambulacral system.

One obtains true appreciation of the size of this capturing device only if one considers the total length of all the ambulacral grooves in one animal. As an extreme example we shall consider a feather-star, *Comantheria grandicalyx*, which is found in the Sino-Japanese ocean at a depth of about 50 to 70 m. This has sixty-eight arms, each of which measures 12.5 cm. The total length of this crinoid's ambulacral system is 102.7 m. A ten-armed subspecies, *Tropiometra afra macrodiscus*, found along the Japanese coast, is characterized by extremely long arms, measuring 25.5 cm each. Its total ambulacral length is 47.6 m.

The extent of the capturing apparatus

Although the free-swimming feather-stars are able to change their location at will, they nevertheless always have to be in a place with a minimal water current which will sweep the necessary food particles along. Some feather-stars may alternately beat the water with their outstretched arms or brush through it, although this procedure probably does not significantly increase the amount of available food. Since water currents play a great role in the feeding of feather-stars and sea lilies, it is not surprising that these animals are extremely sensitive to water currents and respond to them in a very characteristic manner. Consequently, feather-stars tend to avoid still waters, preferring locations with gentle currents. They also avoid water eddies and strong currents, since these would damage these extremely fragile animals, or sediments stirred up from the ocean floor would interfere with the efficiency of the food-catching mechanism.

Feather-stars are dioecious. Strands of sex cells penetrate along the hemal lacunae into the arms and pinnules. However, the sex cells reach a stage of maturity only in the pinnules near the crown, where they form extensive gonads. Eggs and sperms are released to the outside when the walls

Spawning periods

of the genital pinnules rupture. In regions where living conditions remain relatively constant throughout the year, spawning is not limited to specific periods. However, individual species usually vary in their spawning periods. In areas where the climate fluctuates and the temperature, as well as the plankton growth, changes, spawning usually coincides with the time of greatest planktonic density. Along western European coasts this optimum time falls in the late spring. *Antedon bifida*, the most frequently found feather-star along Europe's Atlantic coast, spawns from the end of May to the beginning of July, and the Mediterranean feather-star *Antedon mediterranea* spawns in April and May. Usually the males discharge their gametes first, which seems to trigger the release of the eggs by the females.

The Japanese feather-star *Comanthus japonicus* is characterized by a well-defined spawning period. This species releases its gametes in the first half of October and always on those days when the moon is either in the first or last quarter. All animals of this species spawn within a time interval of two hours, even those in aquaria. Even arms that have become detached (autotomy) spawn within this particular period. How can such a unique event be explained? Probably the amount of light present during the different day lengths and moon phases, influences the maturing of the gametes and triggers the exact timing of this brief spawning period.

Usually all eggs or sperms are discharged from all the genital pinnules. When several neighboring animals discharge their gametes at the same time, the water becomes clouded. Of course, eggs and sperms become intermingled and fertilization takes place. However, there are many species where the eggs protrude from the pinnule but do not detach from the maternal body. This progeny only leaves the mother as fully developed swimming larvae.

Only a few species from Antarctica have special brood-care structures such as brood pouches. In these species the ovary-bearing pinnules have deep, pocketlike skin invaginations which reach close to the ovaries. The ripe eggs are probably discharged directly into the brood pouches, where they develop directly into the fixed crinoid or "sea lily" stage (pentacrinoid larva) without going through the free-swimming doliolaria stage. At the end of the larval development the young feather-stars of these species will protrude out of the brood pouch with their arm-bearing crowns while they are still anchored inside with the stalks.

The embryology of the sessile sea lilies is still unknown, since these forms are primarily deep-sea inhabitants, and have not yet been induced to reproduce in an aquarium. The feather-star eggs which have been discharged into the water usually measure 0.15 to 0.5 mm in diameter. Within a few hours these eggs will have developed into primary doliolaria larvae. This stage is characterized by four or five transverse ciliary bands and a sensitive ciliary tuft at the top. The larva moves by beating the ciliary bands, and swims by spinning around its longitudinal axis, sometimes for only a few hours or for days, depending on the species. Some-

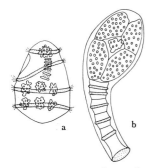

Fig. 11–7. Embryology of a feather-star: a. Doliolaria larva, b. Formation of the stalk and plates of calyx.

times these swimming larvae may be caught in a water current and carried over long distances. Naturally, some larvae are unfortunate and drift to unsuitable habitats, but for the species as a whole this dispersal of the larvae is of great advantage, enlarging the range of distribution by exploiting new habitats. During its swimming existence the larva lives off the yolk stored in the egg.

At the end of the free-swimming stage, the larva settles on a solid substrate, by means of an adhesive pit, and then starts to metamorphose. First the now-useless ciliary bands and ciliary tuft disappear. Then the larval body begins to stretch away from the place of attachment; concurrently the appearance of the crown and stalk becomes apparent. The development of the skeletal elements, which had already begun at the end of the swimming phase, now becomes accelerated. For the first time the transformation from the bilaterally symmetrical larva to the pentamerous, radially symmetrical feather-star becomes noticeable. Along the outer wall of the goblet-shaped crown three overlapping zones of five skeletal plates each develop. These plates make up the protective calyx around the animal's body. The first supporting plates, looking like a roll of coins (Fig. 11-7), also develop along the stalks.

The development of the internal organs progresses at the same time as the skeletal growth. In the doliolaria larva the archenteron has already divided into three secondary coeloms (see Chapter 10). One of these coeloms, the long mesocoel, at first is curved into a horseshoe shape, but later forms a complete ring which comes to lie on top of the archenteron. (The concepts "top" and "bottom" are used in relation to the longitudinal axis of the adult animal.) There now appears a depression in the body, which closes over again and forms the vestibule. This cavity is pushed on top of the enteron and ring-shaped mesocoel. The vestibule connects with the enteron through a hollow, rod-shaped process that penetrates through the mesocoel. At a later stage this process penetrates to the intestine and becomes the mouth.

In the meantime, the first five arms have developed. The arms erupt from five spaces between five skeletal plates, which had covered the vestibule in which the arms developed. By this time the tegmen with the mouth have also become exposed. The arms can now be extended into the surrounding water. The totally metamorphosed feather-star larva is now twice as long as the doliolaria larva, and is now capable of trapping and eating food particles. The number of arms increases because of the development of lateral arms at the base of the first five arms. The eruption of the vestibule creates five triangular epidermal flaps which become supported by skeletal plates, the deltoid plates. These five deltoid plates cover up the oral disc, and serve as a protective cover for the retracted tentacles.

The transition to the fixed crinoid stage is characterized by the development of the arms. This stage is initiated by the growth of a new skeletal

Fig. 11–8. Fixed crinoid stage of a feather-star.

plate at five points (in each interradius) between the deltoid plates and the subsequent circlet of plates. This second radial series of five radial plates forms the "basis" for the arms. A grooved arm bud grows out on top of each radial plate and encloses its corresponding primary arm, which in turn becomes the radial canal of the water-vascular system. The first vertebralike skeletal plates appear in these elongating small arms. During the course of further development the growth rate accelerates. The arms fork, and pinnules develop. The lower arm plates become incorporated into the calyx, and the deltoid plates become reduced. It seems as if the calyx shrinks in comparison to the arms, and the goblet shape of the calyx becomes apparent. The upper stalk segments (columnae) start to sprout cirri.

Fixed crinoid stage

The duration of the fixed crinoid stage in the young feather-stars is not known; it probably varies in the different species. Nevertheless, some forms attain a stalk length of 4–5 cm and an arm length of slightly over 1 cm. It is most likely that the feather-stars maintain this sessile stage, which is a recapitulation of phylogeny, for several months, until they detach from the stalk and start their free-swimming existence. *Promachocrinus kerguelensis* is a feather-star species with the longest known fixed crinoid stage. It stays attached to the stalk up to two and one-half years. The North Atlantic feather-star *Antedon bifida*, on the other hand, is already

Life span

full grown and sexually mature at one year of age. Exact dates concerning the age of feather-stars and sea lilies are not available, but it is assumed that none exceeds a maximum of twenty years.

Aside from the occasional animal which temporarily settles on a feather-star without forming any association with it, there are many animals that form a living association with feather-stars for longer periods of time. For example, there are many protozoans, usually ciliates, which

Commensals and parasitee

inhabit the crinoid's intestinal tract or crawl about on the body surface. The exact nature of this association with the host animal is not known. In the majority of cases these protozoans may only be harmless commensals.

Feather-stars often serve as a "substrate" for sessile organisms, such as hydrozoans. Many of these organisms take on the coloration and design of the host animal. This phenomenon is particularly evident in numerous small polychaetes, crustaceans, and some shrimps. All these creatures live on their host's body surfaces, where they find ample protection among the feather-star's arms and pinnules. These tiny animals can also share the captured food particles, and in return free their host from foreign organisms. However, some crustaceans are true parasites, for example isopods, copepods, and amphipods.

Myzostomes which are related to the polychaetes (see Vol. I) live almost exclusively on echinoderms, and primarily on feather-stars. These organisms dip their proboscis into the food strings flowing toward the mouth in the ambulacral grooves. Some myzostomes even penetrate into the body wall of the feather-star near these food grooves. The host's tissue reacts by growing a characteristic cyst or gall around the intruder. Only the proboscis protrudes from these cysts; it is dipped into the food groove. Some other myzostomes have become true endoparasites. In some feather-star populations almost every animal is infested with myzostomes. Occasionally such an infestation of myzostomes can assume alarming proportions. For example, a single feather-star of the genus *Antedon* harbored 300 to 400 of these parasites.

Some small snails from the family Melanellidae can be considered as among the most harmful parasites to the feather-stars. The snail bores through the skeletal plates with its proboscis, and feeds on the soft tissues of the host. The most highly developed parasite belongs to the fishes (family Gobiesocidae). This fish species, *Lepadichthys lineatus*, attaches itself to the feather-star with an adhesive disc located between its pectoral fins, and starts to feed on the pinnules.

Aside from these damaging parasites, which, however, are never fatal to their host, the practically defenseless feather-stars are without enemies. This is probably because feather-stars do not make rewarding prey animals: the edible organic body content is minute, compared to the mass of skeletal material. In addition, it appears that the feather-star's secreted mucus is poisonous, or at least irritating, to many animals. Fish food mixed with feather-star fragments will remain untouched by fishes. Even the highly carnivorous starfishes, which usually prey on smaller starfishes, brittle-stars, sea urchins, and sea cucumbers, will avoid feather-stars. Otherwise, no other animals are known to feed regularly on sea lilies or feather-stars. Nevertheless, it can happen that a feather-star colony can capture and eat its own free-swimming larval stages.

Anyone who has tried to transport a feather-star into an aquarium without breaking it knows how delicate and fragile these animals are. Relatively small mechanical stimuli, or just a deterioration in living conditions, prompts the animal to break off different-sized segments of arms or cirri, or detach the tegmen and the adjoining intestinal tract. The arms always break at the syzygies between two rigidly adjoining skeletal elements. The significance of this self-multilation is not quite clear. It does not function as a defensive mechanism against any predators, and does not lead to asexual reproduction as is the case in some starfishes and brittle-stars. However, it is conceivable that these multi-armed animals with their numerous processes may easily get entangled, and that self-multilation serves as a means of breaking free again. In any case, one encounters many free-living feather-stars that are in the process of regenerating some missing part.

The ability to regenerate lost body parts is well developed in the feather-stars. However, the prerequisite for regeneration is the presence of the intact nervous system which lies at the aboral surface of the calyx. If this system is undamaged, missing arms, cirri, the tegmen, and the digestive tract can be regenerated. The feather-star *Antedon* can lose four

▷
Sea cucumbers:
1. *Pelagothuria ludwigi*
2. *Holothuria tubulosa* (see Color plate, p. 312)
3. *Ocnus planci*
4. *Rhopalodina lageniformis*
5. *Ypsilothuria bitentaculata*

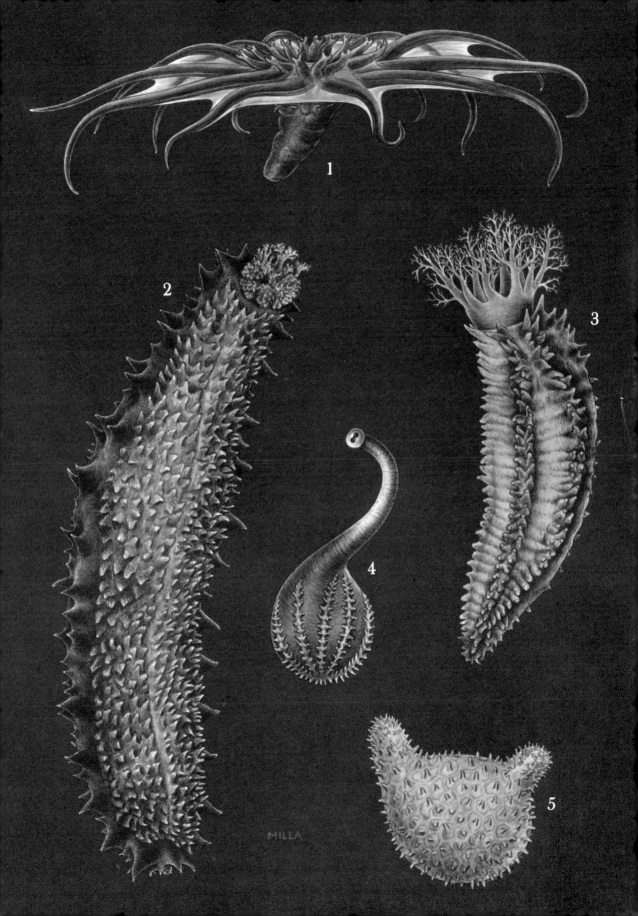

of its five arms, which means eight arm branches, and completely regenerate them all. The reconstruction of the tegmen and the digestive tract requires approximately three weeks. For some time the newly formed arms are considerably thinner and more delicate than the old arm stump from which they grew.

The new formation of a missing body part is initiated by freely moving cells which are present in large numbers in the coelomic fluid and most other tissues. These cells migrate along the brachial nerves to the place of breakage and first of all close over the wound. Later the cells and the epidermis form an elongated regeneration bud, into which the water-vascular system and the brachial nerve grow, to be surrounded in turn by new arm-skeletal elements.

Feather-stars and sea lilies are strictly marine. Their distribution varies greatly in the world's oceans. The highest concentration of feather-star species is found in the shallow coastal regions of the Indian and Pacific Oceans, with particularly high densities in the waters around Borneo, the Philippines, and New Guinea. From this focal point the density of species decreases in all directions. There are no feather-stars in the eastern Pacific Ocean, including along the western coast of the Americas. The species density declines less rapidly toward the west and north. Of approximately 650 known feather-star and sea lily species, only about ninety are found in the Atlantic, while the others are limited to the Indian and Pacific Oceans. In the Atlantic the focal point of feather-star species lies around the Caribbean-West Indian region. Almost half of the Atlantic species are concentrated here. The European coasts of Norway and Iceland, south to the Mediterranean, and further south along Africa to the equator are primarily inhabited by the genus *Antedon*. Although the main range of the feather-stars falls into tropical coastal zones, there are a number of species inhabiting the colder regions and even the polar areas. These species may often occur in large numbers. The antarctic genera *Notocrinus*, *Isometra*, *Phrixometra*, and *Kempometra* practice brood care and "bear" live young.

Only a few members of this class penetrate into the deeper water zones. Nevertheless, there are nineteen species inhabiting depths to 3,000 m, and four species penetrate down to 5,000 m. Only one species, *Bathycrinus australis*, is found at the abyssal depth of 9,000 m. The sea lilies are practically the only crinoids found at great depths.

The stalked larval form (pentacrinoid larva) of the feather-stars is dependent on a solid substrate where it can attach the adhesive disc at the end of its stalk. Frequently the maternal body acts as substrate for attachment, as do small stones, fragments of shell, or tests of polychaetes. Sea lilies possessing an adhesive disc on their stalks also require a solid substrate. Other sea lilies with rootlike processes on the lower end of their stalk are able to anchor these "roots" into soft bottoms such as mud or silt.

Free-swimming feather-stars can settle on many different substrates.

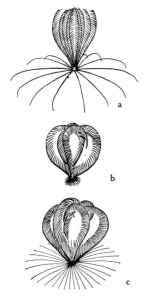

Fig. 11–9. Different feather-stars: The relationship between the length of the cirri and the substrate to which they are attached. a. *Asterometra macropoda*, b. *Pentametrocrinus tuberculatus*, c. *Pentametrocrinus varians*.

Their cirri are correspondingly modified. The majority, however, live on solid ground and on hard vegetable or animal structures to which they can cling. Those feather-stars that are found on cliffs or on rocky, sandy, and "grassy" bottoms usually have long, well-developed cirri with claw-like terminal segments well adapted for clinging. The many feather-stars inhabiting the branches of stony and horny corals or the root tangle of seaweeds have short, stout, curved cirri adapted for clinging to branches and little stems. Only a few feather-stars are adapted for living on muddy bottoms or silt. These forms are characterized by very long, slender cirri that can be spread in large numbers over the substrate. The cirri in this case function in a manner similar to a snowshoe, preventing the animal from sinking into the ground.

Feather-stars avoid direct radiation from the sun and therefore tend to prefer niches protected from sunlight. They shelter along shady walls, underneath overhangs, and in niches, caves, and the labyrinth of seaweed. Generally the animals leave their shelter only at dusk or during the night to seek out places where food availability is greater, returning to their shady sites at daybreak.

Nobody has yet made exact counts of the number of feather-stars and sea lilies inhabiting any specific area. However, catches from various research vessels have shown that the population density of crinoids is very high in certain favorable areas. A single sweep of a net dredging for bottom-dwelling animals along the coast of Massachusetts at a depth of 240 m brought up approximately 10,000 individuals of *Hathrometra tenella*. A similar operation revealed about the same number of sea lilies of the species *Rhizocrinus* dredged from a depth of 550 to 700 m. These sorts of data bring to mind huge feather-star meadows and extensive sea lily forests. Since the advent of underwater cameras which can be lowered to greater depths, photos have revealed that this concept is absolutely true. Photographs taken at a depth of 650 m on the Galicia Bank off Spain's northeastern coast showed that in an area of 95 m² there was an average population density of sixty-five animals per m². Other pictures of less densely populated areas reveal a density of one to five animals per m².

Population density

12   The Sea Cucumbers

Class: Holothuroidea,
by H. Fechter

In form the sea cucumbers (class Holothuroidea) are elongated, worm-shaped or bun-shaped echinoderms (see Color plates, pp. 301 and 302). The mouth is located on the anterior end and the anus at its posterior end. Unlike the other representatives of this phylum (except crinoids), their oral side does not face the substratum; the lateral body side does. The mouth is surrounded by a ring of simple or branched tentacles that usually are retractable. The body wall is elastic and leathery. Calcareous ossicles are embedded in the dermis. Unlike those of other echinoderms, these are almost always microscopic. The largest species, *Synapta maculata* (L 2 m, diameter 5 cm), is vermiform and cylindrical. The bulky species *Stichopus variegatus* (BL 1 m, diameter 21 cm) is bun-shaped.

Sea cucumbers can be surprisingly colorful. This seems hard to believe if one judges them by colder-water species, which are usually uniformly dark or light brown, black, gray, or yellow. Along tropical coral reefs one can encounter brilliant red, sky-blue, and dark green sea cucumbers. Occasionally these forms are vividly patterned by colored warts, or longitudinal or transverse stripes. However, many sea cucumbers that burrow in the sea bottom are just an ordinary dirty-white color. Deep-water forms are often purple or violet.

The thickness of the body wall varies greatly in the individual species. Unlike those of other echinoderms, the sea cucumber's body wall is generally extremely elastic. The microscopic ossicles embedded in the connective tissue layer underneath the thin unciliated epidermis are not interconnected into a rigid exoskeleton. The small ossicles scattered throughout the dermis assume various species-specific shapes, in the form of perforated plates, buckles, crosses, rods, wheels, towers, or anchors. Only a few families, represented by few species, have plate-shaped skeletons consisting of individual scales that touch or overlap and thereby form a scaly armor. Pelagothuriids which drift with the plankton on the water surface have no calcareous ossicles, and several burrowing synaptids and deep-sea holothurians lack ossicles.

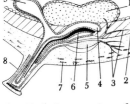

Fig. 12–1. Cross section of the body wall of a sea cucumber in the region of a podium:
1. Longitudinal muscle strand, 2. Circular muscle layer, 3. Nerve endings, 4. Radial water canal, 5. Ectoneural part of radial nerve (the hyponeural part lies above it), 6. Radial hemal canal, 7. Ossicle in the body wall, 8. Podium (tube foot), 9. Podial ampulla.

The absence of a rigid skeleton has been compensated for by a highly developed body musculature. Under the epidemis is a layer of well-developed circular muscles which are interrupted by five bands of longitudinal muscles that run along the ambulacral areas. The longitudinal muscles insert anteriorly to a ring-shaped internal skeleton which usually consists of a ring of ten calcareous plates surrounding the pharynx. The inner surface of the muscles is covered by the peritoneum.

A specimen of the genus "*Cucumaria*" shows (see Color plate, p. 301; compare Color plate, p. 312) five double rows of tubelike processes extending from the body wall, each process ending with a suction disc. These are the tube feet (podia) which adhere to the substratum. Most holothurians always expose the same side of the body to the bottom; this side is usually flattened and solelike. The creeping sole always consists of three ambulacral areas (trivium). The dorsal side has two ambulacral areas (bivium). In those sea cucumbers that are often sessile for long periods of time the creeping sole is difficult to differentiate from the dorsal surface. This condition is exemplified by *Ocnus planci*. In an aquarium such an animal often does not change its location for two years.

Since the podia on the dorsal surface are no longer used for adhesion and movement, they have been modified into globular papillae and warts, or have disappeared altogether. One group of sea cucumbers which is represented by many species has only degenerate podia, even on its ventral side. This order, Apodida (footless), seems to do well even without podia.

The tentacles surrounding the oral region, from ten to thirty, are in reality modified podia. These tentacles vary greatly in the individual groups. Some are fingerlike, forked at the tips, shieldlike, flaplike, dendriform, or pinnate (Fig. 12-3). The anterior part of the body is necklike and is extremely pliable. In the Dendrochirotida this necklike section and the tentacles can be retracted into the pharynx. Additionally, each tentacle can be withdrawn individually by contracting its longitudinal muscles.

In its basic plan, the water-vascular system is similar to that of other echinoderms. It consists of a water ring around the proximal pharynx, and five radial canals running the length of the body wall between the longitudinal and transverse muscles. Each radial canal gives off a closed, tubelike process to the tentacles in its vicinity, and, further along, lateral canals go to the podia and papillae on both sides. At the branching-off points to the podial canals, and frequently also to the tentacles, one finds muscular, bladderlike organs (ampullae) which provide receptacles for the fluid that is pushed back when the podia or tentacles are contracted. When the tube foot or tentacle is extended again, the muscles of the ampullar wall contract and press the fluid into the podial or tentacular canal, thereby re-extending the corresponding organ. In order to prevent a loss of fluid into the radial canal during the hydraulic interplay between ampullae and podia, a check valve in the form of a membranous flap has

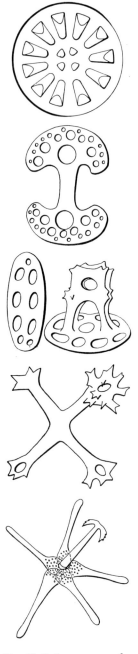

Fig. 12–2. Some types of ossicles from the body wall of a sea cucumber.

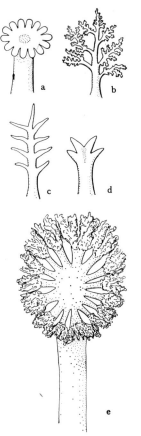

Fig. 12–3. Various types of holothurian tentacles: a. Shield-shaped, b. Tree-shaped, c. Pinnate, d. Digitate, e. Leaf-shaped.

been developed at the joint between the radial canal and podium-ampulla system.

Stone canals, occasionally over 100, but usually only one, arise from the water ring. The stone canals terminate in the coelomic cavity. The water ring also bears one or several ampullar vesicles (polian vesicles) which probably serve as expansion chambers for the maintenance of pressure in the water-vascular system. Representatives of the order Apodida (without podia) do not have radial canals, and the tentacular canals join directly to the ring canal. Tentacular ampullae are also missing, but there may be up to fifty polian vesicles (Fig. 12-4).

The mouth is found at the center of the ring of tentacles. It can be closed by a sphincter muscle. The mouth leads into the short pharynx and a long, usually looped, intestine. The gut is supported by mesenteries that attach it to the coelomic cavity at three different places. The stomach is only vaguely differentiated from the rest of the alimentary canal. The large intestine is usually widened considerably before terminating in the posterior anus. The large intestine—cloaca—is attached to the surrounding body wall by radiating strands of connective tissue and muscle. The wall of the digestive tract consists of longitudinal and circular muscle layers and numerous blood vessels.

In the sea cucumbers the aboral system (see Chapter 10) of the three echinoderm nervous systems is missing. The subdermal ectoneural system is concentrated around a ring-shaped center at the mouth edge and follows the water-vascular system. It supplies the tentacles, podia, and body-wall plexus. The hyponeural system is spread between the ectoneural system and the water-vascular system, and innervates the body-wall muscles (Fig. 12-5).

Sensory cells are distributed over the entire body surface, although they are particularly concentrated at the anterior and posterior ends. Several members of the order Apodida have warty elevations on the body surface and ciliated grooves on the oral side of the tentacles, containing numerous sensory cells. The sensory cells of the skin react partially to chemical and mechanical stimuli; some also react to light. Some sea cucumbers have ten to one hundred statocysts along the nerve ring of the ectoneural system. These tiny fluid-filled spheres consist of innervated cells, and contain one to twenty cells (calcareous statoliths), which in turn contain some inorganic material. The differential movement of the calcareous statoliths is registered by special nerves, and serves to right the animal if its orientation is upset.

In the thin-walled representatives of the orders Apodida and Elasipodida, oxygen-carbon dioxide exchange takes place over the entire surface of the body. Sea cucumbers with thicker body walls breathe with the aid of respiratory trees. Respiratory trees are two evaginations of the digestive tract at the anterior part of the cloaca. These breathing devices are branched like a tree, as the name implies, and they extend anteriorly into the coelomic cavity. Fresh, oxygenated water is sucked in by the cloaca and is forced into the respiratory trees. Water is forced out again by relaxation of the cloaca, contraction of the tree tubules, and opening of the anus. Some species "breathe" rapidly. Water is drawn in and out of the trees several times a minute. Other species draw fresh water in by numerous pumping cycles until the respiratory trees are turgid, and then all the used water is expelled at one time. Oxygen from the inspired water diffuses into the coelomic fluid, which is constantly kept circulating by the ciliary action of the body wall, and oxygen is thereby supplied to the organs.

The hemal system consists mainly of two parts: the food-absorbing network of lacunae in connection with the intestine, and the radial hemal sinuses that parallel the water-vascular system into the smallest branches. These radial hemal strands start at the oral hemal ring around the pharynx and end blindly in the tentacles, podia, and posterior end of the body. This is not a closed circulatory system. The intestinal hemal system joins into the ring canal. This system consists of two hemal sinuses. One sinus runs along the descending intestine and then crosses over (transverse connection) to the sinus running along the ascending intestine. In the large sea cucumbers of the order Aspidochirotida the dorsal sinus is connected to the ascending intestine by a network of lacunae (so-called rete mirabile). Processes of this network run into the left respiratory tree and thereby facilitate favorable conditions for oxygen transport. The intestinal hemal sinuses contract four or five times a minute, forcing the hemal fluid through the hemal system.

Most sea cucumbers are dioecious. Hermaphroditic forms are sometimes found, particularly in the family Synaptidae. Although most other echinoderms have five similar sets of sexual organs, the sea cucumbers have only one gonad, which consists of a tuft or cluster of closed tubules. These tubules may be located on one or both sides of the anteriorly attached mesentery on the dorsal side. The gonoduct passes in the mesentery to the gonopore, at the base of the tentacular ring.

The sea cucumbers are classified into three subclasses: 1. Dendrochirotacea: the front end of the body can be completely withdrawn with the aid of special retractor muscles. Podia and respiratory trees are usually present; unattached tentacular ampullae are absent. There are two tufted gonads, one at each side of the upper mesentery. 2. Aspidochirotacea: ten to thirty shieldlike tentacles. The anterior end lacks retractor muscles. The body is bilaterally symmetrical. Podia are present. 3. Apodacea: simple tentacles, either finger-shaped or pinnate. Podia usually are totally absent or are greatly reduced. The anterior body section lacks retractor muscles.

In the Dendrochirotacea one distinguishes between two orders: A. Dendrochirotida: ten to thirty bush-shaped tentacles. The calcareous ring around the pharynx is simple or is endowed with posteriorly pointing

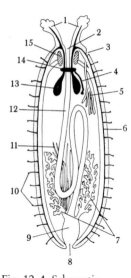

Fig. 12–4. Schematic longitudinal section of a sea cucumber: 1. Oral podium (tentacle), 2. Mouth, 3. Gonopore, 4. Stone canal, 5. Gonad, 6. Radial canal of the water-vascular system, 7. Respiratory trees, 8. Anus, 9. Cloaca, 10. Podia, 11. Cuvierian tubules, 12. Intestine, 13. Tentacle ampullae, 14. Water ring around the pharynx, 15. Calcareous ring around the pharynx.

Fig. 12–5. Cross section of the sea cucumber's body wall (*Holothuria*): 1. Gonad within the mesentery, 2. Longitudinal muscle strand, 3. Circular muscle layer, 4. Respiratory tree, 5. Body wall, 6. Podia, 7. Intestine.

processes. There are seven families. B. Dactylochirotida: eight to thirty finger-shaped tentacles, which occasionally are branched (digitate). The body wall contains closely fitting plates. The calcareous ring is without processes. There are three families.

The order Dendrochirotida encompasses the following families: 1. Placothuriidae: the body wall is covered by overlapping plates. The calcareous ring has long, paired processes. The genus *Placothuria* belongs to this family. 2. Paracucumidae: simple calcareous ring without processes; includes the genus *Paracucumis*. 3. Heterothyonidae: the body wall is covered by overlapping plates and the calcareous ring has short posterior processes; includes the genus *Heterothyone*. 4. Psolidae: the body's dorsal surface is covered by ossicles. The ventral surface is modified as a soft sole. This group includes *Psolus phantapus* (L up to 15 cm). The posterior end is pointed in a taillike fashion, and is directed upward. The tentacles are orange and the body is yellowish-brown to dark brown. It is distributed in the North Sea from Greenland to Spitsbergen at depths ranging from sea level to 380 m. This form practices brood care. The juveniles adhere to rocks by suction. 5. Phyllophoridae: the body wall is naked, with only minute, undefined ossicles. The projections of the calcareous ring consist of several pieces. The following species belong to this family: *Trachythyone elongata* (L up to 15 cm; see Color plate, p. 302); the body is U-shaped, brownish or gray; the species is distributed in the Mediterranean and Atlantic from Norway to Morocco. It is found in sand and mud, from sea level to a depth of 110 m. *Sclerodactyla* ("*Thyone*") *briareus* and *Thyone fusus*; the latter has a length of up to 20 cm and is spindle-shaped or compressed into a pear shape. It is either white, pink, or yellowish. It is distributed in the Mediterranean and the North Atlantic. It is found on reeds, seaweed, and mud flats, at depths from 5 to 150 m. 6. Sclerodactylidae: the calcareous ring projections consists of one undivided section. 7. Cucumariidae: the calcareous ring has no projections. *Thyonidium pellucidum* and *Ocnus planci* (L up to 15 cm; see Color plate, p. 301; compare Color plate, p. 312); the brownish body is pentagonal. The latter species is distributed in the Mediterranean and North Atlantic. It is found at depths from 5 to 250 m on a variety of substrates, including sand, mud, stones, corals, seaweed, and underneath algae.

The three families of the order Dactylochirotida are the following: 1. Ypsilothuriidae: eight to ten tentacles; the plates have small spines; the body is crescent-shaped. The genera *Echinocucumis* and *Ypsilothuria* (compare Color plate, p. 301) belong to this family. 2. Vaneyellidae: ten to twenty tentacles; the plates have only small, thornlike projections or none at all; includes the genera *Vaneyella* and *Mitsukuriella*. 3. Rhopalodinidae: anus and mouth lie side by side; the body is bottle-shaped; includes the genus *Rhopalodina* (see Color plate, p. 301).

The subclass Aspidochirotacea is classified into two orders: A. Aspidochirotida: the respiratory trees are always present; the mesentery of the posterior intestinal loop attaches to the right ventral interradius; tentacular ampullae are present. There are three families. B. Elasipodida: respiratory trees are absent. Mesentery of the posterior loop of the intestine attaches to the right dorsal interradius. Tentacular ampullae absent. There are five families.

Some of the most common and best-known sea cucumbers belong to the order Aspidochirotida. Its three families are: 1. Holothuriidae: only one cluster of gonads on the left side of the dorsal mesentery. The tentacular ampullae are long and slender. The rete mirabile on the left intestinal loop is well developed. Members of the genus *Holothuria*, particularly *Holothuria forskali*, are well known. *H. forskali* (L up to 25 cm, diameter 5 cm): black with white-rimmed papillae on the dorsal surface. Distributed in the Mediterranean and Atlantic up to Scotland at depths from 1 to 100 m. It frequently inhabits seaweed, corals, and muddy sand flats. *Holothuria tubulosa* (L 30 cm, diameter 6 cm; brown to brownish-purple; see Color plates, pp. 301 and 312): found primarily on sand and sandy mud flats, but also seaweed meadows, from sea level to a depth of 100 m. It is one of the most frequently found Mediterranean sea cucumbers. 2. Stichopodidae: two clusters of gonads, one on each side of the dorsal mesentery; long, slender tentacular ampullae; the rete mirabile between the intestinal loops is very well developed; podia are limited to three zones on the ventral surface. The dorsal surface is covered with papillae. *Stichopus regalis* (L up to 35 cm, diameter 6–7 cm; body shape appears flattened; color is light brown with white ringlets on the dorsal surface) is distributed in the Mediterranean and the Atlantic from the Canary Islands to Ireland at depths from 5 to 800 m. It is found on sandy, muddy, and coral substrata. 3. Synallactidae: usually two tufts of gonads lacking tentacular ampullae; the rete mirabile is almost undeveloped; predominantly deep-water forms; includes *Bathyplotes natans* and *Galatheathuria aspera*.

The order Elasipodida is frequently represented by sea cucumbers that are purple, violet, or red. These forms all live at great depths, and, in contrast to shallow-water forms, have usually developed bizarre shapes. 1. Family Deimatidae: podia are present only along the margins of the creeping sole; numerous conical papillae are present on the dorsal surface. This group includes the genera *Deima* (see Color plate, p. 302) and *Oneirophanta*. 2. Laetmogonidae: closely related to the deimatids but distinguished from them by their wheel-shaped ossicles; includes genus *Laetmogone*. 3. Psychropotidae: mouth and anus lie on the creeping sole; the podia have been concentrated in a double row along the center of the creeping sole; frequently there is a large taillike appendage above the posterior end. The genera *Psychropotes* (see Color plate, p. 302) and *Euphronides* (see Color plate, p. 302) belong to this family. 4. Elpidiidae: the calcareous ring around the pharynx consists of only five radial plates; large podia are situated along the margins of the creeping sole and in

▷
The colorful *Paracucumaria tricolor* from the Great Barrier Reef off the northern coast of Australia.

front of the mouth; includes the genera *Peniagone* and *Elpidia*. 5. Pelago-thuriidae: the body wall is thick, gelatinous, completely lacking calcareous deposits. A large brim is developed anteriorly. Bathypelagic. Includes the genera *Pelagothuria* (see Color plate, p. 301) and *Enypniastes*.

The subclass Apodacea encompasses the "footless sea cucumbers." There are two orders: A. Apodida: cylindrical body; respiratory trees, radial canals, and anal papillae are absent; ossicles are wheel-shaped, anchor-shaped, or in the form of anchor plates. Three families. B. Molpadiida: spindle-shaped body, usually with posterior end tapering to form a tail; respiratory trees, radial canals, and anal papillae are present; wheel-shaped ossicles are not present. Four families.

In the order Apodida we already find forms that burrow into the substrate. The three families are characterized by the following features: 1. Synaptidae: pinnate tentacles; wheel-shaped ossicles are absent but anchors and anchor plates are present. Includes the following species: *Leptosynapta inhaerens* (L up to 30 cm, diameter 0.5 to 0.9 cm; see Color plate, p. 302): body color is pink; gonopore is situated on one of the dorsal tentacles; inhabits sandy and muddy substrates but also underneath the roots of seaweed at depths of 2 to 30 m; distributed in the Mediterranean and eastern North Atlantic. *Labidoplax digitata* (L up to 35 cm, diameter 0.8 to 0.9 cm): red or reddish-brown, occasionally also spotted; ventral side is lighter; distributed in the Mediterranean and Atantic up-to the English Channel at depths of 10 to 600 m. The animals burrow in muddy or slimy bottoms. 2. Chiridotidae: tentacles are short, thick, broadened at the tips, and digitate; anchor ossicles are missing; wheel-shaped ossicles with six spokes; includes the genera *Chiridota* and *Taeniogyrus*. 3. Myriotrochidae: no anchor ossicles; wheel-shaped ossicles with eight or more spokes; includes the species *Myriotrochus bruuni*.

Sea cucumbers of the order Molpadida constantly remain buried in the substrate. There are four families: 1. Gephyrothuriidae: fifteen tentacles with two pairs of finger-shaped processes and several whip-shaped papillae on the dorsal surface: ossicles are not present in the body wall; deep-sea forms; genus *Gephyrothuria*. 2. Caudinidae: tentacles with one to two pairs of fingerlike processes (digits); without whip-shaped papillae on the dorsal surface; body wall with ossicles; includes the genera *Caudina* and *Paracaudina*. 3. Molpadiidae: ten to fifteen small, clawlike tentacles with one to three pairs of digits and a terminal digit; ossicles are often degenerated in older animals. *Molpadia musculus* (L averages 6 cm); distributed in temperate oceans at depths of 60 to 80 m; has worldwide distribution. 4. Eupyrgidae; small, arctic forms with fifteen simple, finger-shaped tentacles; includes genus *Eupyrgus*.

Most sea cucumbers are only able to crawl slowly along the ocean floor. Some can also swim for short distances, but only a few have a body structure that permits constant swimming. Forward motion is achieved by a process similar to that of the earthworm. The entire body or succes-

◁
Upper left: *"Cucumaria" miniata* (compare Color plate, p. 301);
Upper right: Indo-Pacific sea cucumber (*Holothuria*);
Center: Indo-Pacific sea cucumber (*Holothuria*);
Below: *Holothuria tubulosa* (see Color plate, p. 301).

sive body sections are repeatedly stretched and contracted in a mouth-anus direction. The most simplified version of this motion is the following: the anterior part of the body firmly adheres to the substratum with the podia. The body contracts by contraction of the longitudinal muscles. Then the more posterior podia anchor themselves slightly in front of the old adhesion place while the anterior podia release themselves from the substratum. Then the circular muscles contract progressively, starting at the posterior end. These muscular contractions slowly ripple over the entire body toward the anterior end. Then the body is stretched and the anterior podia gain a new foothold through this elongation. This interchange of contraction and extension may occur simultaneously in various sections of the body in more complex cases: waves of contraction and elongation follow each other starting from the hind end along the body toward the front, and consequently the animal is pushed forward in little steps. In *Stichopus parvimensis* (L 25 cm) such a locomotive wave lasts about one minute before it has passed over the entire body.

Many of the footless members of the synaptids also crawl in this manner. For example, in *Synaptula hydriformis* (L 4.5 cm) a wave of circular muscle contraction lasts for approximately fifteen seconds. In these sea cucumbers the hooks of the anchor-shaped calcareous plates embedded in the epidermis serve to anchor the various body sections to the substratum. These hooklike projections protrude from the entire body surface. Other footless sea cucumbers additionally use their very sticky tentacles when moving about. These forms extend their long tentacles out in front of them as far as possible, and then fasten the tentacles to the substratum and pull the body up afterward. Almost all sea cucumbers can climb up vertical cliff walls or the glassy sides of an aquarium, although not all are able to move backward. If a sea cucumber is caught in a narrow passage where it cannot move forward, it more or less has to make a "hair-pin" turn. Compared to this maneuver, the motions of *Sclerodactyla briareus* seem very acrobatic. This sea cucumber not only can creep forward and backward, but also sideways with its entire side. The speeds in the different sea cucumbers vary greatly. *Stichopus parvimensis* (L 25 cm) requires fifteen minutes to travel 1 m; *Holothuria suriramensis* (L 10 cm) travels 4 cm in the same time; and *Sclerodactyla briareus* (L 10 cm) takes that long to cover a distance of 7 cm.

When burrowing in the substratum, the major portion of the work is carried out by the tentacles and body musculature. The tentacles clear away slime and mud from the anterior body, and subsequently a small hollow is formed, into which the front part of the body penetrates through contraction of the circular muscles. Then the longitudinal muscles contract and the rest of the body is pulled into the hollow. During this process the front end swells, thereby enlarging the hollow. The tentacles resume their activity and burrow a hollow in front of the old one, so that the body can push onward. It takes less than a minute for *Leptosynapta* (L

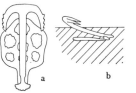

Fig. 12–6. Anchor-shaped endoskeletal ossicle and plate, from the body wall of a synaptid sea cucumber: a. Dorsal view, b. Lateral view, embedded in the body wall, showing how the anchor projects from the skin surface.

Burrowing

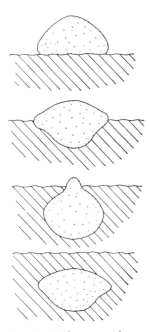

Fig. 12–7. Changes in the body cross section of *Bohadschia vibittata* during burrowing.

Swimming

10 cm) to disappear completely into the substrate. This species does not sweep aside the material in front of its body, but frequently ingests it with its mouth. In a manner of speaking, the sea cucumber eats its way through the substrate.

Other species from the family Holothuriidae use somewhat different burrowing techniques. They push their creeping sole into the substrate like a wedge, and then greatly enlarge it. In this manner, loose material is pushed to the sides, and the entire body sinks in. Constant repetition of this process permits the animal to disappear into the substrate. A sea cucumber measuring 20 cm may take approximately thirty minutes to disappear from the surface. Species of the genera *Molpadia* and *Caudina* simultaneously employ the two digging techniques described above. The alternate shrinking and enlarging of the creeping sole in conjunction with the rippling action of the muscular contractile waves over the animal's body results in relatively fast substrate penetration. Depending on the size of the animal, it may take from ten to forty minutes to disappear completely.

Although sea cucumbers are generally slow, sluggish animals, there are some forms that make quick, undulating motions so that they are actually swimming. Several *Leptosynapta* species, *Astichopus multifidus*, and *Bathyplotes natans* swim by means of rather clumsy vertical S-shaped body twists. *Labidoplax dubia*, found near Japan, is a nocturnal swimmer, but only during the months of June and July. *Leptosynapta inhaerens* (see Color plate, p. 302), found in the Atlantic along European coasts, also makes occasional nocturnal swimming excursions.

True agile swimmers among the sea cucumbers are those forms inhabiting the deep water. These deep-water forms have a gelatinous body with an extremely high water content, just as in the jellyfishes. The genera *Pelagothuria* and *Enypniastes* also have sails and umbrellas like the jellyfishes. *Pelagothuria* (see Color plate, p. 301) swims in a medusoid fashion, by opening and closing its web. *Galatheathuria*, on the other hand, employs a different swimming technique: its flattened body is surrounded by a broad fringe which moves in graceful waves through the water, similar to the cephalopod *Sepia* (cuttlefish).

Except for the few species that are capable of swimming, most sea cucumbers are bottom dwellers. Some prefer rocky backgrounds, where they can hide beneath projecting cliffs, cracks, and hollows. Others are found in the root tangle of seaweed or the jungle of coral thickets. The large holothuriids of the tropical and subtropical coastal regions primarily inhabit sandy bottoms. One frequently finds several hundred animals concentrated on a relatively small area. Since these sandy flats are often fully exposed to the sun, the animals frequently protect themselves with bits of vegetation or pieces of shell stuck to their dorsal surface. This also provides some camouflage.

An astonishingly large number of sea cucumbers, particularly from the subclass Apodacea, live buried in the substrate. This mode of life has resulted in more or less extensive modifications in the body structure of many forms. The relatively thin-skinned members of the family Synaptidae, which lack respiratory trees and breathe predominantly through the epidermis, live totally submerged in sand or mud. These forms dig tunnels, stabilizing the walls somewhat with mucus to prevent their caving in soon after the animal leaves. *Leptosynapta*, for example, is almost buried in the substratum. It digs its tunnels parallel to the upper surface (Fig. 12-8). This form only occasionally leaves its tunnel, for a nocturnal swim or during spawning. During the breeding season the animal may not actually leave its tunnel but merely expose its front end, with the tentacles and gonopore, into the surrounding water. Total burial, as was seen in the synaptids and related forms, is not possible for other sea cucumbers, since they breathe by the respiratory trees. In animals which feed on bottom detritus, the anus, at least, has to be in contact with the fresh water from the bottom surface, since the anus serves as inhalent and exhalent opening. As an adaptation to this living condition (detritus feeding), members of the families Molpadiidae and Caudinidae, i.e., genera *Molpadia*, *Caudina*, and *Paracaudina*, underwent striking modifications in their body structure. The thick, sausage-shaped front end of the body, containing all the viscera, tapers off into a thin, tubelike hind end containing only the large intestine. The posterior end is usually longer than the front end, but has only the function of a respiratory tube which protrudes out of the substratum. The hind end serves to keep the obliquely buried animal in contact with the substrate surface where oxygen-carbon dioxide exchange takes place.

Other forms of sea cucumbers which also feed on plankton prefer to stay buried within the substratum. These forms have to cope with the problem of exposing the tentacular oral region and also the anus to the upper surface. To solve this problem the animals merely had to adjust their body posture. *Cucumaria* and *Thyone*, of the order Dendrochirotida, bend their bodies, which are submerged in either mud or sand, in a crescent shape, so that both the mouth and anus project above the substratum. Other representatives of the subclass Dendrochiroacea have solved this same problem by modifying their body shape, well exemplified by the genera *Ypsilothuria* (compare Color plate, p. 301) and *Echinocucumis*. In these forms, mouth and anus are situated on small, retractile, chimneylike projections on the crescent-shaped body.

However, the most highly specialized modifications occur in the genus *Rhopalodina* (see Color plate, p. 301). As a result of the extreme U-shaped curvature, the mouth and anus projections came to lie side by side, and fused into a single long tube. Mouth and anus are situated side by side at the upper tip of the tube. This mud-submerged animal has thus acquired the appearance of a Chianti bottle. The bulbous section contains the viscera, and the slender neck corresponds to the fused mouth-anus tube.

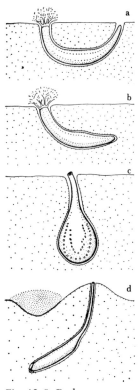

Fig. 12–8. Body posture of various burrowing sea cucumbers: a. *Cucumaria* species, b. *Synapta*, c. *Rhopalodina*, d. *Paracaudina*.

The species *Opheodesoma grisea* can survive long periods of dryness and exposure to the burning sun during low tide. The animal may desiccate to a state where it resembles a piece of dried-out seaweed more than a sea cucumber. However, the animal recovers with the returning tide, reassuming its normal life activities.

Many sea cucumbers are plankton feeders; the microscopic organisms are caught from the surrounding water. Others feed upon small benthic organisms selected from the substrate which passes through the animal's body. Holothurians that catch small planktonic organisms belong primarily to the order Dendrochirotida and the few swimming forms. Their method of capture is based on the principle of the lime twig. The greatly extended tentacles with the more or less highly branched tips are covered with sticky secretions to which minute crustaceans, protozoans, various larval forms, small jellyfishes, and others adhere. In rhythmic succession these tentacles are bent into the mouth, where they penetrate deeply into the pharynx. The tentacle is pulled out again with the mouth closed. The sticky surfaces are regularly sucked, cleaning off the food.

It is far more common for holothurians to feed on deposits that are swallowed as the animals burrow in the substratum. The organic part of the ingested material is digested. There are innumerable ways in which bottom detritus can be picked up. Some members of the order Apodida, for example, pick up individual sand granules with their sticky tentacles, and put them into their mouths. The surface of a sand granule is usually not sterile, but frequently is covered by algae or bacterial growth which can be digested by the sea cucumber. Radiolaria, Foraminifera, and diatoms may flourish between the sand granules. They serve as a food source to the holothurians. Obviously, a few sand granules will not suffice to provide the sea cucumber with an adequate food supply. The animal must be practically stuffed full of sand to obtain sufficient digestible organic matter. It is therefore not surprising that many sea cucumbers must swallow substrate deposits almost without interruption during the day and night.

The more imposing species of the order Aspidochirotida, which includes, for example, the widely distributed genera *Holothuria* and *Stichopus*, have developed very effective methods of devouring as many substrate particles as possible. Each tentacle of these forms has a broad shield at the tip, which function like rakes as the tentacles are pulled over the substrate toward the mouth.

Various other holothurians burrow through the sandy substrate like earthworms, swallowing all the sand lying directly in front of the mouth. Their activity produces the same result as that of the earthworms. These sea cucumbers are responsible for extensive shifting and mixing of the substrate. It is quite amazing how much substrate deposit passes through certain holothurians, as is shown by the following example: an *Isostichopus badionotus* (L 20 cm) living on the ocean floor can contain an average of

60 to 70 g of sand (dry weight). In twenty-four hours the intestinal content is completely renewed two or three times. This means that a daily amount of approximately 160 g of ocean debris passes through the animal's digestive system. If one assumes that this activity proceeds at this pace throughout the year, approximately 45 kg (100 lbs) of material will have been ingested and passed out of the animal annually. This species is very common in certain bays of the Bermuda Islands. It has been estimated that in a certain region of 4.4 km², 500 to 1000 t of sand and slime are passed through these holothurians annually. The apodous species *Paracaudina chilensis*, which lives buried in the substrate, passes out 6–7 g hourly. The annual performance of this species is similar to that of *Isostichopus badionotus*.

Holothurians ingest almost everything that is within the range of their tentacles. Only a few forms are known that select their food. One of these species is the Chilean sea cucumber *Athyonidium chilensis*, which feeds on large quantities of brown algae, apparently its exclusive source of food. It was observed on the coral reefs of the Palau Islands that aside from the constant day-and-night feeders there were also species that kept definite, although long, feeding periods. Such sea cucumbers would spend only two-thirds of the day on feeding, spending the remainder of the time resting under the protection of corals or lying buried underneath algae or sand. Not all individual sea cucumbers would start feeding at the same time. Some began in the morning, others in the afternoon. Several holothurians maintain annual rest periods, either in the summer or during the winter, when they do not feed. These animals resume their normal activities again when the water has reached a specific temperature. **Food selection**

The majority of sea cucumbers spawn into the surrounding water. Almost every species has a specific spawning period, which usually falls within one or two months of every year. Species inhabiting tropical waters are not tied to a specific season of the year, as are those found at higher latitudes. In the tropics, spawning occurs either in the spring or summer. For example, *Cucumaria planci* and *Labidoplax digitata*, found along the European coasts of the Atlantic and Mediterranean, spawn during the months of March and April. *Cucumaria frondosa*, distributed in the North Sea and the Atlantic, spawns from February to March. Arctic species, on the other hand, start spawning in June or July. *Holothuria tubulosa*, from the Mediterranean, discharges its gametes in August or September, and the North Atlantic *Labidoplax buski*, from October to December. **Spawning periods**

Spawning usually occurs late in the afternoon or at night. The males always release their gametes first, and then the females lay the eggs, probably triggered by the presence of pheromones. Hermaphroditic species, like *Labidoplax buski*, never release eggs and sperms simultaneously, but at intervals of one to two days. In this way, self-fertilization is prevented. During the release of the gametes, many free-living holothurians hold the front end of their body erect. Even the synaptids, which are usually submerged in the substratum, project their front end well above the surface of the ground. Some species wave the front end, or at least the tentacles, back and forth while spawning. A fine spray of sperms or eggs is discharged from the gonopore into the water, and the waving motions facilitate greater dispersion.

On the average, spawning lasts for approximately thirty minutes, although it can be as brief as fifteen minutes or as long as four hours. Not all mature gametes are discharged at one time; the process may occur in intervals. Fertilization takes place in the water. The fertilized egg then either sinks to the bottom or rises to the water surface.

More than forty species of holothurians practice brood care, but in very different ways. In *Taeniogyrus contortus*, found in the antarctic oceans, the embryos and young develop within the ovaries. Several species of the synaptids release their eggs into their coelomic cavity, where embryologic development up to the pentactula stage takes place. The young sea cucumbers are "born" through a slit in the anal region. Most broodcare structures, however, are found on the outside surface of the sea cucumber (order Dendrochirotida). Often this external care does not require much of the sea cucumber's energy. For a brief period the eggs adhere to the tentacles, and are released at an early developmental stage. *Psolus antarcticus*, for example, goes a little beyond the previous example. The eggs and larvae are attached to the apodal regions of the sole, where they are protected by the maternal body.

In *Cladodactyla crocea* the first beginnings of simple brood pouches become evident on the external surface of the body. Two thick longitudinal bulges are situated along the two rows of podia. These form a groove into which the eggs are deposited and where the larvae develop. *Lissothuria nutriens*, on the other hand, has several depressions on the dorsal surface of the body, which are known as brood pouches. Here the eggs can develop within the protection of the parent's body.

The means by which the eggs are deposited in the brood pouches is quite remarkable. The eggs are laid on the tentacles, to which they adhere. The tentacles are laid on the dorsal body surface. There the eggs are transferred to the podia, which place them into the brood pouches. Species of the genera *Cucumaria* and "*Thyone*" have only two brood pouches instead of several, although these two pouches are larger. They develop as deep invaginations on either the dorsal or ventral body surface, and open close to the anterior end of the body. Up to 140 young sea cucumbers have been found in such brood pouches.

The *Psolus* species seem to be particularly highly developed brooding forms, containing the only holothurians where one finds almost all the brood-care techniques mentioned above. In addition, these forms have developed a brood pouch of totally different origin. The dorsal surface of some *Psolus* species consists of a regular calcareous carapace constructed

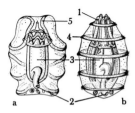

Fig. 12–9. Holothurian larvae: Auricularia (a) in transition to doliolaria stage (b). 1. Primary tentacle, 2. Anus, 3. Stomach, 4. Ciliated rings, 5. Ciliated bands.

of tightly joined ossicles up to 5 mm in length. The brood chamber develops through the elevation of several skin-covered plates by small stalks, which results in a pillar-supported "vault" between body wall and plates. The eggs and juveniles develop within this "nursery."

Depending on the amount of yolk present in the egg, sea cucumber ova may measure 0.1 to 3 mm in diameter. Further embryonic development is closely related to the egg's yolk content. In holothurians that produce a great number of eggs with a low yolk content, the embryological development passes through two free-swimming larval stages. The first stage is the auricularia larva (see Color plate, p. 272), which feeds on plankton, and the second, the barrel-shaped doliolaria larva, which does not feed at all.

Embryology

The bilaterally symmetrical auricularia larva represents a growth stage during which the larva reaches several times its original body volume. The largest known auricularia (of still-unknown affiliation) measures 15 mm. Usually, however, this larval stage does not measure more than 2 mm. The auricularia is characterized by a long, ciliated band that makes several loops around the larva. With the aid of this ciliated band, the animal is able to swim and to trap food. After three to four days of this stage, the bilaterally symmetrical auricularia changes into a radial (5 rays) doliolaria larva. This larval stage appears barrel-shaped, and, in contrast to the auricularia, has only simple ciliated bands which encircle the body and aid in the rotary method of locomotion. The five-rayed body plan reveals the presence of five growing radial canals and five primary tentacles. These initial tentacles penetrate slightly from the vestibulum where the tentaculate anlage originated. During this time period the larva sinks to the bottom.

The appearance of the first pair of podia on the posterior end signals the beginning of the pentactula stage. At this point the sea cucumber is fully differentiated and has merely to increase in size. After the pentactula larva has shed its ciliated bands, it begins to take up food. It crawls about on its podia and tentacles. The latter are instrumental in food capture. In those species that produce few, but very yolky, eggs and that practice brood care, the auricularia stage or even both free-swimming stages, i.e., auricularia and doliolaria, are eliminated. Development is more or less direct up to the pentactula stage. In *Holothuria floridana*, for example, the fully developed pentactula hatches out of the egg shell five days after fertilization. The entire embryonic development takes place within the egg.

Little is known about growth rates in sea cucumbers. However, the growth rate for *Stichopus japonicus* has been determined. This species grows in length from 4 mm to 20 mm within one month. At the end of one year the animal measures 25 cm. It is sexually mature in the third year and probably has a life span of at least five years. *Paracaudina chilensis* requires three to four years before it is fully grown.

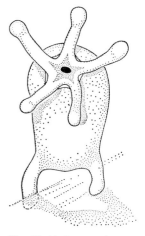

Fig. 12–10. Pentactula stage of a young sea cucumber.

Commensals and parasites

Holothurians have to live with many different parasites and commensals. Ciliates may occupy the oral region, the surface of the intestines, or the inside of the respiratory trees. Gregarine species may swim about freely in the coelomic cavity or may be embedded in the intestinal tract and the hemal system. Flatworms prefer the intestine and coelomic cavity.

The larger and therefore more conspicuous bivalves and gastropods also infest holothurians. These former organisms are found on the external surface or in the intestine, where they pierce the host's tissue and suck its body juices. Certain polychaetes also live in association with holothurians. The pharynx of certain holothurians is occupied by a copepod, and the cloacal region frequently contains small pinnotherid crabs.

Intestinal fishes

The most peculiar parasites to invade the holothurians are fishes of the genus *Carapus* (see Vol. IV, Chapter 18). The juvenile forms of these fishes enter the cloaca of the sea cucumber, and then penetrate through the walls of the respiratory tree into their host's voluminous coelomic cavity. *Carapus* feeds on the host's gonads and parts of the respiratory tree. The young fishes invade the sea cucumber's cloaca headfirst, while the adult fishes enter tailfirst. These fishes do not always return to the same host but may enter other holothurians as well. Adult *Carapus* fishes primarily use their former hosts as an escape location to which they retreat in case of danger.

Aside from the parasites mentioned, it would appear that the holothurians do not have any major specific enemies. There have been reports that sea gulls occasionally have fed on sea cucumbers along the shallow coastal waters in the intertidal zones, but sea cucumbers do not seem to be the "favorite dish" of sea gulls. Certain bottom-dwelling fishes also feed on the lethargic sea cucumbers. On occasion it has been observed that crabs, larger gastropods (*Charonia*; see Chapter 5) or Tonnoidea (see Chapter 5), and certain starfishes attempted to feed on holothurians, but frequently with little success.

Trepang

Along certain coastal stretches, man certainly is these animals' most formidable enemy. In the Indian and Pacific Oceans, primarily in southern Japan, the Philippines, and Indonesia, approximately twenty-five to thirty species from the family Holothuriidae are made into trepang and are eaten. Trepang is a Malayan word, and it refers to cooked, dried, and smoked sea cucumbers, which are considered as delicacies. *Microthele nobilis* is particularly favored. It goes by the commercial name of "mammy fish" or "teat fish." The holothurians *Actinopyga echinites* and *Actinopyga mauritiana* are known as "red fish." *Thelenota ananas* is sold as "prickly fish." The Japanese use *Stichopus japonicus* for making trepang. This sea cucumber is common along the Japanese coast line.

Sea cucumbers are gathered, cut open, and cleaned. The prepared animals are usually boiled to desalt them, and then are dried in the sun.

Often they are also smoked. Well-prepared sea cucumbers have the epidermis and outer layer with the calcareous ossicles removed. Trepang is generally used as a soup base. The dried muscular strands in the sea cucumber's skin are cut into strips or cubes. During boiling, these strips swell greatly, and take on a transparent, slimy, tapiocalike consistency. Samoans eat trepang in the raw state, and in the Philippines it is frequently roasted. Trepang has a protein content of fifty to sixty percent, and therefore is a high quality food. This dish does not appeal greatly to western palates. China annually imports approximately 3000 t of trepang at a value of 5 to 6 million dollars.

The holothurians have certain poisonous substances in the mucous secretions of their skin, which affords excellent protection against eventual enemies. Also, many holothurians, when attacked, can distract their attacker by ejecting their digestive tract, and in particular the Cuvierian tubules. One poison that has been isolated from the West Indian species *Actinopyga agassizi* is known as holothurin. This substance is a steroid-glycoside with saponinlike properties, which means that it is highly hemolytic. At least thirty additional species from four holothurian orders investigated so far have poisonous substances which are lethal to fishes, yet are harmless for man as long as they do not enter his blood stream. It is not known if all thirty species studied have the poison holothurin. Poisonous substances are destroyed during the process of making trepang.

Holothurin

People living along the coasts of the Indian and Pacific Oceans have long been familiar with the lethal properties of the holothurians. Fishermen make a practice of putting mashed or cut-up holothurians into coral reef pools or into the open water. Fishes present in the poisoned water come to the surface in a paralyzed condition, and can easily be gathered. Sea cucumber poison supposedly also harms sharks, and on occasion has been used in controlling them.

Many species of the genera *Holothuria* and *Actinopyga* eject long whitish or reddish threads from their anus when threatened. These threads are unusually tough and sticky and can stretch twenty or thirty times their original length. A predator usually gets entangled in this extremely adhesive thread mass. Struggling merely results in greater entanglement, thoroughly distracting any enemy from additional attacks. These tubules are known as Cuvierian tubules, and have a diameter of 2 to 3 mm. Cuvierian tubules are special attachments at the base of the respiratory trees. Large specimens may have up to 150 organs of Cuvier. Their ejection involves a powerful contraction of the entire skin musculature and pressure produced by a push of water from one of the respiratory trees. The tubules are ejected through a preformed rupture line along the cloacal wall and the anus.

Adaptations for defense

Cuvierian tubules

Most holothurians, however, do not have this defensive mechanism. When severely threatened, these sea cucumbers may achieve a similar reaction by forcing out the gut and associated organs. If this happens,

Evisceration

the digestive tract and respiratory trees will slowly creep about outside the body for some time. These organs are also sticky, and can entangle an enemy. Not all holothurians force their gut through the anus. Some species eject their viscera through the ruptured body wall. Evisceration is not only a response to mechanical stimuli but also to unfavorable environmental conditions, such as a rise or fall in temperature, or foul water. Some species regularly eject their viscera at specific times of the year without any apparent sudden external stimulus. This has been known for a long time about *Stichopus regalis*, which is distributed in the Mediterranean and the Atlantic. Almost all specimens caught in October did not have any viscera. The same has been observed for *Parastichopus californicus*, which ejects its internal organs during the late fall.

True self-mutilation, whereby whole body sections including the body wall are severed, has been observed in members of the family Synaptidae. If one grasps these animals, or when living conditions become unfavorable, they constrict several body sections and break into pieces. All sections die except the anterior end, which contains the tentacles. The animal regenerates from this anterior portion. Self-mutilation can also function as a defensive mechanism. The predator is left with a section of its prey, and, if it does not eat the front end, the survival of the animal is ensured.

Regeneration of lost body parts

If the digestive tract has been shed or ejected, regeneration starts at the remnants of the mesentery or the stubs of esophagus and cloaca. Tropical holothurians can regenerate their entire digestive tract within nine days. *Stichopus regalis* takes two to three weeks, and *Parastichopus californicus* takes approximately three months to regenerate its viscera.

Asexual reproduction

Some species of the genera *Cucumaria* and *Psolus*, as well as *Holothuria parvula*, *Holothuria surinamensis*, and probably *Holothuria difficilis*, can reproduce by transverse fission. A segment will regenerate the missing part and will grow into a fully developed individual by the asexual method. Segmentation, as has been shown in the self-mutilation process of the synaptids, is not related to asexual reproduction. Here only the front end with the tentacles regenerates. There is no further reproduction.

Distribution

Holothurians are strictly marine. In this environment they have occupied almost all possible habitats. Only a few sea cucumbers can tolerate brackish water, including *Thyone*, which is found near estuaries, or *Thyonidium pellucidum*, which occurs in regions of the North Sea and Baltic Sea that have a lower salinity concentration than the average. This group also includes *Protankyra similis*, which is found in mangrove swamps.

As with the feather-stars, most of the species of holothurians are from the Indo-Pacific area. Some species found in this area enjoy worldwide distribution in the tropical and subtropical oceans. Representatives of the order Aspidochirotida, which are found in shallow coastal waters, and which usually are large forms with shieldlike tentacles, greatly decrease in numbers toward the colder, northern latitudes. In these regions the plankton-feeding members of the order Dendrochirotida and the Synaptidae increase in numbers.

Off Africa's eastern coast one generally encounters the same holothurian species that are common in the West Pacific area. South Africa, however, is an exception, because its coastal holothurian fauna is unique. A parallel situation is found in New Zealand. Its holothurian fauna is unlike that of Australia.

Just as in the Indo-Pacific region, there is a reduction in the number of species in the more northern latitudes in the Atlantic Ocean. The only area in the Atlantic influenced by tropical conditions is the West Indian-Caribbean region, which has a characteristic fauna. Within this area the large species of the order Aspidochirotida predominate. There are few sea cucumbers in the North Atlantic that are found on the western European as well as on the eastern North American coasts. Some of these species are *Cucumaria frondosa* and *Leptosynapta inhaerens*. In the Mediterranean one finds many species that are also common to the Atlantic; but some species are indigenous only to the Mediterranean, such as *Holothuria tubulosa*.

The polar regions have a unique holothurian fauna characterized by cold-adapted forms. Some of these species are otherwise found only in deeper zones in the ocean. With the exception of two very closely related *Psolus* species, the Arctic and Antarctic do not have similar holothurian faunas. The South Polar regions are characterized by particularly interesting holothurians. Not only are there many more holothurians around Antarctica than in the Arctic, the antarctic forms are very interesting biologically. For example, there are more than fifteen different brood-caring species around the Antarctic.

An oceanographic expedition in the late 19th Century discovered, among other fascinating surprises, that a great number of holothurian species formed a large proportion of the deep-sea fauna. When evaluating the data of modern deep-sea expeditions, it was found that at a depth of 4000 m holothurians comprised fifty percent of the living organisms, while at 8500 m they made up ninety percent of the mass of all living organisms. The dredging operations of the *Galathea* Expedition during 1950–1952 in the abysses of the world's oceans proved finally that the holothurians are among the few animals that can penetrate to the greatest depths and survive there. One entire group of bizarre-looking sea cucumbers is almost exclusively limited to great depths. These are representatives of the order Elasipodida. Other abyssal sea cucumbers belong to the aspidochirotid family Synallactidae, and are found primarily at depths between 1000 and 5000 m, as are a few species from the orders Molpadida and Apodida. *Myriotrochus bruuni* belongs to the order Apodida. This sea cucumber, according to the *Galathea* Expedition research, was found to live at the greatest depth, in a region of the Philippine Trench at 10,200 m.

Depth distribution

One wonders what animals at these desolate, sparsely inhabited deep-sea trenches and plateaus use as food. Modern deep-sea research has shown, however, that the deep-sea floor is not so desolate as had been assumed. The water layers above the ocean bottom release sufficient organic debris to support the growth of bacteria and nematodes. Animals that feed on the ocean substrate, such as the holothurians, ingest these bacteria and nematodes in turn.

Density of animals

Exact data concerning the population density of holothurians is available in only a very few cases. Nevertheless, there are some reports of more or less dense concentrations in specific regions. In sandflats of an area of 5000 m² in shallow water along the coast of Bermuda, 675 *Isostichopus badionotus* were counted. This means that each holothurian of this species occupied an area averaging 7–8 m². *Cucumaria curata* and *Holothuria glaberrima* are frequently found crowded together. Several dozen animals may lie side by side. *Trachythyone elongata*, which inhabits certain soft bottoms of the North Sea, may reach a density of four or five animals per m² in certain areas. The density of holothurians in a certain area is greatly determined by the nutritional value of the substrate.

The high percentage of holothurians living at the bottom of the deep seas has already been mentioned. If one considers that a single dredging operation in the Sunda Trench, at a depth of 7150 m, yielded approximately 3000 individuals of *Elpidia glacialis*, then it must become evident that this is an appreciable population density under such extraordinary living conditions.

13 The Echinoids

The ECHINOIDS (class Echinoidea; see Color plates, pp. 333 ff. and 345 ff.) are globose, heart-shaped, or disc-shaped echinoderms whose oral side faces the substratum. The body is enclosed by an endoskeletal test or shell usually composed of immovable fused plates. These skeletal plates bear spines and pedicellariae, and are arranged in ten double rows extending from the mouth on the oral surface to the center of the aboral surface. Each double row alternates with double columns of ambulacra and interambulacra. The ambulacral plates are distinguished by having pores, which the interambulacral ones lack. The largest echinoid (*Sperosoma giganteum*) attains a test diameter of 32 cm. Coloration in many species is dark, from black to dark brown or olive-brown, reddish-brown, purple, violet; others are lighter in color, including muddy yellow, green, and white. Aside from the uniformly colored species there are numerous species that are patterned with light or contrastingly colored banded spines. The test may have blue, white, or red stripes and spots (see Color plate, p. 336). There are two subclasses: 1. Perischoechinoidea, with four orders and eleven families, 2. Euechinoidea, with eighteen orders and eighty-three families.

Nearly half of the echinoid species are characterized by a more or less flattened apple shape. The mouth is located in the center of the lower (ventral) body surface, and the anus is in the center of the upper (dorsal) surface. Echinoids that fulfill this description are known as "regular" sea urchins. The other half of the echinoids have undergone changes (after metamorphosis) in the regular body structure. In these forms the anus has shifted from the center of the upper surface to the outer margin of the lower surface. This shift has resulted in a secondary bilateral symmetry. In this connection there is almost always an elongation in the animal's body in the direction of the anus, so that the animal has an oval outline (see Color plate, p. 348). Sea urchins characterized by such a body shape are known as "irregular" echinoids. In contrast to the regular sea urchins, which can move forward with any side of their body, the irregular sea urchins are

Class: Echinoidea, by H. Fechter

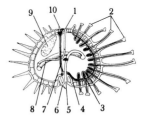

Fig. 13–1. Schematic section through a sea urchin: 1. Anus, 2. Tube-feet, 3. Ampullae of tube-feet, 4. Radial canal of water-vascular system, 5. Mouth, 6. Tooth, 7. Stone canal, 8. Intestine, 9. Gonad, 10. Madreporite.

Fig. 13-2. Sea urchin cut through the middle, exposing the two halves: 1. Tube-foot, 2. Intestine, 3. Gonad, 4. Hindgut, which leads to the outside via the anus, 5. Stone canal, 6. Siphon, 7. Esophagus, 8. Row of podial ampullae, 9. Masticatory apparatus.

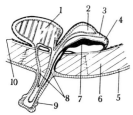

Fig. 13-3. Cross section of the body wall of a sea urchin in the region of a tube-foot: 1. Podial ampulla, 2. Radial canal of water-vascular system, 3. Perihemal canal, 4. Epineural sinus, 5. Epithelium, 6. Ossicle, 7. Ectoneural nerve strand, 8. Podial pores, 9. Tube-foot, 10. Coelomic lining.

characterized by definite anterior and posterior ends. The anus is located posteriorly and the anterior end is opposite. In the heart urchins not only the anus is at the posterior end; the mouth has also moved from the center of the ventral surface toward the anterior end. In addition, these animals have a groovelike depression in the anterior end of the test, giving these echinoids a heart-shaped appearance. Many of the irregular sea urchins have a strongly flattened, disc-shaped body. They are commonly referred to as sand dollars (see Color plate, p. 348). Many sand dollars have lobed test margins, while others are characterized by slitlike openings (lunules) in the test margin (see Color plate, p. 348).

As is characteristic for the echinoderms, the body wall is made up of three layers: the external epidermis, a dermis of connective tissue, and a coelomic lining. The epidermis and coelomic lining are almost always ciliated. The epithelium covers all external appendages of the body wall, but is rubbed off at the tips of the spines. The coelomic lining covers the interior of the body wall and the organs within the coelom.

The well-developed connective-tissue layer contains the supporting endoskeleton, composed of calcareous plates. In the primitive Echinothuriidae and a few Diadematidae the plates can rub against each other, but otherwise they are fused and immovable. The plates bear small hemispherical tubercles forming a ball-and-socket joint with the spines. The spine has a milled ring at its base which serves for attachment of muscles which originate from a marginal depression around the tubercle. This musculature consists of an external layer of ordinary, quickly responding smooth-muscle fibers, and an inner layer of "catch" muscles that hold the spine erect for longer periods of time (Fig. 13-4).

Usually the spines vary greatly in size and shape. Spines are classified into primary and secondary types. Primary spines are long and better developed than the others. In certain species these spines can measure over 30 cm, which is several times more than the diameter of the shell. The secondary spines are short, and at times look like scaly platelets. They may function as a protection for the muscles around the base of the primary spines (see Color plate, p. 336), as in the Cidaroida. Aside from the bulky and stout spines of *Heterocentrotus mammillatus* and *Cidaris cidaris* (see Color plate, p. 336) and the knitting-needlelike long spines of *Diadema* (see Color plate, p. 336), there are short, prismatic spines, as illustrated by *Colobocentrotus* (see Color plate, p. 336). The sand-burrowing irregular echinoids would be hindered by long spines; therefore their spines are short and bristly, and lie flat against the body. These spines are not rigid, as are those in the regular sea urchins (see Color plates, pp. 346 and 348).

In between the large and small spines, little stalked spines project out of the body wall. These are known as pedicellariae. They consist of a head, usually composed of three jaws which can open and close. There are many different types of pedicellariae that perform different functions. The tridentate is characterized by long, slender blades, and the ophicephalous by broad, leaflike blades with serrated edges. The functions of these pedicellariae are not clearly understood, but they are used to discourage intruders. The triphyllous pedicellariae have small, broad, unserrated blades that function in keeping the body surface free of settling particles. The globiferous (gemmiform or glandular) pedicellariae have poisonous glands that serve as defensive organs against predators. The external surfaces of the blades bear contractile poisonous glands which discharge through the tip of the toothlike blade. Certain glandular pedicellaria have three additional globose poisonous glands near the neck of the stem, with ducts lying against the blades (Figs. 13-7, 13-8).

The echinoids also have five double rows of perforated plates arranged as meridians between oral and aboral poles. These radii bear the tubular and highly movable podia which usually terminate in a suction cup. The sea urchins use the podia for adhering to the substratum and pulling themselves up walls. The extremely flexible podia have a long range of extension and contraction. The ambulacral plates bearing the podia are characterized by a pair of pores at the base of each podium. These pores serve as the terminus for the lateral branches of the water-vascular system, which unite into a single canal on the outer surface of the test to form the lumen of the podium (Fig. 13-11).

In the regular sea urchins the ambulacral plates and their podia are arranged in five double rows as meridians between the oral and aboral poles. The plates are approximately of equal size, but decrease in size toward the poles. The double rows of podia alternate with a double row of interambulacral plates. In the irregular sea urchins, however, the aboral ambulacral area is petallike in arrangement (petaloid; Fig. 13-16). The oral ambulacral plates, known as phyllodes, are leaflike in arrangement. In the sand dollars the interambulacral plates frequently bear podia.

The water-vascular system

Basically the structure of the echinoid water-vascular system follows the general echinoderm scheme. The water ring surrounding the esophagus gives rise to five radial canals that run underneath the test and ascend to the center of the aboral surface. Each radial canal gives off lateral branches that enter each podium through two small canals. Prior to penetrating through the skeletal plates, the lateral canals widen to form a pocketlike ampulla controlled by muscular action, fulfilling a function similar to that in the sea cucumbers (see Chapter 12). A stone canal ascends from the ring canal to the aboral surface, where it is covered by a madreporite (Fig. 13-1).

Digestive system

The mouth is located in the center of the membranous peristome on the oral surface. It is surrounded by oral podia. The mouth is often provided with bulging, circular lips, from which projects a cone of five teeth which meet at their tips. This is found in the regular sea urchins and sand dollars. The projecting end of each tooth is only the tip of an ever-growing structure, each of which is anchored to its own wedge-shaped jaw apparatus. These five jaws surround the esophagus. Muscular action integrates

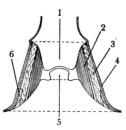

Fig. 13–4. Longitudinal section through the basal part of a sea urchin's spine: 1. Concave proximal part of the spine, 2. Cog muscle (acts like a clamp), 3. Outer muscle (moves the spine), 4. Epithelial layer, 5. Mamelon (central part of the tubercle which articulates with the proximal part of the spine), 6. Nerve ring.

Aristotle's lantern

the function of this complex opening-and-closing masticatory apparatus. Pliny the Elder (A.D. 23–79) called this apparatus "Aristotle's lantern" (Figs. 13-12, 13-13); it still bears that name.

In the sea urchins the esophagus leads into a long intestinal tract which makes two opposing horizontal loops. In the first circuit a siphon runs parallel to it up to the point of doubling (to the second part of the gut). The rectum ascends to the anus, located on the periproct at the aboral surface on a small, plated anal cone.

Nervous system

In the echinoids only two nervous systems are well developed. Since the skeletal musculature is missing, the hyponeural system is correspondingly greatly reduced and is limited to the masticatory apparatus. The ectoneural system is the best developed. It follows the same pattern as the water-vascular system, running between the radial canal and the coelomic lining. This system innervates the tube feet and forms a nervous network below the skin. The aboral nervous system is situated within the gonoduct, from which it sends out branches to the gonads. The subepidermal plexus terminates in many sensory cells. These cells are highly concentrated along the inside of the blades of the tridentate pedicellariae, the tube-feet, and the spines. These areas are receptive to mechanical, chemical, and optical stimuli. Sensory organs are absent in echinoids.

Oxygen uptake takes place at all thin-walled regions of the body surface. The numerous tube-feet are particularly well suited for this function. This oxygen is passed to the gill-shaped ampullae via the podial epidermis, and from the ampullae to the coelomic fluid. The cilia of the coelomic epithelium keep this fluid in constant motion, and the dissolved oxygen thereby comes in contact with the various organs. In many regular sea urchins the peristomial membrane is surrounded by ten slits in the shell, through which ten bushy epidermal sacs can be pushed. Gas exchange takes place in these areas, and these sacs are referred to as gills (peristomial gills).

An adequate supply of oxygen poses a problem for the burrowing irregular sea urchins. To adapt to this condition, the tube-feet projecting from the rosette have become greatly enlarged. These structures resemble leaves, and have been modified into gills. Freshly oxygenated water is constantly fanned to the gills by the action of cilia lining the body surface, and by small paddle-shaped spines which are densely covered by cilia. These spines surround the rosette of petaloid ambulacra, or are located within a ring (fasciole) inside the latter (Figs. 13-14, 13-16).

The intestinal tract is associated with a nutrient-gathering plexus of hemal lacunae. The hemal fluid from the lacunae concentrates in ventral and dorsal marginal sinuses located respectively on the ventral and dorsal sides of the intestine. The sinuses end in a hemal ring around the esophagus. The inner sinus leads directly into the hemal ring, but the outer one follows the axial organ extending along the stone canal. Sections of the axial organ and certain segments of the sinuses pulsate, thereby circulating the hemal fluid. Radial vessels branch out from the ring lacuna which follow the scheme of the water-vascular system. The radial vessels supply the podia. The aboral end of the pulsating axial organ is in contact with the aboral hemal ring which supplies the sexual organs.

Echinoids are dioecious. Very rarely can one differentiate externally between males and females. The gonads are long saccate structures that develop underneath the interambulacral plates and hang suspended into the coelomic cavity. Five gonads are always developed in the regular echinoids, but in the irregular echinoids there are usually only four gonads, and occasionally only two or three because of the migration of the anus (periproct). The gonoducts always terminate in an interambulacral plate of the apical system at the aboral end. This plate is therefore known as the genital plate.

Almost all members of the subclass PERISCHOECHINOIDEA became extinct during the Paleozoic era. Only the order Cidaroida, with five families (three of them extinct), survived into present times. These forms are characterized by simple ambulacral and interambulacral plates that cover the oral surface right up to the mouth opening. Each interambulacral plate bears one well-developed spine, with a correspondingly well-developed tubercle in the test.

One well-known representative of this group is *Cidaris cidaris*. The diameter of the test is up to 6.5 cm; height 4 cm. The thick primary spines of the interambulacral plates measure up to 13 cm. The secondary spines are small and flat, forming a protective ring around the base of the primary spines and also covering the rows of tube-feet. Coloration is a dirty yellow, greenish, or reddish. The bases of the spines are pink. The species is distributed in the Mediterranean and Atlantic from Norway to Cape Verde, at depths of 50 to 2000 m. *Stylocidaris affinis* is similar to *Cidaris*, but is somewhat smaller. Its primary spines are gray and covered with thorns. The secondary spines are reddish. It is very common in the Mediterranean and the eastern Atlantic, occurring at depths of 30 to 1000 m.

All other echinoids that evolved after the Paleozoic belong to the subclass EUECHINOIDEA, which includes the great majority of today's echinoids. The ambulacral and interambulacral plates are always arranged in five double rows. There are four superorders: 1. Diadematacea: regular echinoids lacking the longitudinal keel on the inside of the teeth. There are four orders. 2. Echinacea: regular echinoids with the longitudinal keel on the inside of the teeth; seven orders. 3. Gnathostomata: irregular echinoids with a masticatory apparatus present throughout life; two orders. 4. Atelostomata: irregular echinoids lacking the masticatory apparatus in the adult stage; four orders.

The three surviving orders of the superorder DIADEMATACEA are characterized in the following manner:

a) Order Echinothurioida: the skeletal plates overlap like shingles,

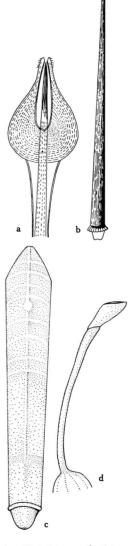

Fig. 13–5. Types of echinoid spines: a. Club-shaped poisonous spine of an echinothuriid, b. Common primary spine of a sea urchin, c. Primary spine of *Heterocentrotus mammillatus*, d. Primary spine of the oral side of an echinothuriid. The tip is broadened, resembling a horse's hoof.

and are movable against each other. The ambulacral plates extend over the oral surface. The spines are hollow. There is a single family: Echinothuriidae.

b) Order Diadematoida: the ambulacral plates do not extend over the oral surface; the spines are hollow. There are four families.

c) Order Pedinoida: the ambulacral plates do not extend over the oral surface; the primary spines are solid and the secondary ones are hollow. The single family, PEDINIDAE, is primarily represented by deep-water forms distributed in the Indian and Pacific Oceans.

The family Echinothuriidae is characterized by ten muscular fans that extend into the coelomic cavity from the test, like septa. The action of these muscles can cause the body to be blown up like a globe or flattened like a cake, and can cause undulations in the body wall. The animals are frequently very colorful and equipped with poisonous pedicellaria (Fig. 13-5, a). They are primarily deep-water inhabitants. The species *Asthenosoma varium* (see Color plate, p. 336) is also found in the tidal zone in the Indo-Malayan archipelago.

The species of the family ASPIDODIADEMATIDAE belonging to the order Diadematoida are characterized by long, slender, outrigger-type spines, curved downward, and some are broadened, horseshoelike, at the ends (see Color plate, p. 336). The widely distributed species *Diadema setosum* (see Color plate, p. 336) represents the Diadematidae in the Red Sea and the Indian and Pacific Oceans. The extremely long, needlelike spines of the Diadematidae bear little thorns. The animal is black with white spots, numerous blue dots, and an orange ring around the anus. The diameter of the test is up to 9 cm. This echinoid often occurs in great numbers around coastal areas. The only European species of the Diadematidae is *Centrostephanus longispinus*. Its test diameter is 4 to 6 cm. The brownish-violet animal with its lighter-banded spines inhabits the Mediterranean and Atlantic at depths of 40 to 200 m. This group also includes the genus *Echinothrix*. The family Micropygidae consists of only two deep-water species, inhabiting the Indo-Malayan region. The family Lissodiadematidae includes shallow-water species, also from the Indo-Malayan region and Hawaii.

The superorder ECHINACEA is classified in the following manner (extinct order Hemicidaroida omitted):

a) Order SALENIOIDA: one or more large, polygonal plates in the periproct. Each interambulacral plate bears one long, slender, thorny spine. One family: SALENIIDAE.

b) Order PHYMOSOMATOIDA: lacking large, angular plates in the periproct; interambulacral plates each also have one large spine, with a corresponding large tubercle. Two families: PHYMOSOMATIDAE and STOMECHINIDAE, each with only one species.

c) Order ARBACIOIDA: four or five large plates in the periproct. One family: ARBACIIDAE. This group includes *Arbacia lixula* (test diameter up to

Family:
Echinothuriidae

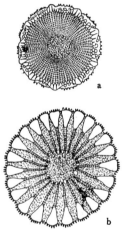

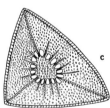

Fig. 13–6. Cross sections through various echinoid spines: a. *Cidaris cidaris*, b. *Arbacia*, c. *Coelopleurus*.

5.8 cm, height 2.5 cm, spines equally long), frequently found on hard substrates exposed to the surf. In contrast to other sea urchins, *Arabacia* does not cover itself with foreign particles. It moves about at night and likes to feed on calcareous algae. Another species: *Coelopleurus floridanus*.

d) Order TEMNOPLEUROIDA: the aperture in the masticatory apparatus is bridged by bars; the surface of the test is characterized by numerous depressions or by widely notched gill slits along the peristomial margin. Two surviving families. Some representatives of the family TEMNO-PLEURIDAE are the smallest sea urchins, with test diameter 6–7 mm. These forms usually bear short, lightly banded spines, and are primarily inhabitants of tropical oceans. In the family TOXOPNEUSTIDAE the test is unsculptured but has deep gill slits. This group includes *Sphaerechinus granularis* (test diameter 13 cm; see Color plate, p. 345), which is a deep purple. The short, stout spines have white tips. This echinoid is found in the Atlantic from the English Channel to the Gulf of Guinea at depths of 3 to 100 m. In some regions it is very common. *Lytechinus variegatus* (test diameter 8.5 cm) usually has a greenish-white coloration, but is also represented by brownish and purple forms. *Toxopneustes pileolus* (test diameter up to 15 cm), distributed in the Indo-Pacific region, has extremely large poisonous pedicellaria. This family also includes *Tripneustes gratilla*.

e) Order ECHINOIDA: the aperture in the masticatory apparatus is bridged by bars; there are no depressions in the surface of the test; the gill slits are shallow and often indistinguishable. Four families: ECHINIDAE, the most frequently found sea urchins along the European coasts, includes the edible sea urchins. *Echinus esculentus* (see Color plate, p. 336; compare Color plate, p. 333) is found from Norway and Iceland to Portugal, from the ocean's surface to a depth of 1200 m. The largest test diameter is 17.5 cm. This sea urchin is flesh colored or green. The most frequently found European sea urchin is the common sea urchin *Paracentrotus lividus* (test diameter up to 7 cm). This dark purple, brownish, or greenish sea urchin is distributed from Ireland to the Canary Islands and the Mediterranean from the ocean's surface down to 80 m.

The second family of Echinoida, the ECHINOMETRIDAE, frequently have an oval test. The following genera and species typify this family: *Echinometra mathaei* (purple, brownish, or greenish; the largest test diameter is 7.3 cm); *Heterocentrotus mammillatus* (see Color plate, p. 336), found in the Indian and Pacific Oceans, with spines that are 8–12 cm long and 1.3 cm thick; *Colobocentrotus* (see Color plate, p. 336), with conical spines; *Echinostrephus*. The only living genus of the family PARASALENIIDAE is *Parasalenia*, with two species in the Indian and western Pacific Oceans. The family STRONGYLOCENTROTIDAE includes the species *Strongylocentrotus droebachiensis*, widely distributed in the colder northern oceans. The purplish-colored species *Strongylocentrotus purpuratus* and *Strongylocentrotus franciscanus* are characteristic of the North American coast of the Pacific Ocean. Their test diameters are 18 and 25 cm respectively.

▷
The sea urchin *Echinus melo* (compare Color plate, p. 366) inhabits the Atlantic and the Mediterranean.

▷▷
Eucidaris tribuloides is found in West Indian waters as well as in the eastern Atlantic coastal regions around the Cape Verde Islands and the Gulf of Guinea.

▷▷▷
Strongylocentrotus purpuratus, from the coast of California.

MILLA

With the superorder GNATHOSTOMATA we introduce the irregular sea urchins. One of the two orders of this group, the HOLECTYPOIDA, forms the transition between the regular and irregular sea urchins. These oval sea urchins lack the petaloid ambulacra (petals) on the aboral surface. The ambulacral plates on the oral surface are smaller than the interambulacral plates. There is only one surviving family, ECHINONEIDAE, with four living species.

The second order, CLYPEASTEROIDA, includes the forms known as SAND DOLLARS. As the common name suggests, these echinoids are disc-shaped and flattened sea urchins. They have five petals. The spines are very short and very dense. The mouth is in the center of the oral surface, where the ambulacral plates are broader than the interambulacral ones. We differentiate between nine surviving families:

The CLYPEASTERIDAE, with the genus *Clypeaster*, are represented by many species that inhabit the sandy bottoms of the warmer oceans; they are not present along European shores. ARACHNOIDIDAE has a limited distribution around Australia, New Zealand, and the Indo-Malayan region. The LAGANIDAE are sand dollars with thick margins, found in the shallow waters of the Indian and western Pacific Oceans. The ASTRICLYPEIDAE have lunules in the shell. *Echinodiscus auritus* is a typical species of the Indian and western Pacific Oceans. It has two lunules on the posterior test margin. The DENDRASTERIDAE include *Dendraster excentricus*, a very common sand dollar species along North America's Pacific coast. The ECHINARACHNIIDAE includes *Echinarachnius parma* (see Color plate, p. 348), which is very widely distributed along North America's Atlantic coast. The ROTULIDAE are characterized by a lobed posterior test margin. *Rotula orbiculus* (test diameter 7.5 cm; see Color plate, p. 348) occurs along Africa's western coast. The so-called "KEYHOLE SAND DOLLARS" (family Mellitidae) are characterized by more than two lunules. *Leodia sexiesperforata* (test diameter 9 cm, height 0.6 cm; see Color plate, p. 348) has six lunules. This species is distributed in the western Atlantic from South Carolina to the Rio de la Plata.

Within the family FIBULARIIDAE there is a noteworthy exception among the irregular echinoids. This is *Echinocyamus pusillus*, which, in contrast to other irregular echinoids, usually does not dig itself in, but rather pushes itself in between coarse objects lying on the ocean floor, covering its exposed side with foreign objects. This gray, oval echinoid has a test diameter of 8–10 mm and a height of 4 mm. It inhabits the European coast from Norway to northwestern Africa, and also the Mediterranean. This form is found on sandy and gravel substrates from the ocean's surface down to 1250 m.

The last superorder, the ATELOSTOMATA, is also composed of irregular echinoids. This group is subdivided into four orders:

a) CASSIDULOIDA: the ambulacral plates near the mouth are arranged in a leaflike arrangement (phyllodes); the interambulacral plates in the

◁
Echinoids:
1. *Prionocidaris baculosa*
2. Long-spined sea urchin (*Diadema setosum*; compare Color plate, p. 347)
3. Common or edible sea urchin (*Echinus esculentus*; compare Color plate, p. 333)
4. *Colobocentrotus pedifer*
5. *Heterocentrotus mammillatus*
6. *Asthenosoma varium* (seen from below)
7. *Plesiodiadema indicum*

proximity of the mouth are conspicuously arched. The genital plates are closely spaced together. The test is round or oval. Cassiduloids inhabit sand. Five extant families, represented by very few species: Neolampadidae, Apatopygidae, Pliolampadidae, Cassidulidae, and Echinolampadidae.

b) HOLASTEROIDA: the anterior and posterior genital plates are separated or are pushed forward greatly; at least one large unpaired interambulacral plate lies behind the lower lip. Four extant families, represented by very few species; primarily deep-water inhabitants. The families are Holasteridae, Urechinidae, Calymnidae, and Pourtalesiidae. The last families include some of the most aberrant irregular echinoids, such as *Pourtalesia jeffreysi*, which has a bottle-shaped test, 6 cm in diameter; purplish-violet; found at depths of 50 to 2450 m around Iceland, the Faeroes, and Spitsbergen.

c) HEART URCHINS (Spatangoida): genital plates are close together; the posterior margin of the lower lip borders on large, paired interambulacral plates. Represented by eleven living families, some containing only a few species, and some of limited distribution: Palaeopneustidae, Palaeostomatidae, Asterostomatidae, Aeropsidae, Toxasteridae, Hemiasteridae, Spatangidae, Loveniidae, Pericosmidae, Schizasteridae, Brissidae.

The VIOLET HEART URCHIN (*Spatangus purpureus*) is a member of the family SPATANGIDAE. This form is characterized by a large, heart-shaped test measuring up to 12 cm; notched at the anterior margin. There are four petals on the aboral surface. This species is purple or violet, with lighter colored, sometimes even white, spines. The heart urchin is buried in sand, mud, or gravel. It is distributed from North Cape (Norway) to Senegal and the Mediterranean at depths of 5 to 900 m.

The LOVENIIDAE includes the HEART URCHIN (*Echinocardium cordatum*; see Color plate, p. 348), the only echinoid enjoying a worldwide distribution. The heartshaped, anteriorly grooved test is characterized by four petals, and measures up to 9 cm. This grayish to yellowish-gray echinoid is found from the ocean's surface to a depth of 230 m. These animals often occur in high concentrations, buried in the sand.

The family SCHIZASTERIDAE is primarily distributed in subantarctic and antarctic waters, but some genera, such as *Schizaster*, *Moira*, and *Faorina*, occur in tropical areas.

Two species from the family BRISSIDAE shall be mentioned. *Brissus unicolor* is found in the Mediterranean and the Atlantic Ocean. Its test has a diameter of up to 13 cm, and is 5 cm high. The dark gray test is oval. The species inhabits sandy and muddy substrates at depths from the ocean's surface to 250 m. The gray or grayish-green *Brissopsis lyrifera* is very common. Test diameter up to 7.5 cm. The anterior body section is characterized by a groovelike depression. The species is found in the Mediterranean and Atlantic from Iceland and Norway to South Africa at depths of 5 to 1500 m, sometimes occurring in great concentrations, buried in the sand.

d) NEOLAMPADOIDA: the ambulacra are not in the forms of petals.

Fig. 13 7. Various types of pedicellariae: a. Triphyllous pedicellaria, b. Tridentate pedicellaria, c. Globiferous poisonous pedicellaria. 1. Valves with externally located poisonous glands, 2. Additional poisonous glands, 3. Stem support.

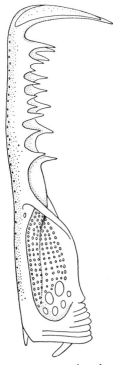

Fig. 13–8. A single valve of a poisonous globiferous pedicellaria of *Psammechinus miliaris.*

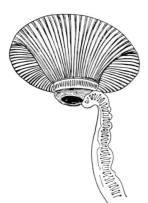

Fig. 13–9. A podium of *Micropyga tuberculata*, with a particularly large suction cup.

Two to four genital pores. One family, Neolampadidae. A poorly known group from depths of 130 to 1300 m.

A tendency to occupy extreme habitats is usually associated with special adjustments in the animal's body structure. This phenomenon also applies to the sea urchins. Regular sea urchins frequently inhabit hard substrates. There are noticeable differences between regular urchins that inhabit quiet bays and those that are exposed to strong surfs. The species inhabiting the quiet waters usually have domed tests in contrast to those in the surf zones, characterized by more or less flattened tests with short, strong spines and powerful tube feet on the aboral surface. The genus *Colobocentrotus* (see Color plate, p. 336) demonstrates the extreme extent to which such adaptations evolved in response to living in the surf zone of tropical coral reefs. The test of this genus is flattened and bears short spines of equal size, which are pyramid-shaped from the base up. The thick, blunted ends of the spines are in very close contact with one another, forming an additional armor. These flattened echinoids adhere to the hard substrate like suction cups.

Other echinoids have adapted to the danger of being rolled over the surface by strong water currents by an elongation and thickening of the lateral spines. This modification results in a shift of the center of gravity in relationship to the substrate, and also decreases the possibility of being tipped over. This situation is exemplified by the genus *Heterocentrotus* (see Color plate, p. 336), which has long, laterally thickened, club-shaped spines, and the sand-inhabiting species of the genus *Coelopleurus*, characterized by outriggerlike spines that bend downward.

However, much more comprehensive modifications in the body structure were necessary to enable the echinoids to inhabit the extensive sandy and muddy areas. These changes finally led to the completely new irregular echinoids, which in many ways are bilaterally symmetrical. Since the tube-feet are unable to anchor themselves in the loose texture of the sand, the animals would have been at the mercy of strong tidal surges. By burrowing into the soft substrate, they overcame this situation. Life beneath the ocean floor, and above all the directive movements within the sand, resulted first in a shortening of the spines, which no longer protrude in all directions, but are flattened against the test, almost all of them appearing to be "combed" in the posterior direction. In most cases the body became streamlined, facilitating better forward movement within the substrate. This leads to a situation where a specific body section always moves forward and thereby becomes the anterior end. Thus, in contrast to most regular sea urchins, the irregular ones have now lost the ability to move in every direction. Logically the irregular echinoid has also developed a posterior end and symmetrical body flanks. Since the substrate offers a sufficient degree of protection, the selective forces for the calcareous armature are decreased; consequently it becomes thinner and at times extremely fragile (Fig. 13-16).

The regular echinoids living on soft substrates, and the irregular ones living in it exclusively, use the spines on the aboral surface for movements. This makes sense if one considers that the tube-feet cannot gain a foothold in the loose ground. When moving, the echinoid repeatedly places the spines, on which it usually rests, in one direction. These animals move over the ground in this stiltlike manner. Depending on the length of the spines, short or long steps will result from the same degree of bending. The long-spined *Centrostephanus longispinus* undoubtedly belongs to those echinoids that can move the fastest. This Mediterranean and Atlantic species can cover a distance of 3.5 cm in one second, while shorter-spined species only manage 0.5 cm in the same time span.

When moving on a hard substrate, the regular echinoids use the tube-feet in addition to the spines. The podia are greatly extended, adhere, and pull the body toward them by contracting. The tube-feet become indispensable when navigating over slanted, uneven terrain. When the animal climbs up steep walls, the tube-feet alone take over the locomotory function and ensure a secure hold. *Psammechinus microtuberculatus* is decidedly a climbing expert. This species is specialized for dangling and climbing along thin objects, such as coral or vegetative stalks and branches. In an experiment it was shown that this echinoid could even climb along a violin string.

In the course of time, some regular echinoids form a depression in the substrate. These animals tend to retreat to these holes. The details of the burrowing activity are unknown, but the traces on the substrate seem to indicate that the animal uses its teeth in this process. The secretion of certain chemical substances to dissolve away the underlying material has to be discounted, since the echinoids dig themselves into a great variety of substrates. It is astonishing that these echinoids can erode not only on sandstone, calcareous stone, slate, or other soft rocks, but also hard lava, granite, and even iron pillars.

The purple sea urchin *Strongylocentrotus purpuratus* (see Color plate, p. 335), *Paracentrotus lividus*, *Psammechinus miliaris*, and *Heterocentrotus mammillatus* (see Color plate, p. 336) are among the best-known "diggers." In addition to these species, which are found primarily in the surf zones, another burrowing species, *Allocentrotus fragilis*, was recently discovered at depths from 100 to 150 m. Without doubt the two species of the genus *Echinostrephus* have the distinction of being the most proficient "stone borers," boring deep, cylindrical channels of lengths from 7 to 10 cm vertically into the substratum. These hollows, with smoothly polished walls, provide the animals with protection against predators. Usually the echinoid sits in the entrance of its channel, but when attacked it drops to the bottom and then extends a bundle of extremely long spines in the direction of the intruder (Fig. 13-19).

The irregular echinoids inhabiting the sand and mud carry out more extensive burrowing feats. When burrowing, the sand dollars push the anterior margin of their body through the substrate at a slight angle, and

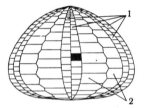

Fig. 13–10. Arrangement of the skeletal plates in a regular sea urchin: 1. Ambulacral plates, 2. Interambulacral plates.

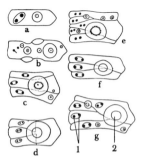

Fig. 13–11. Various types of ambulacral plates: a. *Cidaris cidaris*, b. Echinothuriid, c. *Diadema*, d. *Arbacia*, e. Polyporous *Diadema*-type, f. Simple, g. Polyporous echinoid type. 1. Pores through which the water-vascular system enters the tube-foot, 2. Mamelon of a primary spine.

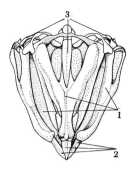

Fig. 13–12. Masticatory apparatus ("Aristotle's lantern") of *Strongylocentrotus droebachiensis*: 1. Pyramids, which enclose teeth for most of their length, 2. Pointed, feeding ends of teeth, 3. Growing ends of teeth.

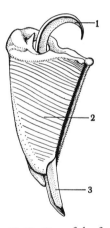

Fig. 13–13. One of the five jaws of the masticatory apparatus of *Strongylocentrotus droelachiersis*: 1. Growing end of tooth, 2, Pyramid, 3. Tooth.

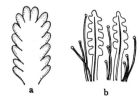

Fig. 13–14. Respiratory podia of a heart urchin (a) and a sand dollar (b).

plow forward and upward beneath the upper surface. Sand dollars penetrate only slightly beneath the surface, using their spines exclusively for digging, in the same manner as in the capture of food. The forward push is brought about by the action of the spines on the aboral surface. *Mellita lata*, which measures 2.5 cm, requires about eighty seconds to bury itself. Another specimen of the same species, measuring 7 cm, takes about three minutes, and *Echinarachnius parma*, also 2.5 cm, takes approximately ten minutes.

While sand dollars can penetrate the substrate by forward motion, only the heart urchins can do this by vertical digging. Members of the genera *Spatangus*, *Echinocardium*, and *Brissopsis* primarily use the lateral spines for burrowing. The animal rests on its walking spines and then pivots the marginal spines, thereby removing the sand from underneath the body. The echinoid sinks deeper and deeper, with sand accumulating at the sides and at the front end; finally the sand walls collapse on top of the animal, burying it.

Many heart urchins have evolved special means of burrowing considerably deeper than the sand dollars. However, the depth to which the animal can penetrate is limited, because the animal requires a constant supply of oxygenated water. Sand dollars and also *Spatangus purpureus* inhabit only shallow depths, and are seldom found deeper than 2 cm below the ocean's floor. Additionally, these forms prefer areas of coarse sand and reeds, where extensive air pockets are found. In these locations it is possible to have a sufficient supply of fresh water, which swirls around the tube-feet by the action of the cilia. The heart urchin *Echinocardium cordatum*, which is buried from 5 to 20 cm in the sand, or the mud-dwelling *Brissopsis lyrifera*, on the other hand, construct special inhalent and exhalent canals that have to be exposed, and that supply the essential respiratory current. The animals produce this structure in the following manner: When burrowing into the substratum, the animal erects two rows of long spines on its aboral surface in such a fashion that no substance can fall into the slit they make in the substratum. Specially modified podia with many secretory cells can extend a shaft which can measure up to 20 cm and have a terminal disc consisting of numerous petaloid lobes. These special tube-feet produce copious amounts of mucus, which they spread over the spines. This hardened mucus is pressed against the walls of the slit, cementing the substrate particles together, preventing immediate cave-ins. As the animal burrows deeper, it forms a canal which is constantly kept open and cemented with mucus by the action of the tube-feet. Fresh water from the surface is constantly sucked into this canal by the action of the fascioles, and swirls past the respiratory tube-feet. From the fasciole the water is distributed over the entire upper body half toward the oral surface, and the anal region toward the posterior end. In the anal region, excretory products become mixed with the water that has been depleted of oxygen but has picked up carbon dioxide.

This ingenious method of fresh-water supply and distribution could not function, however, if there were no concurrent provision for the removal of deoxygenated water and waste products. To prevent congestion which would hinder an inflow of fresh water, a drainage system has been developed at the posterior end of the body. *Echinocardium* solved this problem by constructing a blindly ending drainage canal into which the used water collects and then slowly seeps through the canal walls. A circle of densely ciliated spines (fasciole; Fig. 13-16, b), located underneath the anus, sweeps away the exhalent current. *Brissopsis*, on the other hand, is characterized by a more efficient inhalent-and-exhalent system. This form constructs two ascending sanitary funnels to the substrate surface, facilitating a more evenly flowing water current. The construction of the sanitary funnels follows the same pattern as that of the respiratory funnel, mainly using the action of the spine rows and the funnel-building tube-feet on the posterior margin of the body.

These echinoids do not merely burrow into the ground and stay in one place; they move about below the surface, although not at a great speed, since that would expend a great deal of energy. *Echinocardium* pushes through the ground like a milling machine. It can cover a distance of 4 cm in fifteen to twenty minutes. The spines along the anterior margin scrape the material from the walls of the funnel, and the lateral spines throw it to the posterior end. As the funnel is extended anteriorly, the respiratory water system regularly collapses but is quickly reconstructed by the funnel-building tube-feet.

All regular echinoids are grazers. The animals slowly push through the substratum, scraping off the organic material. Apparently the animals do not seem to differentiate between organisms of vegetable or animal origin. Experiments in the laboratory have shown that echinoids will eat almost anything if not given a choice. Aside from dead fishes and other carrion, they will consume amphipods, barnacles, small snails and bivalves, their own conspecifics, as well as algae, lettuce, and even wood and peanut butter!

If the echinoids are offered a great variety of foods, they seem to have preferences. From a great choice of various algae, certain echinoid species preferred one algal species above all others. The animals have a greater appetite in the summer than in the winter. When observed in its natural habitat, the edible sea urchin (*Echinus esculentus*) would, within an hour, graze off everything from a rocky substratum in an area of 5.8 cm². Six of these sea urchins, trapped underneath a nylon net in an area of 1 m², ate everything that grew there within one month. This included various blue, green, red, and brown algae, all hydrozoans, bryozoans, ascidians, barnacles, as well as the tubes of polychaetes and amphipods plus the contents of the tubes. It seems probable that in regions where algae represent the main food for the echinoids, the growth of the algae and their consumption by the echinoderms are kept in balance.

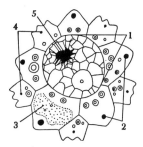

Fig. 13–15. Periproct of the sea urchin *Strongylocentrotus droebachiensis*: 1. Periproct, 2. Genital openings, 3. Madreporite, 4. The first ambulacral plates of each ambulacral radius, 5. Anus.

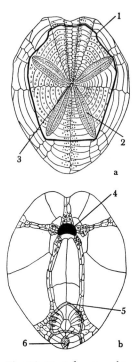

Fig. 13–16. A heart urchin (a) aboral surface, (b) oral surface, with the spines removed: 1. Peripetalous fasciole, 2. Petal, 3. Genital opening, 4. Mouth, 5. Fasciole situated below the anus, 6. Anus.

Fig. 13–17. The brushlike tip of a mucus-producing papilla of the heart urchin.

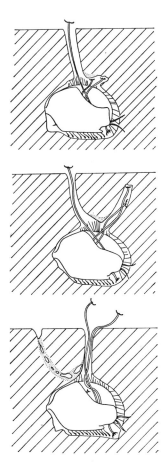

Fig. 13–18. The construction of a new funnel for oxygenated water and the collapse of an old respiratory funnel while the heart urchin (*Moira atropos*) moves through the substrate.

Although the regular echinoids prefer to settle on hard substrata, quite a number of them are also found in sand or mud. The intestine of these echinoids is filled with ocean-floor materials, but it is still unknown whether these animals pick up individual substances or ingest these materials indiscriminately. Echinoids which feed on substances found on hard substrates can occasionally be observed in their feeding activity on algae-covered aquarium walls. Usually these animals press the circular lips surrounding the masticatory apparatus against the substrate before biting. Then they open the jaws, thereby parting the five teeth. In the process of closing the teeth over the substratum, they scrape off the nutritive cover. In *Sphaerechinus granularis* one such bite lasts approximately thirty seconds. The five padded lips of the oral cavity grasp the scraped-off material and conduct it down the pharynx with a type of swallowing motion.

In contrast to the regular forms, the irregular echinoids are characterized by a very different method of obtaining food. Since these latter forms inhabit soft substrata, they have become specialized in feeding on organic particles. They obtain their food particles by dabbing, licking, or straining it from the substratum. Unfortunately, the precise sequence of this feeding method is known from only a few representatives of the sand dollars and heart urchins. In the sand dollar *Leodia sexiesperforata* this process occurs in the following sequence:

Just below the surface of the ocean floor the animal plows through the sand, thereby lifting off the upper sand layer, which contains the highest concentration of living organisms. During this process the sand particles come to rest on top of the first of the numerous and closely spaced club-shaped spines; from there the particles are conducted backward from the top of one spine to the top of the next, and so on. Thus the plowed-up sand mass rolls over the dorsal side of the body like a rug and is deposited at the posterior end. All particles that are smaller than 20 microns in diameter (one micron = one millionth of a meter) are filtered from this sand carpet, falling between the spines onto the skin. These minute particles are swept along by the centrifugal ciliary current at the base of the spines, toward the test margin or lunules, and from there to the ventral side. Finally this sifted material reaches the network of ambulacral grooves, where it is covered by mucus and conducted toward the mouth. Aside from numerous mineral particles, this strained material contains many micro-organisms; these constitute the food source for the echinoids. In addition, extended tube-feet also grope for food particles among the sand grains, and conduct them to the mouth via the ambulacral grooves.

Echinocyamus pusillus, for example, removes nutritive material that adheres to the sand granules, using its five pairs of oral tube-feet and adjacent podia, which are richly endowed with sensory cells, to test individual sand granules for adhering organic growth. When the oral tube-feet detect edible particles, the adjacent tube-feet with suckers transport these to the mouth. Here the substance is wedged against some spines in close proximity to the mouth, and the growth is scraped off with the teeth. The intestinal contents of *E. pusillus* may include diatoms, radiolarians, and also small sections of sponge. The heart urchins, for example members of the genera *Spatangus*, *Echinocardium*, or *Brissopsis*, constantly feel about with their oral tube-feet as the animals plow through the mud and sand. As soon as these tube-feet touch sand granules which have adhering growths, or small snails and bivalves, crustaceans, worms, foraminiferans, or even dwarf sea urchins, the heart urchins grasp and swallow their prey.

It is known that certain sea urchins form larger aggregations during the reproductive phase. The West Indian sea urchin *Lytechinus variegatus* is known to do this. *Psammechinus miliaris*, found along the European coast of the Atlantic, reportedly forms pairs—usually one male and one female—during the spawning period. Every spring, the edible sea urchin *Echinus esculentus* migrates from deeper zones to more shallow coastal regions. The spawning period is greatly dependent on temperature. In the northern hemisphere, spawning usually takes place in the spring or summer. Many tropical species, on the other hand, are able to spawn during most of the year. The ovaries and testes of these species are mature during all seasons. Members of the same species may have completely different spawning periods, depending on whether they inhabit warm or cold water.

In some instances, synchronized spawning of an entire colony of sea urchins has been observed, for example in *Paracentrotus lividus* and *Strongylocentrotus purpuratus*. Presumably the males release the sperms first, thereby inducing the females to discharge the ova. In certain species, such as *Diadema setosum* and *Lytechinus* spawning seems to be triggered by the rhythm of the lunar phases; however, this phenomenon occurs only in specific geographic regions, where the life habits of these animals are strongly influenced by tidal conditions. The eggs, which possess a low yolk content, usually have a diameter of 0.19 mm, rarely up to 0.22 mm. Fertilization takes place in the open sea. Sea urchins are frequently used for research purposes. Basic knowledge of embryological physiology has advanced through research on sea urchin eggs.

Already the first developmental stages of the egg, the blastula and gastrula (see Vol. I), swim independently, by the action of numerous cilia. These forms usually swim close to the surface of the water. The gastrula elongates, forming four to six pairs of slender arms supported by skeletal rods and bearing well-developed ciliary bands (see Color plate, p. 272). This larval stage, known as an echinopluteus, usually lasts for four to six weeks, rarely up to ten weeks. During this planktonic existence, the larvae can drift over long distances, depending on the water current. The echinopluteus larvae feed at this stage, eating various diatoms and other small organisms found in the plankton. They catch this food with the aid of the ciliary apparatus.

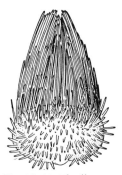

Fig. 13–19. The "stone driller" *Echinostrephus aciculatus*.

▷
Color variant of the purple sea urchin (*Sphaerechinus granularis*), with white spines.

▷▷
Sand dollars.

▷▷▷
A colony of long-spined sea urchins (*Diadema setosum*; compare Color plate, p. 336).

MILLA

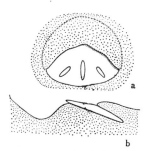

Fig. 13–20. A keyhole sand dollar of the genus *Mellita* in the process of burrowing into the substrate: (a) seen from above, (b) in cross section.

◁

Sea urchins:
1. *Rotula orbiculus*
2. to 4. *Echinosigra paradoxa*:
2. Without spines,
3. Oral side,
4. Aboral side.
5. Heart urchin (*Echinocardium cordatum*)
6. Keyhole sand dollar (*Leodia sexiesperforata*)
7. and 8. A sand dollar (*Echinarachnius parma*):
7. Aboral side,
8. Lateral view.

In certain species, mainly the Cidaroida, the long larval arms are movable, endowed with a relatively well-developed musculature. It is not known, however, to what degree the larvae take advantage of this particular attribute. When touched, the pluteus larvae of *Prionocidaris baculosa* start to beat with two of the longer arms, which makes them seem to "jump away." One can also interpret this motion as a flight reaction. However, this situation seems to differ in the long-spined sea urchin larvae which perform six to eight short, rhythmic strokes with their two long arms, and then extend the arms horizontally. After some time, this whole sequence is repeated.

During this planktonic existence, which may vary in duration, a vestibule on the left side of the echinopluteus larva usually develops. The floor of the vestibule develops into the ventral (oral) side of the future sea urchin. Primary tentacles grow out of the vestibule. The tips of the tentacles develop suction discs and thereby metamorphose into primary podia. These finally erupt through the outer wall of the vestibule. At this stage the larvae sink to the ocean floor and start to crawl about, using their primary podia.

During the course of metamorphosis, the larval organs degenerate, and the organs of the young sea urchin start to develop. The arms with the ciliary bands and calcareous spicules retrogress, as do sections of the larval intestine. The dorsal side starts to develop opposite the ventral side, which is already endowed with podia and which has the first spines appearing along its margins. Aside from several pedicellariae, a circle of five skeletal plates and, later, genital plates develop on the aboral surface. A skeletal plate also forms around each primary podium. These so-called terminal plates represent the first five unpaired ambulacral plates. The ambulacral plates, which are always paired, and their tube feet, develop along the lower margin of the terminal plates. The primary podia become the blind sacs of the water-vascular system. The interambulacral plates develop alongside the ambulacral plates. The terminal plates are progressively pushed up to the aboral surface, until they come to lie beside or between the genital plates. At this stage the young sea urchin is fully developed. It usually measures less than 1 mm in diameter.

The early development of species characterized by eggs with a high yolk content, eggs measuring up to 0.5 mm, and brood care, is not well known. The scant information that is available seems to indicate that these forms undergo direct development, with only a brief or completely suppressed free-swimming larval stage. Approximately eighteen species of sea urchins which practice brood care are known, fifteen of which inhabit antarctic and subantarctic waters. With a few exceptions, all brood-caring forms belong to the Cidaroida and Spatangoida.

The cidaroids retain their eggs and also the subsequently developing young forms near the anal region or the mouth regions (Fig. 13-22). The spines in these areas are bent protectively over the brood. Females in the species *Rhynchocidaris triplopora* develop a ring-shaped groove in the peristome. Up to thirty-five young sea urchins have been counted in such a depression. *Hypsiechinus coronatus*, from the upper North Atlantic, is an exception among the brood-caring species, since it belongs to the family Temnopleuridae. Three to seven eggs are deposited along the external wall of the female's globular, protruding anal region, protected by the adjacent spines. The young stay on the maternal body until fully developed. Females of brood-caring heart urchins develop sunken petals covered by bent spines, thereby forming regular brood chambers. The eggs are directly transported from the genital openings into these brood chambers, remaining there until they have developed into complete sea urchins (Fig. 13-23).

The sea urchin's rate of development and life span are related in a complex way to temperature variations and food availability. Generally speaking, sea urchins grow faster in warmer regions, but they have a shorter life span there than do those in colder oceans. *Psammechinus miliaris*, which has a test diameter of approximately 1 mm after metamorphosis, grows to 20 mm within one year, also reaching sexual maturity during this time. In the course of six years this animal can grow to 38–39 mm. The largest known *P. miliaris* have a diameter of 47 mm. After metamorphosis, *Strongylocentrotus droebachiensis* measures 0.5 mm, but within the first year it can grow to 5–6 mm, and by eight years of age can measure 78 mm. Generally speaking, sea urchins have a life span of between four and eight years.

Numerous animals utilize the spine forest of the sea urchin as escape terrain. The commensal relationship of certain cardinal fishes (Apogonidae; see Vol. IV) with long-spined sea urchins is well known. The fishes often form small swarms between the long and extremely sharp spines which offer them protection and camouflage. From within this spinal thicket, these fishes prey on organisms drifting past the sea urchin. The crustacean *Saron marmoratus* exhibits similar behavior. During the day it rests near the long-spined sea urchins, accompanying its host on its excursions during the night.

Numerous small sessile organisms, such as barnacles, tube-worms, sponges, bryozoans, and hydrozoans, grow on the surface of the thick, skinless spines of many cidaroid sea urchins. Other animals, such as certain species of ctenophores, polychaetes, bivalves, gastropods, copepods, ostracods, amphipods, shrimps, crabs, small sea cucumbers, and brittle stars, may live on the surface of the sea urchin's shell. The exact nature of their association with the individual host is usually unknown.

Numerous ciliates occupy the sea urchin's intestinal tract. Up to 1000 of these organisms per 0.1 cc of intestinal content were once counted. Various gregarines and flat-worms also parasitize the sea urchin's intestine. Parasitic gastropods cause grotesque galls on the spines of certain long-spined sea urchins. Other snails live on the body surface of many species

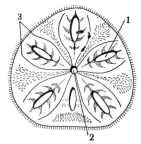

Fig. 13–21. The oral surface of a keyhole sand dollar (*Leodia sexiesperforata*). The arrows indicate the flow of food particles toward the mouth along the ciliated bands. 1. Lunule, 2. Mouth, 3. Ambulacral groove.

Commensals and parasites

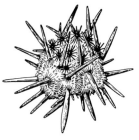

Fig. 13–22. Brood care in *Austrocidaris canaliculata*.

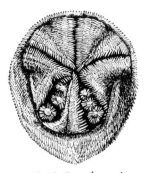

Fig. 13-23. Brood care in the heart urchin *Abatus philippii*.

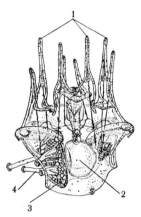

Fig. 13-24. Metamorphosis of the echinopluteus larva of *Psammechinus miliaris*: 1. Larval arms, 2. Stomach, 3. Vestibule, where the future oral side of the sea urchin will develop, 4. Protruding primary tube-feet.

of common sea urchins, punching out the host's skin or boring their proboscis through the test into the host's body and feeding on the internal organs. Parasitic barnacles and copepods also infest the sea urchin.

Although the sea urchins are endowed with numerous defense mechanisms in the form of poisonous and non-poisonous spines, and triphyllous and globiferous pedicellariae, they still are victims to many enemies. Not even their test armor offers the necessary protection. The sea urchin's ovaries and testes are a favorite food source of many fishes such as plaice, sole, turbot, flying gurnards, seawolf, cod, certain graylings and trigger-fishes, sculpins, and grunt, as well as the Greenland shark. *Balistes vetula* overpowers the sea urchin *Diadema antillarum*, which is covered by long, poisonous spines, by getting hold of one spine and lifting the sea urchin from the ground and dropping it again. The fish repeats this procedure until its victim's oral side faces up, and then it attacks the soft-skinned peristome.

Sea urchins that inhabit the tidal zones are often attacked and eaten by marine birds. The glaucous gulls are particularly rapacious in this regard. They will grasp a sea urchin, fly up with it, and let it drop from the air. In case the sea-urchin test does not break after the first fall, the gull will repeat the process until it does and the inside of the prey is exposed. In the northern hemisphere, where sea urchins become exposed during low tide, the arctic fox will feed on them. Cannibalism among sea urchins is known to occur. The starfish (*Marthasterias glacialis*) is a predator that cannot be kept at bay by the sea urchin's spiny cover. It merely everts its highly extensible stomach. This structure's pliable mucous epithelium glides between the spines and invades the sea urchin's body, which is then slowly digested by the asteroid.

Last, but not least, man is also on the list of predators of the sea urchin. In many coastal regions, sea urchin gonads are considered a delicacy. People living around the Mediterranean prefer the ovaries of the sea urchin *Paracentrotus lividus*, eating them raw. The coastal populations of Peru, Chile, and Ecuador enjoy eating the gonads of *Loxechinus albus*, roasting them in one half of the sea urchin's test. In New Zealand, *Evechinus chloroticus* is very frequently consumed.

The calcareous test, which varies in thickness, offers a certain degree of protection for the sea urchin, but by itself it does not act as a deterrent against attackers. The most effective defensive organs are the two differently developed appendages of the test, the previously mentioned spines and poisonous pedicellariae. However, both weapons are not equally well developed in all sea urchins. Vacationers that have spent a holiday along the seashore may have experienced the effectiveness of the sea urchin's spines on their own body. In contrast to the heavily armed forms, the spiny cover of the sand dollars appears velvety or furlike. The sand dollars derive sufficient protection from the substrate in which they are buried.

Of the regular sea urchins, the diadematoids and echinothuriids have developed particularly effective defense mechanisms. The spines of the genera *Diadema* and *Echinothrix* can reach a length of 25 cm. They resemble long, thin needles, and have extremely fine tips that are frequently colorless and are invisible in the water. If a shadow falls on a long-spined sea urchin, all spines are jerked in the direction of the shadow-covered body section. The inner muscle (cog muscle), located around the mamelon, clamps the spine rigid like a pointed bayonet. If the shadow changes its position or something touches the sea urchin, a frightening rattling of the spines results, since they often touch. In the water this noise sounds like the clattering of knitting needles. The delicate tips of the spines can easily penetrate deep into the skin and then break off. The puncture can hurt like a painful burn for some time, but usually does not have any serious consequences.

Fig. 13–25. Young echinoi after metamorphosis.

The species of the genus *Echinothrix* have evolved a very effective double coat of armor. If an attacker penetrates beyond the first wall of long spines, it will encounter a second layer of much denser and more numerous, extremely fine and relatively short secondary spines. A substance similar to noradrenalin (a hormone, which, among other effects, causes constriction of blood vessels and an increase in the heart beat) has been extracted from the skin covering these secondary spines. However, this substance alone is not responsible for the pain they cause. The spines of many echinothuriid sea urchins are considerably more toxic than those of *Echinothrix*. The epithelium covering the echinothuriid spines forms a poisonous blister at the spine tip. The contents of this blister are injected into the skin of the attacker as the spine punctures it.

As we have already seen with the sand dollars, not all echinoids are equipped with such a dangerous spiny defense mechanism. Most of these forms have only short, and not always pointed, spines, in contrast to the long-spined echinoids. Frequently the spiny covering merely functions to protect the delicate tube-feet. The short-spined sea urchins, on the other hand, have more highly evolved globiferous pedicellariae, which is the secondary defense weapon. These pedicellariae are particularly highly specialized in the families Echinidae and Toxopneustidae.

Poisonous globiferous pedicellaria

The teeth of the globiferous pedicellariae open when a chemical stimulus affects the sea urchin's skin. Such a stimulus could be a substance secreted by a predator, or merely its "scent," or a touch on the outside of the pedicellaria. This reaction is particularly evident when a predaceous starfish approaches the sea urchin. The sea urchin is already aware of the attacker before contact is made, and its spines are pointed in the direction of the enemy. If a tactile stimulus follows, the spines are directed away from the touched area to allow the pecidellariae free grasp. For example, if a starfish's tube-foot contacts a pedicellaria, the latter will inject its poisonous contents into the puncture wound. Several such bites can drive a starfish away. Since the pedicellariae adhere to the attacker's body surface, they are torn away from the sea urchin when the predator retreats. Repeated attacks by the predator deplete the supply of pedicellariae, and the sea urchin becomes increasingly more defenseless.

The effect of the injected poison varies in the different species. Young eels of 2–3 cm that have been envenomated by poison from *Sphaerechinus granularis* die from the effects of the poison. Dilutions of this toxin, which have been injected into snails, cuttlefish, shrimps, and fish, have produced paralyzing effects. Specific concentrations of this poison can be lethal. The poison extracted from forty globiferous pedicellariae is sufficient to kill a rat within two to three minutes.

Generally, the globiferous pedicellariae are too weak to penetrate human skin. Up to now, only *Toxopneustes pileolus* and *Tripneustes gratilla* have been known to be harmful to man. The Japanese scientist Fujiwara was bitten on the middle finger by seven pedicellariae from *Toxopneustes pileolus*. At first he felt severe pain, followed later by very grave symptoms including difficulty in breathing and paralysis of all facial muscles including the lips, tongue, and eyelids. His leg muscles also felt numb. The pain subsided after an hour, but the paralysis of the facial musculature persisted for six hours. C. B. Alender, an American scientists who worked with sea urchin poisons, was frequently bitten by *Tripneustes gratilla*, but always by only one pedicellaria. The pain resembled that of a bee sting. After three minutes the region around the bite swelled somewhat. Subsequent bites resulted in increasingly larger swellings, but general toxic effects throughout his body were never evident.

Sand dollars escape the approaches of predaceous starfishes by quickly burying themselves into the sand. Many regular echinoids try to avoid an attacker by fleeing. Since these forms move very slowly, they can escape only from those enemies that are slower than themselves.

Sea urchins have a great tolerance to injuries and damage, due to their ability to regenerate. Skin injuries heal very quickly. Broken skeletal plates, lost tube-feet and pedicellariae, as well as lost spines, can be com-

Regeneration

pletely regenerated. In this context, it should be mentioned that the process of regeneration is very well developed in the sea urchins, and the growth rate of the regenerating tissue is very rapid. Migrating coelomocytes are responsible for the healing process. They are particularly numerous in the large coelom and the spaces between the skeletal plates. After an injury has occurred, these cells concentrate at the wound and close it by forming threadlike plasma processes which join with those of other cells. In this manner, a network forms over the injured body section, reminiscent of the action of fibrinogen in the coagulation of human blood.

Habitats

Sea urchins are essentially bottom-dwellers. The majority of the regular sea urchins settle on hard substrates. Only a few forms inhabit sand and mud, where their tube-feet cannot gain a foothold. For this reason these species select only soft substrates in quiet bays and deeper water zones, where there is little possibility that the "unsteady" animals can be rolled back and forth by wave action. These echinoids frequently hide in rocky shelters or coral reefs during the day. They disappear in cracks, hollows, or niches which have openings that seem too small to permit the entry of a sea urchin. However, actual experiments have shown that a sea urchin can move through an opening that is only 4 mm larger than the animal's diameter. A sea urchin that, with its extended spines, will fill a hole of 3.5 cm, requires an opening that measuring only 1 cm to enter and leave its hiding place effortlessly. The species *Heterocentrotus mammillatus*, an inhabitant of tropical coral reefs, which is endowed with bulky spines, exhibits great agility. During the day this species is rarely seen, but at night it crawls out of unbelievably narrow crevices, and moves about on the reef.

However, there is a group of species that prefer to retreat after their noctural wandering to their daily resting places that are frequently exposed to strong sunlight. The long-spined sea urchins of the tropical coastal zones, and the sea urchins of the Mediterranean and Atlantic coasts belong to this group. These forms partially cover their aboral surface with bits of bivalve shells, plant sections, or flat stones, affording some protection against the intense radiation. This also provides a certain amount of camouflage, but this is probably only a welcome side-effect.

The previously discussed burrowing activity of certain sea urchins is practiced mainly by those forms that live along coastal cliffs that are exposed to strong surfs. By burrowing into the substrate, these sea urchins avoid being washed away by the surging water.

Burrowing

Certain species, for example *Heliocidaris erythrogramma*, burrow into the substrate while still very young. As their body size increases, these forms can enlarge their dwelling hole, but not its entrance. In a way they have built themselves a jail for life. Nevertheless, these sea urchins can maintain themselves, because the surf always splashes enough plankton into their hollows for them to survive. Many other species, usually small forms including *Psammechinus microtuberculatus* and *Psammechinus miliaris*, inhabit thickets of seaweed or kelp.

The famous echinoderm student, Theodore Mortensen, has described how, along certain coastal areas on the Kei Islands of Indonesia, the long-spined sea urchins are settled in such densities that neighboring individuals penetrate each other's spiny cover. The whole ocean floor looks black. Swimmers in the Mediterranean have probably often attempted to swim to certain desolate rock cliffs, hoping to stay there for awhile, but after several painful trials have had to give up the idea of finding a gap in the "defense chain" of the sea urchins, and have had to approach the land on foot. At favorable locations one can find more than ten animals of this species (*Paracentrotus lividus*) per m². Approximately fifteen adults and 250 juveniles of *Lytechinus variegatus* have been counted per m², while along Europe's northern coasts only one common sea urchin occurs in 4.7 to 5.5 m². Up to now the largest population density has been observed in the sand dollars. In an area of 0.835 m², sixty-seven *Dendraster excentricus*

Population density

were counted; and in another spot 486 animals were crowded so closely together that they partially overlapped with each other. The heart urchins *Echinocardium cordatum* and *Brissopsis lyrifera*, which can bury themselves as deep as 20 cm in sand or mud, can also reach densities of twenty animals per m². Several sea urchins, for example *Moira atropos*, tend to form larger aggregations during the reproductive phase.

Sea urchins are strictly marine. Only a few species have penetrated into brackish water, including *Psammechinus miliaris*, *Echinocyamus pusillus*, and *Strongylocentrotus droebachiensis*, which also occurs in the western part of the Baltic Sea. Just as with the feather-stars and sea cucumbers, the greatest number of species of sea urchins is found in the Indo-Malayan region. From this center, numerous species are distributed over a wide area in the Indian and Pacific Oceans. Along the various coastal regions, the indigenous sea urchins form part of the local fauna. Another concentration of sea urchin species is found along the coasts of Australia, where 135 of an approximate total of 850 known species of sea urchins live.

The numbers of sea urchin species typical of the tropical fauna of the Indo-Malayan region soon decrease toward the north and south. However, there is a series of species concentrated along the equator which are widely distributed toward the east and west. Some of these species extend from the coasts of eastern Africa to the eastern Pacific. This group of sea urchins includes what is probably most frequently encountered species, *Echinometra mathaei*. This species inhabits the coastal regions of the Red Sea, eastern Africa, and Madagascar, and the island complex of the entire Indian and western Pacific Oceans, including Japan, Australia, and Hawaii. Furthermore, *E. mathaei* spreads eastward to Clarion Island, only 1,000 km from Mexico. It has not yet reached the Pacific coast of North America. No species of sea urchin is distributed throughout the entire region of the Indian and Pacific Ocean as well as eastern Africa and western North America, for it is extremely difficult to bridge the extensive deepwater barrier of the eastern Pacific, which lacks islands which could serve as footholds for sea urchins as they penetrate into new regions.

The Caribbean-West Indian region is another center for an abundance of species, comparable to the Indo-Pacific concentration. From this tropical region in the Atlantic, a number of species penetrate southward to Brazil and northward to the coasts of the U.S.A. The great range of the sand dollar family Scutellidae is particularly noticeable.

Europe's Atlantic coasts are characterized by a relatively uniform sea urchin fauna stretching from Norway and Iceland to the British Isles, and along the continent's Atlantic coast to Morocco and into the Mediterranean. Representatives of this fauna are *Echinus esculentus*, *Psammechinus miliaris*, *Sphaerechinus granularis*, the heart urchin *Brissopsis lyrifera*, *Echinocyamus pusillus*, and *Cidaris cidaris*. Sand dollars are absent, but the Echinidae are represented in large numbers, as are the heart urchins, which are not present in the corresponding latitudes of North America.

The Arctic and Antarctic do not have the same species of sea urchins. Relatively many cidarids and heart urchins inhabit Antarctica. Many forms are of special interest because they practice brood care.

There are no sea urchins with a worldwide distribution. The small heart urchin (*Echinocardium cordatum*) has the most extensive range of distribution. It is found in Norway and Iceland, along Europe's Atlantic coast, in the Mediterranean, and south from Morocco to South Africa, but also along Japan, Australia, Tasmania, and New Zealand to the east, and the coasts of Brazil and North Carolina to the west.

Although the majority of sea urchins inhabit shallow coastal waters from the high-water mark on the beach out to the edge of the continental shelf, at an average depth of 200 m, there are some species that are predominantly or exclusively found at the deeper zones, between 1000 and 4000 m. These deep-sea forms primarily include the echinothuriids, for example *Kamptosoma asterias*, found in the central Pacific Ocean at depths of 6000 m, and the pourtalesiids. The bottle-shaped *Pourtalesia* occurs to depths of 7200 m. The remaining deep-sea sea urchin families, on the other hand, have maximum depths of around 3000 m.

Vertical distribution

▷
Allopatiria ocellifera inhabits the coasts of Australia.

14   The Starfishes

Class: Asteroidea,
by H. Fechter

The starfishes (class Asteroidea: see Color Plates, pp. 357ff, 368ff, and 379ff) are star-shaped echinoderms characterized by a pliable body. Their oral surface faces the substratum. The main trunk of the body is disc-shaped. Usually five, and rarely up to fifty, unbranched arms of varying lengths project from the body. The coelom of each arm contains a pair of digestive caeca and a pair of gonads. In contrast to all other free-living echinoderms, the radial canals of the water-vascular system are located on the outside of the endoskeleton. The largest starfish, *Midgardia xandaros*, can reach an arm length of 68 cm. The entire animal has the remarkable span of approximately 1 m. Carotenoid pigments which are prevalent in the animal and plant world are present in the body wall of starfishes. The majority of starfishes are red, orange, or yellowish, although green, violet, purple, blue, brownish, and multi-colored forms also occur.

The body shape of the starfish is somewhat influenced by the varying degrees of dorso-ventral flattening, but considerably more greatly influenced by the length of the arms and their size relationship with the central disc. The short and only slightly indicated arms of the thick *Culcita* (see Color plates, pp. 379 and 386) or the thin (flat) duck's foot (*Anseropoda placenta*; see Color plate, 385) result in a roughly pentagonal body outline. The families Linckiidae (see Color plate, 386) and Brisingidae (see Color plate, p. 387), on the other hand, are characterized by an extremely small central disc and very long, narrow, cylindrical arms which give the animal an extreme star-shape. In between these two extreme forms there are many transitional shapes. Aside from the majority of five-armed species, one also finds forms with six, seven, and even fifteen to fifty arms. Four-armed animals are not common.

The body wall of an asteroid consists of an outer epithelium or epidermis, a connective tissue layer made up of muscle fibers and skeletal elements, and an inner coelomic lining. Almost all epithelial cells are ciliated, and many are mucous-gland cells. The epidermis varies in thickness in the different starfishes. The arrangement and shape of the

◁
Pisaster ochraceus, from the coast of California.

◁◁
Luidia ciliaris, from the Mediterranean and eastern Atlantic.

◁◁◁
Astropecten aranciacus.

embedded calcareous ossicles on the oral and aboral surfaces also differs greatly.

Below each arm, and bordering the ambulacral groove, are the two longitudinal rows of ambulacral plates. In between the series of plates of each row there are pores which take the canals from the tube-feet to the ampullae associated with them. A single column of small adambulacral plates runs along each side of the ambulacral plates. The adambulacral plates and ambulacral plates in the vicinity of the mouth form a skeletal ring around it. The sides of the arms are covered by a longitudinal row of inferior and superior marginals. These plates, however, are clearly developed only in the more primitive starfish groups. In all others they have been reduced to small, calcareous pieces. The upper surface of the body and the regions between the arms are supported by a more or less regular network of rodlike or scalelike skeletal elements.

All calcareous ossicles are bound together by muscles, and are flexible. A well-developed muscular strand that can bend the arm runs along the inner wall of the upper side of the arm from the distal end to the middle of the central disc. The central disc is surrounded by a layer of circular and longitudinal muscles. The simultaneous contraction of all muscles results in the stiffening of the entire body. This frequently occurs when one lifts a starfish out of the water.

Movable and rigid spines, with their corresponding tubercles, are hard skeletal appendages projecting from the body surface. Spines can be present on all calcareous ossicles. Several species are entirely covered by spines (see Color plate, p. 386). Usually, however, the spiny armature is localized on the adambulacral plates along the ambulacral groove. In this area the spines can form a protective cover for the row of tube-feet if the need arises. The Astropectinidae bear large spines along the margins of the arms, and in the order Forcipulatida the entire upper side of the animal is covered with spines, which, however, are frequently not readily visible. The family Oreasteridae is characterized by well-developed tubercles.

Many forms in the order Phanerozonida have specific arrangements of small spines. These structures, termed paxillae, consist of special body ossicles, each of which has a crown of spines that can be raised to form a covering for the aboral surface. This arrangement of the "second roof" has an advantage for sand-burrowing forms, because the sand will settle on the paxillae, thereby permitting an unobstructed flow of respiratory water over the aboral surface.

Two types of pedicellariae are another skeletal appendage element. In contrast to the pedicellariae in the sea urchins, those of the asteroids consist of only two blades. Only the Forcipulatida have stalked pedicellariae which function like scissors or forceps. These structures may be scattered over the entire aboral surface or they may be arranged in thick bunches around the spines. Unstalked, sessile forms with steel-trap-

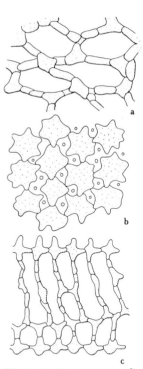

Fig. 14-1. Various types of skeletal plates that are embedded in the aboral body wall: a. The netlike skeletal plates of *Henricia*; b. Lobed, overlapping skeletal plates of an astropectinid; c. Section from half of an asteroid arm, showing the arrangement of the skeletal plates.

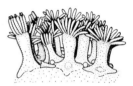

Fig. 14-2. Paxillae are specialized types of spines found on the aboral surface of many phanerozonids.

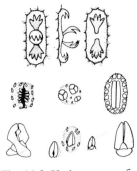

Fig. 14-3. Various types of pedicellariae found in the asteroids.

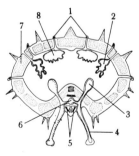

Fig. 14-4. Cross section of an asteroid arm: 1. Papulae, 2. Gonad, 3. Podial ampulla, 4. Podium, 5. Hyponeural sinus, containing strands of the hyponeural nervous system (black), 6. Radial water canal, 7. Ossicle —bearing spine—is embedded in the body wall, 8. Pyloric caecum.

like blades are chiefly found in the Phanerozonida and some representatives of the order Spinulosida (Fig. 14-3).

The most conspicuous soft body appendages are without doubt the tubular podia, which are usually found as double rows, rarely in rows of four, along the ambulacral grooves. The tube-feet usually terminate in a large suction disc. Some Phanerozonida (e.g., Astropectinidae) have club-shaped podia with adhesive terminal mantles. Well-developed longitudinal muscle fibers extend along the wall of the tube-foot. They exert a considerable suction pull. Less conspicuous than the tube-feet are the numerous small, extremely thin blisters on the body wall. These are known as papulae. They may be scattered over the entire aboral surface or may be restricted to isolated regions known as papularia. Gaseous exchange takes place at these bulges in the body wall. Thus, the papulae function as gills.

The structure of the water-vascular system in the asteroids follows the general echinoderm pattern. Multi-rayed starfishes have a correspondingly higher number of radial canals. In contrast to the remaining echinoderms, with the exception of the crinoids, the radial canals extend along the outer side of the ossicles. Two short branches from the lateral canal, which leads to the tube-foot, penetrate through pores between the ambulacral plates into the coelom, forming an extended muscular bulb (ampulla). The function of the ampullae is similar to those in the holothuroids (see Chapter 12). A stone canal ascends from the ring canal to the aboral surface, where it ends in the madreporite.

The star-shaped mouth, which is surrounded by ossicles, leads to a very short esophagus and into a voluminous (lower cardiac) stomach. The upper (aboral) region of the stomach (pyloric) branches out into five pairs of caeca, which are greatly lobed laterally. One pair of pyloric caeca penetrates right to the distal end within the coelomic cavity of each arm. The walls of these intestinal diverticula contain secretory cells which produce digestive juices, and storage cells where fat and glycogen (an animal carbohydrate which acts as food reserve in a similar manner to starch in plants) are deposited. A short intestine leads from the pyloric stomach to the anus on the aboral surface. The intestine is provided with a varying number of relatively small rectal caeca. Many primitive starfishes do not have an anus.

Furthermore, only two of the three nervous systems mentioned in the introduction are developed in the starfishes. The ectoneural system just beneath the epidermis forms a dense subepidermal plexus connected with numerous sensory cells scattered over the outer surface. The ramifications of the major nerve strands follow the water-vascular system. The hyponeural system innervates only the musculature. This nervous tissue (Hange's nerve) is located in the lateral oral wall of the hyponeural sinus (see Chapter 10). Such a hyponeural sinus, which is divided in the middle, extends underneath the radial canal of the water-vascular system. On each side there is an accompanying lateral canal that has already branched off in proximity to the ring canal encircling the mouth.

The neurosensory cells scattered over the asteroid body respond to mechanical, chemical, and optical stimuli. Sensory organs are developed only at the base of each terminal tentacle. At this location a great number of light-sensitive cells form an optic cushion which contains several ocelli. In its most evolved form, such an optic cushion may have a modified lens, as is the case in *Marthasterias glacialis*.

The food-adsorbing hemal system passes through the upper walls of the pyloric caeca. Its principle hemal strands run in the mesenteries of the pyloric caeca to the pyloric stomach, where they pass into a hemal ring connected with the axial organ. Another hemal strand connects to the oral hemal ring. The oral hemal ring gives off hemal radial strands, which extend along the septum which pass up each arm oral to the radial water vessels.

The numerous papulae (see Chapter 8) and the tube-feet function as respiratory organs. The papulae are constantly flushed by the coelomic fluid which absorbs the oxygen and releases the carbon dioxide. Similar processes occur in the tube-feet. The epithelial cilia provide an uninterrupted flow of fresh water, which swirls around the respiratory organs. The Pterasterids have evolved a special "respiratory apparatus" associated with brood care.

Asteroids are chiefly dioecious. *Asterina gibbosa* is one of the few starfishes that are hermaphroditic. At first the male gametes develop, but later only female ones are produced. During a transitional period, both eggs and sperms are produced. A pair of gonads branches into each arm off a circular genital strand located along the oral inner surface of the disc. Each gonad looks like a feathery cluster of tubules. During maturation of the gametes, the gonads greatly increase in size, pushing into the perivisceral cavity of the arms, often right up to the ends of the arms. The gonopores of the individual gonads open at the bases of the arms. Several species have several gonopores along the arm margins.

The single primitive asteroid species (*Platasterias latiradiata*) is classified by some in a separate subclass, the Somasteroidea. This form is oral-aborally flattened, with a small central disc and petaloid arms which are small at the base. The arm skeleton is featherlike, and consists of rod-shaped skeletal elements. Pedicellariae, anus, and suction discs on the podia are absent. It occurs off the western coast of Central America.

All other asteroids are grouped in the subclass ASTEROIDEA. The arms are predominantly elongated. There is no featherlike arm skeleton. The shinglelike arrangement of the ambulacral plates forms an ambulacral groove. There are seven orders (Platyasterida, Paxillosida, Valvatida, Spinulosida, Euclasterida, Forcipulatida, and Zorocallida), with a total of twenty-seven families.

The PLATYASTERIDA are characterized by flat, leaflike to straplike arms

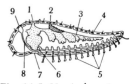

Fig. 14-5. Vertical section through the disc and arm of a starfish: 1. Anus, 2. Genital opening with a gonad behind it, 3. Pyloric caecum, 4. Ossicle embedded in body wall, bearing a spine, 5. Podium, 6. Water-vascular ring canal giving off radial canals which connect with the tube-feet, 7. Mouth opening, 8. Stone canal, 9. Madreporite.

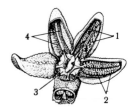

Fig. 14-6. *Asterias rubens* with exposed internal structures; portions of the aboral body wall have been dissected away: 1. Gonads, 2. Pyloric caecum, 3. Cardiac stomach, 4. Rows of podia.

Fig. 14-7. Cribriform organs between two arms of a porcellanasterid.

with transverse rows of many elongated skeletal plates. In the family LUIDIIDAE, the arms are narrow and straplike, without aboral marginal plates. The tube-feet are club-shaped. There is no anus or pyloric caeca. The SEVEN-ARMED STARFISH (*Luidia ciliaris*; see Color plate, p. 359) is found in the Mediterranean and in the Atlantic from the Faroes to Cape Verde. This species is orange-red. Its arms measure 7 to 30 cm. It inhabits sandy ground at depths from 4 to 400 m. The five-armed *Luidia sarsi* is smaller, and reddish-brown. It has a distribution similar to that of *L. ciliaris*, but is found on muddy flats at depths of 10 to 1,300 m.

Family: Astropectinidae

The order PAXILLOSIDA (aboral surface endowed with paxillae, tube-feet without suction cups, paired ampullae) includes the family Astropectinidae, represented by many species. The marginal plates are large, and frequently have long, well-developed spines. The anus is absent. *Astropecten aranciacus* (see Color plates, pp. 358 and 386) is orange-red, and yellow on the oral side. The arms can measure up to 28 cm. It is found in the Mediterranean and in the Atlantic from Portugal to Angola, at depths of 1–20 m. It frequently inhabits sandy flats or seaweed meadows. *A. irregularis* is widely distributed in the Atlantic from the Lofoten to northwestern Africa, and in the Mediterranean. There are several sub-species. The arm length is up to 16 cm. The color is reddish-brown to reddish-violet. It occurs at depths of 10–1000 m. The genus *Leptychaster* practices brood care. This order also includes the deep-sea-dwelling porcellarasterids, including *Porcellanaster caeruleus* (see Color plate, p. 385).

Family: Oreasteridae

The representatives of the order VALVATIDA are characterized by tube-feet with suction cups. The aboral surface has paxillae. There may or may not be spines. The marginal plates are prominent. Of the eight families (see systematic survey), the following are of special interest: 1. ARCHASTERIDAE, including *Archaster typicus*, one of the most common starfishes of the Indo-Pacific. (Its pair formation is described later). 2. GONIASTERIDAE. Almost pentagonal starfishes with very short arms, thick, massive marginal plates, and a broad central disc. This group includes *Sphaeriodiscus placenta* (see Color plate, p. 380). It has a diameter of up to 17 cm, is yellow or reddish-brown, and occurs in the Mediterranean and the Atlantic from Archangel to Senegal. 3. OREASTERIDAE. Primarily larger, massive starfishes with broad arm bases. The aboral endoskeleton is netlike, with numerous papulae in the spaces. Found mainly in the Indo-Pacific. This group includes the genus *Culcita* (see Color plates, pp. 379 and 386), characterized by a pentagonal, cushionlike body. It also includes *Protoreaster lincki*. 4. OPHIDIASTERIDAE. Very small central disc; the arms are long, slender, and cylindrical. The marginal plates are barely visible. It is found mainly in tropical oceans. This includes the genus *Linckia* (see Color plate, p. 386) and *Ophidiaster ophidianus*, found in the Mediterranean and in the Atlantic from Portugal to St. Helena; its arm length is up to 17 cm. The color is carmine or orange-red, occasionally with darker spots. It is found from sea level to a depth of 100 m. *Hacelia attenuata* (diameter 27 cm; see Color plate, p. 380) is scarlet-red with yellow feet. Found in the Mediterranean and along Portugal's coast at depths from 1 to 150 m.

In the order SPINULOSIDA, the marginal plates are usually small and indistinguishable. The suckered tube-feet are arranged in two rows. The aboral skeleton is reticulate or overlapping in a shinglelike fashion. Small spines are often present. Pedicellariae are very rare, and are never stalked. There are twelve families in this order (see systematic survey) but only the most important ones are mentioned here: 1. ASTERINIDAE: the body is usually pentagonal, and the arms are very short. The skeletal plates on the aboral surface overlap in shingle fashion. The best-known species is *Asterina gibbosa* (diameter up to 6 cm; greenish-gray, yellowish-green, or reddish-brown), found in the Mediterranean and the Atlantic from sea level to 130 m, underneath rocks and in seaweed meadows. The DUCK'S FOOT (*Anseropoda placenta*; see Color plate, p. 385) is pentagonal, diameter up to 20 cm, thickness 1 cm; scarlet-red, and yellowish or bluish on the oral side. Found in the Mediterranean and Atlantic at depths of 10 to 600 m. It lives buried in the sand. *Patiria miniata* (see Color plate, p. 380). 2. ECHINASTERIDAE: the central disc is small, the arms long and slender. This group includes the brick-red *Echinaster sepositus* (see Color plate, p. 370; arm L up to 15 cm). It has small spines buried in the skin. Common in the Mediterranean and Atlantic from sea level to 250 m. The BLOOD STAR (*Henricia sanguinolenta*) is somewhat smaller, and is blood-red. It is found from sea level to 2400 m. (Its distribution is discussed later.) 3. ACANTHASTERIDAE includes the poisonous Indo-Pacific species *Acanthaster planci* (see Color plate, p. 386) which has eleven to twenty-one short arms, and long spines on the aboral surface. 4. SOLASTERIDAE are characterized by a broad central disc with many short arms. The purplish-red and whitish patterned COMMON SUN-STAR (*Crossaster papposus*; see Color plate, p. 386) has eight to fourteen arms, and has a diameter of up to 34 cm. The species is found in the North Atlantic from sea level to 1200 m. *Solaster endeca* is similar. 5. PTERASTERIDAE have short, broad arms. There are paxillae on the aboral surface which have a muscular membrane. They are notable because they practice brood care (see Chapter 2; and Color plate, p. 377).

Family: Solasteridae

The order EUCLASTERIDA has only one family, BRISINGIDAE. The body is small and disc-shaped. There are six or more elongated, slender arms, bearing many spines. The marginal plates are indistinguishable. The pedicellariae are small and similar to those of the next order, Forcipulatida. These forms include *Odinella nutrix* (its brood care is discussed later) and *Freyella* (see Color plate, p. 386).

In contrast to all other asteroids, members of the order FORCIPULATIDA are characterized by stalked pedicellariae. The body is not disc-shaped. The marginal plates along the arms are reduced to small skeletal pieces. The tube-feet are frequently arranged in four rows. The best-known family with the most species is the Asteriidae. They have slender arms

Order: Forcipulatida

which gradually merge into the massive disc. The aboral surface is covered with spines which are surrounded by a wall of pedicellariae. The BLUE STARFISH (*Coscinasterias tenuispina*), which inhabits the Mediterranean and Atlantic, has six to twelve arms which measure up to 9 cm. They are bluish-white with large brown spots. They are found on rocky substrates at depths from sea level to 50 m, beneath plants and rocks. *Marthasterias glacialis* (see Color plate, p. 380) is distributed from Norway to Cape Verde in the Atlantic, and in the Mediterranean. The species has five arms which may measure 35 cm and exceptionally reach 50 cm. It has well-developed spines on the marginal and aboral plates. There are great color variations, but the predominant colors are greenish, yellowish, reddish, or brownish. It occurs at depths from sea level to 180 m. The COMMON STARFISH (*Asterias rubens*; see Color plate, p. 385) inhabits the Atlantic from the White Sea to Senegal at depths from sea level to 650 m. The arm length is up to 26 cm. The coloration varies greatly, but is chiefly red, reddish-brown, or reddish-violet. The genus *Leptasterias* practices brood care. *Pisaster ochraceus* (compare Color plate, p. 360) is a frequently encountered starfish on North America's Atlantic coast. The coloration is ocher-yellow, and the arms can measure up to 14 cm.

Members of the order ZOROCALLIDA have a domed disc, with a regular arrangement of disc plates. The order comprises a single family, ZOROASTERIDAE.

Locomotion

In contrast to the echinoids, which make effective use of their spines in locomotion, the asteroids move exclusively with their tube-feet. When a starfish moves over the ground, its tube-feet do not drag the body laboriously over the substrate by suction and subsequent contraction to pull the body in, but rather by the stiltlike steps of hundreds of podia. Pressure from the water-vascular system makes the tube-feet become erect. The body is thus lifted from the ground, and the podia proceed in regular steplike motions. All tube-feet move in the same direction, although not in a uniform step order. As a podium makes a step, it bends slightly, thereby exerting a push.

One arm, or rarely two, will usually take over the lead in a specific march direction. If the direction is changed, the leadership is passed on to the arm or pair of arms facing in the new direction. In only very few species does it seem that one specific arm always moves first. Generally speaking, however, asteroids are sluggish animals that are usually anchored to the substratum, and move only at specific times, when searching for food or for other reasons, such as during intense sunlight or an increase in temperature.

The speed of locomotion varies enormously in the different species, depending on whether the animals live in shallow coastal zones that are rich in food, or inhabit the sand and slime and are able to cover great distances. The large species of the genus *Luidia* are undoubtedly the liveliest and fastest starfishes. Their tube-feet measure up to 3 cm. *Luidia sarsi*, for example, can run 75 cm in one minute. The astropectinids are also fast. Quite a few of them can cover 60 cm per minute. Some members of the family Asteriidae, on the other hand, are much slower. *Asterias rubens*, for example, moves only 5–8 cm per minute. *Asterina gibbosa*, which moves approximately 2.5 cm per minute, is considered one of the slowest asteroids.

It is rather difficult to establish the highest possible locomotory speed, since animals move much faster in a dangerous, threatening situation than during an uneventful walk. *Asterias forbesi* moves 15–20 cm per minute while in the ocean and unthreatened. The same animal, timed in the laboratory during an attack, moved 25–35 cm per minute. Naturally, such top speed records cannot be maintained for a great amount of time. In our discussion of food-gathering we shall learn that *Asterias vulgaris* may cover a distance of only 12 m during two days.

Some species seem to conduct regular migrations to deeper water zones at the onset of the cold season, returning to more shallow waters in the spring. While roaming around in search of food, the asteroids may also discover new habitats, to which they obviously seem to migrate until a certain population density has been reached. An area of 60 m² was at one time completely devoid of *Pisaster ochraceus*; within the subsequent six weeks sixty of them claimed the area.

It is probable that almost all asteroids can climb up vertical walls. Many can master this feat not only on natural hard substrate, but also on the smooth glass walls of an aquarium. Some species of the family Linckiidae which inhabit coral reefs seem to be particularly agile climbers. These same animals can also navigate along rods and thick wires. However, even those species that lack suction cups on the tube-feet, for example the astropectinids, can scale glass walls. The tube-feet of these forms secrete a sticky substance that adheres to the substrate for the short duration of the podial contact.

While moving along an even plane, the tube-feet have to relocate the body by pushing, and they have to pull the body weight during climbing. This is done in the following manner. Individual tube-feet temporarily not being used for adhering to the wall extend upward, stick themselves to a higher "foothold," and pull the body up by contracting the podial musculature. A specimen of *Pisaster ochraceus* which was climbing up a pole covered a distance of 2–3 m in twenty-four hours.

Burrowing is an activity which is carried out chiefly by the astropectinid starfishes which inhabit sand and muddy regions. During their rest periods they disappear from the upper surface. The presence of a buried asteroid is revealed only by the star-shaped unevenness on the upper surface. While burrowing, the tube-feet which are underneath all the arms simultaneously push all the sand toward the sides. Thus the starfish sinks progressively deeper, while part of the piled-up sand walls collapse on its aboral surface, covering it. Aside from the astropectinids, some

▷
Fromia ghardaqana is found in the Red Sea.

▷ ▷
Echinaster sepositus inhabits the Mediterranean and eastern Atlantic Ocean.

species of the genus *Luidia* and the duck's foot (*Anseropoda placenta*) also burrow. All forms are able to emerge from the substrate rather quickly by bending the arms up and thereby stripping off the thin sand cover.

If one turns a starfish on its aboral side, it tries to turn over onto its "feet" after a brief period. It turns over more or less rapidly with varying degrees of agility. *Asterias rubens* takes only two to five seconds to right itself, while more bulky species such as *Marthasterias glacialis* take eight to nine, and *Astropecten aranciacus* takes from two to fifteen minutes. There are even some species that require an hour to reattain their proper position.

During the attempt to catch a foothold and to spread out their arms, the asteroids make use of the topography, such as undulations in the ground, depressions, and other irregularities, as well as walls, stones, outgrowths, and other objects. During this process, some species may twist one or several arms around their longitudinal axis until the tube-feet reach solid ground to which to adhere and then to pull the body into the proper position. Others bend two arms until the oral surfaces of the arm tips touch the ground. Once the tube-feet have grasped a hold, they cause a slow somersault of the entire body.

The two other methods of turning over involve a relocation of the center of gravity. In order to achieve this goal, the animal bends all its arms down, and then lifts up at the armtips. This maneuver results in a fall onto the proper side. In the other procedure, all arms bend upward. In this case only a small part of the starfish's curved trunk touches the ground. The animal loses its balance and falls to one side. From there it tried to catch a hold with at least one arm.

Asteroids, however, are not only able to regain their proper position, but many are also able to turn on their back when attacked. By turning over, they can shake off other asteroids which are predaceous.

Details concerning the method of feeding are known only for a few starfishes. It seems that asteroids can be either predators, plant-eaters, or detritus feeders. However, many transitional stages between these categories are also possible. It is not uncommon that the same animal can interchange between two methods of feeding, depending on prevailing conditions.

Those asteroids that feed on microorganisms have a well-developed ciliary apparatus along the ambulacral grooves. These forms grope along the surface of the substrate with their podia, or stir up the upper layers of the substrate and search for microscopic plants, animals, and decomposed organic substances. They use ciliary action to conduct organisms toward the mouth along the ambulacral grooves, where they are coated with mucus. This feeding method is employed primarily by the deep-sea porcellan-asterids of the order Paxillosida. Stomach contents of these forms reveal abundant substrate materials as well as many foraminiferans, diatoms, and radiolarians. Occasionally there are also small gastropods, worm segments, and pieces of echinoids. From this evidence it seems

◁
The Crown of Thorns (*Acanthaster planci*; see Color plates, pp. 380 and 386), a poisonous starfish of the Indo-Pacific that destroys coral polyps.

◁◁
Nardoa variolata adhering to a stony coral.

apparent that these species will not disdain larger particles if they are available. Shallow-water forms, particularly *Linckia guildingi* and representatives of the genera *Henricia* and *Porania*, employ the filter-feeding method. This feeding method is probably wide-spread among members of the family Linckiidae.

The plant-feeding asteroids introduce us to an ability that is remarkably well developed in many asteroids, reaching its peak of specialization in the carnivorous forms. The animals are able to evert their highly folded cardiac stomach through the mouth to the outside. This event is brought about by contraction of the arm and central-disc musculature, resulting in a high degree of pressure in the coelomic cavity. The coelomic fluid forces the cardiac folds out of the mouth in a bladder shape. The plant-feeders, which may be exemplified by *Patiria miniata* and several species of the tropical family Oreasteridae, mold their extremely flexible cardiac mucoid lining against the substrate, and start to digest the plant growth. One can observe this process beautifully when starfishes feed on algal growth covering the glass walls of an aquarium.

These asteroids feed on animal as well as plant growth, and even carrion. It has even been observed that the antarctic *Odontaster validus* will eat seal feces. The semi-digested broth is conducted into the pyloric caeca from the area in front of the mouth by ciliary action. The final digestive process occurs in these caeca. The CROWN OF THORNS (*Acanthaster planci*; see Color plate, p. 386), which inhabits the coral reefs of the Pacific and Indian Oceans, uses this feeding method on the polyps of the reef-building corals, and has frequently caused extensive damage to the reefs. Several plant-feeders are able to clean the surface of a substrate with the aid of the well-developed ciliary regions of the everted cardiac stomach walls, the action of which is quite similar to that of a street-cleaning machine. The food particles are conducted to the pyloric caeca for digestion.

The majority of asteroids are carnivorous, preying mainly on bivalves, gastropods, crustaceans, echinoids, and ophiuroids. Some starfishes devour the entire prey. For example, the stomach content of a specimen of *Astropecten aranciacus* revealed ten scallops, six *Tellina*, five *Dentalium*, and several conid snails. A single *Luidia sarsi* ingested no less than seventy-three young ophiuroids, and another animal ate a small echinoid in addition to fifty-three ophiuroids. *Crossaster papposus* swallows whole *Mytilus*, including the shells, and digests the soft parts of the animal within approximately twenty-four hours. The empty shells are voided by eversion of the cardiac stomach. The duck's foot feeds on various shrimps, amphipods, crabs, and hermit crabs. The prey is conducted to the mouth either from tube-foot to tube-foot or by bending one or more arms.

Certain solasterid species are characterized by a unique mode of feeding. *Crossaster papposus* (see Color plate, p. 386) and *Solaster endeca* prey on the common starfish (*Asterias rubens*; see Color plate, p. 385).

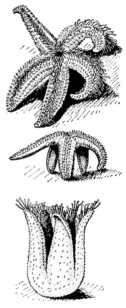

Fig. 14-8. Various "righting" possibilities for a starfish that has fallen on its aboral surface.

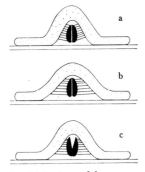

Fig. 14-9. A starfish opening a bivalve: a. Attachment of the tube-feet; b. The stretch phase; c. The pull phase.

Solaster dawsoni prey on *Evasterias troscheli*. When attacked, the victims reject the attacked portion of their body, upon which the solasterid will then feed. In human terms, the action of the small *Solaster endeca* seems "cruel." This species will attach itself to the end of an arm of an *Asterias*, push it into its mouth, and slowly start to nibble on its victim. This process can last for days, until the victim's arm is too large for the small *Solaster*; then it lets go. The comparatively large *Asterias* is not able to shake this "cannibalistic" small rider, and is completely at its mercy.

Among the carnivorous asteroids, there are many forms that do not devour the entire prey, but rather digest their prey in front of their mouth. Just like the plant-feeders, these species evert their cardiac stomach and pry into the victim's body orifices with individual stomach lobes. The everted stomach-folds are able to penetrate into extremely small cracks and holes, for example into narrow slits of certain bivalve shells through which the byssus threads (see Chapter 6) enter the outside, as in the *Mytilus* species, or into many bivalves whose posterior shell region is not entirely closed because of the inhalent and exhalent openings.

The stomach wall is extremely extensible and flexible. Some species are able to extend it up to 10 cm. The stomach wall adheres firmly to the victim's soft parts, but the mucous lining does not secrete digestive juices. These originate, instead, in the pyloric caeca whence they are conducted to the small folds of the everted cardiac stomach, to the place of contact with the prey's tissue, by ciliary action. This method prevents a dilution of the digestive juices, as would be the case if the secretions were discharged freely. The semi-digested tissues of the victim are conducted, along similar ciliary pathways, back to the pyloric caeca for final digestion.

Members of the family Asteriidae have evolved four rows of well-developed tube-feet on each arm, greatly increasing their efficiency in overpowering the prey animal. By skillful maneuvering and the powerful suction of the tube-feet, the asteroid is able to counteract the tension of the bivalve's adductor muscles, and thus to part the two valves. These star-fishes proceed in the following manner: The podia grasp the bivalve and manipulate it underneath the central disc to the mouth, where it is slightly lifted by the arms, and the asteroid hunches over its victim. The bivalve is manipulated by the podia until the hinge faces the substrate and the exposed, firmly oppressed valved margins are located in front of the asteroid's mouth. Once the position of the bivalve is established, a great many tube-feet from the arm-base regions which oppose each other adhere to the two shell valves, while the ends of the arms are anchored to the substrate. Then the hunched central disc relaxes somewhat, and the arm bases spread to the sides, with the result that the podia attached to the bivalve are greatly stretched, often twice as long as their normal size. Subsequently, the entire asteroid freezes in the last position, and the pulling phase sets in. The pull is exerted solely by the contracting tube-feet. The arms do not pull, but merely act as a rigid opposing force.

The tube-feet exert considerable pull. *Asterias forbesi* and *Evasterias troscheli*, for example, exert a pull of 5.5 kg. The common European starfish and *Pisaster ochraceus* exert a 4-kg pull. An asteroid is able to push its everted stomach into a crack between two bivalve shells that measures only 0.1 mm. The predator then begins its extracellular digestion whereby the last resisting force of the bivalve is overcome.

The asteroids also apply this suction mechanism to those animals that are attached to the substratum, such as polychaetes, barnacles, limpets, and chitons. The asteroid tears the victim off the substrate by the method described above. It is even successful in removing the operculum which many marine snails pull over their opening when retreating into the shell. The everted stomach is inserted into the gastropod along all the whorls right up to the tip. If a mollusk is too large or cannot be manipulated into a proper position, these asteroids attempt to penetrate into the prey's natural orifices such as the place where the byssus threads or respiratory siphons emerge from the shell.

Astropecten aranciacus primarily hunts at dawn and dusk. *Luidia sarsi*, *Acanthaster planci*, and *Marthasterias glacialis* are nocturnal. Some species seem to have an excellent sense of smell. A common starfish kept in an aquarium was able to move directly to a dead fish from a distance of 60 cm. If one holds a piece of crab or mussel meat 2–5 cm in front of a starfish's arm tip, one can lead the animal in any direction. Experiments carried out in the ocean have shown that one can attract asteroids by laying out bait. A single specimen of *Asterias vulgaris* was released 12 m from an oyster bed, and unerringly approached its goal at a speed of 6 m per day. However, from a distance of 200 m the starfish was unable to approach the oyster bed correctly.

Asteroids not only perceive prey that is freely exposed, but also animals buried in the substratum. If, for example, *Pisaster brevispinus* senses a bivalve buried in the sand, it will rest on this spot and will start to dig. The tube-feet push the sand toward the arm tips, with the result that the central disc sinks deeper into the sand. It takes the starfish two to three days to burrow to a depth of 10 cm, but there is no hurry, because the bivalve will not run away. As soon as the podia, particularly the well-developed ones in the vicinity of the mouth, have touched the victim, they will grasp it and pull it out of the ground.

Some asteroid species will feed on anything that is available, while others are more fastidious. For example, *Solaster* and *Luidia* species will feed primarily on other asteroids, echinoids, and ophiuroids. Young *Asterias* prefer barnacles to oysters. Many forms may decrease their food intake or stop it altogether if they are offered bivalves that they do not like so well as others. This information may possibly benefit oyster farmers if they could introduce limpets and *Urosalpinx* to their oyster beds. The common European starfish, which usually preys on the oysters, might prefer these other species.

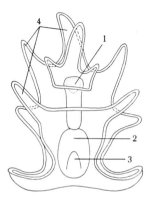

Fig. 14-10. Bipinnaria larva: 1. Mouth, 2. Stomach, 3. Anus, 4. Larval arm with ciliary bands.

Fig. 14-11. Brood care in the asteroids: a. *Leptasterias groenlandica*, b. *Trophodiscus almus*, c. *Leptychaster kerguelensis*, d. *Anasterias antarctica*, e. *Hymenaster pellucidus*.

The stomach content of certain asteroids might give the impression that they are insatiable. However, one has to realize that a large proportion of the ingested bivalves, gastropods, and echinoderms consists of indigestible shells and skeletal elements, and only a small proportion is digestible. For some asteroids the daily food intake is around 2.7 to 3.2 percent of the body weight. Young starfishes, on the other hand, can take in twenty-five to thirty percent of their own body weight.

Spawning usually occurs during specific periods of the year. The reproductive phase, when all gametes are released, usually lasts from one to three months, depending on the species. In the majority of cases the spawning period coincides with the onset of a rise in the water temperatures, usually in springtime. Certain species in the tropical waters do not have specific spawning periods, and others, for example, the Australian *Patiriella exigua*, probably reproduce throughout the year.

Certain asteroid species form aggregations at the onset of the reproductive period. In *Asterina gibbosa* it has even been observed that several males congregated around one female. The most remarkable reproductive behavior, however, is displayed by *Archaster typicus*, from the Indo-Malayan region, which not only forms pairs but also exhibits a kind of copulatory position. The male climbs on the female's aboral surface and bends his arms downward between the female's. The male almost always discharges the sperms first, which probably triggers the release of the eggs by the female. The majority of asteroids, particularly the many shallow-water forms, produce enormous amounts of homolecithal eggs with little yolk—from 2 million up to over 100 million per reproductive period! The diameter of the eggs measures only from 0.1 to 0.2 mm. In contrast, asteroid species inhabiting colder and polar waters, represented by deep-sea forms and brood-caring ones, produce far fewer eggs—several thousand as a maximum—which contain a high percentage of yolk. These eggs have a correspondingly larger diameter—from 0.5 to 1.5 mm.

Fertilization takes place in the open water. The eggs with low yolk content develop into free-swimming larvae within three to four days, and then drift in the plankton. This free-swimming stage is characterized by a larva (bipinnaria; see Color plate, p. 272) with broad, lobed, or slender armlike processes and ribbonlike ciliary bands. Upon full development of the larval intestinal tract, the larvae start to eat. With the aid of the ciliary band in the proximity of the mouth, the bipinnaria larva feeds on unicellular algae, primarily diatoms. The bipinnaria larvae grow quickly, and can reach considerable size, including their arm projections. The dipinnaria larvae of *Luidia sarsi*, for example, reach a length of 35 mm. In the species *Luidia sarsi*, *Luidia ciliaris*, and *Pentaceraster mammillatus* (see Color plate, p. 379), the central arms are elongated and bear well-developed muscles. These arms are used for swimming.

With the exception of astropectinid and *Luidia* larvae, probably all other bipinnaria larvae after three weeks develop three additional lobes in front of the mouth. A suckerlike glandular region develops between the base of these new arms. At this phase, the larva is termed a brachiolaria. The three brachiolaria arms and the suction cup form an adhesive apparatus with which the larvae adhere to the substratum prior to metamorphosis. Only the astropectenid and *Luidia* larvae metamorphose in the swimming, planktonic stage.

Metamorphosis takes place within twenty-four hours. The anterior part of the larval body, which is attached to the substratum, becomes stalklike, and the larval processes and ciliary bands degenerate. The larval mouth and anus are closed over. The oral side of the future asteroid already begins to develop on the left larval side during the swimming stage. The five radial canals begin to form on this surface. In contrast to all other echinoderms, a vestibule does not develop on the oral surface, but rather forms freely on the aboral surface. The first five skeletal plates appear on the right aboral side of the larva, obliquely opposite the oral surface. The first skeletal plates indicate the potential aboral surface of the starfish. During the process of metamorphosis, the rudiments of the aboral and oral surfaces develop toward each other, finally uniting to form the complete young animal.

The anterior arm of the juvenile pushes through the ring canal of the water-vascular system, forming an oral opening on the lower side. At the same time the anus breaks open on the upper side. A pair of disced tube-feet develops behind each of the five primary tentacles previously referred to as the beginnings of the radial canals. Subsequently the first aboral and oral calcareous ossicles develop, and the arms start to extend from the trunk. After the young starfishes have detached themselves from the stalk that anchored them to the substrate, they begin to climb about and to feed on small bivalves of corresponding size.

The development of the brood-caring forms and those asteroids living in cold regions follows a completely different course. Oval larvae hatch from their yolky eggs. The larvae lack ciliary bands, but are completely enclosed by a ciliary covering. In the case of non-brooding species, the larvae persist in the free-swimming stage for a maximum of twenty days. During this time they do not feed, but live off the stored yolk. The only larval organs that develop are two or three arms which correspond to the adhesive papillae of the brachiotaria. Once the larva has sunk to the bottom, it can crawl about with these arms before it attaches itself in preparation for metamorphosis. The larval arms degenerate during this process, while the anterior section of the larva becomes elongated in a stalklike fashion just as in the brachiolaria larvae. The general development of the young starfish proceeds in the same manner as in the majority of species, although there is great variation in the embryology of the internal organs.

Most brood-caring asteroids inhabit the cold and polar oceans, being

▷
Starfishes representing the colorful family Oreasteridae:
Left, from top to bottom: *Asterodiscus truncatus*, *Choriaster granulatus*, *Pentaceraster mammillatus*. Right, from top to bottom: *Culcita schmideliana* (see Color plate, p. 386), *Pentaceraster tuberculatus*, *Protoreaster lincki*.

most prevalent in the antarctic and subantarctic regions. Of a total of approximately 114 antarctic asteroid species, no less than fifty practice brood care. The simplest form of brood care is depositing the eggs in a specific location where they can undergo development, rather than simply releasing them directly into the water. This method is exemplified by *Asterina gibbosa*, found in the Mediterranean and eastern Atlantic. This species carefully attaches its eggs to the substrate, usually the underside of stones, but then leaves. A considerably more advanced form of brood care is practiced by some species found primarily in antarctic waters. These species deposit their eggs on the ground, then cover them with their bodies until the brood has become independent. These eggs are encased in a sticky mucus, which makes them all adhere together. Frequently the young asteroids are still interconnected by these mucous strands. The females hunch over the eggs with a raised body. The group that practices this brood care method also includes *Henricia sanguinolenta* and *Leptasterias mülleri*, found along the coasts of Europe.

In the antarctic species *Odinella nutrix*, from the family Brisingidae, the bases of the numerous arms swell to such an extent that adjacent spines cross over each other, forming a type of spiny basket at the angles between the arms. Five to nine large eggs are deposited in each basket, where they develop. Several species of the genus *Leptychaster* and the species *Ctenodiscus australis*, which also belongs to order Paxillosida, transfer their eggs to the area between the paxillae on the aboral surface from the genital openings, which are located quite a distance from the paxillae. These paxillae consist of special raised body ossicles, each of which has a crown of movable spines. The eggs and young asteroids develop under the protection of this spiny layer. As the young increase in size and outgrow the shallow depression in the maternal central disc, they tend to push the paxillae aside. These females can carry up to thirty young starfish (compare Fig. 14-11, b and c).

The most advanced form of brood care is carried out by members of the pterasterids (Fig. 14-11, e) which have true brood chambers. In these forms the tips of the paxillae are held to each other by a membrane laced with muscle fibers. This provision results in a type of tent roof which is spread over the body proper and is held up by the pillarlike paxillae. The enclosed space is perforated by many pores and one central opening. The eggs are laid directly into the brood chamber from the genital opening which has extended upward. The membrane pulsates, creating currents for the feeding and respiration of the young. Within the brood chamber, the young can develop to a span of 1.5 cm. The most unusual brood-care method is found in *Leptasterias groenlandica*, which belongs to the family Asteriidae. In this species the brood develops in special stomach pouches, which are often turgid with young starfish (Fig. 14-11, a).

The growth rate of asteroids is greatly influenced by the water temperature and the availability of food. In many asteroids, food intake stops

◁
Left, from top to bottom: Color variant of the Crown of Thorns (*Acanthaster planci*; see Color plates, pp. 372 and 386), color variant of *Hacelia attenuata* found in the Mediterranean and eastern Atlantic, *Patiria miniata*.
Right, from top to bottom: *Leiaster leachi*, *Marthasterias glacialis*, *Sphaeriodiscus placenta*.

if the temperature drops below a specific point. For this reason, growth is greatly reduced or arrested during the winter. Some forms can withstand hunger periods of from ten to eighteen months. The common European starfish can reach a span of 4 to 9 cm during the first year. It also becomes sexually mature during this time. In the same period, *Crossaster papposus* can grow to from 4 to 9 cm. On the basis of observations in the aquarium and the ocean, it seems that many asteroids achieve a minimal life span of four to six years, and it is possible that *Pisaster ochraceus* even reaches an age of twenty years.

The asteroids are infested by the following parasites and commensals: ciliates, a ctenophore (*Coeloplana astericola*), and many polychaetes which feed on the asteroid's food supply by inhabiting ambulacral grooves or by approaching the asteroid's mouth-opening within these grooves. Some polychaetes even insert their anterior end into the asteroid's stomach.

Many asteroids are also victimized by parasitic gastropods which we have already encountered in connection with other echinoderm groups. Amphipods and copepods are found among the starfish's spines. The organisms may also feed on tissues which have been tangled in the host's mucus and also properly constitute part of the host's tissue. The Ascothoracica, which are related to the barnacles, are internal parasites of various asteroids. However, there are also vertebrates that parasitize the asteroids. These are various species of carapid fishes (Carapidae; see Vol. IV, Chapter 18) which feed on the internal organs of their hosts. These parasitic fishes were also discussed in connection with the sea cucumbers. The carapids frequently attack various oreasterids.

However, there are no predators that have specialized in asteroids. The young starfishes are most endangered, since they are frequently eaten by conspecifics, i.e. *Luidia* and *Solaster*, or by codfish. Adult starfishes can possibly be overpowered by large gastropods such as *Charonia*. Certain crustaceans may nip off the starfishes' tube-feet or papillae.

The asteroids can employ several techniques to protect their vulnerable body parts against intruders. They can retract such sensitive areas as the podia and skin papillae. Additionally, many starfishes are able to shut the ambulacral grooves which contain the tube-feet, and then spread the spines over them protectively. If a crab or a polychaete succeeds in mounting a starfish, the various pedicellariae start to snap at the intruder. In the Forcipulatida, the pedicellariae are arranged in clumps and are therefore quite effective. These forms may hold onto a crab or polychaete with their pedicellariae for days. In any case the hold is maintained until the opponent has ceased to move.

Another defensive mechanism undoubtedly is the numerous substances contained in various body tissues, which are toxic or even lethal to the animals. The only material which has been isolated and identified is a saponinelike substance with great hemolytic properties. Saponines are any group of glucosides occuring in many plants; they are hemolytic

Defense and protection

even in great dilutions. If some of this asteroid substance is put into water at a specific concentration, fish will die.

Of all asteroids, only *Acanthaster planci* has had a recorded toxic effect on human beings. This multi-armed species, entirely covered by long spines, is found on almost all coasts of the Indian and western Pacific Oceans from the Red Sea to the Great Barrier Reef off Australia. A person pricking himself on a spine of this starfish will almost immediately feel great pain, which may last for hours. Occasionally the area around the wound will swell, lose sensation, and appear to be paralyzed. The symptoms may sometimes be accompanied by nausea and vomiting. It is presumed that the epithelial cells covering the spines contain poisonous glandular cells which release into the skin during the sting.

Some time ago I had the opportunity to observe these symptoms myself. During a diving operation in the Red Sea, my companion had an unpleasant encounter with one of these conspicuous starfish. He touched the asteroid only briefly, but the harm was done. For several hours my companion suffered severe pain and swelling. Local residents of this area were well aware of the starfish's toxicity, and showed great concern. Some species of the genus *Echinaster*, which occur in warm oceans, have short, inconspicuous spines which, however, are poisonous. Exact reports concerning these species are not available.

It seems obvious that asteroids constantly release substances which give them a characteristic odor. Many starfishes make escape motions as soon as they approach another starfish or only touch it with one tube-foot. The escape motions are particularly evident in those forms that are prey to the carnivorous asteroid species. Flight reactions can be artificially elicited with the extracts from the tube-feet or skin of starfishes. Many bivalves and gastropods are extremely sensitive to this extract, as are echinoids and brittle stars. Scallops react particularly dramatically. When approached by certain asteroid species, they will swim away excitedly. The cardiid bivalves and *Spisula* species beat a hasty retreat by repeatedly stretching their curved foot, thereby making wide jumps, often over 10 cm. Some snails try to escape the approach of a carnivorous starfish by turning somersaults; other gastropod species respond by quickly pulling the skin fold of their foot over the entire shell, like a hood. The starfish's tube-feet are unable to gain a foothold on the mucous surface of this covering, and the snail has some chance of escaping its predator.

Many asteroids are able to reject an arm if it has been subjected to severe mechanical or chemical stimuli. Occasionally a starfish may undergo autotomy even when handled carefully. Others react only if injured or severely pinched. Adverse conditions, such as warm or stagnant water or the attack of an enemy, can cause autotomy. This phenomenon is more prevalent in forms possessing a relatively small central disc and long, slender arms. Species characterized by a pentagonal body shape with broad arm bases rarely show this self-mutilating effect.

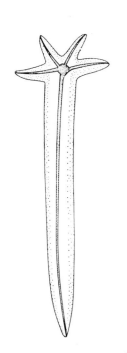

Fig. 14-12. "Comet form" of *Linckia multifora*.

Species like *Luidia sarsi* and *Brisinga coronata* can break their arms off at any point, but most other species possess a definite, pre-designed breaking line, usually close to the central disc. According to existing observations, the break occurs either if the arm segment is anchored to the substrate in front of the suture line and the remaining body moves away, or the rejected arm and the main body move in opposite directions. The rejected arm will almost always deteriorate; in *Linckia*, however, the rejected segment may regenerate into a complete starfish.

The base of the severed arm, at the central disc, regenerates a new arm, although this process is very slow. Some time after the wound is closed over, a bud of tissue develops with the tip of the new arm. The water-vascular system, the pyloric caeca, and radial nerves gradually grow between the end of the arm and the central disc. For a long period of time—frequently longer than a year—the regenerated arm is still recognizable by its relatively smaller size. Not only are entire arms regenerated; smaller injuries and torn-out body sections are also replaced.

Some asteroid species have perfected the ability to autotomize and regenerate to such an extent that they are able to reproduce asexually. Asexual reproduction is known to occur in several species of the genus *Linckia*, particularly *Linckia multifora*, where the rejected arm does not deteriorate as in other asteroids, but regenerates a complete new disc and four arms. This results at first in a so-called "comet form," since the original arm resembles a long tail with a new little four-armed star at its base. Other genera capable of asexual reproduction (*Coscinasterias*, *Sclerasterias*, and *Stephanasterias*), unlike *Linckia*, split in two across the disc, and each half regenerates the arms and disc segment that it lacks. In *Stephanasterias albula*, which is circumpolar in distribution, asexual reproduction seems to be more prevalent than the sexual method.

Asteroids inhabit the ocean floors from the tidal zones to depths of 8,000 m. However, the majority of species are found in shallow coastal waters. Only a few forms have penetrated into brackish water, including the common European starfish, the blood-red starfish, and the common sun-star which are found in the Baltic Sea where the salinity concentration is fifteen parts per thousand. The common European starfish can occur in water where the salinity concentration is only eight parts per thousand. The population density of many species is dependent on the abundance of available prey. *Asterias rubens* and *Asterias forbesi* are equally common on rocks, sand, or mud flats, depending on the availability of bivalves.

Most species, however, show a definite preference for a certain type of substrate. *Echinaster sepositus*, the blood-red starfish, *Asterina gibbosa*, and *Solaster* occur on rocky substrates, where they will be found on cliff faces, underneath stones, or in plant cover. The astropectinids and *Luidia* species are found primarily on sand flats. These forms and the duck's foot frequently burrow into the sand. *Astropecten spinulosus* prefers seaweed

▷
1. Duck's foot (*Anseropoda placenta*), 2. Common European starfish (*Asterias rubens*) in the process of opening a scallop (*Pecten*), 3. *Porcellanaster caeruleus*.

1

2

3

MILLA

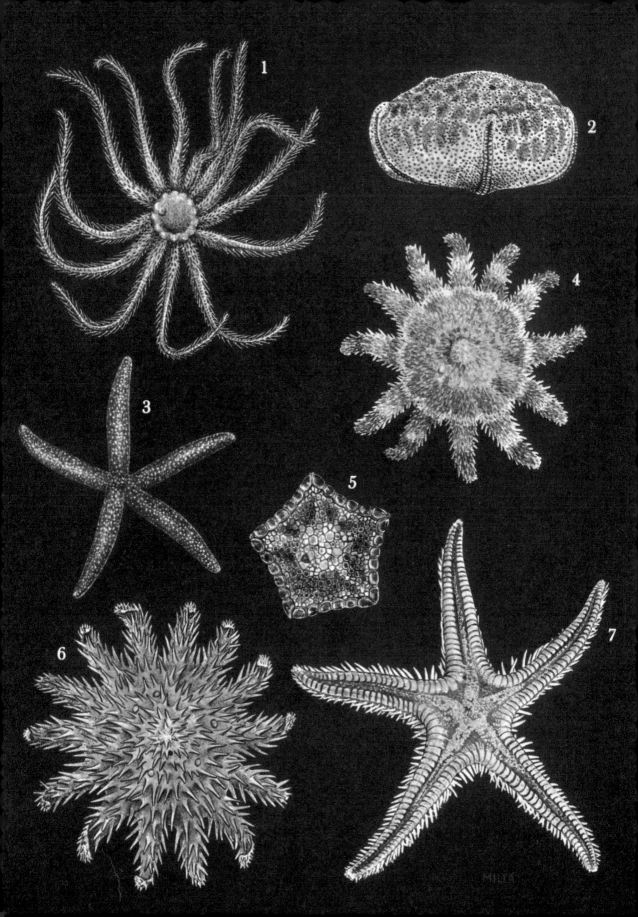

meadows. Coral reefs provide several habitat possibilities to the asteroids. The family Linckiidae in particular has explored this habitat in many ways.

Many starfishes tend to avoid direct sunlight, and during the day they retreat underneath overhangs, or into crevices, hollows, or plant thickets. Other species do not seem to mind strong sunlight, for example numerous tropical species as well as the European starfishes *Asterias rubens* and *Crossaster papposus*. Those species that inhabit the tidal zones or the low-water mark are subjected to adverse living conditions, including strong surges, great temperature fluctuations, dilution by rainfalls, and the threat of desiccation during low tide. *Pisaster ochraceus*, which inhabits Europe's northwestern coast, is very resistant to desiccation. This species can tolerate a loss of thirty percent of its body weight in body fluids.

Time and again one can read reports about extensive concentrations of starfishes, but these reports are rarely accompanied by exact counts and area measurements. *Asterias vulgaris* can reach population densities of up to fifteen animals per square meter. The experiment of resettling *Pisaster ochraceus* described earlier was very illuminating, because it showed quite clearly that there is a definite tendency to maintain the population density at a certain level. The centers for the greatest number of asteroid species do not correspond to the geographical locations of the remaining echinoderm groups. The most significant of these regions is found in the northeastern Pacific Ocean, with the focal point along the coasts of Alaska and western Canada, including Puget Sound. There are more asteroid species there than anywhere else in the world. Several species from this North American center are distributed throughout the entire North Pacific Ocean. The majority of these species, however, are limited in their range to this particular area. In addition, there are several asteroid species with circumpolar distribution, which have penetrated southward into the Pacific and Atlantic Oceans. Some of these forms also occur along European and Asian coasts.

The starfish fauna of the North Atlantic is characterized by polar species which advance southward along the continents, and the intermingling of these species with the local starfishes that occur only around the North Atlantic. Characteristic species for Europe's coasts include the following: *Luidia ciliaris*, *Asterina gibbosa*, *Anseropoda placenta*, *Asterias rubens*, *Marthasterias glacialis*, *Luidia sarsi*, and *Astropecten irregularis*. In the warmer regions of Europe's oceans one also finds *Astropecten aranciacus*, *Ophidiaster ophidianus*, *Echinaster sepositus*, and *Coscinasterias tenuispina*. Almost all these forms, along with the North Atlantic species, are present in the Mediterranean. Several species in the tropical latitudes of the Atlantic are distributed along Old World as well as New World coasts. A unique asteroid fauna has developed in the West Indian-Caribbean region. The tropical coast of western Africa, however, is extremely poor in starfish species.

The number of asteroid species in the Indo-Pacific is not as great as

◁
1. *Freyella echinata*,
2. *Culcita schmideliana* (see Color plate, p. 379),
3. *Linckia laevigata*,
4. Common sun-star (*Crossaster papposus*),
5. *Tosia australis*, 6. Crown of Thorns (*Acanthaster planci*; see Color plates, pp. 372 and 380),
7. *Astropecten aranciacus* (see Color plate, p. 358).

that of the North Atlantic. Nevertheless, the Indonesian-Philippine-Australian region is considered to be another asteroid concentration. Some of these species have spread from the Red Sea, along Africa's eastern coast, including Zanzibar, to Japan and Hawaii. The predominance of the order Phanerozonida, and in particular the Linckiidae and the low number of the order Forcipulatida, is noticeable when the colder zones are approached.

There are also many asteroid species that are limited in their distribution to the Indo-Pacific center. For example, of the 190 asteroid species found in Australia, 140 are limited to this area and are found nowhere else. Some species found in Hawaii spread over the vast area of the Pacific, right to the American continent.

One could consider Antarctica as a third distribution center. It is characterized by numerous brood-caring asteroid species. The asteroids which belong to this group are very large forms, which may reach a span of 70 cm. The species *Ceramaster patagonicus* is found in both polar regions.

Some starfish species enjoy a great range of vertical distribution. For example, *Hymenaster pellucidus* is found from depths of 15 to 2800 m. *Crossaster papposus* is found from the tidal zone to a depth of 1200 m. The families Porcellanasteridae, Benthopectinidae, some Astropectinids, Brisingidae, Zoroasteridae, and many Pterasteridae are limited to the deeper zones of the ocean. Most deep-sea species occur in the North Atlantic. There are only a few in the Indo-Pacific. One hundred and thirty-nine asteroid species are known from below the 3000-m level, only four from below 6000 m, and only *Eremicaster pacificus* is known to occur at 7600 m.

▷
1. Serpent star (*Ophioderma longicauda*), 2. Female of *Amphilycus androphorus*, carrying the much smaller male on its oral surface, 3. Gorgon's head (*Gorgonocephalus caputmedusae*).

MILLA

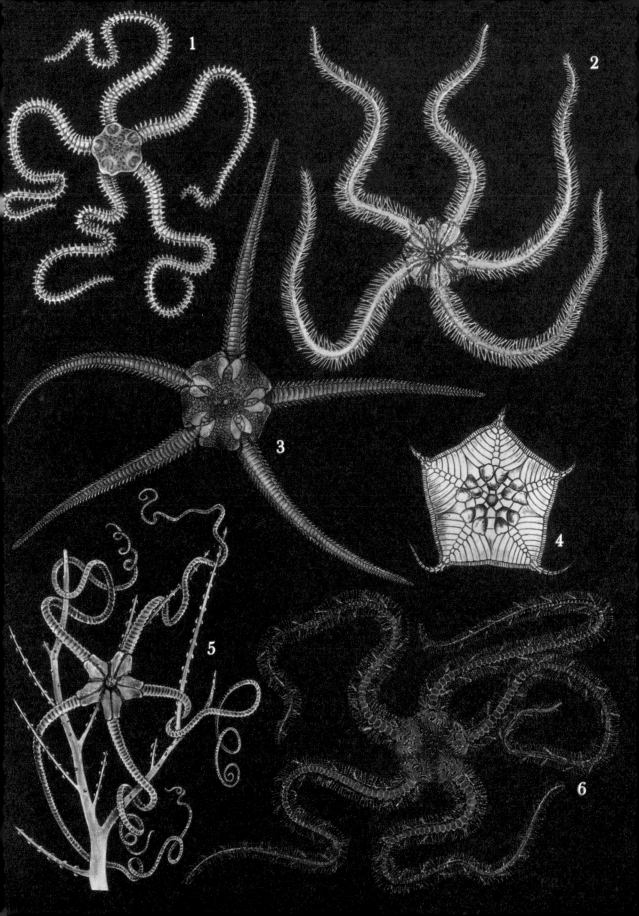

15   The Brittle Stars

Class: Ophiuroidea,
by H. Fechter

The brittle stars (class Ophiuroidea; see Color plates, pp. 389ff and 403ff) are pentagonal echinoderms with slender cylindrical or repeatedly branched arms supported by vertebrae-shaped ossicles. The arms are always sharply demarcated from the central disc. The oral surface faces the substratum. The anus is absent. At both sides of each arm base there is a sac which bulges up into the disc. The tube-feet lack suction discs. The largest species with unbranched arms is *Ophiarachna incrassata*, which has a disc diameter of 5 cm and an arm length of 23 cm. Larger species, however, are found in those groups characterized by branched arms. The largest member of these groups is probably *Gorgonocephalus stimpsoni*, which has a disc diameter of 14.3 cm and an arm length (to the tip) up to 70 cm. Altogether there are three orders with seventeen families.

Brittle stars may be cryptically colored, matching the color of the background. Cryptic colors are grayish-brown or greenish-brown, beige, dark brown, or black. Other brittle stars, however, may be very conspicuous and colorful, such as the white, red, blue, or yellow-patterned species. Coloration is caused by the carotenoid pigments and melanin, which are located in the connective tissue. Many species are characterized by extremely great color variations. In these forms there are hardly two individuals with the same color pattern.

The brittle stars are represented by two different body types. The major difference is in the structure of the arms. The majority of brittle stars have arms that are extremely slender, very flexible, and unbranched. Most of these forms have five arms, rarely six or seven. The slender star-shaped body contour is. clearly recognizable. In several families from the order Phrynophiurida, the arms, well-developed at their bases, divide in a fernlike manner. These forms are more reminiscent of a bush or shrub. In the contracted state, these animals appear like a wild tangle, and it is very difficult to discern the star shape.

Generally, the surface of the body is covered by a thin layer of skin. Unlike the skin of most of the other echinoderms, the epidermis is almost

◁
1. *Amphiura chiajei*,
2. Common brittle star (*Ophiothrix fragilis*; compare Color plates, pp. 403 and 405),
3. *Ophiura texturata* (compare Color plate, p. 404), 4. *Astrophiura permira*, 5. *Asteronyx loveni*, 6. *Ophiomastix annulosa*.

devoid of cilia. The scalelike ossicles which develop underneath the connective-tissue layer penetrate through the epidermal layer. The skeletal plates are particularly evident on the arms. Usually the arms are covered by an upper, a lower, and two lateral rows of plates. Individual ossicles overlap more or less like shingles on a roof. The four series of plates surround the rather massive longitudinal row of ossicles called vertebrae. These articulating ossicles usually have facets and tubercles. Two pairs of muscles move the entire arm. The main portions of the oral and aboral surfaces of the disc are frequently covered by tiny, often granule-sized plates. Only on the oral side are there larger plates, which are located between the bases of the arms. They are known as oral plates.

The only skeletal appendages present in the brittle stars are the spines, which may be borne by the lateral plates on the arms or on the aboral surface of the disc. The spines may be thorny, but often they are modified as cover plates for the tube-feet or as hooks. Along the oral side of the arm there is one pair of suckerless tube-feet for each skeletal vertebra. The podia are not used for locomotion. Their main functions are the capture and transportation of food.

The water-vascular system follows fairly closely the plan of the other echinoderms, although the stone canal terminates on the oral surface. The madreporite usually consists of only a few pores, and is located on one of the oral shields. In the phrynophiurids, such as the basket stars, several stone canals are developed, often up to five. The radial canals extend along the groovelike depressions on the oral side of the vertebrae. In the majority of brittle stars this groove is protectively covered by the oral plates. Only in the phrynophiurids are the grooves merely covered by skin. The canals that branch off from the radial canal and lead to the podia cross through parts of the vertebrae, although these podial canals never expand into ampullae, as in most other echinoderms (Fig. 15-3).

The entrance to the star-shaped mouth is found in the center of the disc on the oral surface. The mouth is surrounded by five interradial jaws which bear spinelike teeth and papillae (Fig. 15-4). The jaw apparatus is manipulated by well-developed concentric series of muscles. Two pairs of buccal podia are located on each side of the jaw at the oral end of the arms. These tube-feet aid in stuffing food particles into the mouth. The mouth leads almost immediately into a voluminous saclike stomach which ends blindly, since there is no anus. Only in the very primitive species *Ophiocanops fugiens* are there still small stomach pouches extending into the coelom of the tapering arms. In all other ophiurids the stomach is limited to the central disc.

The ectoneural nervous system (see Chapter 10) is sunk into a canal which is covered by the aboral arm plates. The radial nerve strands branch off into the tube-feet and spines. The main nerve ring surrounds the mouth, and directly innervates the buccal tube-feet, jaw apparatus, and stomach. The five radial branches of the hyponeural system (see Chapter

Fig. 15-1. Types of ophiuoroid spines:
a. Rings of hooked spines in the basket star,
b. Umbrella-shaped spines of *Ophiohelus*, c. Spine fan of *Ophiopteron*, d. Hook-

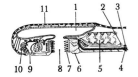

Fig. 15-2. Longitudinal section through disc and arm base of an ophiuroid: 1. Stomach, 2. Radial canal of water-vascular system, 3. Hyponeural sinus with underlying hemal channel, 4. Nerve strands of the hyponeural and ectoneural nervous systems, 5. Vertebral ossicle, 6. Oral podium. 7. Jaw with teeth, 8. Mouth opening, 9. Bursal slit, 10. Gonads, 11. Skeletal plate embedded in the body wall.

10) are paired, and are arranged in ganglionic enlargements. This system extends through each arm beneath the radial hemal channels and along the oral faces of the vertebra ossicles. The aboral nervous system (see Chapter 10) is only weakly developed. It innervates the sexual organs. The brittle stars lack sensory organs. Up to now sensory cells were found only in the podia. These cells responded to chemical, mechanical, and optical stimuli.

The bursae play a great role in the respiratory function. Each arm base is flanked by two narrow slits on the oral side of the central disc. Each bursal slit leads into a sac which bulges up into the body wall. The walls of the bursae are very thin and ciliated. The cilia create water currents which constantly bring fresh, oxygenated water into the bursae, and remove the deoxygenated water. Gas exchange with the coelomic fluid takes place across the bursal walls. In certain species the water exchange in the bursae is supported by the disc musculature. During "inhaling" the aboral disc wall rises, and during "exhaling" it contracts quickly and forcefully, expelling the water. In the phrynophiurids the bursae are fused below the aboral disc wall with the result that the stomach becomes surrounded by bursal cavities.

The hemal channels in the ophiuroids follow the general echinoderm design. However, the nutrient-gathering hemal plexus of the stomach does not terminate in the axial organ, as is the case in other echinoderms, but rather there are five hemal lacunae ending in the aboral hemal ring. The axial organ connects the aboral and oral hemal systems. The oral hemal system consists of the oral hemal ring and the radial hemal channels branching out into the arms.

Most ophiuroids are dioecious. There are only about forty species which are hermaphroditic and which also practice brood care. Except in rare instances, the sexes are alike. Depending on the species, varying numbers of gonads are attached to the lateral walls of the bursae. In a few brittle stars the gonads extend into the coelom of the arms, but in the majority of cases they are confined to the disc. The mature gametes are discharged into the bursae and are then expelled through the bursal slits.

Like all other echinoderms, the brittle stars are marine. Only a few ophiuroids have penetrated into waters of low saline concentration, including *Ophiura albida*, which inhabits the Baltic Sea where the salinity content is ten parts per thousand, and *Ophiophragmus filogranus*, which occurs around Florida where the salinity content is 7.7 parts per thousand.

Surviving members of the class Ophiuroidea are classified into three orders: 1. OEGOPHIURIDA: the most primitive present-day brittle stars. The aboral grooves are covered only by skin. The aboral and oral arm plates are absent. These are paired rows of gonads within the arm coeloms. The stomach pouches also penetrate into the arm coelom. There is one surviving family, the Ophiocanopidae, represented by only one living species, *Ophiocanops fugiens*, from Indonesian waters. 2. BASKET

shaped and thorny spine of *Ophiothrix*.

STARS (Phrynophiurida): the oral grooves are covered by the oral plates. The aboral plates are absent or greatly reduced. Occasionally the gonads, located in the central disc, send out processes into the bursae. There are no stomach pouches in the arms. The body is covered by a soft skin. There are five families (see systematic survey). The Gorgonocephalidae are characterized by extremely long, usually branched arms. Hooked spines girdle the entire arms. This group includes *Gorgonocephalus caput-medusae* (see Color plate, p. 389), which has a disc diameter of 9 cm. It is reddish, yellowish, or white. It is found in the North Atlantic at depths of 150 to 1200 m. 3. OPHIURIDA: the arms are completely covered by aboral, oral, and lateral rows of plates. They have vertebral plates with a type of ball-and-socket joint which permits only lateral articulation. There are eleven families which represent the majority of ophiuroids (see systematic survey). The following families are mentioned here: a. AMPHIURIDAE: the arms are long and thin, with short, protruding spines. The jaw margins are equipped with paired oral papillae. These forms are frequently buried in soft substrates. This family is represented by many species. *Amphiura chiajei* (see Color plate, p. 390) is found in the Mediterranean and Atlantic. Its disc diameter is 1 cm, and arm length is 10 cm. It is reddish or grayish-brown, and occurs at depths from 5 to 1200 m, almost worldwide. *Axiognathus squamata* has a disc diameter of 0.5 cm and an arm length of 2 cm. It is whitish or bluish-gray, and occurs at depths from sea level to 740 m, worldwide. b. OPHIACTIDAE: this group is closely related to the amphiurids, but is represented by only a few species. *Ophiactis virens* and the DAISY BRITTLE STARS (*Ophiopholis aculeata*) belong to this family. *O. aculeata* has a disc diameter of up to 2 cm and an arm length of 8 cm. It is reddish, frequently with dark bands on the arms. c. OPHIOTRICHIDAE: the jaw margins bear spikelike tooth papillae. Oral papillae do not fringe the jaw. The arms usually have long spines. The group is represented by many species. The COMMON BRITTLE-STAR (*Ophiothrix fragilis*; see Color plate, p. 390) inhabits the Mediterranean and Atlantic, often in huge aggregations. Its disc diameter measures up to 2 cm, and its arm length up to 10 cm. The central disc is usually very colorful (although very rarely including blue). The arms have contrasting light and dark bands. With an increase in the depth of the habitat, the central disc becomes more colorful and the spines become longer and more fragile. This species can be found from sea level to 475 m, underneath stones or in plant growth on hard and muddy substrates. *Ophiothrix quinquemaculata* (see Color plate, p. 403) inhabits the Mediterranean at depths from 40 to 250 m. Its disc diameter is 1.5 cm and the arm length is 15 cm. It is greenish-gray or pinkish-gray with purplish banded arms. d. OPHIOLEPIDIDAE: the aboral surface of the disc is covered by well-defined plates. The arms are moderately long with short clinging spines. The jaw is characterized by unpaired rows of teeth and lateral oral papillae. The family is represented by many species.

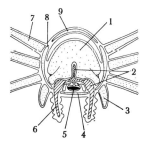

Fig. 15-3. Cross section of arm of ophiuroid: 1. Vertebral ossicle, 2. Water-vascular system, 3. Scale which can cover the pore of the retracted tube-foot, 4. Oral arm plate, 5. Ectoneural nervous system, 6. Podium, 7. Lateral spine, 8. Lateral arm plates, 9. Aboral arm plate.

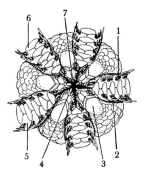

Fig. 15-4. Oral disc surface of *Ophiura sarsi*: 1. Pores through which the tube-feet are protruded, 2. Bursal slit, 3. Oral shield, 4. Jaw with teeth, 5. Lateral arm plates, 6. Oral arm plate, 7. Mouth opening.

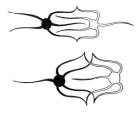

Fig. 15-5. Schematic drawing of a moving ophiuroid.

Ophiura texturata (see Color plate, p. 390) is a common species in the Mediterranean and Atlantic. Its central disc measures 3.5 cm, and the arms 12 cm. Its color is reddish or grayish-brown, with a yellowish oral surface. It inhabits coral and muddy substrates at depths from sea level to 300 m. The smaller *Ophiura albida* (see Color plate, p. 404), reddish-brown, is distributed in the Mediterranean, Atlantic, and the western Baltic Sea, from sea level to 850 m. e. OPHIODERMATIDAE: the aboral and oral surfaces of the disc are covered by granules. The arms are almost smooth, since the spines are very short and appressed. *Ophioderma longi-cauda* (see Color plate, p. 389), blackish-brown, occurs in the Mediterranean and Atlantic. Its disc diameter is 2.5 cm, and the arm length is 15 cm. The species is common, and is found from sea level to 70 m. f. OPHIO-COMIDAE: the jaw margin bears spikelike tooth papillae, and the lateral sides of the jaw are fringed with oral papillae. The arms usually have long, well-developed spines. This family includes *Ophiocomina nigra*, which has a disc diameter of 2.5 cm and an arm length of up to 12 cm. The coloration is black, brown, gray, or reddish. The species inhabits the Mediterranean and Atlantic from sea level to 400 m. *Ophiocoma scolopendrina* feeds on a great variety of substances. It is black, and has arms that measure up to 14 cm. It is a common species in the Red Sea and the coastal zones of the Indo-Pacific.

In comparison to other echinoderms, the brittle stars are extremely active and mobile. Only a very few species use their tube-feet for moving along on a flat plane, as do most other members of this animal phylum. Only *Ophionereis reticulata* is known to move along on its erected podia. *Ophiactis arenosa* can effectively make use of its podia, since they bear suckers.

All other ophiuroids move by powerful arm strokes. Many brittle stars move their arms rather like a swimmer doing the butterfly stroke. The animals swing one or two pairs of opposing arms to the front, and anchor the arm tips to the substratum. The slightly raised disc is thrust forward by the other arms. By constantly repeating this process, the brittle star moves forward rapidly. *Ophiura texturata*, with an arm length of 10 cm, propels its central disc forward by 5.5 cm with each "step." In one minute it can cover a distance of 1.8 m. If only one pair of arms moves, the arm in the middle will point in the direction of movement. If, however, two pairs of arms are active, the fifth one points backward. In either case, the inactive arm does not take part in the forward thrust. Its function is to feel out the terrain.

Slithering

Many brittle stars, especially those with extremely long arms, are able to move with a slithering type of motion. These forms do not flex their arms forward, but pull them along the substrate. This process consists of the animal stretching out one arm, which anchors at its tip to the substrate, and then contracting the arm muscles, resulting in an undulation of the arm. The distance between the arm tip and the central disc decreases, and a frontward pull results. This method of movement is particularly evident in members of the family Amphiuridae. These forms are usually buried in the substratum, and very rarely move long distances, since they are semi-sessile. However, other genera, like *Ophiothrix*, *Ophiocoma*, and *Ophiomyxa*, may also use the slow, "slithering" motions in addition to the forward-thrusting oarlike movements.

Some species are able to climb up the smooth glass walls of an aquarium. These brittle stars secrete a sticky mucus from their podia, enabling them to adhere to glass walls. Others require a rough surface with fine cracks for climbing. The cracks serve as insertion points for the tube-feet. Some of the smaller brittle stars are particularly agile climbers, such as *Nannophiura* and *Amphilycus*, which live as harmless commensals on the spiny tests of certain irregular echinoids. These species have hooked spines at the extreme tip of their highly flexible arms, which are highly adapted for clinging to the sea urchin's spines. The small brittle stars twist around in the echinoid spines, but always turn their mouth to the outside in order to catch food particles which the echinoid stirs up from the ocean floor. The phrynophiuirds (order Phrynophiurida) are also very agile climbers when they feed on the polyps of the coelenterate colonies. These forms bear many movable hooked spines arranged in rings around the arms. These structures are highly adapted for grasping. The animals pull themselves up by bending the hooked or entwined arms. *Climbing*

Two ophiuroid species, *Ophiacantha eurythyra* and the young of *Amphiodia psara*, have been observed swimming short distances. The animals move two pairs of opposing arms very forcefully, while the fifth arm is stretched out and remains inactive. *Swimming*

Several members of the order Ophiurida, notably the long-armed representatives of the family Amphiuridae, burrow into sand and mud. Some burrow into the substratum to such an extent that only the arm tips protrude. Burrowing is primarily the work of the tube-feet, which constantly move from the center of the arm to the sides and in this manner push the underlying material away from the disc and the arms. As the animal sinks, a system of cavities in the shape of the brittle star forms around the arms and the disc. The walls of this cavity are cemented with secreted mucus to prevent cave-ins whenever the animal moves. *Burrowing*

If a brittle star falls on its aboral surface, it behaves in a manner similar to that of a starfish under similar circumstances: The five arms prop up the disc; two adjacent arms spread out at an obtuse angle and support themselves on the substrate; the other arms push the disc over until it somersaults to the proper position. "Righting" can take up to three minutes. Most brittle stars, however, can turn around within forty-five seconds; some need only two seconds. *"Righting"*

The ophiuroids have developed a great variety of feeding methods. Generally one species can employ several techniques, depending on the prevailing environmental conditions. The animal is able to pick up food particles from the surface of the ocean floor and the upper substrate layers, as well as from the surrounding water. Food-gathering is mainly the function of the more or less extensible, sticky tube-feet, which secrete a viscous mucus, and the extremely flexible arm tips.

"Dabbling technique"

The most prevalent method of food-gathering is the "dabbling technique," utilized primarily by the detritus-feeders. The brittle star moves along the substrate, its tube-feet groping for the scattered food particles, which stick to them and are conducted from podium to podium in the direction of the mouth. These food particles may consist of diatoms, foraminiferans, rotifers, and dead plankton. It is highly likely that all ophiuroids can obtain food in this manner.

In order to capture larger prey (small polychaetes, gastropods, bivalves, echinoderms, and crustaceans), the brittle star wraps its pliable arm tips around the victim and conducts it to the mouth by a spiral bending of the entire arm. This grasping method is practiced by numerous detritus-feeders, including many representatives of the family Ophiolepididae. Those species that feed on covering growth or carrion simply place the oral surface of the disc on the food and chew it off with the jaws, without the aid of the arms and the tube-feet except the oral podia. This method of feeding is employed by many omnivorous species, and to an extent also by the basket stars which feed of polyps of various sessile coelenterates.

Trapping suspended particles

The methods of trapping small organisms and particles suspended in the water are extremely varied and adaptable. The most primitive way of doing this is for the brittle star to anchor two of its arms on the substratum or its home burrow, letting the other arms drift in the water. The greatly extended arms are covered by quantities of sticky mucus, to which various suspended particles will adhere. These particles are then conducted to the mouth by the rows of podia. On the way to the mouth, a separation of edible and unedible particles takes place. Useless particles are either dropped or hurled away. This selective process prior to feeding is significant, since the brittle stars have neither an anus nor an intestine. Indiscriminate ingestion of indigestible material would prove problematic in the short alimentary canal, unlike the situation in the sea cucumbers or echinoids.

Many species improved their trapping of suspended particles by secreting copious amounts of viscous mucus, which is suspended between the lateral spines of the arms and hangs down from them in shreds. Of course, this greatly enlarges the trapping surface. Other species sweep their stretched arms through the water, thereby making contact with more food particles and enjoying an increase in the amount of food. A few species use a unique method of brushing particles off the water surface. Near the coast, the water's surface film consists of, among other things, pollen or small bits of plants. In addition, diatoms and dried food particles are picked up from the shore by the rising tide, and for a period of time these particles are also suspended as a film on the water surface. If certain ophiuroids are in a favorable position, they may stretch two or three arms parallel to each other, over the surface, raking in this food and consuming it.

This feeding behavior has been observed in *Ophiocoma scolopendrina* in its natural habitat. If these forms occupy the tidal zone, they will retreat to their hiding places during low tide. They almost always reappear with the rising tide, and turn two or three arms, with the oral side up, toward the water surface. Then they sweep horizontally along the surface film so that the suspended "dust" is pushed between two arms and is conducted to the central disc. Concomitantly, the greatly stretched podia ladle food particles from this concentrated dust film to the arm grooves, where the particles are coated by mucous strings and are swept toward the mouth. It should be noted, though, that *Ophiocoma scolopendrina* belongs to those brittle stars that are highly diverse in their feeding habits and always employ that method which is most advantageous in a particular environmental condition. In undisturbed waters this form dabs for food particles on the ocean floor with its sticky tube-feet, and in moving water it suspends its mucous capture net from its stretched arms.

Ophiocomina nigra, from Europe's coast, is also quite versatile in its feeding methods. This form will graze on algae and various other plant growth, will feed on carrion, overpower smaller prey by the arm-coil method, dab micro-organisms from the ocean floor, suspend mucous capture nets, and on occasion also rake the water's surface film. The common brittle star (*Ophiothrix fragilis*) is also reported to employ to all these techniques.

The amphiurids, which may be buried down to 10 cm in the sand or mud, practice different feeding habits. These forms extend their long arms from their burrow so that they protrude 2–3 cm above the ocean floor. The arms grope around for food particles, but they may also be erected to catch passing material. The gathered or captured particles are wrapped into strings of mucus, and are conducted to the mouth on the oral surface, deeply buried in the substratum, with the aid of ciliary currents and the podia. It was observed that such a brittle star returns the indigestible substances to the outside of the burrow by a different arm.

The basket stars seem to feed primarily on plankton. *Asteronyx loveni* (see Color plate, p. 390), for example, climbs up on the whip-shaped sea plumes or horny corals, clinging to these with two or three arms while letting the others drift in the water. *Astrotoma agassizi*, from Antarctica, lies on its "back" and sweeps its arms through the water. The basket stars have developed very intricate capture nets with the multibranched arms. *Astroboa*, for example, leaves its hiding place at night, taking up a favorable position in the coral reef and then unfolding its fernlike, branched arms, holding them erect into the water current. At first glance this structure gives the appearance of an impenetrable

Fig. 15-6. *Amphiura* buried in the substratum.

bush, because the capture net consists of the many ramifying arms. The appearance of an *Astroboa* colony in capture position is always a memorable experience for a nocturnal diver. A medium-sized *Astroboa* can spread its capture fan over an area of three-quarters of a square meter (see Color plate, p. 406).

Basket stars employ the grasping method while feeding on plankton. The extremely flexible, whiplike ends of the arms instantly coil around planktonic organisms, usually copepods and other small crustaceans, when they touch them. However, the captured prey is not immediately conducted to the mouth, because the animal must roll in the entire arm before it reaches the mouth. The animal waits until sufficient prey organisms are captured before all the arms are rolled successively to the mouth. The rolling process for each captured organism would require too much energy. Usually all the captured food particles are stuffed into the mouth during the early morning, after the night's capture period, when there is a maximum of food particles adhering to all the arm tips. The food particles are stripped off by the jaw apparatus. Some basket stars, for example *Astrophyton*, are not only plankton-feeders but also graze for food along the ocean floor.

Luminescent brittle stars

Presently, seven species of ophiuroids are known to produce luminescence if stimulated either mechanically or chemically. Light-production is always limited to the arms. The yellowish-green glow and blinking is produced within glandular cells in the connective tissue of the dermis. This phenomenon has been known longest for *Axiognathus squamata*, which has a worldwide distribution. In this species, only the bases of the spines are luminescent, even in the young that are within the bursae of the brood-protecting mother. Mother and brood produce a peculiar light pattern. In *Amphiura filiformis*, on the other hand, the entire lateral spines are luminescent. *Ophiopsila annulosa* has light-producing pigment in the aboral plates of the arms as well as the lateral plates. One can well imagine that this brittle star, with its 12-cm-long arms, bearing approximately 20,000 spines, produces quite a brilliant light.

Reproduction

The spawning period for most ophiuroids occurs during the late summer or early autumn, with a peak period in August. Of course, there are also species, particularly those in the tropics, that reproduce in the spring and summer or in the winter. *Amphiura chiajei*, for instance, which occurs in the northern Atlantic, breeds in the winter. Usually the males release their gametes first, which stimulates adjacent females to discharge their eggs. Many species are able to discharge gametes over a period of one month, until the gonad supply is depleted. Not all individuals of one species are sexually mature at the same time, and for this reason it is not uncommon that several species have long spawning periods, frequently up to three months. In *Amphiura filiformis* the reproductive period may be spread over six months.

Many brittle stars rise up on their arms when releasing the gametes.

In some species, pair formation takes place, as in the antarctic basket star—*Astrochlamys bruneus*—and *Astrothorax waitei* of New Zealand. In these species the dwarf male clings to the female's aboral surface. The male coils its arms between those of the female along the disc margin. The females of *Ophiosphaera insignis*, *Ophiodaphne materna*, and *Amphilycus androphorus* carry their dwarf males in a similar manner, but on the oral surface. The oral sides of the female and male face each other. It seems that in *Amphilycus* (see Color plate, p. 389) the partnership forms at birth. In other cases, little information is available concerning the duration of the pair formation.

Most ophiuroids produce very small eggs with a low yolk content and a diameter of 0.1 to 0.18 mm. The eggs are discharged into the surrounding water, where they undergo further development in a swimming existence. However, approximately sixty species deviate from this technique. These forms are brood-caring or produce live young which undergo their development within the bursae or the ovaries. One New Zealand representative of the order Ophiurida lays its 20 to 200 eggs arranged in a row underneath stones, in crevices, or on algae.

The eggs which are suspended in the water develop first into the ciliated blastula (final stage in cleavage during the embryological development; see Vol. I) which in the course of three to four weeks will develop into the so-called ophiopluteus (see Color plate, p. 272). This larva is characteristic for the ophiuroids, and is similar to the pluteus larva of the echinoids. The ophiopluteus larva is usually characterized by four pairs of arms with supporting skeletal rods and ciliary bands which fringe the arms. The most posterior pair of arms is often unusually long, particularly in the Ophiotrichidae.

The ophiopluteus larvae swim around and ingest microscopic organisms. During this stage they grow considerably. The planktonic phase, lasting from two to five weeks, is followed by metamorphosis which proceeds during the swimming stage. The first skeletal elements of the future brittle star are developed to the side of the larval stomach. The oral field becomes depressed into a keylike vestibule, which in contrast to the holothuroids and echinoids does not form into a chamber. Five finger-shaped processes push from the anlage of the water-vascular canal, which encircles the esophagus, to the floor of the vestibule. The tissue of the vestibule and the processes fuse, forming the five primary tentacles. The first pairs of tube-feet develop from the beginnings of the radial canals behind these structures. The larval arms are progressively re-developed. In some species this is a long process. Correspondingly, the animals may drift in the plankton for quite some time, and may be swept over great distances. The species of the genera *Amphiura* and *Ophiura* are even capable of arresting development periodically if they are unable to find a favorable habitat during the swimming stage.

Subsequently, the larval body becomes flattened and the aboral surface

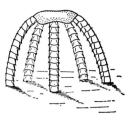

Fig. 15-7. *Amphiura filiformis* in spawning position.

Development

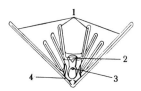

Fig. 15-8. Schematic drawing of an ophiopluteus larva: 1. Larval arms, 2. Mouth, 3. Anus, 4. Skeletal rods supporting the arms.

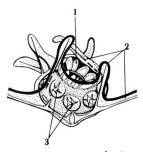

Fig. 15-9. Metamorphosis of the ophiopluteus larva of *Ophiothrix*: 1. Primary tube-foot in the vestibule, 2. Larval arms, 3. First skeletal plates of the developing brittle star.

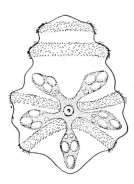

Fig. 15-10. Vitellaria larva of *Ophioderma brevispinum*.

Fig. 15-11. A young brittle star.

of the young brittle star develops opposite the vestibule. The larval gut closes, and a new anus is not formed. With increasing development of the skeleton, the young brittle star grows heavier. It sinks to the bottom and crawls around on its tube-feet, an activity it loses in the adult stage.

Aside from the ophiopluteus larva, there is also the vitellaria. This is a barrel-shaped larva which is similarly ciliated in several species, but in other forms may bear four or five ciliary rings. The vitellaria originates from eggs with high yolk content, which measure from 0.1 to 0.3 mm in diameter. The mouth, devoid of a depression or vestibule, opens on the ventral side of the posterior body. It is surrounded by five pairs of podia. The central disc develops on the dorsal side of the larva, with a ring of five additional skeletal plates surrounding it. Approximately six days after hatching from the egg, the jaw and the first spines start to develop. The young brittle star is fully developed after twelve days (Fig. 15-11). Finally, there are also species whose eggs undergo a direct development. Each distinct larval stage is suppressed, and a flattened, pentagonal embryo emerges immediately after the gastrula stage.

Approximately sixty ophiuroid species practice brood care. These are usually hermaphroditic forms that primarily inhabit colder polar oceans. The eggs are retained in the maternal body, where they undergo development to the complete young brittle-star stage. Generally, the bursae which also are in contact with the gonads serve as brood chambers. Only in rare instances are the gonads functional as brood chambers. Each bursa may enclose from 1 to 200 embryos. A gonadal brood chamber usually houses only one or rarely three young brittle stars. At the time of birth, which actually means the time of departure from the bursae or the gonadal brood chamber, the young brittle stars may already be quite large; some may have an arm length of 2 cm. The young brittle stars crawl out of the bursae with help of their arms. It is not known how the young escape from the gonadal brood chambers.

Occasionally the embryos are fed by the mother, as exemplified by *Amphipholis squamata*. This species carries only one embryo in each bursa. The embryo is attached to the wall of the bursa by a stalk. The food substances for the embryo are probably supplied by a dense network of hemal vessels within the bursal wall, which closely embrace the embryo. In *Ophionotus hexactis*, where the young develop in the gonads, the brood cavities are closely molded to the stomach wall, and probably obtain food substances directly from it. Frequently, immature eggs may disintegrate and thereby furnish nutriments for the embryos. Under certain circumstances, older embryos may feed on younger ones within the same bursa. This situation was observed in *Ophiomitrella falklandica*.

The rate of growth in the ophiuroids is also greatly influenced by the abundance of food and the temperature. *Ophiopholis aculeata* attaines a disc diameter of 5–7 mm in two years. *Ophiura texturata* grows to 7–11 mm within three years. Most ophiuroids are sexually mature in the second or third year. *Ophiura texturata* probably reaches its maximal disc size of 30–35 mm in five to six years. *Amphiura chiajei* reaches a disc diameter of 7.5 mm in ten to fifteen years, which probably corresponds to its life span. Young basket stars start to branch the arm tips when their disc measures only 1 mm. When the disc diameter is 5 mm, each branch again subdivides. Some experts assume that the large basket stars have a life span of twenty to thirty years.

When compared to other echinoderm groups, the brittle stars and basket stars seem to be plagued by relatively few parasites and commensals. This may be due partly to their high level of activity, which seems to discourage "settlers" from landing on the host and staying there. A unicellular alga and a sponge seem to settle in the network of the skeletal plates in many brittle stars, causing the disintegration of the calcareous ossicles. If the infestation is heavy, the brittle star will die.

Ophiura albida, from the North Atlantic, serves as intermediate for the fluke *Fellodistomum fellis*. The larvae of this parasite form capsules on the outer stomach wall of this brittle star. The adult form of *F. fellis* parasitizes the gall bladder of a fish, *Anarhichas lupus*, which preys heavily on various echinoderms. Primarily, however, the ophiuroids are plagued by parasitic copepods. These organisms may infest the body surface of the ophiuroid as ectoparasites or as endoparasites, mainly within the bursae. The major predators of the ophiuroids seem to be various flatfish and haddocks, as well as various asteroids.

The only defense mechanisms of the ophiuroids are their spines, and in many forms these structures are poorly developed or absent. The skeletal armor may provide some protection. Nevertheless, in contrast to other echinoderms, the ophiuroids can escape slower predators because of their great agility. A certain amount of protection may be achieved by the ophiuroid's well-developed ability for self-mutilation (autotomy). The description "brittle star" is indicative of this phenomenon. Autotomy occurs in response to injuries, adverse living conditions, and attempts to hold on to the animal. In some species, like *Ophiopsila aranea* and *Ophiothrix fragilis*, a mere touch brings about self-mutilation. Usually the ophiuroid rejects parts of the arms or entire arms. The arms always break off between two vertebrae at any location on the arm. Several members of the families Amphiuridae, Ophiacanthidae, and Ophiocomidae are even able to reject the entire upper half of the disc including the stomach, bursae, and sexual organs, leaving only the arms and jaw apparatus.

An ophiuroid has a great tendency not only to autotomy but also to regeneration. Both phenomena are generally widespread among the ophiuroids. Complete regeneration of rejected parts takes place even in those cases where the aboral surface of the disc is lost. Only when all of the arms have been broken off up to the base on the disc does regeneration cease, and the animal dies.

Parasites

▷
Ophiothrix quinquemaculata (compare Color plates, pp. 390 and 405) sitting on red ascidian (*Halocynthia papillosa*; see Color plate, p. 428).

▷ ▷
Ophiura albida (compare Color plate, p. 390) is found in the Mediterranean, eastern Atlantic, the North Sea, and the western Baltic Sea.

The wound which results from autotomy is closed off by amoebocytes and coelomic fluid, a process similar to the coagulation of blood. The regenerative process starts with an epidermal bud at the arm stump. Later the radial nerve and the water-vascular system penetrate into this bud. Subsequently, new vertebrae are formed. Some species regenerate lost parts faster when more arm tips have been lost. Regenerated parts are easily recognizable by their generally weaker appearance.

This well-developed ability for regeneration enables certain ophiuroids to replace missing parts which were lost by a transverse division through the disc. This phenomenon occurs primarily in smaller species of the family Ophiactidae, characterized by six or seven arms and a disc diameter of 3 mm at the most. Usually, transverse division results in two parts which have the same number of arms and intact jaws. The two new individuals appear asymmetrical for quite some time.

Most ophiuroids inhabit the coastal zones. Usually their life habits are cryptic. One rarely sees brittle stars running around freely, and for this reason their frequently large concentrations go unnoticed. The animals are extremely sensitive to light, and are active mainly at night. During the day they hide underneath stones, in cracks, hollows, crevices, empty bivalve and gastropod shells, seaweed, between corals, in the holes of sponges, or buried in sand or mud. Ophiuroids prefer contact with a solid surface (stereotropism). This becomes obvious if one exposes a brittle star and observes how it quickly tries to make contact with the walls of a niche. It is therefore not surprising to find ophiuroids concealed in the smallest and narrowest cracks.

There are ophiuroid forms, such as the common brittle star and the daisy brittle star, which prefer hard substrates. Many basket stars also prefer this habitat, and they occupy rocky ground which has been cleared by the water currents. These forms capture the plankton swept along by the current. Generally, currents which carry food particles are an important prerequisite in the habitat selection by the brittle stars that feed on suspended particles. Various species of the Ophiurida therefore settle on hard substrates as well as on soft ones.

Strict sand and mud inhabitants are found among the Amphiurida, many of which bury themselves right up to the arm tips. Many other forms live in close association with other animals. Some ophiuroids are found only in sponges; others cling to soft and horny corals, sit on sea feathers, rhizostomes, and feather stars, or inhabit the spiny forest of sand dollars. Young brittle stars often penetrate into the bursa of an adult star.

◁
Basket star (*Astroboa nuda*) in capture position.

◁ ◁
The common brittle star (*Ophiothrix fragilis*; compare Color plates, pp. 390 and 403).

In areas with an abundant amount of food, one can frequently encounter immense concentrations of ophiuroids. Such concentrations may often persist for years. Members of the Amphiuridae are occasionally crowded to such an extent that a dense tangle of overlapping arms is spread over the ocean floor. Under certain circumstances, 300 to 500

Amphiura filiformis are found on 1 m². A sponge the size of a human hand housed seventy-five *Ophiactis savignyi*. Seven hundred individuals of the species *Ophiura texturata* were counted per m². Up to approximately 340 animals of the common brittle star (*Ophiothrix fragilis*) can congregate per m². In this situation the animals are found in layers; they have been found to cover an area of 15 to 30 km² at a density of 100 per m². On one occasion 100,000 individuals of the small *Ophiactis virens* were found per m² within a cave in the Mediterranean. On the basis of underwater photos made in certain regions of the Antarctic, experts have estimated a population density of 10 million brittle stars per km².

Presently, the ophiuroids are the most successful group among all echinoderms. This is expressed in the number of species as well as in the high population densities found in the world's oceans. Some species have achieved an unusually wide distribution. This is examplified by *Axiognathus squamata*, which is found in all oceans including the subarctic and subantarctic regions. *Ophiactis savignyi* also enjoys a worldwide distribution in the tropical belt.

Like the other echinoderm groups, the ophiuroids also have a central point in species density in the Indonesian-Philippine region. From here the number of species decreases toward the west to Africa's eastern coast and the Red Sea, toward the east to Hawaii, and toward the north in the direction of Japan. At least seven ophiuroid species were successful in bridging the watery desert of the eastern Pacific Ocean, settling on the coast of the Americas. These species and ophiuroid species endemic to western American waters are found along the coasts of Mexico, Panama, Peru, and Chile. Unique ophiuroid forms are also found along the coasts of Australia, New Zealand, Japan, and South Africa.

Distribution

In the tropical Atlantic certain forms occur both in the West Indian section and the western African region, which is represented by few species. The large West Indian region is characterized by a substantial number of native brittle star species which spread southward as far as Brazil and northward as far as the Carolinas.

In the temperate latitudes, tropical species which have penetrated into the colder waters overlap with species that have spread to the subtropics from the subpolar regions. The extensive coastal regions of Europe are characterized by ophiuroids that have originated in arctic regions and migrated southward along the coast of Norway, and by some species that inhabit the North Atlantic and which are found from Norway to the Canaries or the Cape Verde Islands and the Azores. These forms also occur in the Mediterranean. There are also species that are unique to the Mediterranean. Many of these species also occur on the other side of the Strait of Gibraltar on the coasts of Morocco and Portugal.

With the exception of the widely distributed *Amphipholis squamata*, the two polar regions do not share the same species. As with the other echinoderm groups, the brittle stars are far more widely represented around Antarctica than in the Arctic. It is notable that there is a predominance of brood-caring species in Antarctica and the subantarctic regions.

Vertical distribution

The almost uniform living conditions in the deep-water zones have probably contributed to the fact that many deep-sea forms enjoy a worldwide distribution. *Ophiomusium lymani* is one of the most commonly encountered species at the deeper zones, and has the greatest range of distribution. It is found at depths of 700 to 4000 m. Below the 4000 m depth, one still encounters eleven ophiuroid species which belong primarily to the genera *Ophiura*, *Amphiophiura*, and *Ophiacantha*, which may penetrate to depths of 6800 m.

16 Hemichordates and Beard Worms

Only a few decades ago it would have been difficult to find a discussion of the hemichordates (acorn worms and pterobranchs) and beard worms in any volume dealing with the animal kingdom, with the possible exception of *Brehm's Tierleben* (*Brehm's Animal Life Encyclopedia*), which, in 1922, described the acorn worms as an "isolated group." Some experts placed the acorn worms in close proximity with the echinoderms, others with the chordates.

We now know a great deal more about these three unusual groups of animals, but even so, their systematic position remains the subject of much discussion, speculation, and disagreement.

On the basis of embryological processes, and, in certain cases, the possession of external openings (coelomopores) of the coelom, the acorn worms, pterobranchs, and beard worms seem to show some affinities with the echinoderms, with which they have occasionally been united as the phylum Coelomopora. On the other hand, certain characteristics of the hemichordates are reminiscent of the chordates, and they are frequently regarded as a side branch of common ancestry. Gutmann recently referred to them as degenerated chordates. Some uncertainty still surrounds the classification of the beard worms (pogonophorans).

Common anatomical features in these three animal groups have led some scientists to classify them as a single phylum of deuterostomes (Pentacoela), characterized by three body divisions and a five-part secondary body cavity. Some recent investigations have pointed out the several similarities between the beard worms and annelids, and have called into question the inclusion of the beard worms with the deuterostomes. Strong evidence still remains for inclusion of the beard worms with the deuterostomes, however, and similarities with some tube-dwelling annelids (segmentally arranged setae, posterior growth zone, and others) perhaps should be regarded as analogies rather than homologies. In view of the uncertain position of the beard worms, and in spite of the several features shared with hemichordates, most authorities would prefer to regard them as separate phyla.

Hemichordates and pogonophorans, by L. von Salvini-Plawen

Distinguishing characteristics

Phylum:
Hemichordata,
by E. Wawra

The hemichordates (phylum Hemichordata) are closely affiliated with the chordates (Chordata) on the basis of certain common characteristics. The first two body segments are short only in comparison to the posterior segment which represents the animal's long trunk. The alimentary canal is well developed. There is a buccal diverticulum anterior to the esophagus. The epidermis is highly glandular and devoid of a cuticle. In the colonial forms the epidermis secretes substances that are incorporated into the tubular structures. There are two classes: 1. Acorn worms (Enteropneusta), 2. Pterobranchs (Pterobranchia). There are about 100 species.

Although the acorn worms and pterobranchs are very diverse in their external appearances, they share many common characteristics, with the result that they were recognized as one supergroup very early. The body of these forms consists of two short anterior segments and one long posterior segment. The front end or protosome, also known as the "proboscis" or "buccal shield," is followed by the "collar," which represents the central portion (mesosome). In the burrowing acorn worms this section looks like a broad bulge, and in the pterobranchs this section bears the tentacles. The posterior body segment or trunk (metasome) houses the digestive organs as well as the gonads. This segment varies greatly in its shape.

The dorsal nerve cord and ventral nerve cords, which extend along the body midline of the hemichordates, show certain resemblances to corresponding structures in the chordates, the phylum that also gave rise to the vertebrates, including man. The hemichordate dorsal nerve cord is already a type of spinal cord. In addition, the hemichordates have a "stomochord," an outgrowth from the anterior end of the digestive tract, which might correspond to the palatine invagination—the indentation which develops into the mouth—of a vertebrate embryo, or the pineal gland in the vertebrates. The characteristic gill clefts in the acorn worms, which are also present as rudiments in the pterobranchs, serve as further evidence of a close link between the hemichordates and chordates.

Class: Enteropneusta

The acorn worms (class Enteropneusta), are vermiform ciliated hemichordates with numerous gill slits. They have a straight digestive system and a posterior anus. Body length ranges from 3 to 250 cm. The anterior body segment is also known as the proboscis. It is short and conical, and functions as a muscular burrowing organ. The shape of the proboscis changes with the degree of contraction. A stalk connects the proboscis to the equally short and muscular mesosome or collar. The trunk section (metasome) is elongated and vermiform, and is broadly attached to the collar. Internally the trunk is divided into branchial, genital, hepatic, and caudal regions, which are also recognizable externally. The branchial region is characterized by round or cleftlike gill pores, arranged in pairs on the dorsal side. The genital region is externally recognizable by the dorsal genital ridges (as in the genus *Saccoglossus*) or else as broad lateral lamellae or bulging genital wings, as shown in the ptychoderids. The hepatic section is characterized by deep outpocketings on the midventral ridge. The last body segment (abdominal or caudal) is tubular, with an anus at the end (Fig. 16-6). There are three families: Harrimaniids (Harrimaniidae), Spengeliids (Spengeliidae), and the Ptychoderids (Ptychoderidae); there are approximately seventy living species.

The grayish-yellow to dull brown animals have a reddish to orange proboscis. The primary locomotive apparatus is the proboscis and collar region, which is well endowed with muscles, particularly longitudinal muscle fibers. The proboscis stalk includes a rectangular skeletal plate on the ventral side. The two posterior processes of this plate fork around the collar intestine. Observations on burrowing acorn worms of the genus *Ptychodera* and *Saccoglossus* have revealed that the elongated proboscis penetrates into the substratum at an oblique angle. The proboscis then contracts, and a wavelike bulging motion ripples along the entire body. During the next thrust, the remainder of the body is pulled in after the proboscis has anchored. The internal pressure of the body and the muscular action of the collar act as opposing forces to the proboscis. The skeleton of the proboscis serves as an advantageous support for the proboscis stalk.

The lateral mouth opening is located at the constricted border-line between the proboscis and the collar. The mouth is closed by contraction of the proboscis and by the action of a sphincter muscle. A narrow buccal diverticulum arises from the dorsal part of the buccal cavity and extends anteriorly through the stalk into the proboscis. In cross section this "stomochord" shows a certain resemblance to the notochord (Chorda dorsalis) of the chordates. The resemblance, however is superficial; similar structures are also found in planarians (see Vol I).

The simple buccal tube is followed by the complex branchial pharynx, which connects dorsally with the gill pouches although the ventral part functions as part of the digestive tract. The gills, actually a paired row of narrow slits in the ciliated pharyngeal wall, merge into widened branchial sacs which arose from outpocketings of the intestine and depressions in the body wall. The internal gills communicate with the outside through gill pores. Occasionally the branchial sacs are fused into a single cavity on one side, with only a single gill pore. Depending on the size of the species, acorn worms usually have 40 to 200 pairs of gill slits, with an apparent maximum of 700 pairs.

Independent of the number of gill slits, the downgrowth of the tongue bar into the branchial sac provides an increased surface area for gas exchange to take place. In most acorn worms, the tongue bar hangs freely suspended in its gill slit. In the ptychoderids, however, the tongue bar is connected to the branchial septa by transverse rods (synapticles). Skeletal elements, as well as the blood, penetrate into these transverse rods and prevent collapse of the gill. Just like the skeleton of the proboscis stalk, the skeletal elements of the tongue bars and skeletal rods are localized

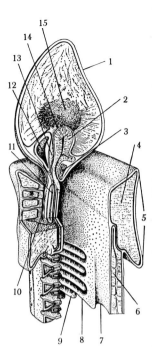

Fig. 16-1. Part of a longitudinal section of an acorn worm (ventral side = right): 1. Proboscis, 2. Buccal diverticulum, 3. Proboscis skeletal plate, 4. Muscles, 5. Collar, 6. Blood vessel, 7. Digestive section of buccal tube, 8. Branchial septa and cross ridges, 9. Pharynx with tongue bars, 10. Collar cord with canal, 11. Blood vessel, 12. Pericardium, 13. Heart, 14. Glomerulus, 15. Proboscis coelom.

Fig. 16-2. Food current along the anterior end of *Protoglossus koehleri*: 1. Gill pores, 2. Respiratory current, 3. Mouth opening, 4. Mucous string, 5. Ciliary tract.

structures or enlargements of the basement membrane. The structure of the pharynx strongly resembles that of *Amphioxus* (see Chapter 18).

The acorn worms feed primarily on microscopic organisms which they ingest along with other material from the ocean bottom. Frequently the intestine of an acorn worm is turgid with substrate particles, and when one picks up the animal, the trunk will tear off because of its weight. Several species of acorn worms are known, therefore, only on the basis of the anterior section of the body. Adult acorn worms feed by burrowing with the proboscis (Fig. 16-5) into the substrate surrounding their tube. The animals do not swallow particles indiscriminately; instead, ciliary and mucous activities separate food particles out. These particles adhere to the mucus secreted by the numerous glandular cells of the proboscis. A ciliary current sweeps the mucous stream toward the posterior part of the proboscis in the direction of the collar. At this point the mucous bands merge through the action of the ciliary organ, and are transported to the buccal cavity by ciliary action. Part of the ingested material is conducted further down the collar and trunk. During feeding, the collar functions like a funnel to the proboscis. This is the most efficient way to widen the mouth.

Food particles and bottom material are conducted along the digestive part of the pharynx, while the water current flows along the branchial pharynx on the dorsal side and is discharged via the branchial slits and pores (Fig. 16-1). The ciliary areas of the branchial sacs produce suction by beating the cilia vigorously, resulting in a respiratory water current. The food clumps are digested in the hepatic segment of the animal. While traveling down the digestive tract, the food clumps are constantly rotated. Digestive juices for starches, sugar, and protein are known to occur within the hepatic section. Indigestible particles are passed along the hindgut and out of the anus as thin coils of feces. When defecating, the acorn worm pushes the posterior end out of its tube and moves it in a circular pattern until a hollow coil of feces, measuring from 3 to 6 cm, has been deposited. Presence of these little fecal heaps always characterizes the burrows of the acorn worms (Fig. 16-3). The intestinal contents of several species, however, may include only plankton, with no bottom material.

The acorn worms lack true excretory organs. It is generally assumed that excretory products are voided through the coelomic lining into the coelomic fluid. A lobed enlargement of the blood vessel in front of the heart, known as the glomerulus, seems to have an excretory function. The glomerulus develops in the following manner: the inner lining (peritoneum) of the protocoel, highly folded to provide a greater surface area, encloses the dorsal surface of the buccal diverticulum. The resulting space is filled with blood from the central sinus. This organ may free the blood from waste. The excretory products penetrate the wall of the glomerulus and also the peritoneum, and are finally discharged, via the coelomic cavity, through the proboscis pore.

Fig. 16-3. Tube of an acorn worm with the characteristic fecal heaps.

Only part of the circulatory system, or none at all (open circulatory system), consists of blood vessels. In the acorn worms the blood flows through sinuses within the tissues, and in this manner bathes the individual organs. The main circulatory sinuses consist of fused coelomic lining and independently evolved circulatory lining, as is found in the heart. The extra lining developed in response to the heavy activity of the heart. The gently flowing blood is pumped through the body by the action of the muscular, rhythmically beating heart, located above the buccal diverticulum and supported by the slightly muscular central sinus. The heart originates as a ventral, groovelike invagination of the pericardium. Depending on the species, this groovelike depression may be closed off as a tube. In the mollusks (see Chapter 1), a comparable heart formation takes place. In the small circulatory system the blood flows from the heart to the glomerulus in the proboscis and back again; in the large system the blood flows backward along the ventral vessel located along the midline of the trunk. Some of the blood flows along the gills and some along the digestive organs, returning to the heart via the dorsal vessel.

Just as in the echinoderms, the true epidermal nervous system consists of a scattered network of fibers underneath the epidermis. Along the dorsal and ventral surfaces of the body this net is thickened into two longitudinal nerve trunks. The dorsal nerve trunk extends from the proboscis to the end of the body. This nerve strand is recognizable externally as a narrow ridge, although in the collar region it is depressed to protect it from harm during the burrowing activity of the collar. The genus *Ptychodera* is characterized by a primitive nerve cord which has a continuous channel throughout its life. The species *Saccoglossus pusillus* also has an opening (neuropore) on the dorsal side, indicating that the dorsal nerve trunk in the collar originated as a groovelike depression in the tissue which gradually fused into a hollow tube. The dorsal nerve trunk is connected with an anterior nerve ring which sends a dorsal and ventral branch out into the proboscis. A second nerve ring is located between the collar and proboscis. The central nerve trunk continues from here.

Epidermal nervous system

The preoral ciliary organ on the ventral side of the proboscis, with its many sensory cells and nervous pad, is probably the only true sensory organ. Neurosensory cells are scattered over the entire epidermis of the acorn worms, but in higher concentration along the proboscis. Consequently, the entire body surface reacts to tactile or light stimuli, although at various reaction times. The proboscis seems to be sensitive to various degrees to light intensity, because when stimulated, the animal turns away and immediately attempts to burrow, as is indicated by the repeated contractions of the proboscis. If the proboscis is severed from the rest of the body, it continues to react in the same manner toward this stimulus, while the remainder of the body reacts only when stimulated excessively. If the dorsal nerve trunk of the proboscis is cut, the response to a stimulus is much slower, since the excitation is conducted over the scattered nerve

Sensory perception

net without direction, unlike the directed circuit along the fibers of the main trunk.

Acorn worms are very sensitive to disturbances, particularly vibrations. They respond by rapidly contracting the longitudinal muscle fibers and disappearing into the burrow. This response is primarily effective against fishes which feed along the ocean floor. It is probable that the iodoform-like secretions of the skin glands also afford some protection. The epidermis of the ptychoderids also secretes a luminescent substance of still unknown chemical composition. In species of the genus *Ptychodera*, from Bermuda, there is a day-night rhythm in the luminescence. Other species produced light only in response to vibrations, some only at night and some even during daylight.

Regenerative capacities

These delicate animals can fragment easily, and can readily regenerate missing parts. The central point for regeneration is the abdominal segment of the body. If one severs the two anterior body segments from *Glossobalanus minutus*, the trunk will develop a bud, which eventually grows into a new fore body, and finally a new collar.

The only way in which one can distinguish the sexes in the acorn worms is by the differently colored gonads, which can be seen through the body wall in the gonadal region, but which frequently extend to the branchial and hepatic segments.

The ptychoderids also have epidermal structures, known as genital wings or lamellae. The gonads terminate in an opening. Copulatory organs are absent. Fertilization takes place in the surrounding water. The female may discharge 2000 to 3000 eggs, usually in gelatinous strings. The male may release the sperm at the same time that the female lays the eggs, or a little later. Spawning usually takes place at low tide. High tide distributes the fertilized eggs.

There are two ways in which the eggs can develop, depending on the amount of yolk present. The yolky eggs of the harrimaniids undergo direct development. The ciliated gastrula elongates along the longitudinal axis, and two circular grooves indicate the body segments of the adult animal. After hatching, the larva is characterized by a ciliary tuft and a telotroch in front of the anus. During its brief planktonic phase, which may last for only one day, the larva develops a mouth and the first gill slit (compare Fig. 16-4). Soon after this free-swimming stage, the larva sinks to the bottom and burrows into the substrate with the aid of the proboscis and a post-larval adhesive papilla posterior to the anus. At this point the young animal starts to feed in the same manner as the adult. It exposes the proboscis and mouth opening to the surrounding water, and, with the aid of the respiratory current, extracts plankton.

The eggs with a low yolk content, found in the spengeliids and ptychoderids, undergo an indirect development. The gastrula develops along the longitudinal axis, the primitive mouth closes, and the embyro becomes a barrel-shaped tornaria larva which hatches after twenty-four to thirty-

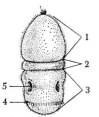

Fig. 16-4. A stage in the direct development of *Saccoglossus kowalevskii*: 1. Proboscis, 2. Collar, 3. Trunk, 4. Telotroch, 5. Gill pore.

six hours. The protocoel is formed by a constriction of the archenteron roof. The new mouth opening develops at the ventral surface. The anus is formed at the location of the primitive mouth. The tornaria larva is entirely covered by cilia which, however, disappear when the ciliary bands are formed. At this stage the tornaria is characterized by a ciliary band anterior and posterior to the mouth, and an apical plate. This developmental stage is known as the Müllerian stage, which is very similar to the bi-pinnaria larvae found in the starfishes and sea cucumbers. This larval resemblance has also been used as proof for an affiliation between the echinoderms and acorn worms. The third ciliary band (telotroch) is located anterior to the anus of the tornaria larva. The telotroch helps to rotate the animal around its longitudinal axis. The longitudinal bands of cilia in the oral region, which extend like loops, conduct microscopic organisms to the mouth like conveyor belts.

After a few months the tornaria has reached its peak of development. During this period it grows to a length of 12 mm, from an egg measuring only 0.12 mm. Now metamorphosis commences. The body size becomes reduced because of loss of water, the ciliary bands degenerate, and the epidermis becomes thickened. Subsequently the body again increases in size, elongates, and two constricted areas in the body form the anlage for the three body segments. At this point the animal sinks to the bottom, the branchial sacs break open to the outside, and the collar cord makes a groovelike depression into the collar. Presently, approximately sixty species of tornaria larvae and seventy species of adult acorn worms have been identified. Many of these, however, undergo direct development, so the discovery of several new species is still probable.

Acorn worms are found in all oceans. They are frequently present in shallow water, particularly in the tidal zones. A fragment of an acorn worm of the species *Glandiceps abyssicola* was found at a depth of 4500 m. However, it is entirely possible that acorn worms have more frequently been collected at depths under 50 m because the sampling techniques are so much easier. All acorn worms inhabit the bottom substrate. The ptychoderids live above the ground among rocks, roots of mangroves, or seaweed. The genera *Glandiceps* and *Glossobalanus* are also found in this type of habitat. These forms construct burrows out of sand granules which they cement with mucus. Another group, consisting of the genera *Balanoglossus* and *Saccoglossus*, burrows vertical U-shaped tubes into the substrate (Fig. 16-5). The walls of these tubes are supported by rapidly hardening mucus secreted by the skin of the acorn worms. Some of these tube complexes, with associated feeding tubes, can reach a depth of 50 cm.

Of the three families, the harrimaniids (Harrimaniidae) are the most primitive group, with few specializations. Their branchial septa are not cross-connected by transverse rods. The genital segment lacks genital wings, and the hepatic segment lacks outpocketings. As has already been mentioned, the yolky eggs undergo direct development. The genus

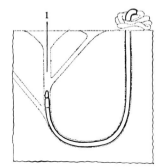

Fig. 16-5. U-shaped burrow of an acorn worm. 1. Tube.

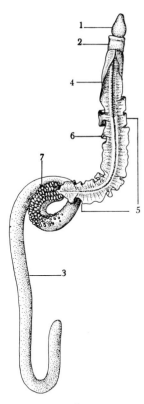

Fig. 16-6. Body segment-
ation of *Balanoglossus
clavigerus*: 1. Proboscis,
2. Collar, 3. Trunk,
4. Branchial region,
5. Genital region,
6. Genital wings,
7. Hepatic region.

Class: Pterobranchia

Protoglossus is particularly primitive, because the coelomic cavities in all three body segments are spaciously developed. The heart vesicle, glomerulus, and buccal diverticulum are simple, and the gill clefts are short. *Protoglossus koehleri* (BL 5 to 7 cm) is characterized by fifty gill slits; it occurs in the English Channel. The genus *Saccoglossus* (see Color plate, p. 427) has an exceptionally long proboscis. This group includes the smallest species of this class, *Saccoglossus pygmaeus* (BL 25 mm), found in Heligoland Bay. Other species include *Saccoglossus horsti* and *Saccoglossus pusillus*. *Harrimania kupfferi* (BL up to 9 cm; see Color plate, p. 427) is characteristic of the North and Baltic Seas, but it is also found along the coast of Greenland at depths of 20 to 1000 m.

The second family, the SPENGELIIDS (Spengeliidae), have many characteristics which are intermediate between those of the other families. In this group the buccal diverticulum is elongated and vermiform. The genital ridges are present, and usually there are transverse rods between the branchial septa. The genera *Glandiceps* (with *Glandiceps abyssicola*) and *Spengelia* (BL up to 56 cm) live on the substrate underneath stones or roots.

The PTYCHODERIDS (Ptychoderidae) are the most highly evolved group of the acorn worms. They are characterized by long genital ridges or genital lamellae, a well-developed tuberculate hepatic region, and gill slits that always have transverse rods. Their eggs are minute, and undergo indirect development through the tornaria stage. Adult animals often grow to a large size. In the genus *Ptychodera* the broad genital wings cover the genital body segment like two mantle lobes. The gill pores are widely open. The species *Ptychodera flava*, found in the Indian and Pacific Oceans, was first discovered by the scientist Eschscholtz in 1825 in the vicinity of the Marshall Islands, although he thought that this animal was a sea cucumber. *Balanoglossus* has broad genital wings but narrow gill pores. The proboscis is often very short. *Balanoglossus clavigerus* (see Color plate, p. 427) is found along the coast lines of Naples and Trieste. This species seems to leave its tube without difficulty, to feed on detritus lying on the ocean bottom. The longest acorn worm, *Balanoglossus gigas* (BL 1.8 to 2.5 m), occurs in Brazil. Of the specimen with a length of 1.8 m, the caudal segment occupied 1.77 m. This clearly shows the relationship of the two anterior body segments to the trunk. The branchial segment is purple and the hepatic region is yellowish-green. In order to capture this acorn worm, one has to dig a trench measuring 50 cm deep and 2 m in diameter in shallow water. *Glossobalanus* is characterized by bulging genital wings. *Glossobalanus minutus* (BL 10 to 15 cm; see Color plate, p. 427) is found in the northern Adriatic and along the coasts of Naples and Toulon.

The pterobranchs (class Pterobranchia) are sessile animals that are usually aggregated or colonial in secreted tubes. The animals have a disc-shaped anterior body. The central body has one or several pairs of tentaculated arms. BL 0.2 to 14 mm. The gill clefts are almost always absent. The intestine is U-shaped. The coenecia (secreted tubes) may measure up to 25 cm. The coloration of the animals is dark brown to reddish. There are two orders and three families: 1. Cephalodiscids (Cephalodiscidae) and 2. Atubariids (Atubariidae) forming the order Cephalodiscida; and 3. Rhabdopleurids (Rhabdopleuridae) forming the order Rhabdopleurida. Altogether there are approximately twenty species.

Just as in the acorn worms, the body of the pterobranchs is divided into three regions. The double tilt in the body axis is explicable on the basis of its sessile mode of life. The anterior body (buccal shield) consists of a well-developed, flattened adhesive disc well supplied with muscles and glands. The ventral side of this shield partially hides the crescent-shaped mouth. A reddish-brown band separates this overlapping fold from the rest of the shield. The highly movable collar or neck, the mesosome, connects the buccal shield with the longitudinal trunk, which has the shape of a bent cylinder. The mouth opening is located on the anterior side of the shorter ventral surface of the collar. The curved dorsal portion of the neck bears paired arms on its lateral ridges. The genus *Rhabdopleura* has only one pair of arms, while the cephalodiscids usually have four to nine pairs. Each arm is composed of two rows of twenty to fifty ciliated tentacles each. A ciliated groove runs between the tentacles. It extends from the mouth opening and occasionally up into the buccal cavity.

The third body segment, also known as the trunk or metasome, is a saclike structure which tapers posteriorly into a tubular stalk. In the cephalodiscids this stalk can measure several times the body length. The V-shaped intestine terminates in the anus, located on the dorsal anterior end of the trunk.

The pterobranchs live in aggregations or colonies which originated by asexual reproduction. In the cephalodiscids the daughter animals bud from the stalk of the mother. In *Rhabdopleura* new individuals, created by budding, stay together. The composition of the substance secreted by the pterobranchs for their coenecia is still unknown, although it does not consist of chitin. The stalks of the genus *Rhabdopleura* consist of irregularly branching stolons which adhere to a hard substratum. Such a colony usually measures only 1 to 1.5 cm, but it can also expand to 5 cm. However, cephalodiscid colonies can grow to over 25 cm.

Many small, vertical secreted tubes, each with a zooid, branch off from one or several "main stolons." The tubular substance is secreted by an epidermal gland in the buccal shield. As the animal brushes new secretions along its tube, an additional ring is deposited; this causes the banded appearance of the tubes.

Generally the epidermis of the pterobranchs is very thin and is only thickened along the lower wall of the buccal shield. It is assumed that the animals are entirely covered by cilia. Glandular cells are frequently found

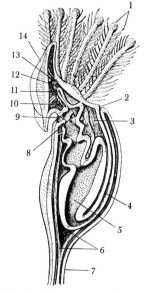

Fig. 16-7. Anatomy of *Cephalodiscus*, longitudinal section: 1. Lophophore arms, 2. Anus, 3. Gonad, 4. Hindgut, 5. Stomach, 6. Muscles, 7. Peduncle, 8. Gill pore, 9. Mouth opening, 10. Glomerulus, 11. Buccal diverticulum, 12. Heart, 13. Pericardium, 14. Buccal shield.

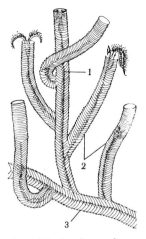

Fig. 16-8. A colony of *Rhabdopleura normani*: 1. Stalk, 2. Vertical tubes of the zooids, 3. Horizontal main tube.

along the tips of the tentacles and in the ventral body wall of the buccal shield (which provides the tube secretion). The musculature is only weakly developed, except for a few delicate longitudinal fibers.

The coelomic cavity has three main divisions, two of which are paired. The unpaired protocoel is represented as a flat cleft in the center of the buccal shield. The protocoel terminates in two pores (buccal pores) on the dorsal side. The mesocoel in the collar is paired, and circles around the buccal cavity giving off a blind sac into each tentacle. The mesocoel terminates in an opening at the base of each arm. Like the mesocoel, the paired metacoel extends into the stalk.

The ciliated epidermis creates a constant water current flowing from the stalk over the trunk, up to the tips of the tentacles. Species like *Rhabdopleura* capture diatoms, rotifers, and crustacean larvae (see Vol. I) with the aid of their ciliated tentacles. The captured food is coated by a film of mucus and is conveyed along two lateral ciliary streams to the crescent-shaped mouth opening, which is surrounded by lips. Ciliary currents continue to conduct the food along its path. The pharynx forms two lateral diverticula at the border between the collar and trunk, which in *Cephalodiscus* and *Atubaria* open to the outside via a gill pore. In *Rhabdopleura* these diverticula remain as two pouches. A diverticulum (stomochord) extends from the buccal cavity into the buccal shield in the direction of the heart vesicle. The gill anlage and the presence of the stomochord demonstrate the close relationship between the pterobranchs and acorn worms. The mucous food string is turned into balls by the action of the cilia within the small saccular stomach. Constant turning propels the indigestible particles to the small intestine, which makes a sharp turn upward and extends upward along the dorsal side toward the anterior margin of the trunk, where the anus is located.

The tentacles function as primary respiratory organs. These structures are well adapted for this function because of their large surface area and the presence of many cilia to produce a steady stream of water. Only rudimentary gill fragments are present. The pterobranch vascular system resembles that of the acorn worms, but is less well developed because of the small size of the pterobranchs. A net of nerve fibers extends within the epidermis of the entire animal. The chief nerve center is located at the midline of the collar. This center sends one nerve strand out along the dorsal surface to the anus, one dorsal and two lateral nerve strands to the buccal shield, and a pair of strands to the arms, where they branch off to supply each arm. One final pair of nerve strands encircles the trunk at the junction with the collar, joining up with the ventral nerve trunk which extends from the tip of the stalk and continues in the direction of the anus. No one has yet been able to demonstrate the presence of sensory organs. The animals do not respond to light and dark stimuli.

The simple saccular gonads are located in the trunk, and terminate in the vicinity of the posterior end of the collar. Most zooids within a

Rhabdopleura stolon do not develop gonads. They remain infertile and reproduce by budding. In *Cephalodiscus* one frequently finds colonies which are either entirely male, female, or hermaphroditic. New stolons or colonies are founded by sexual reproduction. Pterobranch eggs are large and yolky. There is only fragmentary information available concerning their embryological development. There is a yolky ciliated larva characterized by an apical tuft and ventral glandular field (Fig. 16-10), and another type of acorn worm larva which is elongated and in which tilting has not yet occurred. The intermediate stages between these two developmental forms remain unknown.

In contrast to the beginning of a new colony, the expansion of a colony occurs by asexual reproduction, namely by budding from the main stalk. In *Cephalodiscus*, one to fourteen vaselike bulges form at the base of a stalk. The anterior portion of such a bud starts to swell, and separates by constriction. This develops into the buccal shield. Subsequently, two more outpocketings form, from which the arms and tentacles develop. In the meantime the coelomic cavity has subdivided into three segments. Finally the collar becomes demarcated from the trunk. In the cephalodiscids the stalk of the former bud separates from the parent and begins to secrete its own tube, while in *Rhabdopleura* the daughter stays in organic continuity with the parent via the stalk (black stolon).

The majority of pterobranchs occur at depths of 100–600 m, usually on stones, shells, and other hard substrate objects. They are more rarely found on sponges or ascidians. *Cephalodiscus indicus*, found in the Indian and western Pacific Oceans, occurs along coastal cliffs at depths of 10–20 m. *Cephalodiscus gracilis* and *Rhabdopleura striata* are found around Ceylon at 2 to 3 m in the tidal zone. The cephalodiscids seem to be concentrated in the southern hemisphere, with the focal point around Antarctica, and the rhabodopleurids are distributed in the northern hemisphere, especially around the coasts of Greenland and northern Europe, and also in the South Atlantic. *Atubaria heterolopha* lives on hydroid colonies along the Japanese coast.

Up to now very few pterobranchs have been observed alive. They protrude their front end out of their tubes if left undisturbed in a basin for some time. *Rhabdopleura* crawls up in its tube, with the aid of the buccal shield, until the mouth opening is even with the edge of the tube. Then the arms are bent upward. When disturbed, the animal slides down into its tube by contracting the stalk. The arms return to their original position along the main axis of the body. *Cephalodiscus hodgsoni* (Fig. 16-9) can leave its tube completely, except for the bottom of the stalk, which remains attached inside. While the animal crawls around, the stalk becomes greatly stretched. *Cephalodiscus gilchristi* can detach itself completely from its tube. The species adheres to the walls of the tubes with the buccal shields of its many zooids.

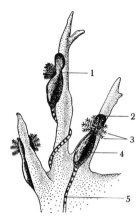

Fig. 16-9. *Cephalodiscus hodgsoni* climbs around on the branches of its tubes: 1. Collar, 2. Buccal shield, 3. Lophophore arms, 4. Trunk, 5. Stalk.

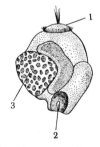

Fig. 16-10. Lateral view of the larva of *Cephalodiscus indicus*: 1. Apical plate, 2. Depression, 3. Glandular area.

System

The CEPHALODISCIDS (family Cephalodiscidae) have one pair of gill clefts and more than one pair of arms. The daughter animals separate from the parent, but all the tubes containing individual zooids are interconnected by a common stalk (coenoecium). The shape of the coenoecium varies in the individual species. The genus *Cephalodiscus* consists of several subgenera: *Orthoecus, Idiothecia, Cephalodiscus* in the narrow sense, *Demiothecia, Acoelothecia*. The subgenus *Orthoecus* is characterized by tubes arranged like organ pipes. In *Cephalodiscus densus* the tubes reach a height of 13 cm and a similar diameter. The tube length of a zooid is approximately 3 cm, body length without stalk is 0.7 cm, and length with stalk up to 4 cm. The subgenus *Idiothecia* constructs treelike tubes with a height of over 19 cm and a diameter up to 11 cm. A zooid of *Cephalodiscus gilchristi* with its stalk measures only 1.6 mm. The colony of the subgenus *Demiothecia* consists of a common, richly branched living space for all zooids. The height of the structure is 25 cm, and the diameter 19 cm. *Cephalodiscus dodecalophus* has a body length of 0.2 cm. The subgenus *Acoelothecia* secretes a netshaped colony with cavities in which the zooids can climb about. The height of the solid structure is 8 cm and the diameter is 3.5 cm. *Cephalodiscus kempi* has a body length of 2 mm. In the ATUBARIIDS (family Atubariidae), characterized by a pair of gill slits and eight pairs of arms, the second pair of arms has been transformed into grasping organs. These animals do not live in secreted tubules, but crawl about on polyp stalks, which possibly serve as a substitute for the missing tube. There is only one species, *Atubaria heterolopha* (L without stalk 1 mm, with stalk 3 mm).

The RHABDOPLEURIDS (family Rhabdopleuridae) lack gill slits and have only one pair of arms. The daughter (zooid) animals are in organic continuity with the parent animal by means of a thin black tissue. The tubes are very delicate. There is an external difference in the sexes: males are spindle-shaped and are twice as large as the females. The tubes of *Rhabdopleura normani* usually measure 1–2 cm, and the maximum length is 10 cm. Zooids without stalks measure 0.3 mm; the tube is approximately 3 mm high.

It is highly probable that aside from the acorn worms and pterobranchs, several extinct animal groups also belonged to this phylum. The graptoliths (class Graptolithida) from the Silurian (approximately 400 million years ago), which frequently have left their petrified scriptlike or sawbladelike imprints of delicate "polyps" on dark slate, can probably be classified with the Pterobranchia.

In the depths of the Bay of Biscay a transparent larva, the size of a gooseberry, was discovered and classified as *Planctosphaera pelagica*. This larva could be derived from a larval form (tornaria) with three ciliary bands, but is of much greater complexity. However, *P. pelagica*, which up to now has only been found in a few areas, shows such great differences from the known hemichordite larvae that it must be regarded as representing a highly distinctive adult stage, which is still unknown to us. For this reason, this particular form is frequently classified as a representative of its own class (Planctosphaerida).

A peculiar, stout vermiform animal (*Xenoturbella bocki*; L approximately 3 cm) was found in the muddy flats of Skagerrak and the North Sea. Some of its characteristics, and particularly its glandular columnar epithelium, are indicative of a distant relationship with the hemichordates. According to Reisinger, *Xenoturbella* could also represent a primitive larval form of this particular group, that reaches sexual maturity prematurely.

The second phylum to be dealt with in this chapter is the phylum Pogonophora (beard worms). These threadlike animals are sessile, and live in cylindrical tubes. Each individual animal lives within one tube. They are devoid of an intestine, and are marine. Body length is 4–40 cm, usually 10–20 cm, body width often only 0.3 mm. The anterior body bears from 1 to 200 tentacles which have the dual function of respiratory and food-gathering organs, since the alimentary canal is degenerate. The central area represents the major part of the body. The short posterior body is segmented. The greater part of the epidermis is covered by a cuticle. The epidermal layer of each individual animal secretes a tube that is usually longer than the actual body. There are two classes: 1. Frenulata, with the orders Athecanephria and Thecanephria and the extinct Hyolithellida. 2. Afrenulata, with the single order Vestimentifera. Altogether there are probably over 100 species.

The beard worms (Pogonophora) have been known since early in this century, although their characteristics were not described until around 1950. It is likely that the beard worms were discovered at an earlier date but were mistaken for threads from a net or some similar object. It has been no easy task to clarify the morphology of these animals. Some scientists assume that the tentacles originate on the ventral side of the body. This assumption was responsible for the name "beard worms." However, others believe that the tentacles are carried on the dorsal surface.

The beard worms resemble the hemichordates not only in the development of the coelomic cavities and its associated openings, but also in the vascular system, heart, and direction of blood flow. Also, the ciliary field which is close to the origin of the gonads—especially if viewed in connection with the partially paired nature of the ventral nerve strand—can be interpreted as a vestigial locomotive organ located on the ventral surface. Larvae with a similar organ on their ventral surface have been known to use it for locomotion. The genital wings of numerous acorn worms are not comparable with the papillose ridges of the beard worms. In the latter case these structures represent various external features in different body segments. The nervous system of the pogonophorans and hemichordates show many common features. In the beard worms the nervous network is concentrated only along the ventral side, while in the

Phylum: Pogonophora, by E. Wawra

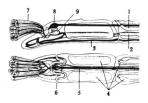

Fig. 16-11. Anatomy of a beard worm—lateral view (above) and dorsal view (below): 1. Coelomoduct = oviduct (female), 2. Ventral blood vessel, 3. Nerve cord, 4. Paired coelomic cavities of the mesosome, 5. Dorsal blood vessel, 6. Coelomopore, 7. Tentacles with blood vessels, 8. Heart, 9. Protocoel with pore.

Fig. 16-12. *Polybrachia annulata*: 1. Tentacular crown, 2. Protosome, 3. Frenulum, 4. Part of the mesosome, 5. Papillae.

hemichordates there is an upper as well as a lower nerve trunk. It is logical to assume that the tentacles have developed on the dorsal body surface, since this is the side that is exposed to freshly circulated water (respiration) and food particles (feed-gathering). Respiration and food-gathering are two functions carried out by the tentacles.

When considering all these factors, it appears that the beard worms show some relationships with the hemichordates. Recently, certain researchers have pointed out a possible phylogenetic relationship to the annelids; such a relationship is not supported by sufficient evidence, especially when the subdivisions of the body are considered.

The class FRENULATA is made up of the orders Athecanephria and Thecanephria. The athecanephrids have a pericardium. The protosome and mesosome are separated on the outside by a distinct groove. The tentacles are attached individually, and are not fused at their bases. The family OLIGOBRACHIIDAE is characterized by two to twelve tentacles. *Oligobrachia webbi* was discovered at a depth of 260 m off northern Norway. The SIBOGLINIDS (family Siboglinidae) have only one or two tentacles. *Siboglinum ekmani* (BL approximately 5 cm, diameter 0.1 mm, tube L 9 cm) was found in the Skagerrak, and *Siboglinum caulleryi* was found at a depth of 8164 m east of the Kuril Islands (north of Japan). The family SCLEROLINIDAE includes *Sclerolinum brattstromi*, which lives in the interstices of rotten wood, paper, and leather on the bottom of some Norwegian fjords. Sclerolinids have two tentacles without pinnules.

The THECANEPHRIDS (order Thecanephria) lack a pericardium. The protosome is only occasionally separated from the mesosome by a groove. The coelomic ducts are surrounded by saclike bulges from the ventral vessel. The POLYBRACHIIDS (family Polybrachiidae) have three tentacles. This group includes *Zenkevitchiana longissima* (BL without tentacles, 36 cm) which occurs in the region east of the Kuril Islands at 4000 to 9000 m. The genus *Galanthealinum*, found in northern Canadian waters, is significant because of its resemblance to the extinct Cambrian HYO-LITHELLIDS (Hyolithellida). The LAMELLISABELLIDS (family Lamellisabellidae) have ten to thirty-one tentacles with processes (pinnules) fused at the bases. Representatives of the SPIROBRACHIIDS (family Spirobrachiidae) are characterized by 39 to 223 tentacles, attached to a spiral tissue.

In 1969 Webb discovered a new species, *Lamellibrachia barhami*, in the northeastern Pacific Ocean. This form was referred to the new class AFRENULATA, order Vestimentifera, family Lamellibrachiidae. This beard worm has many—approximately 2000—fused tentacles. The mesosome is devoid of the frenulum. The protosome is characterized by a paired fold. The excretory organ is not connected to the coelomic cavity.

Beard worms are threadlike in appearance, and their tubes are pale. It is therefore no surprise that they remained unknown for such a long time. Externally, such a creature is characterized by a short anterior segment which bears the tentacles and is clearly differentiated from the long trunk by a groove. The anterior segment consists of the true anterior section (protosome) and the mesosome. This segmentation is not always distinguishable from the outside. In the more primitive families, a shallow depression between the protosome and mesosome is still recognizable, although this has disappeared in the more highly evolved families. The conspicuous tentacles are borne by the short anterior body segment. The basic anatomical arrangement of the tentacles is very uniform in the beard worms, but the total appearance of the tentacles often varies greatly. They grow from the dorsal surface, and form into a tube by touching sides. They do not form a funnellike tube. In the case where the tentacles have been reduced to one, as in the genus *Siboglinum*, the one arm coils into a spiral. This arrangement also produces a tubular cavity (inter-tentacular space). In forms with a low number of tentacles, for example *Oligobrachia*, the tentacular base is horseshoe-shaped, while in those forms with numerous tentacles it is ringshaped. Species with a very high number of tentacles have a lophophore. In the spirobrachiids this structure is a drawn-out spiral. This enlarged area of attachment can bear numerous tentacles.

The somewhat longer cylindrical mesosome of almost all known species is characterized by two ridges on its anterior part, that encircle it in a V-shape and are therefore known as reins (frenulum). The frenulum can vary greatly in appearance. It is assumed that this rigid structure serves for holding the animal in its tube or to support the animal when it crawls out. The mesosome has well-developed glands which probably function in constructing the tube.

The disproportionally long trunk is subdivided into approximately two equal segments by girdles or belts (annuli). The anterior part (pre-annular section) is again subdivided into two segments: the most anterior one (serial or metameric part) and posterior part (ametameric part). The most anterior part is characterized by a ciliary band on its ventral surface, which is considered a vestige of an entire ciliary cover which covered the primitive adult animal. The dorsal surface bears a double band of papillose ridges, usually arranged in rows and separated by a longitudinal groove. In the male the gonopores open into the first two papillae. The other papillae are usually equipped with hard structures in the form of hooks or pincers. This segmentation is not a true metameric division, as in the annelids (see Vol. I), since the internal morphology does not correspond to the external segmentation.

The regular arrangement of the papillae phases out in the posterior section of the anterior preannular segment. *Oligobrachia* is the only genus where the papillae are never arranged in rows. This condition is considered primitive. The papillose hooks and platelets serve as locomotive aids within the tube. The girdle separating the two regions of the trunk consists of a double belt bearing two to five rows of toothed platelets (uncini). In the athecanephrids the last and longest trunk segment has

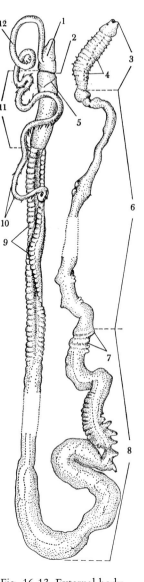

Fig. 16-13. External body features of *Siboglinum fiordicum*: 1. Protosome, 2. Groove which separates the anterior body segment from the trunk, 3. Anchor of the hind body, 4. Bristles, 5. Frenulum, 6. Post-annular body segment, 7. Annulus with uncini, 8. Preannular body segment, 9. Dorsal papillae, 10. Tentacular processes (pinnules), 11. Area formerly regarded as mesosome, 12. Tentacles.

Fig. 16-14. Cross section of a tentacular crown of *Spirobrachia grandis*: 1. Tentacular processes (pinnules), 2. Tentacles.

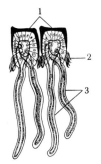

Fig. 16-15. Cross section through the tentacle of *Spirobrachia grandis*: 1. Cuticle, 2. Lateral cilia, 3. Pinnules.

regularly spaced glandular bulges along its ventral surface. In the thecanephrids, on the other hand, there are transverse rows of adhesive papillae on the dorsal surface, which aid in maintaining a hold within the tube.

It is not surprising that these delicate animals are often found only as fragments; the trunk region in particular is rarely recovered in its intact state. Recently, the scientists Webb and Bubko have discovered an additional body region which attaches to the postannular segment. This segment has been found in a few beard worms, such as the species *Siboglinum fiordicum* and *Siboglinum ekmani*. It is characterized by a series of rings and usually also bristles. It serves to maintain the animal in its tube, and was therefore named the "anchor." It is interesting that the external segmentation corresponds to an internal one in this body segment. This information led Webb to the conclusion that the anchor represents the actual trunk (metasome) and that the two portions anterior and posterior to the annulus, which up to then had been referred to as the "trunk," really belong to the central body (mesosome). Webb regards the former mesosome as the gonadless section and the former "trunk" as the gonadal section.

The epidermis of the beard worms consists of a single layer of columnar epithelium with glands, which secretes the elastic double layer of cuticle. Only the inner surface of the tentacles and the ventral ciliary strip on the anterior section of the mesosome are ciliated. In contrast to the bristles on the annulus, the frenulum and papillose platelets are epidermal thickenings. The bristles arise from the epidermis and penetrate through the cuticle. Internally the epidermis is bordered by a basal membrane. The smooth musculature consists of two layers: a fine external layer of circular muscles, and a somewhat more highly developed layer of longitudinal fibers.

The unpaired coelomic cavity of the protosome is extremely small, but the protocoel sends off a saclike projection into each tentacle. The protosome opens to the surface at a slightly lateral location on the dorsal surface via two ciliated coelomoducts. The coelomic cavity in the gonadless segment lacks openings. The second part of the mesocoel is also paired, and opens to the surface via two dorsal openings which function as gonopores for the females. The coelom of the anchor is internally segmented by means of separating walls (dissepiments) which correspond to the external segmentation.

Since the alimentary canal is absent in the beard worms, the tentacles not only function in food-gathering but also in digestion. The inner surface of the tentacles is lined by numerous long, single-celled processes or pinnulae. These pinnulae project freely into the space between the tentacles. The former serve as an excellent filter for plankton and detritus. Lateral cilia situated at the base of the tentacles produce the necessary water currents. This satisfies the food requirements for the beard worm; but how is the food digested? In 1965, with the aid of an electron microscope, Nørrevang discovered that sections of the tentacular epidermis and the cells of the pinnulae from *Siboglinum ekmani* are lined with tiny plasmic projections (microvilli) which, of course, would provide a favorable increase in surface area for digestion to take place. In man, for example, the duodenum is lined with many intestinal villi. These microscopic processes rupture the cuticle at these locations, and absorb food substances. It is not quite clear if the food is broken down by digestive enzymes secreted by the pinnulae or by bacteria, or if digestion takes place extracellularly or intracellularly.

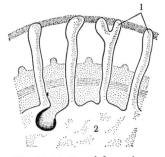

Fig. 16-16. Detail from the cross section of a pinnule of *Siboglinum ekmani*. Part of the striated border: 1. Microvilli, 2. Cell body.

The beard worms have a closed circulatory system. There is a dorsal and a ventral blood vessel running the length of the body. These well-developed vessels also contain muscular components. The muscular ventral heart is found in the mesosome. A pericardium is still located near the heart in the more primitive forms. A vessel extends anteriorly from the heart, branching off into the individual tentacles with their pinnulae. Gas exchange, greatly facilitated by a flowing water current and the great surface area of the pinnulae, takes place at the highly vascular pinnulae. The capillaries flowing away from the pinnulae join into a draining tentacular vessel, and these fuse into the ventral vessel which brings the oxygenated blood to the individual organs, and in particular to the gonads. From here the blood flows backward in the dorsal vessel.

Generally, excretory products pass from the blood through the coelomic lining and into the coelomic cavity. The ciliated coelomoducts from the protocoel, which is in close contact with the abdominal cavity, serve as excretory ducts. In the athecanephrids the area of contact between the two coelomic cavities is greatly increased by the folding of the coelomic lining. Finally the excretions are passed to the outside through the coelomopores.

The beard worms seem to be devoid of sensory organs. Some scientists suspect that the ventral ciliary area fulfills some type of "chemosensory" function, although this is still questionable. There are a few sensory cells on the cephalic lobes which might possibly be light-sensitive. The nervous system consists of a loose network of nerve fibers within the epidermis, which does not follow a definite arrangement. There are a few main trunks. There is a ventral ganglionic mass with two flaps forming a type of circle sending off numerous lateral tentacular nerves in the cephalic lobes. Posteriorly this nervous center continues as a broad net with a pair of giant nerve fibers. Two nerve circles in the gonadless mesosome are significant. These rings encircle the central body from the ventral to the dorsal side (compare acorn worms and pterobranchs). A ventral nerve trunk has not been found in the anchor.

The beard worms are dioecious, and only reproduce sexually. The only way one can differentiate the sexes is by the position of the male and female gonopores. The paired ovaries are located on the outside of the coelomic cavity, which they deeply indent in the preannular section. During the

▷
Acorn worms: 1. *Balanoglossus clavigerus*, 2 and 3. *Glossobalanus minutus*, 4. *Saccoglossus kowalevskii*, 5. *Saccoglossus mereschkowskyi*, 6. *Harrimania kupfferi*.

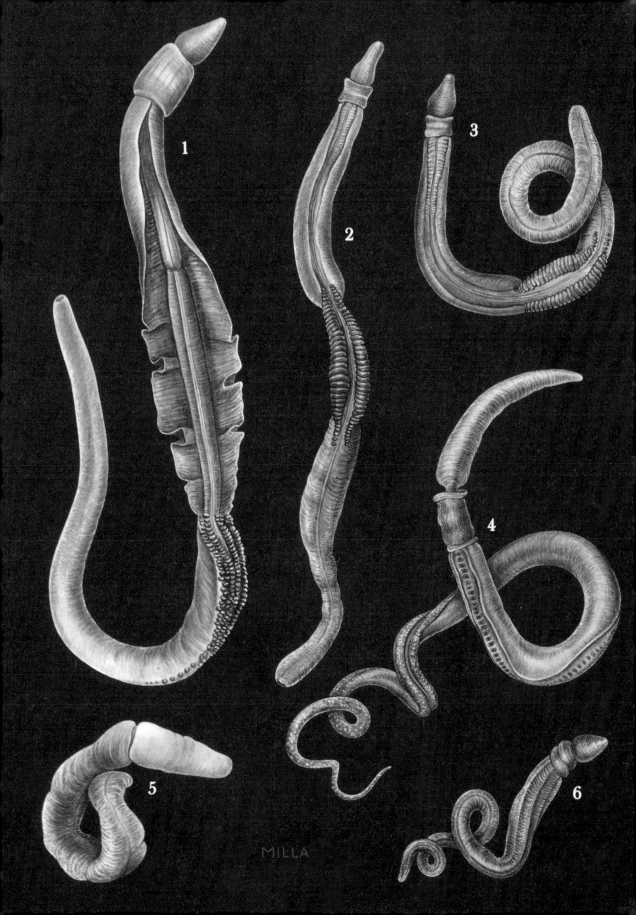

MILLA

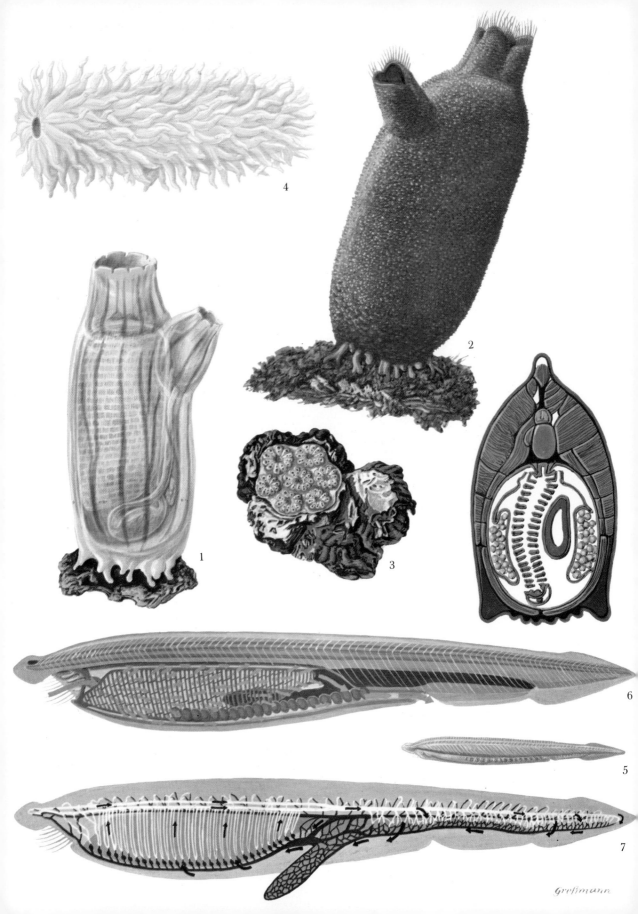

4

2

1

3

5

6

7

Greßmann

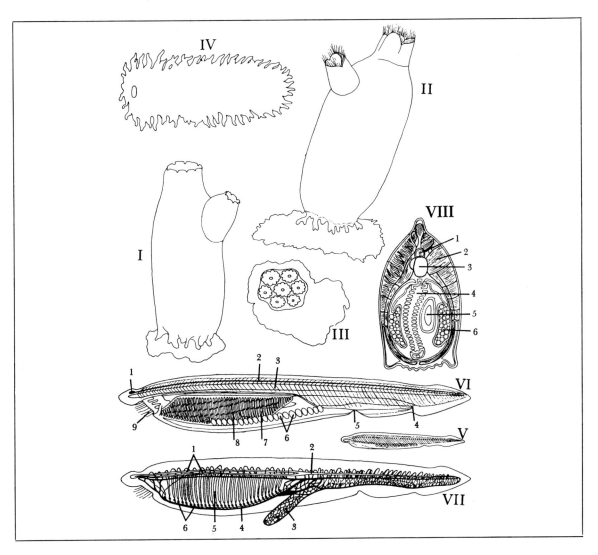

△
◁

Tunicates:
I. *Ciona intestinalis.*
II. *Halocynthia papillosa*
(see Color plates, pp. 403
and 449).
III. Star ascidian
(*Botryllus schlosseri*)
attached to an oyster shell.
IV. *Pyrosoma atlanticum*
Acrania:
Amphioxus (*Branchiostoma
lanceolatum*).

V. Natural size.
VI. Anatomy (semi-
schematic). The blue
arrows indicate the
direction of the water
current as it passes through
the mouth opening,
branchial basket with gill
slits and atriopore.
1. Cerebral vesicle,
2. Neural cord, 3. Noto-
cord (Chorda dorsalis),

4. Anus, 5. Atriopore,
6. Gonads, 7. Liver,
8. Pharyngeal pouch with
gill slits, 9. Mouth with
buccal cirri.
VIII. Anatomy (semi-
schematic). Circulatory
system: arterial system in
pale color and venous
system in dark color.
1. Paired dorsal aorta,
2. Descending aorta,

3. Liver, 4. Ventral aorta,
5. Afferent branchial
arteries, 6. Branchial
hearts.
VIII. Cross section
(schematic). 1. Nerve cord,
2. Trunk muscles,
3. Notochord, 4. Pharynx,
5. Liver, 6. Gonads.

course of maturation the eggs migrate posteriorly, and, following the rupture of the coelomic lining, drop to the oviducts, which are actually a pair of U-shaped coelomoducts with a lateral opening (coelomopore) to the outside. The testes are located from almost the middle of the preannular region to the end of the next segment. The sperm ducts extend to the first papilla of the preannular section without coming into contact with the coelomic cavity. In the last part of the sperm duct the mature sperms are packed in spermatophores (each up to 2.5 mm). Each beard worm species is characterized by a distinct spermatophore. The eggs stay in the oviduct, where they are fertilized. It is still unknown how the sperm is transferred to the oviduct, but the tentacles probably could perform this function.

Little information is available about the direct development of beard worms. The yolky eggs are usually oval. The clutch of *Siboglinum caulleryi* consists of ten to thirty eggs. Cleavage is total and almost even. There is an endodermal gut primordium, but never with an internal cavity. Cells from the anterior portion of this cell ball separate and undergo rapid division to form the protocoel of the protosome. The coelomic cavities of the mesosome and metasome arise in a similar manner, but each body segment has two lateral pouches. Later, ciliated bands develop at opposite ends of the body; two grooves divide the larva into three segments. At this point the larva shows great similarity to the larvae of the acorn worms. The larva also grows bristles which are similar to those of the annelids (see Vol. I) but have a different origin. In the adult animal the ciliary bands and bristles disappear, with the exception of the ventral ciliary tract which aids the larvae in their gliding motions. A depression, still evident on the anchor of the adult animal, forms on the hind body of the larva.

The unbranched tubes of the beard worms are often much longer than the animals themselves, enabling them to climb up and down in them. They are usually delicate and ringed; more rarely, they are rigid. The rigid tubes look like stalked funnels. The tubes are buried in fine deposits on the surface of the ocean floor. Only the uppermost sections of the tubes of many beard worms seem to protrude erectly from the substrate; at least this is what one assumes, since in many tubes only the upper part is covered by other growing organisms. The tubes are formed from the secretions of the club-shaped glands in the gonadless section of the mesosome. The tubes of *Siboglinum* and *Zenkevitchiana* consist of proteins and chitin. In certain species the tubular rings are alternately dark and light.

Beard worms inhabit muddy substrates in deep water, to a known maximum depth of 9900 m. Occasionally they occur in masses. Beard worms seem to enjoy a worldwide distribution. A wealth of species is found in the Okhotsk Sea, the Bering Sea, the Kuril Trench, and the Pommerellen Trench. More than fifty percent of all species inhabit a depth between 3000 and 8000 m.

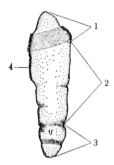

Fig. 16-17. Larva of the genus *Siboglinum*: 1. Prososoma, 2. Mesosoma, 2. Anchor, 4. Ventral ciliary tract.

Distribution

17   The Chordates

Phylum: Chordata,
by O. Kraus

The chordates (phylum Chordata) are probably the most significant evolutionary line in the entire animal kingdom. This group gave rise to the vertebrates, including man, who has been and still is a major force in changing the face of the earth. The chordates represent a very diverse group; this becomes evident when we look not only at all the vertebrates (subphylum Vertebrata) but also at the marine amphioxus (subphylum Cephalochordata) and the tunicates (subphylum Urochordata). It is entirely possible that several further groups have yet to be included with the chordates.

Because of their great diversity, it is impossible to characterize the chordates on the basis of general external properties. Creatures like the delicate, transparent marine appendicularans, which belong to the tunicates, as well as the fishes, amphibians, reptiles, birds, and mammals, all belong to this phylum. Body size alone may vary from 1 mm, as in some tunicates, to approximately 30 m in the blue whale. The unifying characteristics of the chordates are of an anatomical and phylogenetic nature.

The chordate body is always bilaterally symmetrical. The epidermis does not give rise to any supporting skeletal elements; instead there is an internal skeleton. The secondary coelom is always developed, although due to degeneration it is present only as the pericardium in the tunicates. These characteristics and the type of embryological development would, however, only classify the chordates as deuterostomes (supergroup Deuterostomia; see Chapter 9). Additional common characteristics prove that the chordates are a unique group within the deuterostomes.

The following distinguishing characteristics are found in all chordates, even if they are only restricted to some embryonic stage: The development of a rodlike supporting organ extending along the longitudinal body axis—the notochord (Chorda dorsalis). This characteristic is stressed in the scientific name of the group—the Chordata. The nervous centrum is always located dorsally to the notochord. In the primitive forms the hollow nerve chord is long and tubular. The central germ layer (meso-blast), which originated from cells from the primitive gut wall, is paired. These longitudinal mesodermal layers running along the sides of the archenteron later give rise to musculature, the coelom, and various organ systems.

A comparison of the feeding methods in the various subgroups of this phylum reveals an additional common characteristic: Tunicates and amphioxus feed exclusively on plankton and microscopic organisms. This is also true for the primitive vertebrates. The jawless ostracoderms (Ostracodermata; see Vol. IV), from the early Paleozoic, were also filter feeders, as was proved by Stensiö. Some jawless forms, the cyclostomes (class Cyclostomata; see Vol. IV), have survived into recent times. Although these modern representatives are carnivorous or else feed on carrion, their larval forms—particularly the Ammocoetes larvae of the lamprey—still employ the primitive filter-feeding mechanism. This shows that all chordate animals shared the same primitive food-gathering method. Only in the vertebrates has the manner of food-gathering evolved along very diverse paths.

Large-sized animals can subsist on small organisms only if these are available in correspondingly large quantities. It would then follow that one would expect to find suitable trapping and filtering apparatus. This is indeed true: In principle the corresponding organ is similar to the tunicates, lancelets, and vertebrates. In all cases the pharynx consists of lateral openings in contact with the external environment, namely the water surrounding the animal. In the German language this structure is aptly called *Kiemendarm*, meaning "gill gut." The water that enters through the mouth is expelled through these pharyngeal slits. During this process, suspended food particles are filtered out of the water current, and gaseous exchange or respiration takes place along the pharyngeal slits. Only in the vertebrates have these structures undergone drastic changes. One can still discern the basic structural components without great difficulty in the gill apparatus of the fishes, but in the higher verte-brates, including man, only the phylogenetic development bears witness to these primitive pharyngeal elements as fragments in the hyoid and laryngeal apparatus.

And yet there is still another special feature in the chordate animals which may appear odd at first glance but which has some connection to the primitive filter-feeding method. This is the initially small tendency for cephalic and brain development. Examples of filter-feeding in other animal phyla have shown that an organism can feed efficiently without such a center at the anterior end of the body, as do, for example, the mollusks. The lancelets also lack a true head region with a brain and large sensory organs. The situation is not much different in the tunicates, but there is a slight indication of a head region in their larval forms. Only in those vertebrates which lost the filter-feeding ability very early

Fig. 17–1. Comparison of the position of the noto-chord, musculature (axial musculature), and branchial area (digestive tract respec-tively) in a *Doliolum* larva and a cartilaginous fish (shark).

in their phylogenetic history did a distinct head region develop, reaching its highest peak of organization in the higher vertebrates.

Subphyla: Tunicata, Acrania, Vertebrata

We differentiate between three "undisputed" groups of chordates, the subphyla of tunicates, amphioxus, and vertebrates. Certain anatomical and embryological characteristics found in a few other animal groups lead to the possibility that these are also true chordate animals. This applies particularly to the acorn worms (Enteropneusta; see Chapter 16), which also have pharyngeal gill slits, a central nervous system similar in its morphology and embryology to the chordates, and finally a supposed notochord in the proboscis that, according to some authors, is comparable with a true notochord. Despite this seemingly convincing evidence, the controversy about classifying the acorn worms with the chordate animals has not yet been settled. If this should come about one day, then scientists will also have to examine the pterobranchs (Pterobranchia; see Chapter 16), which are closely related to the acorn worms and should also be included in the chordates. Some experts even classify the beard worms (Pogonophora; see Chapter 16) with the chordates. The beard worms have been recognized as a distinct group only since 1937. Presently, no one can settle this question beyond doubt.

Several uncertainties still cloud the concept of "chordate animals." It is therefore not surprising that one cannot and probably never will be able to give an undisputed view about the origin of the chordates and their subgroups. Many scholarly theories about chordate origin have been expounded, but in the final analysis most reports have to substantiate their writings with data based on comparative anatomy and embryology. Some fossil animals have been regarded as primitive chordates by some scientists, but their chordate affinities have been disputed by others. Nevertheless, the phylum must have originated long before the time of the earliest known records of animal life.

During the course of evolutionary development, which encompasses many hundreds of million years, only the vertebrates, of all the chordates, were able to conquer practically all habitats. All other groups from this phylum were limited to a marine environment despite their infinitely "long run."

18 Tunicates and Acrania

Nobody would suspect, when first glancing at a tunicate, that this animal is in fact a chordate and a close relative to the vertebrates. This is one reason why this long-familiar creature occupied a doubtful taxonomic position for a long time. It was only in 1866 that the well-known Russian zoologist Alexander Kowalewskij could demonstrate that the tunicates (subphylum Tunicata) belonged to the chordates. In many respects the tunicates seem unique and highly specialized, but at the same time many characteristics appear to be simplified. The loss of progressive structural features in the tunicates also resulted in the degeneration of characteristic chordate properties. The true phylogenetic relationship of the tunicates with the other chordate animals becomes more evident if one examines the phylogeny and anatomy of their larval forms.

Some tunicates are pelagic throughout their life, and others are sessile in their adult stage. The notochord is rarely preserved in the fully developed form. Frequently the notochord is degenerate, like the entire posterior body segment. Length in the colony-building species is 0.5–2 mm. Solitary sea squirts can reach a height of up to approximately 33 cm. The body is saclike or barrel-shaped; only the appendicularians retain the larval tail segment throughout their life. The body is protected from the outside environment by a thick secretion of the epidermis which is known as a tunic or test. The majority of the animal's body is made up of a well-developed pharynx which functions as food filter and a respiratory organ. Tunicates are hermaphroditic. There are three classes: 1. Ascidians or sea squirts (Ascidiacea), sessile in the adult phase; 2. Thaliaceans (Thaliacea), pelagic; 3. Appendicularians (Larvacea), also pelagic oceanic forms.

The unique enclosure of the sea squirts, the tunic, is the feature responsible for the name TUNICATA. In some cases the tunic is gelatinous, but in others it is quite solid and of a tough leathery consistency. Since the tunic is an epidermal secretion, one could be tempted to call it cuticle. However, a closer examination of the morphology of this structure

Subphylum: Tunicata, by O. Kraus

Distinguishing characteristics

reveals that it does not consist of dead material, but is a living tissue. The basic substance is made up of organic and inorganic nitrogenous bonds and, above all, tunicin. In the ascidian of the genus *Ciona*, tunicin makes up approximately sixty percent of the dry weight. Tunicin can also be referred to as "animal cellulose." Among all the multicellular animals, only the tunicates contain this substance, which is so widely distributed in the plant kingdom. The basic substance of the tunic is interspersed with living cells that have migrated from the deeper body layers. Even tactile and nervous cells have been found within the tunic. In some forms there might even be color pigments and calcareous elements. In the ascidians this entire structure is additionally interlaced with a branching vascular system. The tunic has a protective function, but also serves as an antagonistic force for the trunk musculature, because of its own elasticity.

The anterior part of the body is characterized by a wide mouth opening passing into a short buccal cavity and then directly into the pharynx, which is really the largest organ of the tunicates (see Color plate, p. 428). Commensurate with the anatomy of the pharynx in all chordate animals, the lateral gill slits are also developed in the tunicates. The interior of the gill clefts is in contact with the surrounding water. Some forms, such as the appendicularians and several thaliaceans (see Color plate, p. 437), have only a single gill slit. In other forms, particularly the ascidians (see Color plate, p. 437), the number of successive gill slits is high. The horizontal rows of slits are separated by horizontal and vertical bars, resulting in a gridlike "branchial basket." The beating of the cilia lining the gill slits produces a current enabling water to flow through the mouth opening, into the pharynx, and out via the gill slits.

Branchial basket

These special adaptations serve not only to filter out the planktonic organisms, but also to conduct the food particles to the subsequent digestive segments. The ciliated glandular endostyle (that is, the hypobranchial groove) produces mucus which coats the food particles. The endostyle is a narrow longitudinal band extending along the entire ventral side of the pharynx. A similarly developed endostyle is also found in the amphioxus. The secreted mucus resembles an endless string. In the ascidians the string of mucus has a thickness of only one micron (one millionth of a meter). The mucous string is conducted along the wall of the pharynx toward the dorsal side by ciliary action. As a result, the filtered food particles flow toward the middle of the dorsal pharynx. At this surface, and opposite the endostyle, is another longitudinal organ known as an epibranchial groove. Here the mucus, laden with food, is rolled into sausagelike strands that are passed posteriorly in the direction of the subsequent sections of the digestive canal. Depending on the developmental stage of the various forms, one can differentiate between an esophagus, stomach with glandular outgrowths, an intestine, and a hindgut. These features are most clearly seen in the ascidians.

The circulatory system, and particularly the heart of the tunicates,

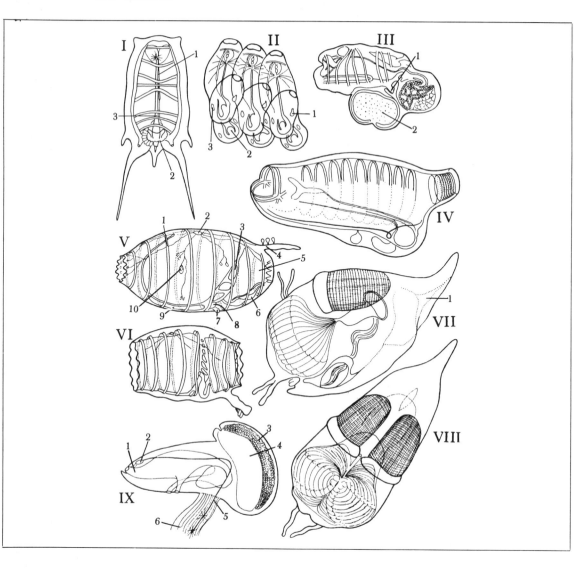

Salps:
Thalia democratia
I. Nurse (oozooid):
1. Endostyle, 2. Stolon,
3. Gill ridge.
II. End of a young salp
stolon: 1. Ovary,

2. Nucleus with testes,
3. Heart.
III. Developing oozooid:
1. Stolon, 2. Placentalike
organ.
IV. *Salpa maxima*, oozooid

(nurse).
V. *Doliolum rarum*
(Doliolida), nurse
(oozooid): 1. Muscle ring,
2. Nerve ganglion, 3. Gill
slit, 4. Dorsal spur with

the first buds, 5. Cloaca
(atrium), 6. Intestine,
7. Stolon, 8. Heart,
9. Endostyle, 10. Statocyst.
VI. *Doliolum muelleri*
(Doliolida), Phorozooid.

Appendicularians:
Oikopleura albicans
VII. Appendicularian
within its house (lateral
view):

1. Emergency exit.
VIII. Abandoned house
(seen from above).
IX. Animal without house;

basal part of tail is shown:
1. Mouth opening,
2. Nerve ganglion,
3. Ovary, 4. Testes,

5. Nerve strand, 6. Musculature. The blue arrows
indicate the direction of
the water current.

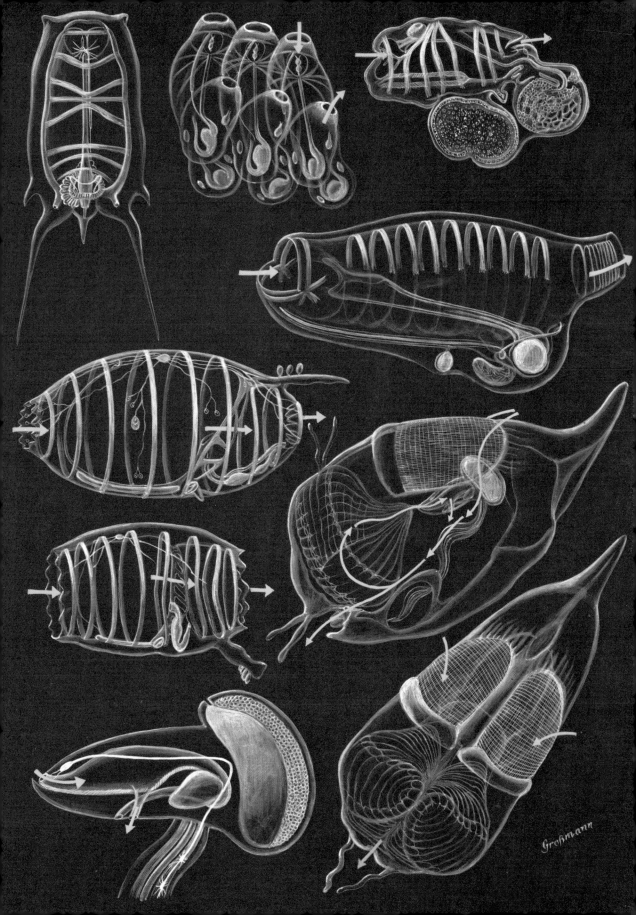

Großmann

show certain special features which, because of their uniqueness, have repeatedly fascinated zoologists. Blood vessels as such are not present in the tunicates. Except for a few blood channels, the blood flows through spaces in the tissue. The blood contains blood cells which are quite colorful: red, green, brown, or white. Oddly enough, the blood fluid and specific blood cells contain a high percentage of the element vanadium. It does not have the oxygen-carrying capacity of the iron in vertebrate blood. The tubular heart is located on the ventral side posterior to the pharynx. The heart is surrounded by the pericardium, which in the tunicates represents a vestige of the otherwise unrecognizable secondary coelom. Waves of contractions pass back and forth from one end of the heart to the other, thereby pumping the blood through the body. The periodic reversal of blood flow is unique in the entire animal kingdom. Pacemakers in the heart are responsible for the alternation in blood flow. According to observations on sea squirts, the pumping activity of the heart slows down gradually after approximately one hundred beats, and then stops altogether. Then the heart resumes pumping, but in the opposite direction, and the entire process is repeated.

The nervous system of a fully developed tunicate appears reduced, particularly in the sessile sea squirts. They are characterized by an elongated nerve ganglion which lies anterior to the branchial basket and sends out several nerves. In the pelagic forms, a statocyst is usually developed in the vicinity of this cerebral ganglion. The thaliaceans also have a simple eye, the ocellus.

When mature, practically all tunicates are hermaphroditic; often, however, the male and female gonads mature at different times within the same animal (as will be noted in the sections on the individual groups). A tadpolelike larva (Fig. 17-1) hatches from the eggs. The tail end of the larva contains the notochord, and, totally in agreement with the chordate "blueprint," the nerve cord is found dorsally to this notochord. The nerve cord extends into the anterior body part, where it enlarges into a cerebral vesicle, in some cases with a statocyst and cerebral eye. After a brief larval phase, metamorphosis sets in. Only the appendicularians retain the tail throughout their life. In all other forms the tail is either reabsorbed, as in the ascidians, or else it was not even present in the larval form, as in the thaliaceans. Aside from sexual reproduction, the animals can also propagate by asexual means, and this frequently results in the formation of colonies. The thaliaceans are characterized by a regular alternation of generations (metagenesis).

Even though the tunicates have been classified with a degree of certainty in the phylum Chordata, their origin and phylogenetic position within this group are still largely uncertain. Today some scientists still assume that the sessile forms are the more primitive ones, but this is highly unlikely. Embryological studies have shown that free-swimming larvae appear first, becoming sedentary after undergoing metamorphosis.

◁
The ascidian *Phallusia mammillata* is easily mistaken for a sea cucumber.

Specializations found in sessile tunicates can therefore be interpreted only as newly evolved adaptations. This view is substantiated by the anatomy of the appendicularians whose tail section with the notochord and nerve cord is retained into the adult stage.

More recently the view has gained weight that the tunicates are a side branch but are secondarily simplified (degenerate) chordates. In 1964 Rendel compared the DNA content (DNA = deoxyribonucleic acid) of cells from the major groups of the animal kingdom. He found that there was a general tendency of increase in the DNA content going from the poriferans and coelenterates to the vertebrates. Only the tunicates were exceptional. Although they are chordates, their DNA content (i.e., genetic coding) was comparable only to the values determined for the coelenterates (Coelenterata; see Vol. I). This can be interpreted as a phylogenetic "leap backwards"; it is possible, however, that although tunicates have less DNA, they use more of it.

In contrast to the pelagic appendicularians and thaliaceans, the sea squirts or ascidians (class Ascidiacea) are sessile, although only in the adult phase. The ascidians are the most successful group of the tunicates. They are represented by approximately 2000 species. Some of these are quite familiar to us, and have quite regularly been raised in aquaria at marine biological stations.

Class: Ascidiacea

The most important characteristics of the ascidians are closely related to their sedentary mode of life. Their body shape is more or less plump, saccular, or budlike. The incurrent and excurrent siphons usually are in close proximity, although rarely they are widely separated. The branchial basket is surrounded by an atrium. The notochord is totally reduced, but is still present in the larval stage. The tunic is often thick, gelatinous, cartilaginous, or leathery. Some ascidians are solitary, but others live in colonies or groups produced through budding.

Formerly this animal class was divided into three groups: 1. Solitary ascidians (Monascidians); 2. Social forms which live in groups but are in organic continuity via stolons; 3. Synascidians which live in dense "systems" created by budding. Classification now is based on anatomical differences, particularly the detailed structure of the branchial basket and the position of the gonads. According to this classification, two major orders emerge: 1. Enterogona: gonads unpaired, and located either anterior or posterior to the intestinal loop; 2. Pleurogona: gonads paired, and located on either side of the body wall.

Many ascidian species are notable because of their bright coloration. There are yellow, orange, red, blue, and green forms. The color is produced by pigment cells which may be located within the tunic or in both the tunic and the deeper tissues. Other ascidians are a dull white or are more or less transparent. In some species the surface of the tunic is quite slippery and smooth, in others, relatively solid and leathery. The tunic may be bumpy or full of folds, or even covered with spines and hairy

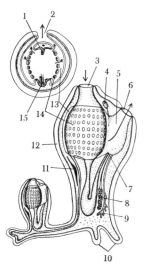

Fig. 18-1. Anatomy of an ascidian (*Clavelina*): 1. Epibranchial (epipharyngeal) groove, 2. Exhalent opening, 3. Inhalent opening, 4. Nerve ganglion, 5. Cloaca (atrium), 6. Exhalent opening, 7. Anus, 8. Ovary, 9. Testes, 10. Stolon, 11. Heart, 12. Tunic (test), 13. Atrial cavity, 14. Pharynx (branchial basket), 15. Endostyle.

processes. Such surface appendages may serve as attachment areas for foreign objects or other organisms which might provide an excellent camouflage cover for the ascidians. The posterior end of the animal frequently bears rootlike processes on its tunic, which serve as an anchor.

The striated musculature is relatively weakly developed. A layer of longitudinal muscles lies next to the external membranous tunic which is followed by a layer of circular muscles. The ascidians are able to change the shape of their body to a certain degree by the antagonistic action between body musculature and the elasticity of the tunic. They are able to contract the body and to bend it to one side. The two body openings are endowed with their own musculature which facilitates the opening, closing, and even pulling in of these siphonal structures. In the colonial forms there are usually only a few longitudinal strands and sphincter muscles associated with the body openings.

The mode of feeding generally follows the described chordate pattern. Nevertheless, there are certain specializations. There is usually a short incurrent siphon which protrudes like a chimney. The edge of this siphon is often lobed or serrated (see Color plate, p. 428). *Halocynthia papillosa* (see Color plates, pp. 403 and 449) has bristlelike structures around the siphonal openings. These siphonal accessories vary in the individual species, and therefore play an important role in the systematics. The internal lining of the inhalent siphon has the same embryonic origin as the epidermis. Along the border of the pharynx, which originated from the innermost germ layer (endoblast), there is a tentacular ring which prevents the entry of coarser foreign particles. The branchial basket, which takes up the better part of the animal's body, is perforated by many—often extremely numerous—slits arranged in circles and in horizontal rows. Among the various genera and species there are notable differences in the shape and arrangement of these pharyngeal openings. In order to classify an ascidian properly, it is frequently necessary to examine the characteristics of the branchial basket.

The water which leaves through the pharyngeal slits first passes into a paired atrial cavity which encloses the branchial basket laterally. These two cavities fuse along the dorsal side of the ascidian, where the single subsequent cavity merges into the excurrent siphon. The inhalent and exhalent siphons are usually similar. However, it is easy to differentiate between the two, since the inhalent siphon is always at the animal's upper pole, while the exhalent opening is almost always in a lateral position. The atrium also receives the discharged products of the hindgut and the gonads, because their openings terminate within it. For this reason the atrium is also referred to as a "cloaca." A modification of this basic plan is found in the colonial ascidians which have a common cloaca (atrium) and consequently are also grouped around a common exhalent siphon.

Digestion takes place in the segments of the alimentary channel that follow the branchial basket, particularly in the more or less saccular stomach, which frequently is associated with many diverticula. Digestive enzymes are secreted by the pyloric gland, which enters the stomach via a duct, and by the stomach lining itself. The midgut and hindgut make a loop so that the anus comes to lie at the animal's dorsal side, where it empties into the atrium. Excrement leaves the atrium through the exhalent siphon. Other metabolic by-products (excretory products) accumulate in "excretory storage organs."

Excretory storage organs

The vascular system is characterized by certain modifications that deviate from the overall chordate scheme. The hypobranchial vessel, next to the heart, extends to the branchial basket. This structure is supplied by numerous circular and vertical blood channels. A visceral vessel from the opposite end supplies the digestive and genital organs. In accordance with the rhythmical reversal of the blood flow, the blood is pumped alternately through these two major vessels.

The nervous system and the sensory organs are only weakly developed. This is obviously a by-product of its sedentary mode of life. The nerve cord is still quite evident in the free-swimming larval form, although in the fully developed animals this cord has hypertrophied during the course of metamorphosis. The central nervous system in the adults consists of a long neural ganglion located between the inhalent and exhalent siphons. Several nerves extend from this ganglion. Larger sensory organs are totally absent, and yet the animals respond to light stimuli. Numerous neurosensory cells are embedded in the body surface, concentrated most frequently around the two siphonal openings.

The gonads consist of closely located but separate male and female units. In many species the gonads are located in the loop of the intestine, whence the two separate gonoducts ascend side by side to the gonopores, which empty into the cloaca. In most forms, the eggs are discharged directly into the surrounding water through the exhalent siphon. The eggs are fertilized in the sea by sperms discharged from the same species. To ensure fertilization, spawning occurs only at very specific times. Synchronization of this process is probably increased by the fact that chemosensory organs respond to the presence of species-specific discharged sperms by releasing their own eggs. Self-fertilization is prevented by two methods. Sperms of one individual do not fertilize their own eggs (self-sterilization); or the ova and sperms in the same animal mature at different times.

However, eggs are not discharged to the outside in all cases. Mainly in the colonial species, and also with a few solitary ascidians, the sperm is swept to the parental animal by the water current. After fertilization, which takes place in the atrium, the eggs are retained and undergo development within the cloaca. Some forms may have a special brood pouch (a lateral cavity off the cloaca) in the atrium or partially located in the tunic. In these species the fully developed larva with tail (tadpole form) leaves the parental animal.

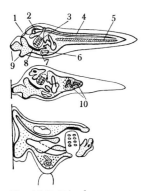

Fig. 18-2. Development of an ascidian. Above: Larva. Center: Transitional stage from larva to sedentary ascidian. Below: Young ascidian. 1. Mouth, 2. Cerebral vesicle with ocellus and statocyst, 3. Branchial basket, 4. Neural cord, 5. Notochord, 6. Heart, 7. Left atrial cavity. 8. Endostyle, 9. Adhesive papillae, 10. Notochord vestiges.

Only after zoologists became familiar with the anatomy of the ascidian larvae and the modifications which occurred during metamorphoses toward a sedentary mode of life (Fig. 18-2) did the anatomy of the fully developed ascidian become understandable. The free-swimming larva, which may persist in this state for only a few hours, does not feed. These ascidian larvae are superficially reminiscent of tadpoles. Their plump anterior body merges into the slender tail segment which functions as a locomotive organ. During active swimming, interrupted periodically by resting phases, the animals rotate around their longitudinal axis. Larval anatomy reveals the distinguishing chordate characteristics. A well-developed nerve cord extends from the anterior body to the end of the tail; anteriorly the cord is enlarged into a cerebral vesicle with an associated ocellus and statocyst. The notochord, however, is limited to the tail section. A heart is also already present, taking up the typical chordate position along the ventral body side. The mouth opening is found in the body's anterior pole. It leads into an already conspicuously saccular branchial basket. The paired atrial cavities, which surround the branchial basket symmetrically, have also been developed. They are ectoblastic in origin. During the course of embryological development, a left and a right invagination are formed on the dorsal side of the embryo. These paired structures become pocket-shaped, so that they surround the anterior digestive tract from two sides. The interior of the branchial basket is in contact with these invaginations via the first pharyngeal slits. At a later period the two atrial cavities fuse into the unpaired cloacal cavity, which is in contact with the external environment through a common opening, the rudimentary exhalent siphon opening.

The posterior digestive tract in the larva is only a closed saccular structure. The tail fin in the larva consists of the same substances as the tunic in the fully developed animals. There is a strand of closely woven tissue, of endoblastic origin, below the notochord. This tissue probably represents an extension of the digestive tract which became hypertrophied due to some unknown phylogenetic process. This would indicate that the hindgut of present-day ascidians is shorter than that of their primitive ancestors. The nerve cord, notochord, and endoblastic strand are enclosed on both sides by tail musculature.

After a certain time period, the larva attaches itself to the substratum with the aid of the adhesive papillae located on the anterior end. At this point, far-reaching transformations take place within the animal. The tail and the notochord degenerate, although their fragments remain recognizable for some time. The nerve cord, cerebral vesicle, and larval sensory organs are reabsorbed. Concomitantly, the adult cerebral ganglion and the remainder of the larval nervous tissue are newly developed and reorganized. A change in body shape is also evident at this point. Rapid unilateral growth processes have caused the larval mouth opening, which originally was located beside the adhesive papillae, to move to the opposite side. In a similar manner, the exhalent opening and the internal organs are also shifted.

In addition to sexual reproduction, asexual propagation also plays a great role. There are various types of bud formation, depending on the family or genus. The most familiar type is the so-called "stolon budding," where tubular buds (stolons) form at the animal's base. Budding leads to colony formation. Loosely arranged associations are referred to as social ascidians, while densely formed colonies are known as compound ascidians or synascidians (see Color plate, p. 455). Some ascidians are characterized by a more or less well-regulated alteration between asexual and sexual reproduction in alternate generations (generation change). Thus sexually produced individuals of the genus *Perophora* are incapable of sexual production, although their buds are able to produce eggs and sperms, and so on.

Colony formation

The life span of ascidians seems to be about one year. Colonial forms may live for several years. The genus *Diazona* is said to have a life span of three to four years. Ascidians are cosmopolitan in distribution. Most species are found at the inter-tidal zone from the surface to 400 m. Isolated deep-sea forms have been found at 5000 m.

We shall mention *Ciona intestinalis* (height 8 cm; see Color plate, p. 428) as a representative of the order ENTEROGONA. This form is solitary, but is often found in groups. Its tunic is quite transparent, making the longitudinal muscle strands clearly visible. This common species occurs down to 500 m. *C. intestinalis* enjoys a wide distribution, including European oceans.

The order PLEUROGONA includes the species *Halocynthia papillosa* (height 10 cm; see Color plates, pp. 403 and 428). This form also inhabits European oceans, where it prefers sandy substrates. *H. papillosa* is conspicuous because of its brilliant red color. Its coarse tunic bears numerous papillae. Both siphonal openings are lined by long, brownish, bristlelike structures. The inhalent opening has four lobes; the lateral exhalent opening has one, or rarely two. This feature permits easy identification.

Botryllus schlosseri (L 2–2.5 mm; see Color plate, p. 455) is a species that is quite common in the North Sea. It is a compound ascidian that belongs to the Pleurogona. The animals are embedded in a gelatinous mass. They are arranged around a common cloacal cavity, in a nearly circular shape.

The thaliaceans (class Thaliacea) are pelagic and planktonic; occasionally they occur in high concentrations. A series of characteristics distinguishes this group from the remaining tunicates: the body is barrel-shaped, and the inhalent and exhalent openings are at opposite ends of the body. In some forms a complicated change of generations takes place, but at some time, or during an individual's entire life, the animals form a colony from a stolon. In contrast to the equally pelagic appendicularians, the fully developed thaliaceans always lack both a tail segment and a tract

Class: Thaliacea

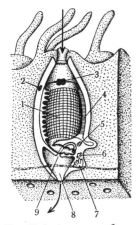

Fig. 18-3. Anatomy of *Pyrosoma*. Direction of water current: black arrows. 1. Branchial basket, 2. Nerve ganglion, 3. Luminescent organ, 4. Endostyle, 5. Heart, 6. Ovary, 7. Testes, 8. Cloaca, 9. Anus.

Luminescent organs

of notochord, even in the larval stage (exception: larvae of Doliolida). There are great variations in the length of adult animals. Individuals may measure from 1 mm to more than 10 cm, rarely 19 cm. Thaliaceans form colonies. Some form long chains by budding. Riedl described how one could pull in a chain of thaliaceans almost like pulling "a line into a boat." Many of the smaller forms appear clear and transparent, because of their high water content. Larger species have a bluish or yellowish color. The group is represented by just over forty species.

There are three orders: 1. Pyrosomatida, with the single genus *Pyrosoma* and approximately ten species; 2. Salps (Salpida), with barely more than twenty species; 3. Doliolida, with approximately eleven species.

The PYROSOMATIDS (order Pyrosomatida) form colonies reminiscent of a thick-walled tube closed on one end. The length is usually between 10 and 20 cm, although *Pyrosoma spinosum* can reach a length of 10 m. From the outside, the tube has an uneven surface that bears numerous tentacular processes and is pierced by many openings which are also evident on the inner wall. These pores represent the inhalent and exhalent openings respectively of the individual animals. Only a longitudinal section will reveal the anatomy of the colony.

The zooid measures only 4–5 mm. It is vertically embedded in a common gelatinous mantle. As in the ascidians, the pyrosomatids have a spacious branchial basket, perforated by numerous slits and surrounded by atrial cavities. Cilia lining the gill slits produce water currents facilitating gaseous exchange (respiration) and the straining of food particles. Formerly, the pyrosomatids were considered as free-swimming ascidian colonies, since the anatomy of these two groups is so similar.

The branchial basket is followed by a compressed digestive segment. The anus and gonoducts terminate in the cloacal cavity. The primitive ventral side of the zooids is characterized by an extensive endostyle. The central nervous system, represented by a nerve ganglion, is located opposite the endostyle, but slightly more anterior. A number of nerves branch off this ganglion. There is an acellular mass on each side of the esophagus just anterior to the branchial basket. These are luminescent organs, containing luminescent bacteria. The scientific name in fact means "fire body" (*Pyrosoma* = fire body, from the Greek πῦρ, πυρός = fire, σῶμα = body). One can elicit a yellowish to bluish-green luminescence by mechanical or other stimuli. The nature of the mechanism of light-production is still unknown.

Animals (zooids) within one colony are uniformly aligned. Their endostyles always face the closed body pole. The common mantle contains muscle fibers which interconnect the cloacal muscles of the zooids; this has resulted in muscular tunic strands. Although other muscles can contract the individual animals, it is these tunic fiber strands that contract the entire colony. Water is expelled from the common atrium, permitting the colony a limited amount of active mobility through "jet-propulsion."

All zooids gathered in a colony are sexually reproducing individuals, or gonozooids. The male and female gonads are located in a bulge in the vicinity of the ventral side of the cloacal cavity. The ovary produces only one yolky egg which is fertilized *in situ* by sperms that have entered the oviduct. The fertilized egg develops within the parental body. With increased growth, the primary individual breaks through the body wall into the cloacal cavity. An individual that arose from a fertilized egg is known as an oozoid. This form is short-lived, and gives rise to four buds (ascidiozooids). This little tetrazooid colony is set free from the parental cloaca, and establishes a new colony through asexual budding. This is a well-developed change of generations, with the gonozooids playing the principal role.

In the single genus *Pyrosoma*, the species *Pyrosoma atlanticum* is widely distributed and is not uncommon in the Mediterranean. With advancing age, the animals acquire a cloudy, milky color which gradually turns yellow.

The TRUE SALPS (order Salpida) are characterized by a system of four to nine well-developed transverse muscle bands surrounding the body like rings, which are open along the ventral surface (see Color plate, p. 437). The salps are best understood if one first discusses the solitary forms. These have originated from fertilized eggs, and are regarded as oozooids.

Order: true salps

The nearly barrel-shaped animals are usually colorless. Length is between 1.5 and 10 cm, and in extreme cases up to 19 cm. The body is transparent, and is enclosed by a relatively solid test which occasionally is endowed with species-specific processes. The mouth and cloacal openings are opposite each other, although they are frequently shifted slightly to the dorsal side. The branchial basket is large, occupying the better part of the animal. Two large and long slanted gill slits are always in contact with the cloacal cavity. These gill slits (stigmata) are located to the left and right sides of an unpaired ciliated gill ridge which is on the same slant as the stigmata on each of its sides and extends from the anterior dorsal side to the posterior ventral side.

The functions of respiration and filter-feeding in the salps are similar to those of the other chordates mentioned previously, except that the necessary water current is produced mainly by the action of muscles, and less by ciliary beating in the gill slits. In a manner of speaking, the animals "pump" their way through the water. The digestive tube, which lies behind the posterior base of the gill ridge, is relatively small and is compactly curved as the "nucleus." The anus terminates in the cloacal cavity, the second largest body cavity in the salps. Figure V of the color plate on page 437 illustrates the location of the heart and central nervous system (cerebral ganglion). Even more conspicuous than the oral and atrial sphincter muscles, which are responsible for closing these openings, are the body muscles which surround the body like bands. These muscle

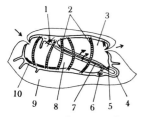

Fig. 18-4. Salp oozooid: 1. Nerve mass, 2. Muscle bands, 3. Cloaca (atrial cavity), 4. Intestine, 5. Anus, 6. Heart, 7. Gill ridge between the two gill slits, 8. Branchial basket, 9. Tunic (test), 10. Endostyle.

bands are open on the ventral side, the nature of the opening varying in the different species.

Oozooids could also be called "nannies" or "nursemaids," since they are devoid of gonads. They reproduce asexually, by means of a budding stolon. This structure extends along the ventral surface of the salp's body midline, just posterior to the end of the endostyle. Numerous buds sprout out from this stolon, developing into sexually reproducing animals (gonozooids). These gonozooids are arranged in two rows, which are mirror images of each other. Usually these zooids remain attached to each other by adhesive papillae even after the stolon has been lost. This is actually the manner in which the well-known salp chains (L can exceed 25 m; see Color plate, p. 437) are formed.

Basically the body structure of the gonozooid is the same as that of the oozooid, but there are differences in size, body shape, arrangement and number of the muscular bands, and other features within the same species. Naturally, the gonozooids possess the male and female gonads. The testes are part of the nucleus, while the ovary is located laterally along the cloacal wall. The true salps, like the pyrosomas, produce only one egg, which is fertilized *in situ*, and develops there. An oozooid (see Color plate, p. 437) is formed which is connected to the parental hemolymphatic fluid by a placenta located on its ventral side. As the size of the embryo and the tissues surrounding it increases, they bulge toward the cloaca until the cloacal wall bursts; the oozooid is released into the cloaca and then out the atrial opening. Gonozooids are therefore viviparous.

The first man to demonstrate the generation change between oozooids and gonozooids was actually much better known as a poet. He was Adelbert von Chamisso, who took part, as a scientist, in a voyage around the world on the Russian brig *Ruric*, between 1815 and 1818. Before von Chamisso studied the salps, the connection between the solitary animal and the salp chains was unknown, and these creatures were often described by two different names.

The true salps are primarily found in the warmer oceans, although they do penetrate into more temperate latitudes.

Order: Doliolida

The third and last group of pelagic, tailless tunicates to be discussed is the DOLIOLIDS (order Doliolida). They are characterized by closed, hooplike muscle rings, arranged vertically to the body's longitudinal axis. The distinction between the salps in the narrow sense and the doliolids will become apparent after discussing the oozooids, or "nurses."

The shape of the doliolid nurse is less cylindrical, but decidedly more barrel-shaped. The oozooids are far smaller than in the salps; length is approximately 10 mm. The mantle is transparent and delicate, and surrounds the equally transparent body. The mouth and cloacal openings are almost at opposite ends. They are surrounded by a larger number of rounded lappets (closing apparatus). The spacious cavity of the branchial basket borders posteriorly on the cloacal cavity. There are 8 to 200 horizontal stigmata, limited to the posterior wall of the branchial basket, where they are usually arranged in two vertical rows from the dorsal to the ventral side of the body. The modes of feeding and respiration are the same as in the salps. Doliolids move by jerky thrusts brought about by closing one body opening and contracting the eight or nine regularly arranged muscle rings so that the water current is jetted out through one opening. Figure V of the color plate on page 437 shows the position of the digestive tract, the heart, and the ganglion. The statocyst is a special structure, located between the third and fourth ring muscles.

The oozooids of the Doliolida also have a stolon from which the buds arise through asexual propagation. The stolon is at the same location as in the salps, but, unlike that of the salps, it does not produce salp chains. The stolon in the doliolids constricts off a continuous series of buds. As the buds constrict off, they are aided by migratory amoeboid phorocytes. Peculiarly, the phorocytes and the buds always migrate toward the right side, and posteriorly up onto the dorsal surface to the base of the nurse's unpaired dorsal spur. From this point the buds become arranged by the phorocytes in a specific order:

a. A pair of lateral rows of buds is arranged on each side of the spur. These develop into trophozooids that supply the nutritional needs of the nurse and other parts of the colony that have developed. The trophozooids are simplified and yet highly specialized in their body structure. In the meantime, a degeneration of the branchial basket and gills has taken place in the nurse. On the other hand, the nurse's muscle bands have strongly widened, and fulfill the function of "motor" for the entire colony.

b. Buds also migrate to the median line between the two rows of lateral trophozooids. These buds develop into normal doliolids, but, like the nurse, without gonads. These buds have a stalk of attachment along their posterior end, connecting them with the nurse's dorsal spur.

c. Additional buds transported by the phorocytes attach to these stalks. These buds are the precursors of the gonozooids.

d. The precursor gonozooids in turn divide, thereby producing a larger number of buds of the second order. During this process the gonadless buds mentioned in section "b" become detached from the nurse's ever-increasing dorsal spur. They are known as phorozooids. When the phorozooids are free-swimming, the buds of the second order finally develop into gonozooids that detach from the nurse.

The body structure of the gonozooid (see Color plate, p. 437) is similar to that of the oozooid, except that the gonozooid has gonads. In contrast to the pyrosomas and similar salps, the doliolids release their eggs into the surrounding water, where they are fertilized. The fertilized ovum develops into a tailed larva with a notochord. After hatching and the hypertrophy of the tail, the larva develops into an oozooid, the doliolid form without the gonads. This extremely complicated change of generations thereby comes full circle. To recapitulate the individual steps

▷
The conspicuous red sea squirt (*Halocynthis papillosa*; Color plates, pp. 403 and 428).

of the various developmental types that appear in the change of generations: 1. Oozooids; 2. Trophozooids, Phorozooids, and probuds (precursor to gonozooid); all three forms are of nearly equal rank but are differently constructed as they arise from the stolon of the oozooid; 3. Gonozooids.

The Doliolida also prefer the upper water layers of tropical and subtropical oceans. Occasionally they occur in great concentrations. *Doliolum muelleri* and *Doliolum denticulatum* are quite frequently found in the Mediterranean (compare Color plate, p. 437).

Class: Larvacea

The APPENDICULARIANS (class Larvacea) are pelagic. Except for a few forms, they inhabit the upper water layers from the surface to 200 m. Particularly in tropical oceans, they occur in unbelievably high concentrations. Lohmann, who is responsible for most of the information on these animals, described the appendicularians as the most frequently encountered planktonic animals after the copepods (Copepoda; see Vol. I). The appendicularians are surrounded by a clear test, which is considerably larger than their actual body. This surrounding house (test) is very complex, and shall only be discussed after the body structure has been described.

In contrast to all other tunicates, the tail present in the appendicularian's larval stage is not reduced, but is retained in a somewhat modified form throughout its entire life. The tail is more or less twisted into a horizontal position and directed forward under the body, giving the entire animal a hammer-shaped form. The anterior body is oval to irregularly cylindrical. Length is usually 1–2 mm, minimally 0.5 mm and maximally 8 mm. It contains the branchial basket and the digestive tract, the heart (which is almost always present), the central nervous system (ganglion) and branching nerve strand, and the testes and ovaries (see Color plate, p. 437 for position of organs). There is only a single gill slit on one side of the branchial basket, in direct contact with the external environment. In contrast to the other subgroups of the Tunicata, an "atrium" or cloaca is not present. Only the testes are connected to the outside via a gonoduct. The sperms mature considerably earlier than the eggs (protandry). The small eggs, on the other hand, are released by a rupturing of the body wall, which naturally results in the death of the parental animal.

The tail, which is such a characteristic feature of the appendicularians, attaches to the body ventrally, behind the anus and in front of the gonads. The tail is laterally fringed by a more or less broad, transparent fin containing the notochord, with large muscle cells on each side of it, and a neural cord. On the other side of the neural tube there is a solid strand of tissue of entoblastic origin, which corresponds to the conditions found in the larvae of the ascidians; this might represent the vestige of a hindgut segment which extended through the hind body of the phylogenetic ancestors. The position of the tail is both unique and characteristic. It is not only twisted but turned ninety degrees around its own axis. The

⊲
The ascidian *Clavelina lepadiformis* attached to a horny coral *Paramuricea chamaeleon* (see Color plate, p. 282).

actual left side of the tail points frontward, while the right side faces posteriorly; the fin margins point left and right, respectively.

The eggs are fertilized in the surrounding sea. The resultant larvae are at first very similar to the ascidian larvae. The larval tail is straight, like that of the ascidian larvae, but the nervous system is not dorsal to the notochord, instead lying to the left of it. During metamorphosis the tail becomes twisted and rotated ninety degrees about its own axis.

The specializations in the body structure are functionally associated with the structure of the house. The external house is secreted by sections (oikoplast) of the epithelium of the body (oikothelium), which are concentrated to specific regions in the anterior body. An overall tunic is not produced, on the other hand. The house is extremely complex, but at the same time is of such transparence that it is barely recognizable in the sea water. The only way in which Lohmann was able to study the position, boundaries, and interconnections of the inner cavities in this gelatinous and extremely fragile house was by carefully injecting it with colored sea water. The results of his classic investigations appeared in 1899. Here they are reiterated in the example of the external house of *Oikopleura albicans* (see Color plate, p. 437).

The appendicularians which are embedded deeply within the interior of their houses produce a water current by undulating motions of their tails. The water enters the house through a special sieve with a mesh size of 0.034 by 0.132 mm. In an equally complex manner the water is conducted to a paired food trap which in turn is connected with the animal's mouth opening. From the mouth, the concentrated food particles are swept to the digestive tract by a water current created by the ciliary zones in the branchial basket (pharynx). The water leaves the house through an exhalent aperture. This results in a weak backward thrust which slowly propels the structure forward by rhythmic spiral turns around its own axis. Since the appendicularians are connected with the house only around the oral region, they are remarkably quick in leaving it via an "emergency exit." This phenomenon is not only a response to external stimulation, but is also observed when the house has gradually become uninhabitable because of an accumulation of particles in the sieve and filtering apparatus.

Usually a house is reconstructed within only twenty to thirty minutes. As has become evident from the minute nature of the sieve apparatus, these animals feed on extremely small organisms, the so-called nannoplankton, and in particular on coccolithophorids which belong to the unicellular flagellates (see Vol. I). In some cases the houses of other appendicularians deviate greatly from the anatomical arrangement described here. For a period of time the appendicularians were considered to be primitive tunicates, because of some of their doubtlessly primitive features. This, however, is dubious, since the appendicularians also have significant specializations.

Subphylum:
Cephalochordata

In 1774 the famous scientist Peter Simon Pallas (1741–1811) discovered a nearly colorless marine animal of fishlike appearance, which measured barely 6 cm. He believed that he had found a gastropod, and his discovery was almost forgotten. About sixty years later, the Italian Costa rediscovered Pallas' mollusk, and a little later the renowned anatomist and physiologist Johannes Müller also rediscovered it. At this point the close relationship between this animal—the amphioxus *Branchiostoma*—and the vertebrates was recognized. Since then, this lancelet has played a great role in the discussion of the origin of the vertebrates.

Distinguishing
characteristics

The ACRANIA (subphylum Acrania) are characterized by an elongated, fishlike, laterally compressed body which tapers like a lance anteriorly and posteriorly. Body length is up to 7.5 cm. The head, or rather the cranium, is absent. External extremities are also absent, although fins are present. The general anatomy corresponds to the chordate scheme, deviating less than the extremely specialized tunicates. For this reason the lancelet is more suitable for comparison with the vertebrates. The body is covered by a single layer of epithelium. The notochord is retained throughout the animal's life, and is a well-developed supporting rod extending through the entire body. The trunk musculature is asymmetrical, and is located on both sides of the notochord. These longitudinal muscles consist of chevronlike segments called myomeres. This arrangement is called myometamery. The notochord functions as an elastic axial skeleton in its interrelationship with the body musculature. The central nervous system consists of an elongated neural tube with a narrow central canal dilated anteriorly into a cerebral vesicle. An unpaired eyespot —an eye or a thermal organ?—is present, but there is no indication of an auditory organ. The nerve cord (spinal cord) is lined with numerous visual cells which are capped by pigment on one side (directional sight). In each segment, a dorsal and a ventral spinal nerve branch off the spinal cord. The mouth opening is surrounded by cirri. The buccal cavity is short, and leads into a large pharynx, which takes up approximately the anterior half of the body. There are fewer gill slits in the larval form than in the adult. The gill slits already present are divided by a bar which grows from the top to the bottom. Eventually the total number of gill slits reaches about 180. Only in the larvae do the gill slits open directly to the outside. During early development, the ventral body suface (ectoblast) folds inwards. The edges of the folds fuse, forming an atrial cavity which is in contact with the outside by only one atrio-branchial pore located posteriorly to the ventral midline.

The cilia lining the gill slits produce a water current in the lancelet similar to that in the ascidians. Water enters through the mouth, and flows to the pharynx, whence it enters the atrium through the gill slits, leaving through the atriopore (atriopore; see Color plate, p. 428). The water current supplies the animal with food and the necessary oxygen. Just like the tunicates, the lancelet also filters food from the water. In principle the filtering process is the same in both groups. The lancelet also has an endostyle which secretes mucus to which the food particles adhere. The mucous strings roll upward into the epipharyngeal groove, which is also lined by glands and cilia; the strings are subsequently transported to the digestive section of the alimentary canal. The anus is almost ventral and anterior to the posterior body end, and, peculiarly, is on the body's left side. The excretory organs consist of segmentally arranged protonephridia, short, tubelike structures opening into the atrial cavity.

The sexes are separate in the lancelet. The gonads are arranged on both sides of the atrial cavity as a series of pouches. Eggs and sperms are emptied into the atrium, whence they escape to the outside via the atriopore. It is an astonishing fact that the Acrania already have a circulatory system which in a remarkable way corresponds to the basic pattern of the vertebrate vascular system (see Color plate, p. 428). The blood flows along a ventral aorta to the pharynx, where it distributes blood to the branchial arches of the gill bars. The blood collects in the paired dorsal aortae, which join behind the pharynx to form the single dorsal aorta which furnishes fresh blood to the digestive organs. From capillary regions, particularly the hepatic or "liver" area, a ventral diverticulum of the midgut, the blood returns to the ventral aorta. Unlike the vertebrate circulatory system, the blood in the lancelet is not driven by a central heart but by numerous pulsating "branchial hearts" at the base of each afferent branchial artery. The blood of amphioxus is devoid of red corpuscles.

Lancelets are distributed in temperate to warm oceans. The most common genus is *Branchiostoma*, which formerly was also known as *Amphioxus*. The seven representative species are able to swim freely in the water for short periods of time. They employ undulating lateral body movements. Mostly, however, the lancelet is buried in the surface of semi-coarse sand. They prefer water depths of 4 to 15 m. The only species found in European oceans is the LANCELET (*Branchiostoma lanceolatum*; L up to 6 cm; see Color plate, p. 428). It was described by Pallas. The six species of the genus *Asymmetron* are considerably modified. They have unpaired gonads, developed only on the right side of the body.

It is extremely tempting to assume that the lancelets are a phylogenetic precursor of the vertebrates. Nevertheless, the significant specializations in the Acrania cannot be overlooked, such as the asymmetrical body structure of the genus *Asymmetron*, and, more important yet, the fact that the lancelet develops from an equally asymmetrical larval form. The larva's mouth and anus are on the body's left side, and initially its gill slits are unpaired and are found only on the body's right side. Therefore, it is perhaps best if one considers the Acrania a specialized descendent from the evolutionary line that gave rise both to the vertebrates and to the Acrania. This group, however, separated from the true vertebrate line very early in evolutionary history. Accordingly, the Acrania were able to retain a series of primitive characteristics.

▷
Botryllus schlosseri is a colony-forming ascidian.

▷▷
Microcosmus sulcatus covered with the bryozoan *Retepora beaniana* (see Color plate, p. 233).

Systematic Classification

The page numbers are the main text references. A dagger † signifies a fossil genus or species.

The Phylum Mollusca (Mollusks)

The Solenogasters (Class Aplacophora)

Subclass Neomeniida (or Solenogastres)

Subclass Chaetodermatida (or Caudofoveata)

The Chitons (Class Polyplacophora)

The Ancient Chitons (Order Paleoloricata)

Suborder Lepidopleurina

(Modern Univalves) Order Caenogastropoda

Superfamily Cyclophoracea (Land Operculates)

Superfamily Viviparacea (Apple Pond Snails)

Superfamily Valvatacea

Superfamily Littorinacea

Superfamily Rissoacea

Superfamily Rissoellacea

Superfamily Architectonicacea

Superfamily Cerithiacea

Superfamily Epitoniacea

Superfamily Eulimacea

Subclass Opisthobranchia (Bubble Shells and Sea Hares)

Order Cephalaspidea

Superfamily Acteonacea

Superfamily Bullacea

Superfamily Philinacea

Superfamily Diaphanacea

Family Notodiaphanidae	101	*D. glacilis*	101
		D. minuta (Brown, 1837)	101
Family Diaphanidae	101	Genus *Newnesia*	101
Genus *Diaphana*	101	*N. antarctica* Smith, 1902	101

Superfamily Acochlidiacea

Family Acochlidiidae	101	*M. glandulifera* (Kowalevsky, 1901)	101
Genus *Microhedyle*	101	Genus *Hedylopsis*	101
M. milaschewitchii (Kowalevsky, 1901)	101	*H. spiculifera* (Kowalevsky, 1901)	101

Superfamily Cylindrobullacea

Family Cylindrobullidae	—	*C. fragilis* (Jeffreys, 1855)	101
Genus *Cylindrobulla*	101		

Order Entomotaeniata

Superfamily Pyramidellacea

Family Pyramidellidae	101	Genus *Eulimella*	101
Genus *Odostomia*	101	*E. laevis* (Brown, 1827)	101
O. eulimoides Hanley, 1844	101	Genus *Turbonilla*	101
O. conoidea (Brocchi, 1814)	101	*T. elagantissima* (Montagu, 1803)	101

Order Thecosomata (Pteropods)

Superfamily Spiratellacea

Family Spiratellidae	—	*C. pyramidata* Linné, 1767	102
Genus *Spiratella*	102	Genus *Creseis*	122
S. helicina (Phipps, 1774)	102	*C. acicula* (Rang, 1828)	122
S. retroversa (Fleming, 1823)	—	Genus *Cavolinia*	79
		G. tridentata (Niebuhr, 1775)	79
Family Cavoliniidae	—	Genus *Diacria*	—
Genus *Clio*	102	*D. trispinosa* Lesueur, 1817	—

Superfamily Peraclidacea

Family Peraclididae	—	Genus *Cymbulia*	102
Genus *Peracle*	102	*C. peroni* Blainville, 1818	102
P. reticulata (Orbigny, 1836)	102	Genus *Gleba*	102
		G. cordata Forskäl, 1776	102
Family Procymbuliidae	—		
Genus *Procymbulia*	102		
P. valdiviae Meisenheimer, 1905	102	**Family Desmopteridae**	—
		Genus *Desmopterus*	102
Family Cymbuliidae	102	*D. papilio* Gegenbaur, 1855	102

Order Soleolifera

Superfamily Veronicellacea

Family Rathousiidae	105	**Family Veronicellidae**	105
Genus *Rathousia*	105	Genus *Angustipes*	105
R. *leonina* Heude, 1883	105	A. *plebejus* (Fischer, 1868)	105
Genus *Atopos*	105	Genus *Vaginulus*	105
A. *semperi* Simroth, 1891	105	V. *taunaysi* Férussac, 1821	105

Superfamily Onchididiacea

Family Onchidiidae	—	O. *verruculatum* Cuvier, 1830	105
Genus *Onchidiella*	105	Genus *Onchidina*	105
O. *floridana* (Dall, 1885)	105	O. *australis* (Semper, 1880)	105
O. *celtica* (Cuvier, 1817)	106	Genus *Platevindex*	105
O. *chilensis* (Gay, 1854)	105	P. *montana* (Plate, 1893)	105
Genus *Onchidium*	105	P. *granulosa* (Lesson, 1830)	105
O. *peroni* Cuvier, 1805	113	Genus *Peronina*	—
O. *typhae* Buchanan, 1800	105	P. *alta* Plate, 1893	—

Order Rhodopacea

Family Rhodopidae	106	R. *veranyi* Kolliker, 1847	106
Genus *Rhodope*	106		

Order Anaspidea (Sea hares)

Superfamily Aplysiacea

Family Akeratidae	106	A. *rosea* Rathke, 1799	107
Genus *Akera*	106	A. *dactylomela* (Rang, 1828)	19
Paper bubble shell, A. *bullata* Müller, 1776	106	Genus *Dolabella*	107
		D. *dolabella* Lightfoot, 1786	107
		D. *termidi* (Rang, 1828)	107
Family Aplysiidae (Sea hares)	107	Genus *Phyllaplysia*	107
Genus *Aplysia*	107	P. *depressa* (Cantraine, 1835)	107
European sea hare, A. *depilans* Linné, 1767	107	Genus *Aplysiella*	107
A. *fasciata* Poiret, 1789	107	A. *virescens* (Risso, 1818)	107

Order Gymnosomata (Naked Pteropods)

Family Laginiopsidae	—	**Family Notobranchaeidae**	108
Genus *Laginiopsis*	107	Genus *Notobranchaea*	—
L. *triloba* Pruvot, 1922	107	N. *macdonaldi* Pelseneer, 1886	—
Family Anopsiidae	—		
Genus *Anopsia*	107	**Family Cliopsidae**	108
A. *gaudichaudi* (Souleyet, 1852)	107	Genus *Cliopsis*	—
		C. *krohni* Troschel, 1854	—
Family Thliptodontidae	—		
Genus *Thalassopterus*	108	**Family Pneumodermatidae**	108
T. *zancleus* Kwietnievsky, 1910	108	Genus *Pneumodermon*	108
		P. *mediterraneum* van Beneden, 1838	108
Family Clionidae	108	P. *violaceum* Orbigny, 1840	90★
Genus *Clione*	108	Genus *Pneumodermopsis*	108
Clione, C. *limacina* (Phipps, 1774)	108	P. *ciliata* (Gegenbaur, 1855)	108

Order Sacoglossa

Superfamily Juliacea

Order Notaspidea

Order Nudibranchia (Naked Sea Slugs)

Suborder Doridoidea

Superfamily Bathydoridacea

Superfamily Doridacea

Superfamily Polyceridacea

Superfamily Goniodorididacea

Superfamily Phyllididacea (Perostomata)

Suborder Dendronotoidea

Superfamily Eolidiacea (Cleioprocta)

Family Facelinidae —
Genus *Facelina* 123
  F. drummondi (Thompson, 1843) 123
  F. coronata (Forbes, 1839) 120★
  F. rubrovittata DaCosta, 1866 114★
Genus *Hervia* 123
  H. peregrina (Gmelin, 1789) 123
  H. costai Haefelfinger, 1961 123

Family Caloriidae —
Genus *Caloria* —
  C. maculata Trinchese, 1888 —

Family Favorinidae —
Genus *Favorinus* 123
  F. branchialis (Rathke, 1806) 123
Genus *Dondice* 25★

  D. banyulensis Portmann & Sandmeier, 1960 25★

Family Eolidiidae —
Genus *Eolidia* 123
  Gray sea slug, *E. papillosa* (Linné, 1761) 123
Genus *Spurilla* 123
  S. neapolitana (Delle Chiaje, 1823) 123
Genus *Berghia* 123
  B. coerulescens (Laurillard, 1830) 123

Family Glaucidae —
Genus *Glaucus* 123
  G. atlanticus Forster, 1777 —
  G. marinus 123

Family Myrrhinidae —

Subclass Pulmonata (Lung-bearing Snails)

Order Basommatophora

Superfamily Siphonariacea (False Limpets)

Family Siphonariidae 124
Genus *Siphonaria* 124
  S. pectinata (Linné, 1758) 124

Family Trimusculidae 124
Genus *Trimusculus* 124
  T. garnoti (Payraudeau, 1826) 124

Superfamily Amphibolacea

Family Amphibolidae 124
Genus *Amphibola* 127

  A. crenata (Martyn, 1784) —

Superfamily Lymnaeacea (Pond Snails)

Family Chilinidae 124
Genus *Chilina* 124
  C. fluctuosa Gray, 1828 124

Family Latiidae 124
Genus *Latia* 124
  L. neritoides Gray, 1849 124

Family Acroloxidae 124
Genus *Acroloxus* 124
  Lake limpet, *A. lacustris* (Linné, 1758) 124

Family Lymnaeidae 125
Genus *Lymnaea* 125
  Great pond snail, *L. stagnalis* Linné, 1758 125
Genus *Radix* 125
  Ear pond snail, *R. auricularia* (Linné, 1758) 125
  Wandering snail, *R. peregra* (Müller, 1774) 125
Genus *Galba* 125
  Dwarf pond snail, *G. truncutula* (Müller, 1774) 125
  Bog or marsh snail, *G. palustris* (Müller, 1774) 125

Family Lancidae 125
Genus *Lanx* 125
  L. patelloides (Lea, 1856) 125

Superfamily Ancylacea

Family Physidae (Bladder snails) 125
Genus *Physa* 125
  Bladder snail, *P. fontinalis* (Linné, 1758) 125

  P. acuta Draparnaud, 1805 125
Genus *Aplexa* 125
  A. hypnorum (Linné, 1758) —

Superfamily Ellobiacea

Order Stylommatophora (Land Snails)

Suborder Orthurethra

Superfamily Achatinellacea (Tree Snails)

Superfamily Cionellacea

Superfamily Pupillacea

Suborder Heterurethra

Superfamily Succineacea (Amber Snails)

Superfamily Athoracophoracea

Suborder Mesurethra

Superfamily Clausiliacea

Superfamily Corillacea

Superfamily Strophocheilacea

Suborder Sigmurethra

Superfamily Achatinacea

Superfamily Streptaxacea

Superfamily Ariophantacea

Family Trochomorphidae	—	**Family Ariophantidae**	—
		Genus *Rhinocochilis*	133
Family Euconulidae	—	*R. nasuta* (Metcalfe)	133
Genus *Euconulus*	—		
E. fulvus (Müller, 1774)	—	**Family Urocyclidae**	—
		Genus *Thyrophorella*	133
Family Helicarionidae	—	*T. thomensis* Greeff, 1882	133

Superfamily Oleacinacea

Family Thysanophoridae	—	**Family Oleacinidae**	—
Family Ammonitellidae	—	**Family Sagdidae**	—

Superfamily Testacellacea

Family Testacellidae	—	Shelled slug, *T. haliotidea* Draparnaud, 1801	133
Genus *Testacella*	133		

Superfamily Polygyracea

Family Polygyridae —

Superfamily Helicacea (Helix Snails)

Family Oreohelicidae	133	Genus *Helicella*	134
		Heath snail, *H. itala* (Linné, 1758)	134
Family Camaenidae	133	*H. obvia* (Hartmann, 1842)	134
		Genus *Theba*	134
Family Bradybaenidae (Bush snails)	133	White or sandhill snail, *T. pisana* (Müller, 1774)	134
Genus *Bradybaena*	133	Genus *Helicodonta*	134
Brush snail, *B. fruticum* (Müller, 1774)	133	Cheese snail, *H. obvoluta* (Müller, 1774)	134
		Genus *Trochulus*	134
Family Helminthoglyphidae	—	*T. villosus* (Studer, 1789)	134
Genus *Polymita* (Cuban Tree Snails)	135	*T. unidentatus* (Draparnaud, 1805)	94★
P. picta (Born, 1778)	135	Genus *Isognomostoma*	134
		I. holosericum (Studer, 1820)	134
Family Helicidae	133	Genus *Helicigona*	134
Genus *Helix* (Edible Snails)	134	Lapidary snail, *H. lapicida* (Linné, 1758)	134
Edible snail or "Escargot", *H. pomatia* Linné, 1758	133	Genus *Cepaea*	134
		Grove snail, *C. nemoralis* (Linné, 1758)	134
Common snail, *H. aspersa* Müller, 1774	134	Garden snail, *C. hortensis* (Müller, 1774)	134
Genus *Perforatella*	134	*C. vindobonensis* (Pfeiffer, 1828)	134
P. rubiginosa Schmidt, 1853	134	Genus *Monacha*	134
P. incarnata (Müller, 1774)	89★	*M. cartusiana* (Müller, 1774)	134
Genus *Arianta*	134	Genus *Cylindrus*	134
Copse snail, *A. arbustorum* (Linné, 1758)	134	*C. obtusus* (Draparnaud, 1805)	134

Class Scaphopoda (Tusk Shells)

Family Dentaliidae (Elephant's tusks)	138	Common elephant's tusk, *D. vulgare* DaCosta, 1778	
Genus *Dentalium*	138		138

Class Bivalvia (The Bivalves)

Order Nuculoida

Order Solemyoida

Subclass Filibranchia

Order Arcoida

Sublcass Anisomyaria

Order Mytiloida

Superfamily Tellinacea

Superfamily Solenacea

Superfamily Mactracea

Order Myoida

Superfamily Myacea

Superfamily Pholadacea

Subclass Anomalodesmata

Order Pholadomyoida

Superfamily Pandoracea

Superfamily Clavagellacea

Superfamily Poromyacea

Class Cephalopoda (Squids and Octopods)

Subclass Tetrabranchiata (Nautiluses)

Subclass Dibranchiata

Order Decabrachia

Suborder Sepiodei

Suborder Teuthoidei

Superfamily Loligoidea or Myopsida (Dwarf Squid)

Superfamily Oegopsioidea

Order Vampyromorpha

Order Octobrachia

Suborder Cirrata

Suborder Incirrata

Superfamily Bolitaenoidea

Superfamily Octopodoidea

Superfamily Argonautoidea

LOPHOPHORATES

PHYLUM PHORONIDA

PHYLUM BRYOZOA

Class Phylactolaemata

Class Stenolaemata

Order Cyclostomata

Superfamily Tubuliporoidea

Superfamily Hederelloidea

Superfamily Articuloidea

Superfamily Cancelloidea

Superfamily Cerioporoidea

Superfamily Rectanguloidea

Class Gymnolaemata

Order Ctenostomata

Suborder Stolonifera

Superfamily Walkerioidea

Superfamily Vesicularioidea

Suborder Carnosa

Superfamily Paludicelloidea

Superfamily Halcyonellloidea

Order Cheilostomata

Suborder Anasca

Superfamily Inovicelloidea

Superfamily Scruparioidea

Superfamily Malacostegoidea

Superfamily Cellularioidea

Order Terebratulida

PHYLUM CHAETOGNATHA

SUBPHYLUM CRINOZOA

Class Crinoidea

Order Isocrinida

Order Comatulida

Suborder Comasterina

Suborder Miriametrina

SUBPHYLUM ECHINOZOA

Class Holothuroidea

Subclass Dendrochirotacea

Order Dendrochirotida

Superorder Gnathostomata

Order Holectypoida

Order Clypeasteroida

Superorder Atelostomata

Order Cassiduloida

Order Holasteroida

PHYLUM HEMICHORDATA
Class Enteropneusta

Class Pterobranchia
Order Cephalodiscida

On the Zoological Classification and Names

For many years, zoologists and botanists have tried to classify animals and plants into a system which would be a survey of the abundance of forms in fauna and flora. Such a system, of course, may be established under very different aspects. Since Charles Darwin, his predecessors, and his successors have found that all creatures have evolved out of common ancestors, species of animals and plants have been classified according to their natural relationships. Our knowledge about the phylogeny, and thus the relationship of each living being to the other, is augmented every year by new discoveries and insights. Old ideas are replaced with more recent and more appropriate ones. Therefore, the natural classification of the animal kingdom (and the plant kingdom) is subject to changes. Furthermore, the opinions of zoologists, who are working on the classification of animals into the various groups, are anything but uniform. These differences and changes are usually insignificant. The classification of vertebrates into the classes of fish, amphibia, reptiles, birds, and mammals has been fixed for many decades. Only the Cyclostomata were recently separated from the fish and all other classes of vertebrates as the "jawless" Agnatha (comp. Vol. IV).

The animal kingdom has been split into several sub-kingdoms and these were again divided into further sections, subsections, and so on. The scale of the most important systematic categories follows in a descending rank order:

Kingdom
Sub-kingdom
Phylum
Subphylum
Class
Subclass
Superorder
Order
Suborder
Infraorder
Family
Subfamily
Tribe
Genus
Subgenus
Species
Subspecies

The scientific names of the animals and their spelling follow the international rules for the zoological nomenclature as agreed upon by the XV International Congress for Zoology and are obligatory for all zoological publications. The name of the genus, which is a Latin or Latinized noun, is singular and capitalized. After the name of the genus follows the name of the species and of the subspecies. The names of the species and subspecies may be nouns or adjectives, and they are spelled in the lower case. The name of a subgenus, which is formed in the same manner as a genus, may be added in brackets following the name of the genus. The names of the tribes, subfamilies, families, and superfamilies are plural capitalized nouns. They are formed from the name of a given genus by adding to the principal word the endings -ini for the tribe, -inae for the subfamily, -idae for the family, and -oidea for the superfamily. The names of the authors who were the first to describe and to name a species, subspecies, or group of animals should be cited with the year of this naming at least once in each scientific publication. The name of the author and year are not enclosed in brackets when the species or subspecies is classified as belonging to the same genus with which the author had originally classified it. They are in brackets when another genus name is used in the present publication. The scientific names of the genus, subgenus, species, and subspecies are supposed to be printed with different letters, usually italics.

Animal Dictionary

1. English—German—French—Russian

For scientific names of species see the German-English-French-Russian section of this dictionary or the index.

ENGLISH NAME	GERMAN NAME	FRENCH NAME	RUSSIAN NAME
Argonaut	Papierboot		Обыкновенный аргонавт
Bear's claw	Pferdehufmuschel		
Bivalves	Muscheln	Bivalves	Двустворчатые моллюски
Black fish	Gemeiner Tintenfisch	Sèche	Обыкновенная каракатица
Blood-red starfish	Blutstern		
Blue mussel	Gemeine Miesmuschel	Moule	Съедобная мидия
Boat shell	Pantoffelschnecke		
Brittle stars	Schlangensterne	Ophiurides	Офиуры
Bryozoans	Moostierchen	Bryozaires	Мшанки
Calmary	Gemeiner Kalmar	Encornet	Обыкновенный кальмар
Cask shell	Faßschnecke		
Cephalopods	Kopffüßer	Céphalopodes	Головоногие моллюски
Chambered Nautilus	Gemeines Perlboot		Обыкновенный кораблик
Clam	Pilgermuschel	Coquille de Saint-Jacques	
Cockle	Gewöhnliche Herzmuschel	Coque	Съедобная сердцевидка
– brillon	Klaffmuschel	Mye	Песчаная ракушка
Common brittle-star	Zerbrechlicher Schlangenstern		
– cockle	Gewöhnliche Herzmuschel	Coque	Съедобная сердцевидка
– European starfish	Gemeiner Seestern	Étoile de mer	Красная морская звезда
– gaper shell	Klaffmuschel	Mye	Песчаная ракушка
– limpet	Gemeine, Gewöhnliche Napf-schnecke	Patelle (de la Méditerranée)	
– mussel	– Miesmuschel	Moule	Съедобная мидия
– periwinkle	– Strandschnecke	Bigorneau noir	Обыкновенный берего-вичок
– piddock	– Bohrmuschel	Pholade	Обыкновенный камне-точец
– sea-urchin	Eßbarer Seeigel		Съедовный морской еж
Common sepia	Gemeiner Tintenfisch	Sèche	Обыкновенная каракатица
– squid	Nordamerikanischer Kalmar	Encornet	Североамериканский кальмар
– sun-star	Stachelsonnenstern		
– whelk	Wellhornschnecke	Buccin	Трубач
Cotton-spinner	Röhrenholothurie		Трубчатая голотурия
Crinoids	Seelilien und Haarsterne	Crinoïdes	Морские лилии
Crockling	Gemeine Miesmuschel	Moule	Съедобная мидия
Crow oyster	Sattelmuschel	Rose	Веловатая луковичка
Cuttlefish	Gemeiner Tintenfisch	Sèche	Обыкновенная каракатица
Daisy brittle star	Gänseblümchen-Schlangenstern		
Date shell	Steindattel	Datte de mer	Морской финик
Ear shell	Gemeines Seeohr	Ormeau	Обыкновенное морское ухо
Echinoderms	Stachelhäuter	Échinodermes	Иглокожие
Echinoids	Seeigel	Échinides	Морские ежи
Edible snail	Weinbergschnecke	Vigneron	Виноградная улитка
European rock periwinkle	Gewöhnliche Strandschnecke		
Exotic snail	Weinbergschnecke		Виноградная улитка
Fan shell	Pilgermuschel	Coquille de Saint-Jacques	
Feather stars	Haarsterne		
Flither	Gemeine Napfschnecke	Patelle	
Flithers	Napfschnecken		Морское блюдечко
Flying squid	Gemeiner Kalmar	Encornet	Обыкновенный кальмар
Frill	Pilgermuschel	Coquille de Saint-Jacques	
Furrow shell	Pfeffermuschel	Lavignon	
Gastropods	Schnecken	Gastéropodes	Брюхоногие моллюски
Giant clam	Mördermuschel		Обыкновенная тридакна
– conch	Fechterschnecke		Крылатка великан
– scallop	Große Kammuschel		Большой гребешок
Golden-star tunicate	Sternaskidie		Ботрилл Шлоссера
Green snail	Marmorierte Kreiselschnecke		Мраморная кубарчатка
Heart-urchin	Herzigel		Сердцевидный еж

ENGLISH NAME	GERMAN NAME	FRENCH NAME	RUSSIAN NAME
Hen	Pfeffermuschel	Lavignon	
Holothuroids	Seewalzen	Holothurides	Голотурии
Horse oyster	Sattelmuschel	Rose	Веловатая луковичка
Inkfish	Mittelländischer Zwergkalmar	Petit encornet	Средиземноморский карликовый кальмар
Italian snail	Weinbergschnecke		Виноградная улитка
Larger anomia	Sattelmuschel	Rose	Веловатая луковичка
Limpet	Gemeine Napfschnecke, Schildkrötenschnecke	Patelle	
Limpets	Napfschnecken	Patelles	Морское блюдечко
Long oyster	Gemeine Bohrmuschel	Pholade	Обыкновенный камнеточец
Long-necked clam	Klaffmuschel	Mye	Песчаная ракушка
Long-spinned sea-urchin	Diademseeigel		
Mammy fish			
Molluscs	Weichtiere	Mollusques	Моллюски
Mud-hen	Pfeffermuschel	Lavignon	
Noah's ark shell	Arche Noah		Ноев ковчег
Octopus	Gemeiner Krake	Pieuvre	Обыкновенный осьминог
Old maid	Klaffmuschel	Mye	Песчаная ракушка
Onion-peel anomia	Sattelmuschel	Rose	Веловатая луковичка
Ophiuroids	Schlangensterne	Ophiurides	Офиуры
Ormer	Gemeines Seeohr	Ormeau	Обыкновенное морское ухо
Oyster	Europäische Auster	Huître	Европейская устрица
Oysters	Austern	Huîtres	Устрицы
Pap shells	Napfschnecken		Морское блюдечко
Paper nautilus	Papierboot		Обыкновенный аргонавт
Pearly Nautilus	Gemeines Perlboot	Nautile	Обыкновенный кораблик
Pea-urchin	Schildigel		Щитовидный еж
Penfish	Mittelländischer Zwergkalmar	Petit encornet	Средиземноморский карликовый кальмар
Piddock	Gemeine Bohrmuschel	Pholade	Обыкновенный камнеточец
Pilgrim oyster	Pilgermuschel	Coquille de Saint-Jacques	
Pinpatche	Gemeine Strandschnecke	Bigorneau noir	Обыкновенный береговичок
Poulp	Gemeiner Krake	Pieuvre	Обыкновенный осьминог
Prickly fish			
— piddock	Gemeine Bohrmuschel	Pholade	Обыкновенный камнеточец
Pudworm	— —	—	Обыкновенный камнеточец
Puller	Klaffmuschel	Mye	Песчаная ракушка
Queen	Pilgermuschel	Coquille de Saint-Jacques	
Queen conch	Fechterschnecke		Крылатка великан
Rayed mactra	Große Trogmuschel	Mactre	
Razor shell	— Scheidenmuschel	Couteau droit	Обыкновенный черенок
Red fish			
— slender starfish	Blutstern		
Red-nose clam	Gemeiner Felsenbohrer		
Ribband	Sägezahn	Olive	
Rock-borer clam	Gemeiner Felsenbohrer		
Rock periwinkle	Rauhe Strandschnecke		
Rough periwinkle	— —		
Sabre razor	Schwertmuschel	Couteau courbe	
Saddle oyster	Sattelmuschel	Rose	Веловатая луковичка
Sand clam	Große Scheidenmuschel	Couteau droit	Обыкновенный черенок
— gaper	Klaffmuschel	Mye	Лесчаная ракушка
Sanddollars	Sanddollars		Щитовидные морские ежи
Scallop	Pilgermuschel	Coquille de Saint-Jacques	
Sea arrow	Nordamerikanischer Kalmar	Encornet	Североамериканский кальмар
Sea-cucumbers	Seegurken, Seewalzen	Holothuries	Голотурии
Sea hare	Gefleckter Seehase	Lièvre marin	Пятнистый морской заяц
— lace	Seerinde		
Sea-stars	Seesterne	Stellérides	Морские звезды
Sea-urchin	Steinseeigel	Oursin	
Sea-urchins	Seeigel	Échinides	Морские ежи
Serpent star	Brauner Schlangenstern		
Sheath shell	Große Scheidenmuschel	Couteau droit	Обыкновенный черенок
Ship worm	Gemeine Schiffsbohrmuschel		Корабельный червь
Slipper shell	Pantoffelschnecke		
Smooth periwinkle	Stumpfe Strandschnecke		
Soft-shelled clam	Klaffmuschel	Mye	Песчаная ракушка
Spiked cockle	Stachelige Herzmuschel	Coque épineuse	

ENGLISH NAME	GERMAN NAME	FRENCH NAME	RUSSIAN NAME
Spiny cockle	– –	– –	
Spont fish	Große Scheidenmuschel	Couteau droit	Обыкновенный черенок
Squid	Gemeiner Kalmar, Mittel- ländischer Zwergkalmar	Encornet, Petit encornet	Обыкновенный кальмар, Средиземноморский карликовый кальмар
Sucker	Gemeiner Krake	Pieuvre	Обыкновенный осьминог
Teat fish			
Thin tellen	Platte Tellmuschel		Теллина
Tooth shell	Elefantenzahn		Зубовики
Trough shell	Große Trogmuschel	Mactre	
Tuberculated sea bear	Gemeines Seeohr	Ormeau	Обыкновенное морское ухо
Tunicates	Manteltiere	Tuniciers	Оболочники
Vertrebrates	Wirbeltiere	Vertébrés	Позвоночные
Violet snail	Veilchenschnecke		Янтина
Warted Venus shell	Rauhe Venusmuschel	Praire	
Waved whelk	Wellhornschnecke	Buccin	
Wedge shell	Sägezahn	Olive	
West Indian chiton	Westindischer Chiton		Вестиндский хитон
Wilk	Gemeine Strandschnecke	Bigorneau noir	Обыкновенный берего- вичок

II. German—English—French—Russian

GERMAN NAME	ENGLISH NAME	FRENCH NAME	RUSSIAN NAME
Achtarmige Tintenschnecken		Octopodes	Осьминогие
Achtarmkalmar			Осьминогий кальмар
Aciculidae			Острянковые
Acmaea testudinalis	Limpet		
Acrania			Бесчерепные
Acroloxus lacustris			Озерная чашечка
Actinopyga echinites	Red fish		
– *mauritiana*	– –		
Aeolidia papillosa			Широкососочная эолка
Aeolidioidea			Эолковые
Alloteuthis			Карликовые кальмары
– *media*	Squid	Petit encornet	Средиземноморский кар- ликовый кальмар
– *subulata*			Атлантический карлико- вый кальмар
Amnicola steini			Амникола Штейна
Amphineura		Amphineures	Боконервные моллюски
Amphioxus			Ланцетник
Ampullariidae			Ампуллярии
Ancylidae			Чашечки
Ancylus fluviatilis			Речная чашечка
Anisus vortex			Завернутая катушка
Anodonta cygnaea		Anodonte	Обыкновенная беззубка
Anomia ephippium	Saddle oyster	Rose	Веловатая луковичка
Antedon mediterranea			Средивемноморская волосатка
Aplexa hypnorum			Аплекса
Aplysia dactylomela	Sea hare	Lièvre marin	Пятнистый морской заяц
– *depilans*			Морской заяц
Apoda		Holothuries apods	Безногие голотурии
Arca noae	Noah's ark shell		Ноев ковчег
Arche Noah	– – –		Ноев ковчег
Architeuthidae			Гигантские кальмары
Architeuthoidea			Открытоглазые кальмары
Argonauta argo	Paper nautilus		Обыкновенный аргонавт
Argonautoidea			Аргонавты
Arion hortensis			Садовый слизень
– *rufus*			Большой красный слизень
Armfüßer			Плеченогие
Ascidia		Ascidies	Асцидии
Asterias rubens	Common European starfish	Étoile de mer	Красная морская звезда
Asteroidea	Sea-stars	Stellérides	Морские звезды
Astropecten aranciacus			Оранжевая морская звезда
Atlantacea und Heteropoda			Киленогие моллюски
Atlantische Sepiole			Атлантическая сепиола

GERMAN NAME	ENGLISH NAME	FRENCH NAME	RUSSIAN NAME
Atlantischer Zwergkalmar			Атлантический карликовый кальмар
Austern	Oysters	Huîtres	Устрицы
Baltische Plattmuschel			Балтийская макома
Bankia minima			Малый морской древоточец
Basommatophora		Basommatophores	Сидячеглазые
Bäumchenschnecke			Ветвистая древовидка
Beilschnecke			Филироя
Belemniten		Bélemnites	Белемниты
Belemnoidea		–	Белемниты
Bernsteinschnecken			Янтарки
Binden-Schwimmschnecke			Трехполосая лунка
Birnenschnecken			Грушковые
Bischofsmütze			Епископская митра
Bithynia leachi			Битиния лича
– tentaculata			Щупальцевая битиния
Bivalvia	Bivalves	Bivalves	Двустворчатые моллюски
Blasenschnecken			Ампуллярии, Физы
Blasige Federkiemenschnecke			Гладкая затворка
Blättermoostierchen			Флюстры, Листовидная флюстра
Blutstern	Blood-red starfish		
Borstenkiefer			Щетинкочелюстные
Borysthenia naticina			Гладкая затворка
Botryllus schlosseri	Golden-star tunicate		Ботрилл Шлоссера
Brachiopoda			Плеченогие
Branchiostoma lanceolatum			Ланцетник
Brandhorn		Rocher épineux	Обыкновенная багрянка
Brauner Schlangenstern	Serpent star		
Breitwarzige Fadenschnecke			Широкососочная эолка
Bryozoa	Bryozoans	Bryozoaires	Мшанки
Buccinidae		Buccins	Букциниды
Buccinum undatum	Common whelk	Buccin	Трубач
Bulla striata			Обыкновенный пузырек
Bythinella austriaca			Австрийская битинелла
Calyptraea chinensis			Китайская калиптрея
Capulidae			Колпачковые
Capulus hungaricus			Венгерский колпачок
Cardium aculeatum	Spiked cockle	Coque épineuse	
– edule	Common cockle	Coque	Съедобная сердцевидка
Carychium minimum			Карликовая улитка
Cephalopoda	Cephalopods	Céphalopodes	Головоногие моллюски
Cerithiidae			Игольники
Chaetognatha			Щетинкочелюстные
Chinesenhut			Китайская калиптрея
Chiton olivaceus			Средиземноморский хитон
– tuberculatus	West Indian chiton		Вестиндский хитон
Chordata			Хордовые
Chordatiere			Хордовые
Clausiliidae			Щеминки
Clavagella			Булава
Clione limacina			Клион
Clypeasteroidea	Sanddollars		Щитовидные морские ежи
Comatulida	Feather stars		
Conidae			Конусовые
Conus marmoreus			Мраморный конус
Corbula			Корбулы
– mediterranea			Средиземноморская корбула
Crepidula fornicata	Slipper shell		
Crinoidea	Crinoids	Crinoïdes	Морские лилии
Crossaster papposus	Common sun-star		
Cryptochiton stelleri			Тихоокеанский хитон
Cucumariidae	Sea-cucumbers	Holothuries	
Cypraea tigris			Пятнистая ужовка
Dattelmuschel	Common piddock	Pholade	Обыкновенный камнеточец
Decabrachia		Décapodes	Десятиногие
Dendronotus arborescens			Ветвистая древовидка
Dentaliidae			Зубовики
Dentalium	Tooth shell		Зубовики
–- vulgare			Обыкновенный зубовик
Deuterostomia			Вторичноротые
Diadema antillarum	Long-spinned sea-urchin		
Diademseeigel	– –		

GERMAN NAME	ENGLISH NAME	FRENCH NAME	RUSSIAN NAME
Dibranchiata		Dibranchiaux	Двужаберные моллюски
Dicke Flußmuschel			Толстая перловица
Doliolida			Бочночники
Donax trunculus	Wedge shell	Olive	
Dreiecksmuschel			Речная дрейссена
Dreissena polymorpha			Речная дрейссена
Dreissenidae			Дрейссены
Ecardines			Беззамковые плеченогие
Echinocardium cordatum	Heart-urchin		Сердцевидный еж
Echinocyamus pusillus	Pea-urchin		Щитовидный еж
Echinodermata	Echinoderms	Échinodermes	Иглокожие
Echinoidea	Echinoids	Échinides	Морские ежи
Echinothuridae			Кожистые морские ежи
Echinus exculentus	Common sea-urchin		Съедовный морской еж
Echte Sumpfdeckelschnecke			Речная живородка
Eigentliche Salpen			Сальпы
– Tintenschnecken			Каракатицы
Elefantenzahn	Tooth shell		Зубовики
Elefantenzähne			Зубовики
Elysia viridis			Зеленая элизия
Emarginula huzardi			Средиземноморская вырезка
Ensis ensis	Sabre razor	Couteau courbe	
Erbsenmuscheln			Горошинки
Eßbare Auster	Oyster	Huître	Европейская устрица
– Herzmuschel	Common cockle	Coque	Съедобная сердцевидка
Eßbarer Seeigel	Common sea-urchin		Съедобный морской еж
Europäische Auster	Oyster	Huître	Европейская устрица
– Gastrochaena			Европейская гастрохена
Fadenschnecken i. e. S.			Эолковые
Faßschnecke	Cask shell		
Fechterschnecke	Giant conch		Крылатка великан
Federkiemenschnecken			Затворки
Federsterne	Feather stars		
Felsen-Strandschnecke	Rough periwinkle		
Feuerwalzen			Пирозомы
Fissurella			Дырчатки
Fissurellidae			Дырчатковые
Flache Tellerschnecke			Килеватая катушка
Flügelschnecken		Ptéropodes	Крылоногие моллюски, Крылатковые
Flußkugelmuschel			Речная шаровка
Flußmuscheln			Перловицы и Беззубки
Flußnapfschnecken			Чашечки
Flußperlmuschel			Обыкновенная жемчуж- ница
Flußschwimmschnecke			Речная лунка
Flustra			Флюстры
– *foliacea*			Листовидная флюстра
Furchenfüßer		Solénogastres	Желобобрюхие моллюски
Fußlose			Безногие голотурии
Galba palustris			Болотный прудовик
– *trunculata*			Усеченный прудовик
Gänseblümchen-Schlangenstern	Daisy brittle star		
Gartenwegschnecke			Садовый слизень
Gastrochaena dubia			Европейская гастрохена
Gastropoda	Gastropods	Gastéropodes	Брюхоногие моллюски
Gefleckter Seehase	Sea hare	Lièvre marin	Пятнистый морской заяц
Gekielte Tellerschnecke			Завернутая катушка
Gemeine Auster	Oyster	Huître	Европейская устрица
– Bernsteinschnecke			Обыкновенная янтарка
– Blasenschnecke			Обыкновенный пузырек
– Bohrmuschel	Common piddock	Pholade	Обыкновенный камне- точец
– Erbsenmuschel			Болотная горошинка
– Flußmuschel			Овальная перховица
– Flußnapfschnecke			Речная чашечка
– Landdeckelschnecke			Обыкновенная башне- видка
– Miesmuschel	Common mussel	Moule	Съедобная мидия
– Napfschnecke	– limpet	Patelle	
– Schiffsbohrmuschel	Ship worm		Корабельный червь
– Strandschnecke	Common periwinkle	Bigorneau noir	Обыкновенный берего- вичок
– Sumpfdeckelschnecke			Болотная живородка
– Turmschnecke			Обыкновенная башенка

GERMAN NAME	ENGLISH NAME	FRENCH NAME	RUSSIAN NAME
— Wandermuschel			Речная дрейссена
Gemeiner Elefantenzahn			Обыкновенный зубовик
— Felsenbohrer	Red-nose clam		
— Hakenkalmar			Обыкновенный крючкова-тый кальмар
— Kalmar	Calmary	Encornet	Обыкновенный кальмар
— Krake	Octopus	Pieuvre	Обыкновенный осьминог
— Seestern	Common European starfish	Étoile de mer	Красная морская звезда
— Tintenfisch	— sepia	Sèche	Обыкновенная каракатица
Gemeines Perlboot	Pearly Nautilus	Nautile	Обыкновенный кораблик
— Seeohr	Tuberculated sea bear	Ormeau	Обыкновенное ушко
Gesprenkelte Weinbergschnecke			Крапчатая улитка
Gewöhnliche Herzmuschel	Common cockle	Coque	Съедобная сердцевидка
— Napfschnecke	— limpet	Patelle de la Méditerranée	
— Strandschnecke	European rock periwinkle		
Gewöhnliches Lanzettfischchen			Ланцетник
Glänzende Tellerschnecke			Вывернутая катушка
Glycimeridae			Сердцевидки
Gorgonocephalidae			Головы медузы
Grabfüßer		Scaphopodes	Лопатоногие моллюски
Graptolithen			Граптолитовые
Graptolithida			Граптолитовые
Große Egelschnecke			Большой придорожный слизень
— Erbsenmuschel			Речная горошинка
— Feilenmuschel			Большой напильник
— Kammuschel	Giant scallop		Большой гребешок
— Langfühlerschnecke			Щупальцевая битиния
Große Mantel-Käferschnecke			Тихоокеанский хитон
— Rossie			Большая россия
— Rote Wegschnecke			Большой красный слизень
— Scheidenmuschel	Razor shell	Couteau droit	Обыкновенный черенок
— Schlammschnecke			Озерник
— Strandschnecke	Common periwinkle	Bigorneau noir	Обыкновенный берего-вичок
— Trogmuschel	Trough shell	Mactre	
Grüne Samtschnecke			Зеленая элизия
Gyraulus laevis			Вывернутая катушка
Haarsterne	Feather stars		
Hainschnecken			Улитки
Hakenkalmare			Крючковатые кальмары
Haliotidae			Морские ушки
Haliotis kamtschatkana			Камчатское морское ухо
— tuberculata	Tuberculated sea bear	Ormeau	Обыкновенное морское ухо
Hammermuschel			Молоток
Hängende Wattschnecke			Выпуклая гидробия
Haubenmuschel			Болотная шаровка
Helicidae			Улитки
Helix aspera			Крапчатая улитка
— pomatia	Edible snail	Vigneron	Виноградная улитка
Henricia sanguinolenta	Blood-red starfish		
Herzigel	Heart-urchin		Сердцевидный еж
Herzseeigel			Сердцевидные морские ежи
Heteropoda			Киленогие моллюски
Hiatella arctica	Red-nose clam		
Hinterkiemer		Opisthobranches	Заднежаберные моллюски
Hippipus hippopus	Bear's claw		
Holothuria forskali			Черная голотурия
— tubulosa	Cotton-spinner		Трубчатая голотурия
Holothuroidea	Holothuroids	Holothurides	Голотурии
Hornfarbige Kugelmuschel			Роговая шаровка
Hornschnecken		Buccins	Букциниды
Hutschnecken			Колпачковые
Hydrobia ventrosa			Выпуклая гидробия
Hydrobiidae			Гидробииды
Irreguläre (unregelmäßige) Seeigel		Oursins irréguliers	
Janthina janthina	Violet snail		Янтина
Janthinidae			Янтины
Käferschnecken		Placophores	Бляшконосные
Kahnfüßer			Лопатоногие моллюски
Kalmare		Calmars	Кальмары
Kammerschnecken		Tétrabranchiaux	Четырехжаберные моллюски
Kammuscheln			Гребешки

GERMAN NAME	ENGLISH NAME	FRENCH NAME	RUSSIAN NAME
Kamtschatka-Seeohr			Камчатское морское ухо
Kegelschnecken			Конусовые
Kielfüßer			Киленогие моллюски
Klaffmuschel	Soft-shelled clam	Mye	Песчаная ракушка
Kleine Pfahlmuschel			Малый морской древо-точец
— Schlammschnecke			Усеченный прудовик
Königsholothurie			Королевская голотурия
Kopffüßer	Cephalopods	Céphalopodes	Головоногие моллюски
Korbmuscheln			Корбулы
Kraken		Pieuvres	Осьминоги
Krallenkalmar			Обыкновенный крючкова-тый кальмар
Kugelmuscheln			Шаровки и Горошинки
Landdeckelschnecken			Башневидки
Landlungenschnecken		Stylommatophores	Стебельчатоглазые моллюски
Langarmiger Krake			Длиннощупальцевый осьминог
Lanzettfischchen			Бесчерепные, Ланцетник
Lazarusklappe			Съедобный шарнир
Lederseeigel			Кожистые морские ежи
Leistenschnecken			Багрянки
Leitungsmoos			Болотнянка
Lima hians			Малый напильник
— inflata			Большой напильник
Limax cinereoniger			Черный слизень
— maximus			Большой придорожный слизень
Lingula			Язычки
Lingula unguis			Язычок
Lithophaga mytiloides	Date shell	Datte de mer	Морской финик
Littorina littorea	Common periwinkle	Bigorneau noir	Обыкновенный берегови-чок
— neritoides	European rock periwinkle		
— obtusata	Smooth periwinkle		
— saxatilis	Rough periwinkle		
Littorinidae			Литорины
Lochschnecken			Дырчатковые
— i. e. S.			Дырчатки
Loligo pealei	Common squid	Encornet	Североамериканский кальмар
— vulgaris	Calmary	—	Обыкновенный кальмар
Loligoidea			Закрытоглазые кальмары
Lungenschnecken		Pulmonés	Легочные моллюски
Lymnaea stagnalis			Озерник
Lymnaeidae			Прудовики
Macoma baltica			Балтийская макома
— nasuta			Тихоокеанская макома
Mactra stultorum	Trough shell	Mactre	Обыкновенная перловица
Malermuschel			
Malleus malleus			Молоток
Manteltiere	Tunicates	Tuniciers	Оболочники
Margaritana margaritifera			Обыкновенная жемчуж-ница
Marmorierte Kreiselschnecke	Green snail		Мраморная кубарчатка
Marmorkegel			Мраморный конус
Medusenhäupter			Головы медузы
Meerohren			Морские ушки
Membranipora membranacea	Sea lace		
Messerscheide	Razor shell	Couteau droit	Обыкновенный черенок
Microthele nobilis	Mammy fish		
Mitra episcopalis			Епископская митра
— papalis			Папская митра
Mitraschnecken			Митровые
Mitridae			Митровые
Mittelländische Ausschnitts-schnecke			Средиземноморская вырезка
— Korbmuschel			Средиземноморская корбула
— Miesmuschel		Moule de la Méditerranée	Средиземноморская мидия
Mittelländischer Zwergkalmar	Squid	Petit encornet	Средиземноморский кар-ликовый кальмар
Mittelmeer-Chiton			Средиземноморский хитон
Mittelmeer-Haarstern			Средиземноморская волосатка

GERMAN NAME	ENGLISH NAME	FRENCH NAME	RUSSIAN NAME
Mittelmeer-Schirmschnecke			Средиземноморская зонтичница
Mittelmeersepiole		Sepiole	Сепиола Ронделета
Mollusca	Molluscs	Mollusques	Моллюски
Molluscoidea			Моллюскообразные
Moosblasenschnecke			Аплекса
Moostierchen	Bryozoans	Bryozaires	Мшанки
Mördermuschel	Giant clam		Обыкновенная тридакна
Moschuskrake			Мускусный осьминог
Murex brandaris		Rocher épineux	Обыкновенная багрянка
Muricidae			Багрянки
Muschellinge			Моллюскообразные
Muscheln	Bivalves	Bivalves	Двустворчатые моллюски
Mya arenaria	Soft-shelled clam	Mye	Песчаная ракушка
Mytilus edulis	Common mussel	Moule	Съедобная мидия
– galloprovincialis		– de la Méditerranée	Средиземноморская мидия
Nacktaugenkalmare			Открытоглазые кальмары
Nacktkiemer		Nudibranches	Голожаберные моллюски
Nadelschnecken			Острянковые
Najaden			Наяды
Napfschnecken	Limpets	Patelles	Морское блюдечка, Чашечки
Nautilidae		Nautiles	Кораблики
Nautilus pompilius	Pearly Nautilus	Nautile	Обыкновенный кораблик
Neritacea			Ронковые
Nestbauende Feilenmuschel			Малый напильник
Neumünder			Вторичноротые
Nixenschnecken			Ронковые
Nordamerikanischer Kalmar	Common squid	Encornet	Североамериканский кальмар
Nudibranchia		Nudibranches	Голожаберные моллюски
Octobrachia		Octopodes	Осьминогие
Octopodoidea		Pieuvres	Осьминогие
Octopoteuthis sicula			Осьминогий кальмар
Octopus macropus			Длиннощупальцевый осьминог
– vulgaris	Octopus	Pieuvre	Обыкновенный осьминог
Ohrförmige Schlammschnecke			Ушковый прудовик
Ölkrug			Масляная кубарчатка
Ommatostrephes sagittatus		Seiche	Стреловидный вертиглаз
Onychoteuthidae			Крючковатые кальмары
Onychoteuthis banksi			Обыкновенный крючковатый кальмар
Ophioderma longicauda	Serpent star		
Ophiopholis aculeata	Daisy brittle star		
Ophiothrix fragilis	Common brittle star		
Ophiura texturata			Чешуйчатая офиура
Ophiuroidea	Brittle stars	Ophiurides	Офиуры
Opistobranchia		Opisthobranches	Заднежаберные моллюски
Ostrea edulis	Oyster	Huître	Европейская устрица
Ostreidae	Oysters	Huîtres	Устрицы
Ozaena moschata			Мускусный осьминог
Paludicella articulata			Болотнянка
Pantoffelschnecke	Slipper shell		
Papierboot	Paper nautilus		
Papierbootartige Kraken			Обыкновенный аргонавт
Papstkrone			Аргонавты
Paracentrotus lividus	Sea-urchin	Oursin	Папская митра
Patella	Limpets	Patelles	
– coerulea	Common limpet	Patelle de la Méditerranée	Морское блюдечка
– vulgata	– –	Patelle	
Patellidae			Чашечки
Pazifische Plattmuschel			Тихоокеанская макома
Pecten jacobaeus	Scallop	Coquille de Saint-Jacques	
– maximus	Giant scallop		Большой гребешок
Pectinidae			Гребешки
Perlboote		Tétrabranchiaux, Nautiles	Четырехжаберные моллюски, Кораблики
Pfeffermuschel	Hen	Lavignon	
Pfeilkalmar		Seiche	Стреловидный вертиглаз
Pfeilwürmer			Щетинкочелюстные, Стреловидные черви
Pferdehufmuschel	Bear's claw		
Philine quadripartita			Морская миндалинка
Pholas dactylus	Common piddock	Pholade	Обыкновенный камнеточец

GERMAN NAME	ENGLISH NAME	FRENCH NAME	RUSSIAN NAME
Phylliroe bucephala			Филироя
Physa acuta			Заостренная физа
– fontinalis			Пузырчатая физа
Physidae			Физы
Pilgermuschel	Scallop	Coquille de Saint-Jacques	
Pinna nobilis			Обыкновенная пинна
Pisidium			Горошинки
– amnicum			Речная горошинка
– casertanum			Болотная горошинка
Placophora		Placophores	Бляшконосные
Planorbarius corneus			Роговая катушка
Planorbidae			Катушки
Planorbis carinatus			Килеватая катушка
Platte Tellmuschel	Thin tellen		Теллина
Plattmuscheln			Теллиниды
Pleurotomariidae			Вырезокрайние
Pomatias elegans			Обыкновенная башне-видка
Pomatiasidae			Башневидки
Posthörnchen			Витушки
Posthornschnecke			Роговая катушка
Prosobranchia		Prosobranches	Переднежаберные моллюски
Protostomia			Первичноротые
Pterocera chiragra			Чертов коготь
Pteropoda		Ptéropodes	Крылоногие моллюски
Pulmonata		Pulmonés	Легочные моллюски
Pyrenidae			Грушковые
Pyrosomatida			Пирозомы
Quellenblasenschnecke			Пузырчатая физа
Quellenschnecke			Австрийская битинелла
Radix auricularia			Ушковый прудовик
– peregra			Вытянутый прудовик
Raubschnecken			Хищные слизни
Rauchfangmuscheln			Булава
Rauhe Strandschnecke	Rough periwinkle		
– Venusmuschel	Warted Venus shell	Praire	
Regelmäßige (reguläre) Seeigel		Oursins réguliers	Правильные морские ежи
Reguläre (regelmäßige) Seeigel		– –	Правильные морские ежи
Reihenzähnige Muscheln		Taxodontes	Равнозубые
Riesenkalmare			Гигантские кальмары
Riesenmuscheln			Тридакны
Röhrenholothurie	Cotton-spinner		Трубчатая голотурия
Röhrenschaler			Лопатоногие моллюски
Rossia macrosoma			Большая россия
Rote Bohne			Балтийская макома
Roter Kammstern			Оранжевая морская звезда
Rückgrattiere			Хордовые
Runde Langfühlerschnecke			Битиния лича
Runzeliger Felsenbohrer			Шероховатая бурилка
Sägezahn	Wedge shell	Olive	
Sagittoidea			Стреловидные черви
Salpen		Thaliacés	Сальпы
Salpida			Сальпы
Samtmuscheln			Сердцевидки
Sanddollars			Щитовидные морские ежи
Sarmatische Schwimmschnecke			Дунайская лунка
Sattelmuschel	Saddle oyster	Rose	Веловатая луковичка
Saxicava rugosa			Шероховатая бурилка
Scaphopoda		Scaphopodes	Лопатоногие моллюски
Schädellose			Бесчерепные
Schalenschloß-Armfüßer			Замковые плеченогие
Scheibenförmige Federkiemen-schnecke			Плоская затворка
Scheidenmuscheln			Черенковые
Schiffsboot	Pearly Nautilus	Nautile	Обыкновенный кораблик
Schildigel	Pea-urchin		Щитовидный еж
Schildkrötenschnecke	Limpet		
Schirmschnecken			Зонтичницы
Schlammschnecken			Прудовики
Schlangensterne	Brittle stars	Ophiurides	Офиуры
Schließaugenkalmare			Закрытоглазые кальмары
Schließmundschnecken			Щеминки
Schlitzbandschnecken			Вырезокрайние
Schloßlose Armfüßer			Беззамковые плеченогие
Schnauzenschnecken			Гидробииды

GERMAN NAME	ENGLISH NAME	FRENCH NAME	RUSSIAN NAME
Schnecken	Gastropods	Gastéropodes	Брюхоногие моллюски
Schuppiger Schlangenstern			Чешуйчатая офиура
Schwanenmuschel			Обыкновенная беззубка
Schwarze Egelschnecke			Черный слизень
– Seegurke			Черная голотурия
Schwertmuschel	Sabre razor	Couteau courbe	
Schwimmschnecken			Лунки
Scrobicularia plana	Hen	Lavignon	
Seedattel	Date shell	Datte de mer	Морской финик
Seegurken	Sea-cucumbers	Holothuries	
Seehase			Морской заяц
Seeigel	Echinoids	Échinides	Морские ежи
Seelilien	Sea-lilies		Морские лилии
– und Haarsterne	Crinoids	Crinoïdes	Морская миндалинка
Seemandel			Игольники
Seenadelschnecken			Амникола Штейна
Seenschnecke			
Seerinde	Sea lace		
Seescheiden		Ascidies	Асцидии
Seesterne	Sea-stars	Stellérides	Морские звезды
Seewalzen	Holothuroids	Holothurides	Голотурии
Sepia officinalis	Common sepia	Sèche	Обыкновенная каракатица
Sepie	– –	–	Обыкновенная каракатица
Sepioidei			Каракатицы
Sepiola		Sepioles	Сепиолы
– *atlantica*			Атлантическая сепиола
– *rondeleti*		Sepiole	Сепиола Ронделета
Sepiolen		Sepioles	Сепиолы
Solen vagina	Razor shell	Couteau droit	Обыкновенный черенок
Solenidae			Черенковые
Solenogastres		Solénogastres	Желобобрюхие моллюски
Spatangoida			Сердцевидные морские ежи
Spatangus purpureus			Пупурный сердцевидный еж
Sphaeriidae			Шаровки и Горошинки
Sphaerium corneum			Роговая шаровка
– *lacustre*			Болотная шаровка
– *rivicola*			Речная шаровка
Spiralige Tellerschnecke			Завернутая катушка
Spirulidae			Витушки
Spitze Blasenschnecke			Заостренная физа
– Schlammschnecke			Озерник
Spondylus gaederopus			Съедобный шарнир
Stachelauster			Съедобный шарнир
Stachelhäuter	Echinoderms	Échinodermes	Иглокожие
Stachelige Herzmuschel	Spiked cockle	Coque épineuse	
Stachelschnecken			Багрянки
Stachelsonnenstern	Common sun-star		
Steckmuschel			Обыкновенная пинна
Steindattel	Date shell	Datte de mer	Морской финик
Steinseeigel	Sea-urchin	Oursin	
Sternaskidie	Golden-star tunicate		Ботрилл Шлоссера
Stichopus regalis			Королевская голотурия
Strahlenkorb	Through shell	Mactre	
Strandschnecken			Литорины
Strombacea			Крылатковые
Strombus gigas	Giant conch		Крылатка великан
Stumpfe Strandschnecke	Smooth periwinkle		
– Sumpfdeckelschnecke			Речная живородка
Stylommatophora		Stylommatophores	Стебельчатоглазые моллюски
Succinea putris			Обыкновенная янтарка
Succineidae			Янтарки
Süditalienische Schwimmschnecke			Южноитальянская лунка
Sumpfdeckelschnecken			Живородки
Sumpfschnecke			Болотный прудовик
Tafelauster	Oyster	Huître	Европейская устрица
Taxodonta		Taxodontes	Равнозубые
Teich-Federkiemenschnecke			Обыкновенная затворка
Teichmuschel		Anodonte	Обыкновенная беззубка
Teichnapfschnecke			Озерная чашечка
Tellerschnecken			Катушки
Tellina tenuis	Thin tellen		Теллина
Tellinidae			Теллиниды
Tellmuscheln	– –		Теллина

GERMAN NAME	ENGLISH NAME	FRENCH NAME	RUSSIAN NAME
Teredo navalis	Ship worm		Корабельный червь
Testacellidae			Хищные слизни
Testicardines			Замковые плеченогие
Tetrabranchiata		Tétrabranchiaux	Четырехжаберные моллюски
Teufelskralle			Чертов коготь
Thaliacea		Thaliacés	Сальпы
Thelenota ananas	Prickly fish		
Theodoxus			Лунки
— *danubialis*			Дунайская лунка
— *fluviatilis*			Речная лунка
— *meridionalis*			Южноитальянская лунка
— *transversalis*			Трехполосая лунка
Theutoidei		Calmars	Кальмары
Tigerschnecke			Пятнистая ужовка
Tintenschnecken		Dibranchiaux	Двужаберные моллюски
Tonna galea	Cask shell		
Tonnensalpen			Боченочники
Tridacna gigas	Giant clam		Обыкновенная тридакна
Tridacnidae			Тридакны
Tunicata	Tunicates	Tuniciers	Оболочники
Turbo marmoratus	Green snail		Мраморная кубарчатка
— *olearius*			Масляная кубарчатка
Turmschnecken			Башенковые
Turritella communis			Обыкновенная башенка
Turritellidae			Башенковые
Umbraculidae			Зонтичницы
Umbraculum mediterraneum			Средиземноморская зонтичница
Ungarkappe			Венгерский колпачок
Unio crassus			Толстая перловица
— *pictorum*			Обыкновенная перловица
Unionidae			Перловицы и Беззубки
Unionoidea			Наяды
Unregelmäßige (irreguläre) Seeigel		Oursins irréguliers	
Urmollusken		Amphineures	Боконервные моллюски
Urmünder			Первичноротые
Valvata cristata			Плоская затворка
— *piscinalis*			Обыкновенная затворка
Valvatidae			Затворки
Veilchenschnecke, Veilchen- schnecken	Violet snail		Янтина, Янтины
Veneriidae			Венусы
Venus verrucosa	Warted Venus shell	Praire	
Venusmuscheln			Венусы
Vermetidae			Червячковые
Vertebrata	Vertebrates	Vertébrés	Позвоночные
Vierkiemer		Tétrabranchiaux	Четырехжаберные моллюски
Violetter Herzigel			Пупурный сердцевидный еж
Viviparidae			Живородки
Viviparus contectus			Болотная живородка
— *viviparus*			Речная живородка
Vorderkiemer		Prosobranches	Переднежаберные моллюски
Walaas, Walaat			Клион
Wandermuscheln			Дрейссены
Wandernde Schlammschnecke			Вытянутый прудовик
Wasserlungenschnecken		Basommatophores	Сидячеглазые
Weichtiere	Molluscs	Mollusques	Моллюски
Weinbergschnecke	Edible snail	Vigneron	Виноградная улитка
Wellhornschnecke	Common whelk	Buccin	Трубач
Westindischer Chiton	West Indian chiton		Вестиндский хитон
Wirbeltiere	Vertebrates	Vertébrés	Позвоночные
Wurmschnecken			Червячковые
Zackenmuscheln			Тридакны
Zehnarmige Tintenschnecken		Décapodes	Десятиногие
Zerbrechlicher Schlangenstern	Common brittle-star		
Zungenmuschel, Zungenmuscheln			Язычок, Язычки
Zweikiemer		Dibranchiaux	Двужаберные моллюски
Zwerghornschnecke			Карликовая улитка
Zwergkalmare			Карликовые кальмары
Zwergseeigel	Pea-urchin		Щитовидный еж
Zwergsepia		Sepiole	Сепиола Ронделета
Zwiebelmuschel	Saddle oyster	Rose	Веловатая луковичка

III. French—German—English—Russian

FRENCH NAME	GERMAN NAME	ENGLISH NAME	RUSSIAN NAME
Alisson	Steinseeigel	Sea-urchin	
Amphineures	Urmollusken		Боконервные моллюски
Anodonte	Teichmuschel		Обыкновенная беззубка
Arapède	Gewöhnliche Napfschnecke	Common limpet	
Ascidies	Seescheiden		Асцидии
Astérides	Seesterne	Sea-stars	Морские звезды
Astérie	Gemeiner Seestern	Common European starfish	Красная морская звезда
Basommatophores	Wasserlungenschnecken		Сидячеглазые
Bélemnites	Belemniten		Белемниты
Berlu	Klaffmuschel	Soft-shelled clam	Песчаная ракушка
Bernique	Gemeine Napfschnecke	Common limpet	
Bigorneau	Brandhorn		Обыкновенная багрянка
– de chien	Wellhornschnecke	Common whelk	
– noir	Gemeine Strandschnecke	– periwinkle	Обыкновенный берего-вичок
Biscuit de mer	Gemeiner Tintenfisch	Common sepia	Обыкновенная каракатица
Bivalves	Muscheln	Bivalves	Двустворчатые моллюски
Blanchet	Große Trogmuschel	Trough shell	
Bonne-sœur	Gemeine Bohrmuschel	Common piddock	Обыкновенный камне-точец
Bryozoaires	Moostierchen	Bryozoaires	Мшанки
Bucarde épineuse	Stachelige Herzmuschel	Spiked cockle	
Buccin	Wellhornschnecke	Common whelk	Трубач
Buccins	Hornschnecken		Букциниды
Calmaret	Mittelländischer Zwergkalmar	Squid	Средиземноморский кар-ликовый кальмар
Calmars	Kalmare		Кальмары
Casseron	Gemeiner Tintenfisch	Common sepia	Обыкновенная каракатица
Céphalopodes	Kopffüßer	Cephalopods	Головоногие моллюски
– dibranches	Tintenschnecken		Двужаберные моллюски
– tétrabranches	Perlboote		Четырехжаберные моллюски
Cérites	Hornschnecken		Букциниды
Châtaigne de mer	Steinseeigel	Sea-urchin	
Châtrou	Gemeiner Krake	Octopus	Обыкновенный осьминог
Clanque	Klaffmuschel	Soft-shelled clam	Песчаная ракушка
Coieux	Gemeine Miesmuschel	Common mussel	Съедобная мидия
Compteux	Wellhornschnecke	– whelk	
Coque	Gewöhnliche Herzmuschel	– cockle	Съедобная сердцевидка
Coque épineuse	Stachelige Herzmuschel	Spiked cockle	
– rayée	Rauhe Venusmuschel	Warted Venus shell	
Coquille de Saint-Jacques	Pilgermuschel	Scallop	
Corne de punchas	Brandhorn		Обыкновенная багрянка
Cornet	Gemeiner Kalmar	Calmary	Обыкновенный кальмар
Couteau courbe	Schwertmuschel	Sabre razor	
– droit	Große Scheidenmuschel	Razor shell	Обыкновенный черенок
Crinoïdes	Seelilien und Haarsterne	Crinoids	Морские лилии
Datte de mer	Steindattel	Date shell	Морской финик
Décapodes	Zehnarmige Tintenschnecken		Десятиногие
Derte	Gemeine Bohrmuschel	Common piddock	Обыкновенный камне-точец
Dibranchiaux	Tintenschnecken		Двужаберные моллюски
Échinides	Seeigel	Echinoids	Морские ежи
Échinodermes	Stachelhäuter	Echinoderms	Иглокожие
Encornet	Nordamerikanischer Kalmar, Gemeiner Kalmar	Common squid, Calmary	Североамериканский каль-мар, Обыкноевнный кальмар
Étoile de mer	Gemeiner Seestern	– European starfish	Красная морская звезда
Fausse palourde	Große Trogmuschel	Trough shell	
Feuille de rose	Sattelmuschel	Saddle oyster	Веловатая луковичка
– de salade	Brandhorn		Обыкновенная багрянка
Fleon	Sägezahn	Wedge shell	
Flie	Große Trogmuschel	Trough shell	
Gajin	Gemeine Strandschnecke	Common periwinkle	Обыкновенный берего-вичок
Gastéropodes	Schnecken	Gastropods	Брюхоногие моллюски
Gofiche	Pilgermuschel	Scallop	
Grande pèlerine	–	–	
Gravette	Europäische Auster	Oyster	Европейская устрица
Gros bigorneau	Wellhornschnecke	Common whelk	
Grosse coque	Stachelige Herzmuschel	Spiked cockle	
– palourde	Klaffmuschel	Soft-shelled clam	Песчаная ракушка
Haricot de mer	Sägezahn	Wedge shell	

FRENCH NAME	GERMAN NAME	ENGLISH NAME	RUSSIAN NAME
Hénon	Gewöhnliche Herzmuschel	Common cockle	Съедобная сердцевидка
Hérissons de mer	Seeigel	Echinoids	Морские ежи
Holothurides	Seewalzen	Holothuroids	Голотурии
Holothuries	Seegurken	Sea-cucumbers	
– apods	Fußlose		Безногие голотурии
Homomyaires	Reihenzähnige Muscheln		Равнозубые
Huître	Europäische Auster	Oyster	Европейская устрица
– de roche	Sattelmuschel	Saddle oyster	Веловатая луковичка
Huîtres	Austern	Oysters	Устрицы
Jambe	Gemeine Napfschnecke	Common limpet	
Lamellibranches	Muscheln		Двустворчатые моллюски
Lampot	Gemeine Napfschnecke	Common limpet	
Lavignon	Pfeffermuschel	Hen	
Lièvre marin	Gefleckter Seehase	Sea hare	...нистый морской заяц
Luisette	Sattelmuschel	Saddle oyster	Веловатая луковичка
Mactre	Große Trogmuschel	Trough shell	
Maillot	Gewöhnliche Herzmuschel	Common cockle	Съедобная сердцевидка
Manche de couteau	Große Scheidenmuschel	Razor shell	Обыкновенный черенок
Margade	Gemeiner Tintenfisch	Common sepia	Обыкновенная каракатица
Minard	– Krake	Octopus	Обыкновенный осьминог
Mollusques	Weichtiere	Molluscs	Моллюски
Moule	Gemeine Miesmuschel	Common mussel	Съедобная мидия
– de la Méditerranée	Mittelländische Miesmuschel		Средиземноморская мидия
– – rocher	– –		Средиземноморская мидия
Muscle de Rocco			Средиземноморская мидия
– – vase	Gemeine Miesmuschel	Common mussel	Съедобная мидия
Mye	Klaffmuschel	Soft-shelled clam	Песчаная ракушка
Nautile	Gemeines Perlboot	Pearly Nautilus	Обыкновенный кораблик
Nautiles	Perlboote		Кораблики
Nudibranches	Nacktkiemer		Голожаберные моллюски
Octopodes	Achtarmige Tintenschnecken		Осьминогие
Œil de bouc	Gemeine Napfschnecke	Common limpet	
Olive	Sägezahn	Wedge shell	
Ophiurides	Schlangensterne	Brittle stars	Офиуры
Opisthobranches	Hinterkiemer		Заднежаберные моллюски
Oreille de mer, Ormeau	Gemeines Seeohr	Tuberculated sea bear	Обыкновенное морское ухо
Oursin	Steinseeigel	Sea-urchin	
Oursins irréguliers	Unregelmäßige (irreguläre) Seeigel		
– réguliers	Regelmäßige (reguläre) Seeigel		Правильные морские ежи
Padallida	Gewöhnliche Napfschnecke	Common limpet	
Palourde	Pilgermuschel	Scallop	
– plate	Pfeffermuschel	Hen	
– sauvage	Rauhe Venusmuschel	Warted Venus shell	
Patelle	Gemeine Napfschnecke	Common limpet	
– de la Méditerranée	Gewöhnliche Napfschnecke	– –	
Patelles	Napfschnecken	Limpets	Морское блюдечко
Pélécypodes	Muscheln		Двустворчатые моллюски
Petit encornet	Mittelländischer Zwergkalmar	Squid	Средиземноморский карликовый кальмар
Pholade	Gemeine Bohrmuschel	Common piddock	Обыкновенный камнеточец
Pied de cheval	Europäische Auster	Oyster	Европейская устрица
– – couteau	Große Scheidenmuschel	Razor shell	Обыкновенный черенок
Pieuvre, Pieuvres	Gemeiner Krake, Kraken	Octopus	Обыкновенный осьминог, Осьминогие
Pilau noir	Gemeine Strandschnecke	Common periwinkle	Обыкновенный береговичок
Pilot	Wellhornschnecke	– whelk	
Placophores	Käferschnecken		Бляшконосные
Poulpe	Gemeiner Krake		Обыкновенный осьминог
– sèche	Mittelmeersepiole		Сепиола Ронделета
Praire	Rauhe Venusmuschel	Warted Venus shell	
Ptéropodes	Flügelschnecken		Крылоногие моллюски
Pulmonés	Lungenschnecken		Легочные моллюски
Pupre	Gemeiner Krake	Octopus	Обыкновенный осьминог
Ran à capet	Wellhornschnecke	Common whelk	
Rasoir	Große Scheidenmuschel	Razor shell	Обыкновенный черенок
Religieuse	Gemeine Bohrmuschel	Common piddock	Обыкновенный камнеточец
Rigadeau	Gewöhnliche Herzmuschel	– cockle	Съедобная сердцевидка
Rigadelle	Rauhe Venusmuschel	Warted Venus shell	
Rocher épineux	Brandhorn		Обыкновенная багрянка
Rose	Sattelmuschel	Saddle oyster	Веловатая луковичка
Salpes	Salpen		Сальпы
Scaphopodes	Grabfüßer		Лопатоногие моллюски

FRENCH NAME	GERMAN NAME	ENGLISH NAME	RUSSIAN NAME
Sèche	Gemeiner Tintenfisch	Common sepia	Обыкновенная каракатица
Seiche	Pfeilkalmar		Стреловидный вертиглаз
— anglaise	Gemeiner Kalmar	Calmary	Обыкновенный кальмар
Sepiole	Mittelmeersepiole		Сепиола Ронделета
Sepioles	Sepiolen		Сепиолы
Sépioun	Mittelmeersepiole		Сепиола Ронделета
Silieu	Gemeines Seeohr	Tuberculated sea bear	Обыкновенное морское ухо
Six yeux	— —	— — —	Обыкновенное морское ухо
Socquet	Mittelländischer Zwergkalmar	Squid	Средиземноморский карликовый кальмар
Solénoconches	Grabfüßer		Лопатоногие моллюски
Solénogastres	Furchenfüßer		Желобобрюхие моллюски
Souchot	Mittelmeersepiole		Сепиола Ронделета
Stellérides	Seesterne	Sea-stars	Морские звезды
Stylommatophores	Landlungenschnecken		Стебельчатоглазые моллюски
Supia	Gemeiner Tintenfisch	Common sepia	Обыкновенная каракатица
Taut	— Kalmar	Calmary	Обыкновенный кальмар
Taxodontes	Reihenzähnige Muscheln		Равнозубые
Tête de scorpion	Brandhorn		Обыкновенная багрянка
Tétrabranchiaux	Perlboote		Четырехжаберные моллюски
Thaliacés	Salpen		Сальпы
Tote	Mittelländischer Zwergkalmar	Squid	Средиземноморский карликовый кальмар
Truille	Sägezahn	Wedge shell	
Tuniciers	Manteltiere	Tunicates	Оболочники
Vanne	Pilgermuschel	Scallop	
Vertébrés	Wirbeltiere	Vertebrates	Позвоночные
Vigneau	Gemeine Strandschnecke	Common periwinkle	Обыкновенный берего-вичок
Vigneron	Weinbergschnecke	Edible snail	Виноградная улитка

IV. Russian—German—English—French

RUSSIAN NAME	GERMAN NAME	ENGLISH NAME	FRENCH NAME
Австрийская битинелла	Quellenschnecke		
Амникола Штейна	Seenschnecke		
Ампуллярии	Blasenschnecken		
Амфиокс	Lanzettfischchen		
Амфиокс	Gewöhnliches Lanzet'fischchen		
Аплекса	Moosblasenschnecke		
Аргонавты	Papierbootartige Kraken		
Асцидии	Seescheiden		Ascidies
Атлантическая сепиола	Atlantische Sepiole		
Атлантический карлико-вый кальмар	Atlantischer Zwergkalmar		
Багрянки	Leistenschnecken		
Балтийская макома	Baltische Plattmuschel		
Башенковые	Turmschnecken		
Башневидки	Landdeckelschnecken		
Беззамковые плеченогие	Schloßlose Armfüßer		
Безногие голотурии	Fußlose		Holothuries apods
Белемниты	Belemniten		Bélemnites
Бесчерепные	Lanzettfischchen		
Бесчерепные	Schädellose		
Битиния лича	Runde Langfühlerschnecke		
Бляшконосные	Käferschnecken		Placophores
Боконервные моллюски	Urmollusken		Amphineures
Болотная горошинка	Gemeine Erbsenmuschel		
Болотная живородка	— Sumpfdeckelschnecke		
Болотная шаровка	Haubenmuschel		
Болотный прудовик	Sumpfschnecke		
Болотнянка	Leitungsmoos		
Большая россия	Große Rossie		
Большой гребешок	— Kammuschel	Giant scallop	
Большой красный слизень	— Rote Wegschnecke		
Большой напильник	— Feilenmuschel		

RUSSIAN NAME	GERMAN NAME	ENGLISH NAME	FRENCH NAME
Большой придорожный слизень	— Egelschnecke		
Бороздчатобрюхие моллюски	Furchenfüßer		Solénogastres
Ботрилл Шлоссера	Sternaskidie	Golden-star tunicate	
Боченочники	Tonnensalpen		
Брюхоногие моллюски	Schnecken	Gastropods	Gastéropodes
Букциниды	Hornschnecken		Buccins
Булава	Rauchfangmuscheln		
Веловатая луковичка	Sattelmuschel	Saddle oyster	Rose
Венгерский колпачок	Ungarkappe		
Венусы	Venusmuscheln		
Вестиндский хитон	Westindischer Chiton	West Indian chiton	
Ветвистая древовидка	Bäumchenschnecke		
Виноградная улитка	Weinbergschnecke	Edible snail	Vigneron
Витушки	Posthörnchen		
Вторичноротые	Neumünder		
Вывернутая катушка	Glänzende Tellerschnecke		
Выпуклая гидробия	Hängende Wattschnecke		
Вырезокрайние	Schlitzbandschnecken		
Вытянутый прудовик	Wandernde Schlammschnecke		
Гигантские кальмары	Riesenkalmare		
Гидробииды	Schnauzenschnecken		
Гладкая затворка	Blasige Federkiemenschnecke		
Головые медузы	Medusenhäupter		
Головоногие моллюски	Kopffüßer	Cephalopods	Céphalopodes
Голожаберные моллюски	Nacktkiemer		Nudibranches
Голотурии	Seewalzen	Holothuroids	Holothurides
Горошинки	Erbsenmuscheln		
Граптолитовые	Graptolithen		
Гребешки	Kammuscheln		
Грушковые	Birnenschnecken		
Двужаберные моллюски	Tintenschnecken		Dibranchiaux
Двустворчатые моллюски	Muscheln	Bivalves	Bivalves
Десятиногие	Zehnarmige Tintenschnecken		Décapodes
Длиннощупальцевый осьминог	Langarmiger Krake		
Длиннощупальцевый спрут	— —		
Дрейссены	Wandermuscheln		
Дунайская лунка	Sarmatische Schwimmschnecke		
Дырчатки	Lochschnecken i. e. S.		
Дырчатковые	Lochschnecken		
Европейская гастрохена	Europäische Gastrochaena		
Европейская устрица	Europäische Auster	Oyster	Huître
Епископская митра	Bischofsmütze		
Желобобрюхие моллюски	Furchenfüßer		Solénogastres
Живородки	Sumpfdeckelschnecken		
Завернутая катушка	Gekielte Tellerschnecke		
Завернутая катушка	Spiralige Tellerschnecke		
Заднежаберные моллюски	Hinterkiemer		Opisthobranches
Закрытоглазые кальмары	Schließaugenkalmare		
Замковые плеченогие	Schalenschloß-Armfüßer		
Заостренная физа	Spitze Blasenschnecke		
Затворки	Federkiemenschnecken		
Зеленая элизия	Grüne Samtschnecke		
Змеехвостки	Schlangensterne	Brittle stars	Ophiurides
Зонтичницы	Schirmschnecken		
Зубовики	Elefantenzahn	Tooth shell	
Зубовики	Elefantenzähne		
Иглокожие	Stachelhäuter	Echinoderms	Échinodermes
Игольники	Seenadelschnecken		
Кальмары	Kalmare		Calmars
Камчатское морское ухо	Kamtschatka-Seeohr		
Каракатицы	Eigentliche Tintenschnecken		
Карликовая улитка	Zwerghornschnecke		
Карликовые кальмары	Zwergkalmare		
Катушки	Tellerschnecken		
Килеватая катушка	Flache Tellerschnecke		
Киленогие моллюски	Kielfüßer		
Китайская калиптрея	Chinesenhut		
Клион	Walaas		
Кожистые морские ежи	Lederseeigel		
Колпачковые	Hutschnecken		
Конусовые	Kegelschnecken		
Корабельный червь	Gemeine Schiffsbohrmuschel	Ship worm	

RUSSIAN NAME	GERMAN NAME	ENGLISH NAME	FRENCH NAME
Кораблики	Perlboote		Nautiles
Корбулы	Korbmuscheln		
Королевская голотурия	Königsholothurie		
Крапчатая улитка	Gesprenkelte Weinbergschnecke		
Красная морская звезда	Gemeiner Seestern	Common European starfish	Étoile de mer
Крылатка великан	Fechterschnecke	Giant conch	
Крылатковые	Flügelschnecken		
Крылоногие моллюски	—		Ptéropodes
Крючковатые кальмары	Hakenkalmare		
Ланцетник	Gewöhnliches Lanzettfischchen		
Ланцетник	Lanzettfischchen		
Легочные моллюски	Lungenschnecken		Pulmonés
Листовидная флюстра	Blättermoostierchen		
Литорины	Strandschnecken		
Лопатоногие моллюски	Grabfüßer		Scaphopodes
Лужанки	Sumpfdeckelschnecken		
Лунки	Schwimmschnecken		
Малый морской древо-точец	Kleine Phahlmuschel		
Малый напильник	Nestbauende Feilenmuschel		
Масляная кубарчатка	Ölkrug		
Митровые	Mitraschnecken		
Моллюски	Weichtiere	Molluscs	Mollusques
Моллюскообразные	Muschellinge		
Молоток	Hammermuschel		
Морская миндалинка	Seemandel		
Морское блюдечко	Napfschnecken	Limpets	Patelles
Морской заяц	Seehase		
Морской финик	Steindattel	Date shell	Datte de mer
Морские ежи	Seeigel	Echinoids	Échinides
Морские звезды	Seesterne	Sea-stars	Stellérides
Морские кубышки	Seewalzen	Holothuroids	Holothurides
Морские лилии	Seelilien und Haarsterne	Crinoids	Crinoïdes
Морские огурцы	Seewalzen	Holothuroids	Holothurides
Морские ушки	Meerohren		
Мраморная кубарчатка	Marmorierte Kreiselschnecke	Green snail	
Мраморный конус	Marmorkegel		
Мускусный осьминог	Moschuskrake		
Мускусный спрут	—		
Мшанки	Moostierchen	Bryozoans	Bryozaires
Мягкотелые	Weichtiere	Molluscs	Mollusques
Наутилусы	Perlboote		Nautiles
Наяды	Najaden		
Ноев ковчег	Arche Noah	Noah's ark shell	
Оболочники	Manteltiere	Tunicates	Tuniciers
Обыкновенная багрянка	Brandhorn		Rocher épineux
Обыкновенная башенка	Gemeine Turmschnecke		
Обыкновенная башне-видка	— Landdeckelschnecke		
Обыкновенная беззубка	Teichmuschel		Anodonte
Обыкновенная жемчуж-ница	Flußperlmuschel		
Обыкновенная затворка	Teich-Federkiemenschnecke		
Обыкновенная каракатица	Gemeiner Tintenfisch	Common sepia	Sèche
Обыкновенная перловица	Malermuschel		
Обыкновенная пинна	Steckmuschel		
Обыкновенная тридакна	Mördermuschel	Giant clam	
Обыкновенная янтарка	Gemeine Bernsteinschnecke		
Обыкновенный аргонавт	Papierboot	Paper nautilus	
Обыкновенный бегего-вичок	Gemeine Strandschnecke	Common periwinkle	Bigorneau noir
Обыкновенный ботик	Papierboot	Paper nautilus	
Обыкновенный зубовик	Gemeiner Elefantenzahn		
Обыкновенный кальмар	— Kalmar	Calmary	Encornet
Обыкновенный камне-точец	Gemeine Bohrmuschel	Common piddock	Pholade
Обыкновенный кораблик	Gemeines Perlboot	Pearly Nautilus	Nautile
Обыкновенный крючкова-тый кальмар	Gemeiner Hakenkalmar		
Обыкновенный наутилус	Gemeines Perlboot	Pearly Nautilus	Nautile
Обыкновенный осьминог	Gemeiner Krake	Octopus	Pieuvre
Обыкновенный прудовик	Große Schlammschnecke		
Обыкновенный пузырек	Gemeine Blasenschnecke		
Обыкновенный спрут	Gemeiner Krake	Octopus	Pieuvre
Обыкновенный черенок	Große Scheidenmuschel	Razor shell	Couteau droit
Обыкновенное морское ухо	Gemeines Seeohr	Tuberculated sea bear	Ormeau

RUSSIAN NAME	GERMAN NAME	ENGLISH NAME	FRENCH NAME
Обыкновенное ушко	– –	– – –	–
Овальная перловица	Dicke Flußmuschel		
Озерная чашечка	Teichnapfschnecke		
Озерник	Große Schlammschnecke		
Оранжевая морская звезда	Roter Kammstern		
Острянковые	Nadelschnecken		
Осьминогие	Achtarmige Tintenschnecken		Octopodes
Осьминогие	Kraken		Pieuvres
Осьминогий кальмар	Achtarmkalmar		
Открытоглазые кальмары	Nacktaugenkalmare		
Офиуры	Schlangensterne	Brittle stars	Ophiurides
Папская митра	Papstkrone		
Первичноротые	Urmünder		
Переднежаберные моллюски	Vorderkiemer		Prosobranches
Перловицы и Беззубки	Flußmuscheln		
Песчаная ракушка	Klaffmuschel	Soft-shelled clam	Mye
Пирозомы	Feuerwalzen		
Пластинчатожаберные моллюски	Muscheln	Bivalves	Bivalves
Плеченогие	Armfüßer		
Плоская затворка	Scheibenförmige Federkiemenschnecke		
Позвоночные	Wirbeltiere	Vertebrates	Vertébrés
Правильные морские ежи	Regelmäßige (reguläre) Seeigel		Oursins réguliers
Прудовики	Schlammschnecken		
Пузырчатая физа	Quellenblasenschnecke		
Пупурный сердцевидный еж	Violetter Herzigel		
Пятнистая ужовка	Tigerschnecke		
Пятнистый морской заяц	Gefleckter Seehase	Sea hare	Lièvre marin
Равнозубые	Reihenzähnige Muscheln		Taxodontes
Речная горошинка	Große Erbsenmuschel		
Речная дрейссена	Gemeine Wandermuschel		
Речная живородка	Echte Sumpfdeckelschnecke		
Речная лунка	Flußschwimmschnecke		
Речная чашечка	Gemeine Flußnapfschnecke		
Речная шаровка	Flußkugelmuschel		
Роговая катушка	Posthornschnecke		
Роговая шаровка	Hornfarbige Kugelmuschel		
Ронковые	Nixenschnecken		
Садовый слизень	Gartenwegschnecke		
Сальпы	Salpen		Thaliacés
Сальпы	Eigentliche Salpen		
Североамериканский кальмар	Nordamerikanischer Kalmar	Common squid	Encornet
Сепии	Eigentliche Tintenschnecken		
Сепиола Ронделета	Mittelmeersepiole		Sepiole
Сепиолы	Sepiolen		Sepioles
Сердцевидки	Samtmuscheln		
Сердцевидные морские ежи	Herzseeigel		
Сердцевидный еж	Herzigel	Heart-urchin	
Сидячеглазые	Wasserlungenschnecken		
Средиземноморская волосатка	Mittelmeer-Haarstern		
Средиземноморская вырезка	Mittelländische Ausschnittsschnecke		
Средиземноморская зонтичница	Mittelmeer-Schirmschnecke		
Средиземноморская корбула	Mittelländische Korbmuschel		
Средиземноморская мидия	– Miesmuschel		Moule de la Méditerranée
Средиземноморский карликовый кальмар	Mittelländischer Zwergkalmar	Squid	Petit encornet
Средиземноморский хитон	Mittelmeer-Chiton		
Стебельчатоглазые моллюски	Landlungenschnecken		Stylommatophores
Странствующая ракушка	Gemeine Wandermuschel		
Стреловидные черви	Pfeilwürmer		
Стреловидный вертиглаз	Pfeilkalmar		Seiche
Съедобная мидия	Gemeine Miesmuschel	Common mussel	Moule
Съедобная сердцевидка	Gewöhnliche Herzmuschel	– cockle	Coque
Съедобный шарнир	Lazarusklappe		
Съедовный морской еж	Eßbarer Seeigel	Common sea-urchin	
Теллина	Platte Tellmuschel	Thin tellen	

RUSSIAN NAME	GERMAN NAME	ENGLISH NAME	FRENCH NAME
Теллиниды	Plattmuscheln		
Тихоокеанская макома	Pazifische Plattmuschel		
Тихоокеанский хитон	Große Mantel-Käferschnecke		
Толстая перловица	Dicke Flußmuschel		
Трехполосая лунка	Binden-Schwimmschnecke		
Тридакны	Riesenmuscheln		
Трубач	Wellhornschnecke	Common whelk	Buccin
Трубчатая голотурия	Röhrenholothurie	Cotton-spinner	
Улитки	Hainschnecken		
Улитки	Schnecken	Gastropods	Gastéropodes
Усеченный прудовик	Kleine Schlammschnecke		
Устрицы	Austern	Oysters	Huîtres
Ушковый прудовик	Ohrförmige Schlammschnecke		
Физы	Blasenschnecken		
Филироя	Beilschnecke		
Флюстры	Blättermoostierchen		
Хищные слизни	Raubschnecken		
Хордовые	Chordatiere		
Хордовые	Rückgrattiere		
Чашечки	Flußnapfschnecken		
Чашечки	Napfschnecken		
Червячковые	Wurmschnecken		
Черенковые	Scheidenmuscheln		
Черная голотурия	Schwarze Seegurke		
Черный слизень	– Egelschnecke		
Чертов коготь	Teufelskralle		
Четырехжаберные моллюски	Perlboote		Tétrabranchiaux
Чешуйчатая офиура	Schuppiger Schlangenstern		
Шаровки и Горошинки	Kugelmuscheln		
Шашень	Gemeine Schiffsbohrmuschel	Ship worm	
Шероховатая бурилка	Runzeliger Felsenbohrer		
Широкососочная эолка	Breitwarzige Fadenschnecke		
Щеминки	Schließmundschnecken		
Щетинкочелюстные	Pfeilwürmer		
Щитовидные морские ежи	Sanddollars	Sanddollars	
Щитовидный еж	Schildigel	Pea-urchin	
Щупальцевая битиния	Große Langfühlerschnecke		
Эолковые	Fadenschnecken i. e. S.		
Южноитальянская лунка	Süditalienische Schwimmschnecke		
Язычки	Zungenmuscheln		
Язычок	Zungenmuschel		
Янтарки	Bernsteinschnecken		
Янтина	Veilchenschnecke	Violet snail	
Янтины	Veilchenschnecken		

Conversion Tables of Metric to U.S. and British Systems

U.S. Customary to Metric		Metric to U.S. Customary	

—— Length ——

To convert	Multiply by	To convert	Multiply by
in. to mm.	25.4	mm. to in.	0.039
in. to cm.	2.54	cm. to in.	0.394
ft. to m.	0.305	m. to ft.	3.281
yd. to m.	0.914	m. to yd.	1.094
mi. to km.	1.609	km. to mi.	0.621

—— Area ——

sq. in. to sq. cm.	6.452	sq. cm. to sq. in.	0.155
sq. ft. to sq. mi.	0.093	sq. m. to sq. ft.	10.764
sq. yd. to sq. m.	0.836	sq. m. to sq. yd.	1.196
sq. mi. to ha.	258.999	ha. to sq. mi.	0.004

—— Volume ——

cu. in. to cc.	16.387	cc. to cu. in.	0.061
cu. ft. to cu. m.	0.028	cu. m. to cu. ft.	35.315
cu. yd. to cu. m.	0.765	cu. m. to cu. yd.	1.308

—— Capacity (liquid) ——

fl. oz. to liter	0.03	liter to fl. oz.	33.815
qt. to liter	0.946	liter to qt.	1.057
gal. to liter	3.785	liter to gal.	0.264

—— Mass (weight) ——

oz. avdp. to g.	28.35	g. to oz. avdp.	0.035
lb. avdp. to kg.	0.454	kg. to lb. avdp.	2.205
ton to t.	0.907	t. to ton	1.102
l. t. to t.	1.016	t. to l. t.	0.984

Abbreviations

U.S. Customary	Metric
avdp.—avoirdupois	cc.—cubic centimeter(s)
ft.—foot, feet	cm.—centimeter(s)
gal.—gallon(s)	cu.—cubic
in.—inch(es)	g.—gram(s)
lb.—pound(s)	ha.—hectare(s)
l. t.—long ton(s)	kg.—kilogram(s)
mi.—mile(s)	m.—meter(s)
oz.—ounce(s)	mm.—millimeter(s)
qt.—quart(s)	t.—metric ton(s)
sq.—square	
yd.—yard(s)	

By kind permission of Walker: Mammals of the World
©1968 Johns Hopkins Press, Baltimore, Md., U.S.A.

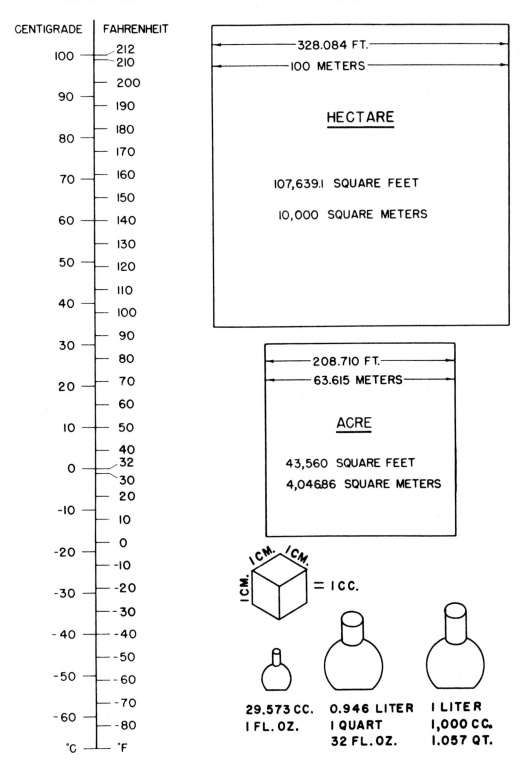

TEMPERATURE

CENTIGRADE   FAHRENHEIT

100 — 212 / 210
90 — 200 / 190
80 — 180 / 170
70 — 160 / 150
60 — 140 / 130
50 — 120 / 110
40 — 100 / 90
30 — 80 / 70
20 — 70 / 60
10 — 50 / 40
0 — 32 / 30 / 20
-10 — 10 / 0
-20 — -10 / -20
-30 — -30 / -40
-40 — -50 / -60
-50 — -70 / -80
-60 — -80
°C — °F

AREA

328.084 FT.
100 METERS

HECTARE

107,639.1 SQUARE FEET

10,000 SQUARE METERS

208.710 FT.
63.615 METERS

ACRE

43,560 SQUARE FEET

4,046.86 SQUARE METERS

1 CM. 1 CM. 1 CM. = 1 CC.

29.573 CC.
1 FL. OZ.

0.946 LITER
1 QUART
32 FL. OZ.

1 LITER
1,000 CC.
1.057 QT.

WEIGHT

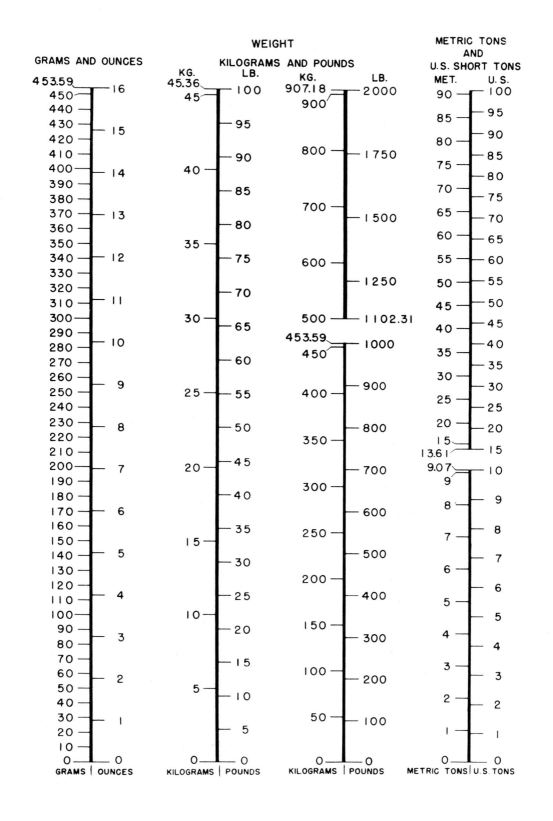

GRAMS AND OUNCES

453.59	
450	16
440	
430	15
420	
410	
400	14
390	
380	
370	13
360	
350	
340	12
330	
320	
310	11
300	
290	
280	10
270	
260	
250	9
240	
230	8
220	
210	
200	7
190	
180	
170	6
160	
150	
140	5
130	
120	
110	4
100	
90	3
80	
70	
60	2
50	
40	
30	1
20	
10	
0	0

GRAMS | OUNCES

KILOGRAMS AND POUNDS

KG. / LB.

45.36	100
45	
	95
40	90
	85
35	80
	75
	70
30	65
	60
25	55
	50
	45
20	40
	35
15	30
	25
10	20
	15
5	10
	5
0	0

KILOGRAMS | POUNDS

KG. / LB.

907.18	2000
900	
800	1750
700	1500
600	1250
500	1102.31
453.59	1000
450	
400	900
	800
350	700
300	600
250	500
200	400
150	300
100	200
50	100
0	0

KILOGRAMS | POUNDS

METRIC TONS
AND
U.S. SHORT TONS

MET. / U.S.

90	100
85	95
80	90
75	85
70	80
65	75
60	70
55	65
50	60
45	55
40	50
35	45
30	40
25	35
20	30
15	25
13.61	20
9.07	15
9	10
8	9
7	8
6	7
5	6
4	5
3	4
2	3
1	2
0	1
	0

METRIC TONS | U.S. TONS

LENGTH: MILLIMETERS AND INCHES

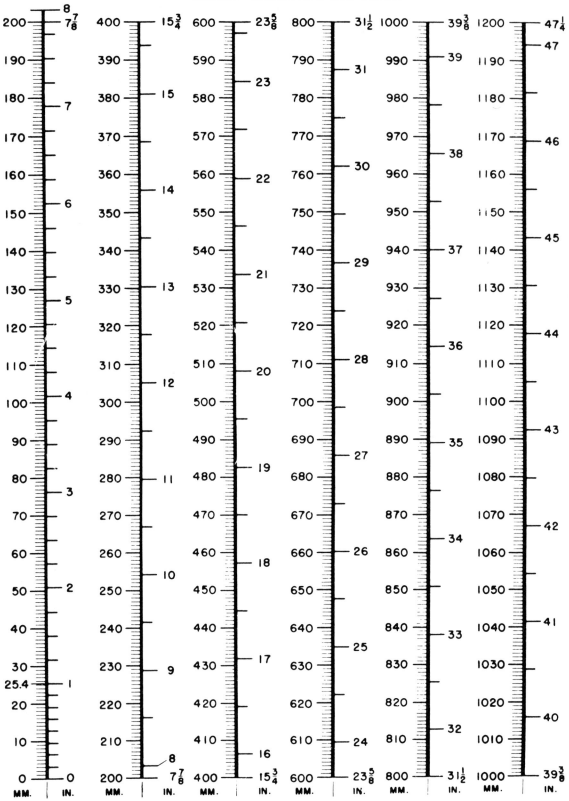

LENGTH

METERS AND FEET

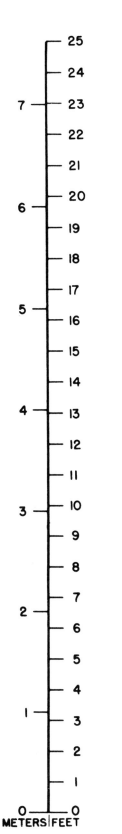

METERS	FEET
7	25
	24
	23
	22
6	21
	20
	19
	18
5	17
	16
	15
	14
4	13
	12
	11
3	10
	9
	8
	7
2	6
	5
	4
1	3
	2
	1
0	0

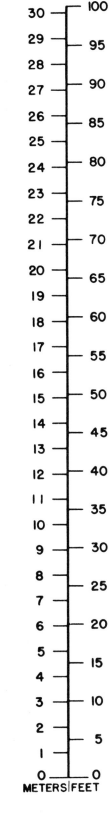

METERS	FEET
30	100
29	95
28	
27	90
26	85
25	
24	80
23	75
22	
21	70
20	65
19	
18	60
17	55
16	
15	50
14	45
13	
12	40
11	35
10	
9	30
8	25
7	
6	20
5	15
4	
3	10
2	5
1	
0	0

KILOMETERS AND MILES

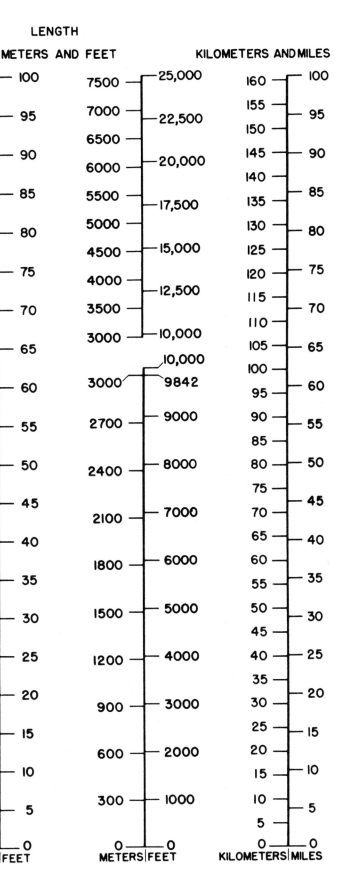

METERS	FEET
7500	25,000
7000	22,500
6500	
6000	20,000
5500	17,500
5000	
4500	15,000
4000	12,500
3500	
3000	10,000

METERS	FEET
3000	10,000 / 9842
2700	9000
2400	8000
2100	7000
1800	6000
1500	5000
1200	4000
900	3000
600	2000
300	1000
0	0

KILOMETERS	MILES
160	100
155	95
150	
145	90
140	85
135	
130	80
125	75
120	
115	70
110	65
105	
100	60
95	
90	55
85	50
80	
75	45
70	40
65	
60	35
55	30
50	
45	25
40	20
35	
30	15
25	10
20	
15	5
10	
5	
0	0

Supplementary Readings

Compiled by John B. Brown

Abbott, R. T. 1968. *Seashells of North America*. Golden Press, New York.

—. 1972. *Kingdom of the Seashell*. Crown Publishers, New York.

—. ed. 1959–74. *Indo-Pacific Mollusca*, Vols. 1–3. Greenville, Del.

—. 1974. *American Seashells*. Van Nostrand Reinhold, New York.

Baker, F. C. 1928. *The Lymnaeidae of North and Middle America*. Chicago Academy of Science, Special Publication 3, Chicago.

—. 1945. *The Molluscan Family Planorbidae*. University of Illinois Press, Urbana.

Barnes, R. D. 1968. *Invertebrate Zoology*. Saunders, Philadelphia.

Barrington, E. J. W. 1965. *The Biology of Hemichordata and Protochordata*. Freeman, San Francisco.

Bequaert, J. C. and W. B. Miller. 1973. *The Mollusks of the Arid Southwest*. University of Arizona Press, Tucson.

Berrill, N. J. 1950. *The Tunicata, with an Account of the British Species*. Ray Society, London.

Boolootian, R. A., ed. 1966. *Physiology of Echinodermata*. Wiley and Sons, New York.

Borradaile, L. A. and F. A. Potts. 1958. *Invertebrata: A Manual for the Use of Students*. 3rd Ed., rev. by G. A. Kerkut. Cambridge University Press, Cambridge, England.

Bousefield, E. L. 1960. *Canadian Atlantic Sea Shells*. National Museum of Canada, Ottawa.

Brusca, R. C. 1973. *A Handbook to the Common Intertidal Invertebrates of the Gulf of California*. University of Arizona Press, Chicago.

Buchsbaum, R. M. 1948. *Animals Without Backbones*. University of Chicago Press, Chicago.

—, and L. J. Milne. 1960. *The Lower. Animals: Living Invertebrates of the World*. Doubleday, New York.

Bullough, W. S. 1956. *Practical Invertebrate Anatomy*. Macmillan Co., New York.

Burgess, C. M. 1970. *The Living Cowries*. A. S. Barnes & Co., New York.

Carson, R. 1955. *The Edge of the Sea*. Houghton Mifflin, Boston.

Carter, G. S. 1951. *A General Zoology of the Invertebrates*. Sedgwick and Jackson, London.

Clench, W. J., ed. 1941–74. *Johnsonia, Monographs of the Marine Mollusca of the Western Atlantic*. Cambridge, Mass.

Cloudsley-Thompson, J. L. 1961. *Land Invertebrates*. Methuen, London.

Coker, R. E. 1962. *This Great and Wide Sea*. Harper Torchbooks, New York.

Cousteau, J. Y. and F. Dumas. 1953. *The Silent World*. Harper, New York.

Ellis, A. E. 1969. *British Snails*. The Clarendon Press, Oxford, England.

Feininger, A. and W. K. Emerson. 1972. *Shells*. Viking Press, New York.

Fretter, V. and A. Graham. 1962. *British Prosobranch Molluscs*. Ray Society, London.

Galtsoff, P. S. 1964. *The American Oyster*. U.S. Bureau of Commercial Fisheries, Bulletin 64, Washington, D.C.

Gardiner, M. S. 1972. *Biology of the Invertebrates*. McGraw-Hill Book Co., New York.

Gosner, K. L. 1971. *Guide to Identification of Marine and Estuarine Invertebrates*. Wiley-Interscience, New York.

Gould, M. C. and others. 1972. *Invertebrate Oogenesis*. MSS Information Corp., New York.

Graham, A. 1971. *British Prosobranch and Other Operculate Gastropod Molluscs*. The Linnean Society of London and Academic Press, London and New York.

Grant, U. S. IV and L. G. Hertlein. 1938. *The West American Cenozoic Echinoidea*. Publications of the University of California, Los Angeles, in Mathematics and Physical Sciences, Vol. 2.

Halstead, B. W. 1965. *Poisonous and Venomous Marine Animals of the World. Vol. 1: Invertebrates*. U.S. Government Printing Office, Washington, D.C.

Hardy, Alister. 1956. *The Open Sea*. The World of Plankton. Collins, London.

Hedgepeth, J., ed. *Treatise on Marine Ecology and Paleoecology. Vol 1: Ecology*. Geological Society of America, New York.

Hegner, R. W. and J. G. Engermann. 1968. *Invertebrate Zoology*. Macmillan Co., New York.

Hickman, C. P. 1967. *Biology of the Invertebrates*. C. Y. Mosby Co., St. Louis.

Hyman, L. H. 1950–67. *The Invertebrates*. 6 vols. McGraw-Hill, New York.

Ivanov, A. V. 1963. *Pogonophora*. Academic Press, London.

Joyce, E. A. Jr. 1972. *A Partial Bibliography of Oysters, With Annotations*. State of Florida, Department of Natural Resources, Report No. 34. St. Petersburg, Fla.

Kaestner, A. 1967. *Invertebrate Zoology. Vol 1: Porifera, Cnidaria, Platyhelminthes, Aschelminthes, Mollusca, Annelida and Related Phyla*. John Wiley & Sons, Interscience Publishers, New York.

Keen, A. M. 1971. *Sea Shells of Tropical West America*. 2nd ed., Stanford University Press, Stanford, Calif.

Knight, J. B. 1952. *Primitive Fossil Gastropods and their Bearing on Gastropod Classification*. Smithsonian Miscellaneous Collections, Washington, D.C.

Kott, P. 1969. *Antarctic Ascidiacea*. Antarctic Research Series. Vol 13. American Geophysical Union, Washington, D.C.

Lane, F. W. 1960. *Kingdom of the Octopus*. Sheridan House, New York.

Macginitie, G. and N. Macginitie. 1949. *Natural History of Marine Animals*. McGraw-Hill, New York.

Marcus, E. and E. Marcus. 1967. *American Opisthobranch Mollusks*. University of Miami Press, Coral Gables, Fla.

McMillan, N. F. 1968. *British Shells*. Frederick Warne, London.

Mead, A. R. *The Giant African Snail*. A Problem in Economic Malacology. University of Chicago Press, Chicago.

Miner, R. W. 1950. *Field Book of Seashore Life*. Putnam, New York.

Moore, R. C., ed. 1965–71. *Treatise on Invertebrate Paleontology*. Geographical Society of America and University of Kansas Press, Lawrence.

Morton, J. E. 1963. *Molluscs*. Hutchinson & Co., Ltd., London.

Nicholds. D, and J. A. L. Cooke. 1971. *The Oxford Book of Invertebrates*. Oxford University Press, Oxford, England.

Nichols, D. 1969. *Echinoderms*. 4th ed. Hutchinson University Library, London.

Orton, J. H. 1937. *Oyster Biology and Oyster Culture*. Arnold.

Pennak, R. W. 1953. *Fresh-water Invertebrates of the United States*. Ronald, New York.

Pilsbry, H. A. 1939–48. *Land Mollusca of North America*. Monograph 3, 2 vols. Academy of Natural Sciences, Philadelphia.

Pratt, H. S. 1935. *A Manual of the Common Invertebrate Animals*. Blakiston Co., Philadelphia.

Purchon, R. D. 1968. *The Biology of the Mollusca*. Pergamon, London.

Raven, C. 1958. *Morphogenesis, the Analysis of Molluscan Development*. Pergamon Press, London.

Ricketts, E. F., and J. Calvin. 1968. *Between Pacific Tides*. 4th ed., revised by J. W. Hedgpeth. Stanford University Press, Stanford, Calif.

Roper, C. F. E., R. E. Young, and G. L. Voss. 1969. *An Illustrated Key to the Families of the Order Teuthoidea*. Smithsonian Contributions to Zoology, Washington, D.C.

Rounds, H. D. 1968. *Invertebrates*. Reinhold Book Corp., New York.

Rudwick, M. S. 1970. *Brachiopods*. Hutchinson University Library, London.

Runham, N. W. and P. J. Hunter. 1971. *Terrestrial Slugs*. Hillary House, New York.

Ryland, J. S. 1970. *Bryozoans*. Hutchinson University Library, London.

Schechter, V. 1959. *Invertebrate Zoology*. Prentice-Hall, Englewood Cliffs, N.J.

Shirai, S. 1970. *The Story of Pearls*. Japan Publ. Trading, San Francisco.

Stasek. C. R. 1972. *The Molluscan Framework in Chemical Zoology*, ed. by M. Florkin and B. T. Scheer, Academic Press, New York.

Stephenson, T. A. and A. Stephenson. 1972. *Life Between Tidemarks on Rocky Shores*. Freeman, San Fancisco.

Taylor, D. W. and N. F. Sohl. 1962. *An Outline of Gastropod Classification*. Ann Arbor, Mich.

Tebble, N. 1966. *British Bivalve Seashells*. British Museum (Natural History), London.

Turner, R. D. *A Survey and Illustrated Catalogue of the Teredinidae*. Museum of Comparative Zoology, Cambridge, England.

VanName, W. G. 1945. *The North and South American Ascidians*. Bulletin 84. American Museum of Natural History, New York.

Wagner, R. J. L. and R. T. Abbott. 1964. *Van Nostrand's Standard Catalogue of Shells*. D. Van Nostrand Co., Princeton, Toronto, New York, London.

Webb, W. F. 1942. *United States Mollusca*. Bookcraft, New York.

Wells, M. J. 1962. *Brain and Behaviour in Cephalopods*. Stanford University Press, Stanford, Calif.

Wilbur, K. M. and C. M. Yonge. 1964–66. *Physiology of Mollusca*. 2 vols. Academic Press, New York.

Wilmoth, J. H. 1967. *Biology of Invertebrata*. Prentice-Hall, Englewood Cliffs, N.J.

Wright, C. A. 1971. *Flukes and Snails*. George Allen and Unwin, London.

Yonge, C. M. 1950. *Oysters*. Collins, London.

German Books and Scientific Journals

Ankel, W. 1936. *Prosobranchia*. Die Tierwelt der Nord- und Ostsee. Becker & Erler, Leipzig.

Beebe, W. 1935. *923 Meter unter dem Meeresspiegel*. Brockhaus, Leipzig.

Benthem-Jutting, V. 1926. *Scaphopoda*. Die Tierwelt der Nord- und Ostsee. Becker & Erler, Leipzig.

Boettger, C. R. 1944. *Basommatophora*. Die Tierwelt der Nord- und Ostsee. Becker & Erler, Leipzig.

Brehms Tierleben. Bd. 1 (Niedere Tiere). Bibliographisches Institut, Leipzig.

Brohmer, P. 1969. *Fauna von Deutschland*. Quelle & Meyer, Heidelberg.

Bronns Klassen und Ordnungen Tierreichs: 4 Bd. Vermes, IV Abteilung Tentaculaten, Chaetognathen u. Hemichordaten. 1. Buch: Phoronoidea, Ektoprokta u. Brachiopoda.

Buchsbaum, R., und L. J. Milne. 1963. *Niedere Tiere*. Knaurs Tierreich in Farben. Droemer-Knaur, Munich/Zürich.

Coker, R. E. 1966. Das Meer—der grösste Lebensraum. Parey, Hamburg. Berlin.

Cori, C. 1937. *Phoronoidea*. Handbuch der Zoologie von Kükenthal/ Krambach. Bd. III, 2. Hälfte, 10. Lieferung. Gruyter, Berlin.

—. 1937. *Tentaculata*. Handbuch der Zoologie von Kükenthal/ Krumbach. Bd. III, 2, Hälfte, 10. Lieferung. Gruyter, Berlin.

—. 1941. *Bryozoa*. Handbuch der Zoologie von Kükenthal/Krumbach. 15 u. 16. Lieferung. Gruyter, Berlin.

Crome, W., R. Gottschalk, H. -J. Hannemann, G. Hartwick, und R. Kilias. 1930. *Urania Tierreich*. Wirbellose Tiere 1. Deutsch, Frankfurt a. M./Zürich.

Dahl, F. v., und H. Bischoff. 1930. *Die Tierwelt Deutschlands*. Fischer Jena.

Eibl-Eibesfeldt, I. 1967. *Grundriss der vergleichenden Verhaltensforschung*, Ethologie. Piper, Munich.

Engelhardt, W. 1955. *Was lebt in Tümpel, Bach und Weiher?* Kosmos, Franckh, Stuttgart.

Floericke, K. 1920. *Schnecken und Muscheln*. Kosmos, Franckh, Stuttgart.

Frey, H. 1957. *Das Aquarium von A-Z*. Neumann, Radebeul.

Geyer, D. 1927. *Unsere Land- und Süsswassermollusken*. Stuttgart.

Hass, F. 1926. *Lamellibranchia*. Die Tierwelt der Nord- und Ostsee. Becker & Erler, Leipzig.

Hafner, F. 1939. *Nordseemuscheln*. Arten und Formen. Kupferberg, Berlin.

Hässlein, L. 1960. *Weichtierfauna der Landschaften an der Pegnitz*. Ein Beitrag zur Ökologie und Soziologie niederer Tiere. Naturhistorische Gesellschaft, Nürnberg.

Helmcke, J. G. 1939. *Brachiopoda*. Handbuch der Zoologie von Kükenthal/Krumbach. 13. Lieferung. Gruyter, Berlin.

Hesse, R., und F. Doflein. 1935. *Tierbau und Tierleben*. Fischer, Jena.

Hoc, S. 1963. *Die Moostiere der deutschen Süss-, Brack-, und Küstengewässer*. Neue Brehm-Bücherei, Ziemsen, Wittenberg Lutherstadt.

Hoffmann, H. 1934. *Oposthobranchia, Pteropoda*. Dir Tierwelt d. Nord- u. Ostsee. Becker & Erler, Leipzig.

—. 1934. *Gastropoda*. Handwörterbuch der Naturwissenschaften. 6. Fischer, Jena.

—. 1950. *Mollusca*. Handbuch der Biologie. VI, H. 6, Potsdam.

Jaeckel, S. H. 1952. *Unsere Süsswassermuscheln*. Neue Brehm-Bücherei, Leipzig.

—. 1953. *Die Schlammschnecken unserer Gewässer*. Neue Brehm-Bücherei, Leipzig.

—. 1954. *Weichtiere*. Das Tierreich, Bd. V. Sammlung Goschen, Gruyter, Berlin.

—. 1955. *Bau und Lebensweise der Tiefseemollusken*. Neue Brehm-Bücherei, Zeimsen, Wittenberg Lutherstadt.

—. 1955. *Stachelhäuter. Tentakulaten, Binnenatmer und Pfeilwürmer*. Das Tierreich, Bd. VI, Sammlung Göschen, Gruyter, Berlin.

—. 1957. *Kopffüsser (Tintenfische)*. Neue Brehm-Bücherei, Ziemsen, Wittenberg Lutherstadt.

—. 1967. *Die Schlammschnecken unserer Gewässer*. Neue Brehm-Bücherei, Ziemsen, Wittenberg Lutherstadt.

Janus, O. 1968. *Unsere Muscheln und Schnecken*. Kosmos, Franckh, Stuttgart.

Johansson, K. E. 1968. *Pogonophora*. Handbuch der Zoologie von Kükenthal/Krumbach, Bd. 3, Teil 2, 18. Lieferung Gruyter, Berlin.

Kaestner. A. 1963. *Lehrbuch der Speziellen Zoologie*. Teil I: Wirbellose. Fischer, Stuttgart.

Kirsteuer, E. 1963. *Morphologie, Histologie, und Entwicklung der Pogonophora, Hemichordata und Chaetognatha*. Fortschritte der Zoologie, Bd. 20, 2, Fischer, Stuttgart.

Krumbach, T. 1937. *Oligomera*. Handbuch der Zoologie von Kükenthal/ Krumbach. Bd. III, 2. Halfte, 10. Lieferung, Gruyter, Berlin.

Kuckuck, P. 1929. *Der Strandwanderer*. Lehmanns, Munich.

Kuhn, O. 1966. *Die vorzeitlichen Wirbellosen—System und Evolution*. Oeben, Krailling.

Liebmann, H. 1951. *Handbuch der Frischwasser- und Abwasserbiologie*. Oldenbourg, Munich.

Luther, W., und K. Fiedler. 1965. *Die Unterwasserfauna der Mittelmeerküsten*. 2., neubearb. Aufl., Parey, Hamburg/Berlin.

Marcus, E. und E. 1960. *Opisthobranchia aus dem Roten Meer und von den Malediven*. Steiner, Wiesbaden.

Marshall, N. B. 1957. *Tiefseebiologie*. VEB Fischer, Jena.

Mehl, S. 1932. *Die Lebensbedingungen der Leberegelschnecke (Galba truncatula Müller)*. Datterer, Freising-Munich.

Meyer, V. 1913. *Tintenfische*. Leipzig.

Naef, A. 1933. *Cephalopoda*. Handwörterbuch der Naturwissenschaften. 2, Fischer, Jena.

Nordsieck, F. 1968. *Die europäischen Meeres-Gehauseschnecken (Prosobranchia)*. Fischer, Stuttgart.

Riedl, R. 1963. *Fauna und Flora der Adria*. Parey, Hamburg/Berlin.

—. 1966. *Biologie der Meereshöhlen*. Parey, Hamburg/Berlin.

Salvini-Plawen, L. v. 1964. *Schildfüsser und Furchenfüsser—verkannte Weichtiere am Meeresgrund*. Neue Brehm-Bücherei, Ziemsen, Wittenberg Lutherstadt.

Schilder, M. 1952. *Die Kaurischnecke*. Neue Brehm-Bücherei, Ziemsen, Wittenberg Lutherstadt.

Schindewolf, O. H. 1961–1968. *Studien zur Stammesgeschichte der Ammoniten*. Steiner, Wiesbaden.

Steinecke, F. 1940. *Der Süsswassersee*. Studienbucher deutscher Lebensgemeinschaften, Bd. I, Leipzig.

Thiele, J. 1926. *Solenogastres*. Mollusca. Handbuch der Zoologie von Kükenthal/Krumbach, Leipzig.

—. 1931. *Bivalvia*. Handwörterbuch der Noturwissenschaften. 1, Fischer, Jena.

—. 1932. *Loricata*. Handwörterbuch der Naturwissenschaften. 6, Fischer, Jena.

—. 1932. *Mollusca*. Handwörterbuch der Naturwissenschaften. 6, Fischer, Jena.

—. 1933. *Scaphopoda*. Handwörterbuch der Naturwissenschaften. 8, Fischer, Jena.

—. 1934. *Solenogastres*. Handwörterbuch der Naturwissenschaften. 9, Fischer, Jena.

Thienemann, A. 1950. *Verbreitungsgeschichte der Süsswassertierwelt*. Die Binnengewasser, Bd. XVIII, Stuttgart.

Van der Horst, C. J. 1939. *Hemichordata*. Bronns Klassen und Ordnungen des Tierreichs. Bd. IV, Abt. 4, Buch 2, Teil 2.

Wells, M. 1968. *Wunder primitiven Lebens*. Kindler, Munich.

Wesenberg-Lund, C. v. 1939. *Biologie der Süsswassertiere*. Wirbellose Tiere. Springer, Wien.

Wickler, W, 1962. *Das Meeresaquarium*. Kosmos, Franckh, Stuttgart.

—. 1968. *Mimikry*. Kindler, Munich.

Picture Credits

Artists: R. Grossmann (pp. 428 and 437). J. Kühn (p. 65/66). S. Milla (pp. 26, 35, 36, 45, 46, 55, 56, 79, 80, 89, 90, 103, 104, 113, 114, 139, 140, 141, 142, 167, 168, 177, 178, 187, 188, 197, 198, 253, 254, 271, 272, 292, 301, 302, 336, 348, 385, 386, 389, 390, 427). Scientific consultants of the artists: Fechter (Milla). O. Kraus (Grossmann). Propach (Kühn). Salvini-Plawen (Milla).

Color photographs: Ax (pp. 71 lower right, 74 lower left and lower right, 93, 120 upper left, 360, 404). Des Bartlett/Photo Researchers (p. 121). Bisserot/Photo Researchers (p. 74 upper left and middle right). Böck (p. 73 upper right). Burton/Photo Researchers (pp. 71 middle left, 94 upper left and upper right, 223 top, 455). Cropp (p. 438). Fechter (p. 224 upper middle and upper right). Fotosub (pp. 119 upper right, lower left, lower middle and lower right, 161, 312 upper right, 334, 357, 359, 369, 379 lower left, 380 middle left, 405). Haefelfinger (pp. 25, 71 upper right, middle lower left and lower left, 74 middle upper left and middle lower left, 119 middle top and middle lower left, 120 upper right, middle left, middle right, lower left and lower right, 122 top, 159 top, 224 lower left, middle lower and lower right). Harstrick/Münchener Internationaler Fotosalon (p. 94 middle upper left, middle lower right, lower left). Harz (p. 94 middle upper right and lower right). Knorr (pp. 234, 281, 312 middle and lower, 347). Köster (pp. 71 upper left and middle upper right, 73 middle upper and lower, 224 bottom, 233, 282, 291, 311, 333, 345, 379 middle left and middle right, 380 middle right, 450). Moosleitner (pp. 380 lower right and 406). v.d. Nieuwenhuizen (pp. 72 top, lower left and lower right, 73 upper left, 119 middle, 122 lower left and lower right, 159 bottom, 160, 223 bottom, 358, 379 upper left and upper right, 380 upper left and upper right). Östmann (pp. 74 upper right, 335, 346, 380 lower left). Paysan (p. 224 upper left). Pfletschinger (p. 99 lower middle and lower right). Piechatzek-/Bavaria (p. 162). Rozendaal (pp. 372 and 456). v. Salvini-Plawen (p. 114). Sauer (p. 243). Schrempp (pp. 94 middle lower left, 403, 449). Siedel (p. 99 lower left). Siegel (p. 99 lower left). Sillner (pp. 119 upper left, 370, 379 lower right). Suominen/Anthony (p. 72 lower middle). Temme (p. 244). Thau/ZFA (p. 371). V-Dia (p. 119 middle right). Wölk/ZFA (p. 100).

Line drawings after authors' sketches: Diller (pp. 274–408). Engstfeldt (pp. 29–225, 266–273, 409–430). Kühn (pp. 431–457 and all distribution maps). Popp (pp. 226–265).

Index

Abbreviations and Symbols

C, °C . . . .   Celsius, degrees centigrade

C.S.I.R.O. . . . .   Commonwealth Scientific and Industrial Res. Org. (Australia)

f . . . .   following (page)

ff . . . .   following (pages)

L . . . .   total length (from tip of nose [bill] to end of tail)

I.R.S.A.C. . . .   Institute for Scientific Res. in Central Africa, Congo

I.U.C.N. . . . .   Intern. Union for Conserv. of Nature and Natural Resources

BH . . . .   body height

BL . . . .   head-rump length (from nose to base of tail or end of body)

N, N- . . . . . .   North, Northern, North-

NE, NE- . . . .   Northeast, Northeastern, Northeast-

E, E- . . . .   East, Eastern, East-

S, S- . . . .   South, Southern, South-

TL . . . .   tail length

SE, SE- . . . . .   Southeast, Southeastern, Southeast-

SW, SW- . . .   Southwest, Southwestern, Southwest-

W, W- . . . . .   West, Western, West-

♂ . . . .   male

♂♂ . . . .   males

♀ . . . .   female

♀♀ . . . .   females

♂♀ . . . .   pair

+ . . . .   extinct

$\frac{2 \cdot 1 \cdot 2 \cdot 3}{2 \cdot 1 \cdot 2 \cdot 3}$ . . .   tooth formula, explanation in Volume X

▷ . . . .   following (opposite page) color plate

▷▷ . . . .   Color plate or double color plate on the page following the next

▷▷▷ . . . .   Third color plate or double color plate (etc.)

⬦) . . . .   Endangered species and subspecies